Readings in Deviant Behavior

Classic and Contemporary

Readings in Deviant Behavior

Classic and Contemporary

Nathaniel Eugene Terrell

Emporia State University

Robert F. Meier

University of Nebraska–Omaha

Harcourt College Publishers

Fort Worth   Philadelphia   San Diego   New York   Orlando   Austin   San Antonio
Toronto   Montreal   London   Sydney   Tokyo

Publisher	Earl McPeek
Acquisitions Editor	Bryan Leake
Market Strategist	Laura Brennan
Project Manager	Andrea Archer

ISBN: 0-15-506438-X
Library of Congress Catalog Card Number: 00-106856

Address for Domestic Orders
Harcourt College Publishers, 6277 Sea Harbor Drive, Orlando, FL 32887-6777
800-782-4479

Address for International Orders
International Customer Service
Harcourt College Publishers, 6277 Sea Harbor Drive, Orlando, FL 32887-6777
407-345-3800
(fax) 407-345-4060
(e-mail) hbintl@harcourtbrace.com

Address for Editorial Correspondence
Harcourt College Publishers, 301 Commerce Street, Suite 3700, Fort Worth, TX 76102

Web Site Address
http://www.harcourtcollege.com

Harcourt College Publishers will provide complimentary supplements or supplement packages to those adopters qualified under our adoption policy. Please contact your sales representative to learn how you qualify. If as an adopter or potential user you receive supplements you do not need, please return them to your sales representative or send them to:
Attn: Returns Department, Troy Warehouse, 465 South Lincoln Drive, Troy, MO 63379.

Printed in the United States of America

0 1 2 3 4 5 6 7 8 9   039   9 8 7 6 5 4 3 2 1

Harcourt College Publishers

To my parents, wife, and children,
from whom I have learned a great deal.

Nathaniel Sr.
Margaret Elizabeth
Cathy Elizabeth
Cameo Elizabeth
Nathaniel Rasean

Nathaniel Eugene Terrell

To Austin and J. P. and those who may follow.

Robert F. Meier

BRIEF CONTENTS

Preface   xv

Part 1: The Nature of Deviant Behavior   1

Chapter 1:   The Meaning of Deviance   5
Chapter 2:   Deviant Events and Social Control   35
Chapter 3:   Becoming Deviant   65

Part 2: Explaining Deviant Behavior: Macro Perspectives   87

Chapter 4:   Structural-Functional Perspectives   91
Chapter 5:   Conflict Perspectives   107

Part 3: Explaining Deviant Behavior: Micro Perspectives   137

Chapter 6:   Interactionist and Learning Perspectives   141
Chapter 7:   Control Perspectives   159
Chapter 8:   Feminist Perspectives   191

Part 4: Forms of Deviant Behavior   231

Chapter 9:   Violent Behavior   235
Chapter 10:   White-Collar Crime   272
Chapter 11:   Drugs and Deviant Behavior   300
Chapter 12:   Alcohol and Deviant Behavior   342
Chapter 13:   Sexual Deviant Behavior   379
Chapter 14:   Suicide   401

Part 5: Studies in Stigma   425

Chapter 15:   Mental Deviant Behavior   427
Chapter 16:   Homosexual Behavior   456
Chapter 17:   Physical Disabilities   491

DETAILED CONTENTS

Preface   xv

Part 1
The Nature of Deviant Behavior   1

Chapter 1
The Meaning of Deviance   5

Howard S. Becker
Outsiders: Studies in the Sociology of Deviance (1963)   5

Stephen J. Pfohl
The "Discovery" of Child Abuse (1977)   13

Justine Janette
Cult Update (1999)   29

Chapter 2
Deviant Events and Social Control   35

Melody A. Graham, Jennifer Monday, Kimberly O'Brien, Stacey Steffen
*Cheating at Small Colleges: An Examination of Student and Faculty Attitudes
and Behaviors* (1994)   35

Rodney Stark
Deviant Places: A Theory of the Ecology of Crime (1987)   44

Craig J. Forsyth
Parade Strippers: Being Naked in Public (1992)   57

Chapter 3
Becoming Deviant   65

Jeffrey W. Riemer
Deviance as Fun (1981)   65

Jacqueline Lewis
*Learning to Strip: The Socialization Experiences of Exotic
Dancers* (1998)   69

Part 2
Explaining Deviant Behavior: Macro Perspectives   87

Chapter 4
Structural-Functional Perspectives

Emile Durkheim
The Normal and the Pathological (1938)   91

Robert K. Merton
Social Structure and Anomie (1957)   95

Richard A. Cloward and Lloyd E. Ohlin
Illegitimate Means and Delinquent Subcultures (1960)   101

Chapter 5
Conflict Perspectives   107

Steven Spitzer
Toward a Marxian Theory of Deviance (1975)   107

Andrew L. Hochstetler and Neal Shover
Street Crime, Labor Surplus, and Criminal Punishment, 1980–1990 (1997)   121

Richard Quinney
The Social Reality of Crime (1975)   131

Part 3
Explaining Deviant Behavior: Micro Perspectives   137

Chapter 6
Interactionist and Learning Perspectives   141

Edwin M. Lemert
Primary and Secondary Deviation (1951)   141

Edwin H. Sutherland and Donald R. Cressey
Differential Association Theory (1974)   146

Liqun Cao, Anthony Adams, and Vickie J. Jensen
A Test of the Black Subculture of Violence Thesis: A Research Note (1997)   148

Chapter 7
Control Perspectives   159

Travis Hirschi
Control Theory (1969)   159

Michael Gottfredson and Travis Hirschi
  A General Theory of Crime (1990)   162

Kimberly L. Kempf
  The Empirical Status of Hirschi's Control Theory (1993)   173

Chapter 8
Feminist Perspectives   191

Meda Chesney-Lind
  Girls' Crime and Woman's Place: Toward a Feminist Model of Female Delinquency (1989)   191

Lois Copeland and Leslie R. Wolfe
  A Radical Feminist View of Rape (1991)   210

Neil Gilbert
  Was It Rape? An Examination of Sexual Abuse Statistics (1994)   221

Part 4
Forms of Deviant Behavior   231

Chapter 9
Violent Behavior   235

Terance D. Miethe and Richard McCleary
  Homicide and Aggravated Assault (1998)   235

Diana Scully and Joseph Marolla
  Convicted Rapists' Vocabulary of Motive: Excuses and Justifications (1984)   248

Shelly Schaefer Hinck and Richard W. Thomas
  Rape Myth Acceptance in College Students: How Far Have We Come? (1999)   260

Chapter 10
White-Collar Crime   272

Michael L. Benson
  Denying the Guilty Mind: Accounting for Involvement in a White-Collar Crime (1985)   272

Tim Jordan and Paul Taylor
  A Sociology of Hackers (1998)   282

Chapter 11
Drugs and Deviant Behavior   300

Howard S. Becker
  Becoming a Marihuana User (1953)   300

Sheigla Murphy, Dan Waldorf, and Craig Reinarman
  Drifting into Cocaine Dealing (1990)   310

James A. Inciardi, Hilary L. Surratt, Dale D. Chitwood, and Clyde B. McCoy
  The Origins of Crack (1996)   329

Chapter 12
Alcohol and Deviant Behavior   342

Norman K. Denzin
The Alcoholically Divided Self (1987)   342

Craig MacAndrew and Robert B. Edgerton
Some People Can Really Hold Their Liquor (1969)   353

Henry Wechsler, Beth E. Molnar, Andrea E. Davenport, and John S. Baer
College Alcohol Use: A Full or Empty Glass? (1999)   370

Chapter 13
Sexual Deviant Behavior   379

Barbara Sherman Heyl
The Madam as Teacher: The Training of House Prostitutes (1979)   379

Philip Elmer-Dewitt
On a Screen Near You: Cyberporn (1995)   391

Susie Bright
Check Mate: What I Teach My Daughter about Sex (1999)   398

Chapter 14
Suicide   401

Jennifer Langhinrichsen-Rohling, Peter Lewinsohn, Paul Rohde,
John Seeley, Candice M. Monson, Kathryn A. Meyer, and
Richard Langford
*Gender Differences in the Suicide-Related Behaviors of Adolescents
and Young Adults* (1998)   401

Emile Durkheim
Anomic Suicide (1951)   414

Part 5
Studies in Stigma   425

Chapter 15
Mental Deviant Behavior   427

David L. Rosenhan
On Being Sane in Insane Places (1973)   427

Peter S. Jensen, Lori Kettle, Margaret T. Roper, Michael T. Sloan,
Mina K. Dulcan, Christina Hoven, Hector R. Bird, Jose J.
Bauermeister, and Jennifer D. Payne
*Are Stimulants Overprescribed? Treatment of ADHD (Attention-Deficit
Hyperactivity Disorder) in Four U.S. Communities* (1999)   437

Christy L. Picard
*The Level of Competition as a Factor for the Development of Eating Disorders
in Female Collegiate Athletes* (1999)   446

Chapter 16
Homosexual Behavior   456

Robert F. Meier and Gilbert Geis
The Gay Movement and Gay Communities (1997)   456

Robert E. Owens Jr.
Becoming Lesbian, Gay, and Bisexual (1998)   461

Michael J. Bailey
Homosexuality and Mental Illness (1999)   486

Chapter 17
Physical Disabilities   491

Robert A. Scott
The Making of the Blind in Personal Interaction (1969)   491

Natalie Angier
Thinking Twice on Splitting Twins (1998)   503

PREFACE

Like many fields, the sociology of deviance reflects a combination of continuity and change. Many important insights about the meaning, nature, and control of deviance have been developed over the last one hundred years, but what is considered deviant and how best to understand deviance must also take into account how the concept of deviance has changed. Our challenge for this collection of readings has been to reflect both the rich intellectual heritage in the literature on the sociology of deviance and to take serious note of recent development in the field.

We embarked on this project with several objectives in mind. Clearly, our overall intent was to make a sampling of important thinking on deviance more accessible to students. But there were other objectives as well. First, we wanted a reader that included material on the definition and meaning of deviance, both structural and processual theories of deviance, examples of different kinds of deviance, and the importance of stigma in understanding both deviance and the processes of social control.

Second, we felt it necessary to include examples of both the bedrock classics and contemporary articles. The development of traditional theoretical frameworks must apply both to established forms of deviance, for example, drug use, suicide, and crime, as well as to newer forms of deviance, such as computer fraud and hacking. Contemporary articles also help students engage in critical and abstract thinking by understanding and applying different theoretical perspectives to everyday illustrations of deviance. By including articles on cults, stripping, rape myths, hacking, cyberporn, and conjoined twins we are trying to provide students with opportunities to view deviance broadly.

Third, we wanted to stretch students' perception of what is normal and abnormal. Deviance cannot be divorced from social power. Having articles that debunk some widely held perceptions of deviance will, we hope, assist students in extending their knowledge of what should and should not be classified as deviant behavior. We also think that it is important to view deviance not only from the perspective of "normals" but from that of the "outsiders" as well. In balancing these perspectives, we believe that stigma both as a cause and consequence of deviance plays a role that is still understated in the deviance literature. By examining social attitudes of deviant behavior, students can gain a better understanding of how society operates.

Finally, we wanted to produce a book that would be useful as a stand-alone volume or in combination with a deviance textbook. We believe this book is suitable for courses on a number of substantive areas such as crime, drugs, alcohol, sexual and homosexual behavior, and suicide.

Numerous people have helped assemble this reader. We would like to recognize them and communicate our appreciation. First and foremost, we thank our contributors and their representatives for allowing us to reprint their work. We wish to thank the reviewers of this book: Christopher F. Armstrong, Bloomsburg University;

Korni Swaroop Kumar, SUNY at Potsdam; Kevin M. Thompson, North Dakota State University; John Broderick, Stonehill College; and David Struckhoff, Loyola University. We are appreciative to the editorial staff at Harcourt College Publishers. We are especially in debt to Stacy Schoolfield who provided us with just the right amount of carrot and stick to guide this project through to completion. Nate Terrell is also grateful to Brenda Carmichael, Shalon "Lacy" Berry, Tanay Forsberg, Stephanie Siebert, Kelly Ladner, Jill Hutchinson, and especially Rochelle "Shelly" Tucker for their help in the project. Bob Meier wishes to acknowledge the assistance of Geri Murphy, June Turner, Steve Culver, Angela Patton, and Kris Piessig.

PART 1

THE NATURE OF
DEVIANT BEHAVIOR

What is deviance? Certainly, the term deviance implies some kind of difference, but what kind? Are Blacks, women, people receiving welfare, obese people, short people, ugly people, the aged, drug addicts, the mentally ill, criminals, homosexuals, the unemployed, the blind, and the poor all deviant? Are any of them deviant? How do we know? In Part 1, a number of selections are chosen to address some of the elementary questions of deviance: how we know or think that some act or condition is deviant, the nature of the context in which the deviance occurs, and how deviance can be constructed.

Some of the most intense and interesting debates in the sociology of deviance concern the definition of deviance itself. There are alternative definitions and each has implications for what acts and conditions are considered deviant. Some may think that deviance is something that is statistically rare, but while most Americans drink coffee, few would say that drinking tea is deviant. Another view is that deviance is *a violation of a norm,* which is an expectation of behavior in particular situations. When people express the idea that something "should not" occur, they are referring to deviance, something of which they disapprove. But how can we determine the content of norms? There is no agreed-upon procedure and those who hold this view often fail to tell how norms evolved and how we come to subscribe to them.

Howard S. Becker's definition, contained in the first selection, is often called the *reactivist* or *labeling* perspective. This view holds that deviance cannot be determined without looking at the reactions of others. But how many people have to consider something deviant before it is deviant? All? Half? Only a few? Regardless of the specific answers to these questions, there is agreement on the notion that deviance refers to behavior and conduct that is disvalued or disapproved by others. Deviance is also a distinctly social entity because the disapproval is shared to some degree among groups. This sense of disapproval for certain acts is often socially constructed. That is, some people may believe that an act is deviant and they are able to persuade others of their conviction. Becker refers to such people as moral entrepreneurs since they promote to others their beliefs of what should and should not occur.

It is also the case that deviance is constructed more deliberately. Stephen J. Pfohl's discussion of the "discovery" of child abuse provides an example of this process. Clearly, child abuse existed before our ability to document it, but it was not until the medical specialty of pediatric radiology was able to interpret more accurately the nature of broken bones of undetermined origin that child abuse became of medical and national concern.

The term "diversity" is often used to refer to the degree of ethnic composition in the population, but it could also apply to ideas and behavior. Is deviance an instance of behavior or can thoughts also be deviant? There are a number of instances where people can be considered deviant for particular thoughts, including religious ones. Many religious people talk to God, but when God talks back, there may be the suspicion of a mental disturbance. Certain religions themselves (for example, those that depart too much from traditional religions) are considered deviant by some segments of the population. The Justine Janette selection provides an update on some religious groups that are considered cults and therefore suspect.

Deviance is widespread but, as mentioned earlier, even if many people engage in an act, the act can still be considered deviant. Melody A. Graham, Jennifer Monday, Kimberly O'Brien, and Stacey Steffen illustrate this in the selection on college cheating. They discovered that cheating on exams and papers is extremely common; nearly 90 percent of a sample of college students admitted to having cheated. Furthermore, the students understood clearly that cheating was prohibited. Faculty regarded cheating as more serious than did students and this was reflected in the reasons students gave for not cheating. While some regarded cheating as morally wrong, many others avoided it simply because of the risk of getting caught.

Many deviant acts have a specific context. They occur more at some times than others, among some groups more than others, and in some places more than others. Rodney Stark discusses one such important context: place. How, Stark asks, can neighborhoods have high crime and deviance rates despite a complete turnover in population? Recalling a similar question and approach from criminologists Clifford Shaw and Henry McKay in the 1920s, Stark finds that where people live is as important as who the people are. Some locations are associated with deviance, regardless of the specific residents.

Many deviant acts are also highly situational. There are occasions—"moral holidays"—when various rules and mechanisms of social control are suspended. Heavy drinking of alcoholic beverages, for example, is more tolerated on New Year's Eve than at other times. Mardi Gras, which signals the beginning of Lent and abstinence for many, is another. One recent characteristic of Mardi Gras is exhibitionistic behavior during the annual Mardi Gras parade. Craig J. Forsyth's article on parade strippers examines the conditions under which people who might not otherwise do so willingly expose themselves to strangers.

One becomes deviant through a social process. The process includes a motivation to commit a deviant act, a social setting in which that motivation can be expressed, and the interaction of a social audience to the act. One element of deviant motivation that is often ignored is that some deviant acts are, well, fun! Jeffrey W. Riemer reminds us that some deviant acts are personally, socially, and economically rewarding. Some are exhilarating and adventuresome. Some are spontaneous acts tied to particular situations, and many such acts have no negative consequences.

Fun can be coupled with other motivations. Jacqueline Lewis describes the learning experiences of women who become exotic dancers. Many have experiences that help prepare them for dancing. They work in clubs, perhaps as wait staff, or have had previous stage experience. The learning continues in interaction with other dancers about the nature of the job, how to handle the audience, and how to develop attitudes that immunize the dancer from the disapproval of others. Dancers come to justify their work in a number of ways, including the financial rewards of such work.

CHAPTER 1

THE MEANING OF DEVIANCE

Outsiders: Studies in the Sociology of Deviance

Howard S. Becker

All social groups make rules and attempt, at some times and under some circumstances, to enforce them. Social rules define situations and the kinds of behavior appropriate to them, specifying some actions as "right" and forbidding others as "wrong." When a rule is enforced, the person who is supposed to have broken it may be seen as a special kind of person, one who cannot be trusted to live by the rules agreed on by the group. He is regarded as an *outsider.*

But the person who is thus labeled an outsider may have a different view of the matter. He may not accept the rule by which he is being judged and may not regard those who judge him as either competent or legitimately entitled to do so. Hence, a second meaning of the term emerges: the rule-breaker may feel his judges are *outsiders.*

In what follows, I will try to clarify the situation and process pointed to by this double-barreled term: the situations of rule-breaking and rule-enforcement and the processes by which some people come to break rules and others to enforce them.

Some preliminary distinctions are in order. Rules may be of a great many kinds. They may be formally enacted into law, and in this case the police power of the state may be used in enforcing them. In other cases, they represent informal agreement, newly arrived at or encrusted with the sanction of age and tradition; rules of this kind are enforced by informal sanctions of various kinds.

Similarly, whether a rule has the force of law or tradition or is simply the result of consensus, it may be the task of some specialized body, such as the police or the committee on ethics of a professional association, to enforce it; enforcement, on the other hand, may be everyone's job or, at least, the job of everyone in the group to which the rule is meant to apply.

Many rules are not enforced and are not, in any except the most formal sense, the kind of rules with which I am concerned. Blue laws, which remain on the statute books though they have not been enforced for a hundred years, are examples. (It is important to remember, however, that an unenforced law may be reactivated for various reasons and regain all its original force, as occurred with respect to the laws governing the opening of commercial establishments on Sunday in Missouri.) Informal rules may similarly die from lack of enforcement. I shall mainly be concerned with what we can call the actual operating rules of groups, those kept alive through attempts at enforcement.

Finally, just how far "outside" one is, in either of the senses I have mentioned, varies from case to case. We think of the person who commits a traffic violation or gets a little too drunk at a party as being, after all, not very different from the rest of us and treat his infraction tolerantly. We regard the thief as less like us and punish him severely. Crimes such as murder, rape, or treason lead us to view the violator as a true outsider.

In the same way, some rule-breakers do not think they have been unjustly judged. The traffic violator usually subscribes to the very rules he has broken. Alcoholics are often ambivalent, sometimes feeling that those who judge them do not understand them and at other times agreeing that compulsive drinking is a bad thing. At the extreme, some deviants (homosexuals and drug addicts are good examples) develop full-blown ideologies explaining why they are right and why those who disapprove of and punish them are wrong.

DEFINITIONS OF DEVIANCE

The outsider—the deviant from group rules—has been the subject of much speculation, theorizing, and scientific study. What laymen want to know about deviants is: why do they do it? How can we account for their rule-breaking? What is there about them that leads them to do forbidden things? Scientific research has tried to find answers to these questions. In doing so it has accepted the common-sense premise that there is something inherently deviant (qualitatively distinct) about acts that break (or seem to break) social rules. It has also accepted the common-sense assumption that the deviant act occurs because some characteristic of the person who commits it makes it necessary or inevitable that he should. Scientists do not ordinarily question the label "deviant" when it is applied to particular acts of people but rather take it as given. In so doing, they accept the values of the group making the judgment.

It is easily observable that different groups judge things to be deviant. This should alert us to the possibility that the person making the judgment of deviance, the process by which that judgment is arrived at, and the situation in which it is made may all be intimately involved in the phenomenon of deviance. To the degree that the common-sense view of deviance and the scientific theories that begin with its premises assume that acts that break rules are inherently deviant and thus take for granted the situations and processes of judgment, they may leave out an important variable. If scientists ignore the variable character of the process of judgment, they may by that omission limit the kinds of theories that can be developed and the kind of understanding that can be achieved.[1]

Our first problem, then, is to construct a definition of deviance. Before doing this, let us consider some of the definitions scientists now use, seeing what is left out if we take them as a point of departure for the study of outsiders.

The simplest view of deviance is essentially statistical . . . from the average. When a statistician analyzes the results of an agricultural experiment, he describes the stalk of corn that is exceptionally tall and the stalk that is exceptionally short as deviations from the mean or average. Similarly, one can describe anything that differs from what is most common as a deviation. In this view, to be left-handed or redheaded is deviant, because most people are right-handed and brunette.

So stated, the statistical view seems simple-minded, even trivial. Yet it simplifies the problem by doing away with many questions of value that ordinarily arise in discussions of the nature of deviance. In assessing any particular case, all one need do is calculate the distance of the behavior involved from the average. But it is too simple a solution. Hunting with such a definition, we return with a mixed bag—people who are excessively fat or thin, murderers, redheads, homosexuals, and traffic violators. The mixture contains some ordinarily thought of as deviants and others who have broken no rule at all. The statistical definition of deviance, in short, is too far removed from the concern with rule-breaking which prompts scientific study of outsiders.

A less simple but much more common view of deviance identifies it as something essentially pathological, revealing the presence of a "disease." This view rests, obviously, on a medical analogy. The human organism, when it is working efficiently and experiencing no discomfort, is said to be "healthy." When it does not work efficiently, a disease is present. The organ or function that has become deranged is said to be pathological. Of course, there is little disagreement about what constitutes a healthy state of the organism. But there is much less agreement when one uses the notion of pathology analogically, to describe kinds of behavior that are regarded as deviant. For people do not agree on what constitutes healthy behavior. It is difficult to find a definition that will satisfy even such a select and limited group as psychiatrists; it is impossible to find one that people generally accept as they accept criteria of health for the organism.[2]

Sometimes people mean the analogy more strictly, because they think of deviance as the product of mental disease. The behavior of a homosexual or drug addict is regarded as the symptom of a mental disease just as the diabetic's difficulty in getting bruises to heal is regarded as a symptom of his disease. But mental disease resembles physical disease only in metaphor. . . .[3] The medical metaphor limits what we can see much as the statistical view does. It accepts the lay judgment of something as deviant and, by use of analogy, locates its source within the individual, thus preventing us from seeing the judgment itself as a crucial part of the phenomenon.

Some sociologists also use a model of deviance based essentially on the medical notions of health and disease. They look at a society, or some part of a society, and ask whether there are any processes going on in it that tend to reduce its stability, thus lessening its chance of survival. They label such processes deviant or identify them as symptoms of social disorganization. They discriminate between those features of society which promote stability (and thus are "functional") and those which disrupt stability (and thus are "dysfunctional"). Such a view has the great virtue of pointing to areas of possible trouble in a society of which people may not be aware.[4]

But it is harder in practice than it appears to be in theory to specify what is functional and what dysfunctional for a society or social group. The question of what the purpose or goal (function) of a group is and, consequently, what things will help or hinder the achievement of that purpose, is very often a political question. Factions within the group disagree and maneuver to have their own definition of the group's function accepted. The function of the group or organization, then, is decided in political conflict, not given in the nature of the organization. If this is true, then it is likewise true that the questions of what rules are to be enforced, what behavior regarded as deviant, and which people labeled as outsiders must also be regarded as political.[5] The functional view of deviance, by ignoring the political aspect of the phenomenon, limits our understanding.

Another sociological view is more relativistic. It identifies deviance as the failure to obey group rules. Once we have described the rules a group enforces on its members, we can say with some precision whether or not a person has violated them and is thus, on this view, deviant.

This view is closest to my own, but it fails to give sufficient weight to the ambiguities that arise in deciding which rules are to be taken as the yardstick against which behavior is measured and judged deviant. A society has many groups, each with its own set of rules, and people belong to many groups simultaneously. A person may break the rules of one group by the very act of abiding by the rules of another group. Is he, then, deviant? Proponents of this definition may object that while ambiguity may arise with respect to the rules peculiar to one or another group in society, there are some rules that are very generally agreed to by everyone, in which case the difficulty does not arise. This, of course, is a question of fact, to be settled by empirical research. I doubt there are many such areas of consensus and think it wiser to use a definition that allows us to deal with both ambiguous and unambiguous situations.

DEVIANCE AND THE RESPONSES OF OTHERS

The sociological view I have just discussed defines deviance as the infraction of some agreed-upon rule. It then goes on to ask who breaks rules, and to search for the factors in their personalities and life situations that might account for the infractions. This assumes that those who have broken a rule constitute a homogeneous category, because they have committed the same deviant act.

Such an assumption seems to me to ignore the central fact about deviance: it is created by society. I do not mean this in the way it is ordinarily understood, in which the causes of deviance are located in the social situation of the deviant or in "social factors" which prompt his action. I mean, rather, that *social groups create deviance by making the rules whose infraction constitutes deviance,* and by applying those rules to particular people and labeling them as outsiders. From this point of view, deviance is *not* a quality of the act the person commits, but rather a consequence of the application by others of rules and sanctions to an "offender." The deviant is one to whom that label has successfully been applied; deviant behavior is behavior that people so label.[6]

Since deviance is, among other things, a consequence of the responses of others to a person's act, students of deviance cannot assume that they are dealing with a homogeneous category when they study people who have been labeled deviant. That is, they cannot assume that these people have actually committed a deviant act or broken some rule, because the process of labeling may not be infallible; some people may be labeled deviant who in fact have not broken a rule. Furthermore, they cannot assume that the category of those labeled deviant will contain all those who actually have broken a rule, for many offenders may escape apprehension and thus fail to be included in the population of "deviants" they study. Insofar as the category lacks homogeneity and fails to include all the cases that belong in it, one cannot reasonably expect to find common factors of personality or life situation that will account for the supposed deviance.

What, then, do people who have been labeled deviant have in common? At the least, they share the label and the experience of being labeled as outsiders. I will begin my analysis with this basic similarity and view deviance as the product of a transaction that takes place between some social group and one who is viewed by that group as a rule-breaker. I will be less concerned with the personal and social characteristics of deviants than with the process by which they come to be thought of as outsiders and their reactions to that judgment.

Malinowski discovered the usefulness of this view for understanding the nature of deviance many years ago, in his study of the Trobriand Islands. . . .[7] Whether an act is deviant, then, depends on how other people react to it. You can commit clan incest and suffer from no more than gossip as long as no one makes a public accusation; but you will be driven to your death if the accusation is made. The point is that the response of other people has to be regarded as problematic. Just because one has committed an infraction of a rule does not mean that others will respond as though this had happened. (Conversely, just because one has not violated a rule does not mean that he may not be treated, in some circumstances, as though he had.)

The degree to which other people will respond to a given act as deviant varies greatly. Several kinds of variation seem worth noting. First of all, there is variation over time. A person believed to have committed a given "deviant" act may at one time be responded to much more leniently than he would be at some other time. The occurrence of "drives" against various kinds of deviance illustrates this clearly. At various times, enforcement officials may decide to make an all-out attack on some particular kind of deviance, such as gambling, drug addiction, or homosexuality. It is obviously much more dangerous to engage in one of these activities when a drive is on than at any other time. (In a very interesting study of crime news in Colorado newspapers, Davis found that the amount of crime reported in Colorado newspapers showed very little association with actual changes in the amount of crime taking place in Colorado. And, further, that people's estimate of how much increase there had been in crime in Colorado was associated with the increase in the amount of crime news but not with any increase in the amount of crime).[8]

The degree to which an act will be treated as deviant depends also on who commits the act and who feels he has been harmed by it. Rules tend to be applied more to some persons than others. Studies of juvenile delinquency make the point clearly. Boys from middle-class areas do not get as far in the legal process when they are apprehended as do boys from slum areas. The middle-class boy is less likely, when picked up by the police, to be taken to the station; less likely when taken to the station to be booked; and it is extremely unlikely that he will be convicted and sentenced.[9] This variation occurs even though the original infraction of the rule is the same in the two cases. Similarly, the law is differentially applied to Negroes and whites. It is well known that a Negro believed to have attacked a white woman is much more likely to be punished than a white man who commits the same offense; it is only slightly less well known that a Negro who murders another Negro is much less likely to be punished than a white man who commits murder.[10] This, of course, is one of the main points of Sutherland's analysis of white-collar crime: crimes committed by corporations are almost always prosecuted as civil cases, but the same crime committed by an individual is ordinarily treated as a criminal offense.[11]

Some rules are enforced only when they result in certain consequences. The unmarried mother furnishes a clear example. Vincent[12] points out that illicit sexual relations seldom result in severe punishment or social censure for the offenders. If, however, a girl becomes pregnant as a result of such activities the reaction of others is likely to be severe. (The illicit pregnancy is also an interesting example of the differential enforcement of rules on different categories of people. Vincent notes that unmarried fathers escape the severe censure visited on the mother.)

Why repeat these commonplace observations? Because, taken together, they support the proposition that deviance is not a simple quality, present in some kinds of behavior and absent in others. Rather, it is the product of a process which involves responses of other people to the behavior. The same behavior may be an infraction of the rules at one time and not at another; may be an infraction when committed by one person, but not when committed by another; some rules are broken with impunity, others are not. In short, whether a given act is deviant or not depends in part on the nature of the act (that is, whether or not it violates some rule) and in part on what other people do about it.

Some people may object that this is merely a terminological quibble, that one can, after all, define terms any way he wants to and that if some people want to speak of rule-breaking behavior as deviant without reference to the reactions of others they are free to do so. This, of course, is true. Yet it might be worthwhile to refer to such behavior as *rule-breaking behavior* and reserve the term *deviant* for those labeled as deviant by some segment of society. I do not insist that this usage be followed. But it should be clear that insofar as a scientist uses "deviant" to refer to any rule-breaking behavior and takes as his subject of study only those who have been *labeled* deviant, he will be hampered by the disparities between the two categories.

If we take as the object of our attention behavior which comes to be labeled as deviant, we must recognize that we cannot know whether a given act will be categorized as deviant until the response of others has occurred. Deviance is not a quality that lies in behavior itself, but in the interaction between the person who commits an act and those who respond to it.

WHOSE RULES?

I have been using the term "outsiders" to refer to those people who are judged by others to be deviant and thus to stand outside the circle of "normal" members of the group. But the term contains a second meaning, whose analysis leads to another important set of sociological problems: "outsiders," from the point of view of the person who is labeled deviant, may be the people who make the rules he had been found guilty of breaking.

Social rules are the creation of specific social groups. Modern societies are not simple organizations in which everyone agrees on what the rules are and how they are to be applied in specific situations. They are, instead, highly differentiated along social class lines, ethnic lines, occupational lines, and cultural lines. These groups need not and, in fact, often do not share the same rules. The problems they face in dealing with their environment, the history and traditions they carry with them, all lead to the evolution of different sets of rules. Insofar as the rules of various groups conflict and contradict one another, there will be disagreement about the kind of behavior that is proper in any given situation.

Italian immigrants who went on making wine for themselves and their friends during Prohibition were acting properly by Italian immigrant standards, but were breaking the law of their new country (as, of course, were many of their Old American neighbors). Medical patients who shop around for a doctor may, from the perspective of their own group, be doing what is necessary to protect their health by making sure they get what seems to them the best possible doctor; but, from the perspective of the physician, what they do is wrong because it breaks down the trust the patient ought to put in his physician. The lower-class delinquent who fights for his "turf" is only doing what he considers necessary and right, but teachers, social workers, and police see it differently.

While it may be argued that many or most rules are generally agreed to by all members of a society, empirical research on a given rule generally reveals variation in people's attitudes. Formal rules, enforced by some specially constituted group, may differ from those actually thought appropriate by most people.[13] Factions in a group may disagree on what I have called actual operating rules. Most important for the study of behavior ordinarily labeled deviant, the perspectives of the people who engage in the behavior are likely to be quite different from those of the people who condemn it. In this latter situation, a person may feel that he is being judged according to rules he has had no hand in making and does not accept, rules forced on him by outsiders.

To what extent and under what circumstances do people attempt to force their rules on others who do not subscribe to them? Let us distinguish two cases. In the first, only those who are actually members of the group have any interest in making and enforcing certain rules. If an orthodox Jew disobeys the laws of kashruth only other orthodox Jews will regard this as a transgression; Christians or nonorthodox Jews will not consider this deviance and would have no interest in interfering. In the second case, members of a group consider it important to their welfare that members of certain other groups obey certain rules. Thus, people consider it extremely important that those who practice the healing arts abide by certain rules; this is the reason the state licenses physicians, nurses, and others, and forbids anyone who is not licensed to engage in healing activities.

To the extent that a group tries to impose its rules on other groups in the society, we are presented with a second question: Who can, in fact, force others to accept their rules and what are the causes of their success? This is, of course, a question of political and economic power. [The] political and economic process through which rules are created and enforced [must be considered]. Here it is enough to note that people are in fact always *forcing* their rules on others, applying them more or less against the will and without the consent of those others. By and large, for example, rules are made for young people by their elders. Though the youth of this country exert a powerful influence culturally—the mass media of communication are tailored to their interests, for instance—many important kinds of rules are made for our youth by adults. Rules regarding school attendance and sex behavior are not drawn up with regard to the problems of adolescence. Rather, adolescents find themselves surrounded by rules about these matters which have been made by older and more settled people. It is considered legitimate to do this, for youngsters are considered neither wise enough nor responsible enough to make proper rules for themselves.

In the same way, it is true in many respects that men make the rules for women in our society (though in America this is changing rapidly). Negroes find themselves subject to rules made for them by whites. The foreign-born and those otherwise ethnically peculiar often have their rules made for them by the Protestant Anglo-Saxon minority. The middle class makes rules the lower class must obey—in the schools, the courts, and elsewhere.

Differences in the ability to make rules and apply them to other people are essentially power differentials (either legal or extralegal). Those groups whose social position gives them weapons and power are best able to enforce their rules. Distinctions of age, sex, ethnicity, and class are all related to differences in power, which account for differences in the degree to which groups so distinguished can make rules for others.

In addition to recognizing that deviance is created by the responses of people to particular kinds of behavior, by the labeling of that behavior as deviant, we must also keep in mind that the rules created and maintained by such labeling are not universally agreed to. Instead, they are the object of conflict and disagreement, part of the political process of society.

NOTES

1. Cf. Donald R. Cressey, "Criminological Research and the Definition of Crimes," *American Journal of Sociology,* LVI (May, 1951), 546–551.

2. See the discussion in C. Wright Mills, "The Professional Ideology of Social Pathologists," *American Journal of Sociology,* XLIX (September, 1942), 165–180.

3. Thomas Szasz, *The Myth of Mental Illness* (New York: Paul B. Hoeber, Inc., 1961), pp. 44–45; see also Erving Goffman, "The Medical Model and Mental Hospitalization," in *Asylums: Essays on the Social Situation of Mental Patients and Other Inmates* (Garden City: Anchor Books, 1961), pp. 321–386.

4. See Robert K. Merton, "Social Problems and Sociological Theory," in Robert K. Merton and Robert A. Nisbet, editors, *Contemporary Social Problems* (New York: Harcourt, Brace and World, Inc., 1961), pp. 697–737; and Talcott Parsons, *The Social System* (New York: The Free Press of Glencoe, 1951), pp. 249–325.

5. Howard Brotz similarly identifies the question of what phenomena are "functional" or "dysfunctional" as a political one in "Functionalism and Dynamic Analysis," *European Journal of Sociology*, II (1961), 170–179.

6. The most important earlier statements of this view can be found in Frank Tannenbaum, *Crime and the Community* (New York: Ginn, 1938), and E. M. Lemert, *Social Pathology* (New York: McGraw-Hill Book Co., Inc., 1951). A recent article stating a position very similar to mine is John Kitsuse, "Societal Reaction to Deviance: Problems of Theory and Method," *Social Problems,* 9 (Winter, 1962), 247–256.

7. Bronislaw Malinowski, *Crime and Custom in Savage Society* (New York: Humanities Press, 1926), pp. 77–80.

8. F. James Davis, "Crime News in Colorado Newspapers," *American Journal of Sociology,* LVII (January, 1952), 325–330.

9. See Albert K. Cohen and James F. Short, Jr., "Juvenile Delinquency," in Merton and Nisbet, *op. cit.,* p. 87.

10. See Harold Garfinkel, "Research Notes on Inter- and Intra-Racial Homicides," *Social Forces,* 27 (May, 1949), 369–381.

11. Edwin H. Sutherland, "White Collar Criminality," *American Sociological Review,* V (February, 1940), 1–12.

12. Clark Vincent, *Unmarried Mothers* (New York: The Free Press of Glencoe, 1961), pp. 3–5.

13. Arnold M. Rose and Arthur E. Prell, "Does the Punishment Fit the Crime?—A Study in Social Valuation," *American Journal of Sociology,* LXI (November, 1955), 247–259.

QUESTIONS FOR DISCUSSION

1. How does the author define deviance?

2. What does the author mean by the term "outsiders"? Give an example of how this applies to your life.

3. After reading about Malinowski's (1926) research on the Trobriand Islands, give an example of how his research applies today.

The "Discovery" of Child Abuse

Stephen J. Pfohl

Despite documentary evidence of child beating throughout the ages, the "discovery" of child abuse as deviance and its subsequent criminalization are recent phenomena. In a four-year period beginning in 1962, the legislatures of all fifty states passed statutes against the caretaker's abuse of children. This paper is a study of the organization of social forces which gave rise to the deviant labeling of child beating and which promoted speedy and universal enactment of criminal legislation. It is an examination of certain organized medical interests, whose concern in the discovery of the "battered child syndrome" manifestly contributed to the advance of humanitarian pursuits while covertly rewarding the groups themselves.

The structure of the present analysis is fourfold: First, an historical survey of social reaction to abusive behavior prior to the formulation of fixed labels during the early sixties, focussing on the impact of three previous reform movements. These include the nineteenth-century "house-of-refuge" movement, early twentieth century crusades by the Society for the Prevention of Cruelty to Children, and the rise of juvenile courts. The second section concentrates on the web of cultural values related to the protection of children at the time of the "discovery" of abuse as deviance. A third section examines factors associated with the organizational structure of the medical profession conducive to the "discovery" of a particular type of deviant label. The fourth segment discusses social reaction. Finally, the paper provides a sociological interpretation of a particular social-legal development. Generically it gives support for a synthesis of conflict and labeling perspectives in the sociology of deviance and law.

THE HISTORY OF SOCIAL REACTION:
PREVENTATIVE PENOLOGY AND "SOCIETY SAVING"

The purposeful beating of the young has for centuries found legitimacy in beliefs of its necessity for achieving disciplinary, educational or religious obedience (Radbill, 1968). Both the Roman legal code of "Patria Patistas" (Shepard, 1965), and the English common law (Thomas, 1973), gave guardians limitless power over their children who, with chattel-like status, had no legal right to protection.

The common law heritage of America similarly gave rise to a tradition of legitimized violence toward children. Legal guardians had the right to impose any punishment deemed necessary for the child's upbringing. In the seventeenth century, a period dominated by religious values and institutions, severe punishments were considered essential to the "sacred" trust of child-rearing (Earle, 1926: 119–126). Even in the late eighteenth and early nineteenth centuries, a period marked by the decline of religious domination and the rise of rationalism and a proliferation of statutes aimed at codifying unacceptable human behavior, there were no attempts to prevent caretaker abuse of children. A major court in the state of North Carolina declared that the parent's judgment of need for a child's punishment was presumed to be correct. Criminal liability was said to exist only in cases resulting in "permanent injury" (*State v. Pendergass,* in Paulsen, 1966b:686).

I am not suggesting that the American legal tradition failed to recognize any abuse of discipline as something to be negatively sanctioned. A few cases resulting in the legal punishment of parents who murdered their children have been recorded. But prior to the 1960's socio-legal reactions were sporadic, and atypical of sustained reactions against firmly labeled deviance.

Beginning in the early nineteenth century, a series of three reform movements directed attention to the plight of beaten, neglected and delinquent children. These included the nineteenth century "house-of-refuge" movement, the turn of the century crusades by the Society for the Prevention of Cruelty to Children and the early twentieth century rise of juvenile courts. Social response, however, seldom aimed measures at ameliorating abuse or correcting abusive parents. Instead, the child, rather than his or her guardians, became the object of humanitarian reform.

In each case the primary objective was not to save children from cruel or abusive parents, but to save society from future delinquents. Believing that wicked and irresponsible behavior was engendered by the evils of poverty and city life, these movements sought to curb criminal tendencies in poor, urban youths by removing them from corrupt environments and placing them in institutional settings. There they could learn order, regularity and obedience (Rothman, 1970). Thus, it was children, not their abusive guardians, who felt the weight of the moral crusade. They, not their parents, were institutionalized.

The "House of Refuge" Movement

Originating in the reformist dreams of the Jacksonian era, the so-called "House of Refuge Movement" sought to stem the social pathologies of an industrializing nation by removing young people, endangered by "corrupt urban environments," to institutional settings. Neglect statutes providing for the removal of the young from bad home lives were originally enacted to prevent children from mingling freely with society's dregs in alms houses or on the streets. In 1825, the first statute was passed and the first juvenile institution, the New York House of Refuge, was opened. Originally privately endowed, the institution soon received public funds to intervene in neglectful home situations and transplant children to a controlled environment, where they shared a "proper growing up" with other vagrant, abandoned and neglected youths as well as with delinquents who had violated criminal statutes. Similar institutions were established in Philadelphia and Boston a year later, in New Orleans in 1845, and in Rochester and Baltimore in 1849.

The constitutionality of the neglect statutes, which formed the basis for the House of Refuge Movement, was repeatedly challenged on the grounds that it was really imprisonment without due process. With few exceptions court case after court case upheld the policy of social intervention on the Aristotelian principle of "parens patriae." This principle maintained that the State has the responsibility to defend those who cannot defend themselves, as well as to assert its privilege in compelling infants and their guardians to act in ways most beneficial to the State.

The concept of preventive penology emerged in the wording of these court decisions. A distinction between "delinquency" (the actual violation of criminal codes) and "dependency" (being born into a poor home with neglectful or abusive parents) was considered irrelevant for "child saving." The two were believed to be intertwined in poverty and desolation. If not stopped, both would perpetuate themselves. For the future good of both child and society, "parens patriae" justified the removal of the young before they became irreparably tainted (Thomas, 1972:322–323).

The underlying concept of the House of Refuge Movement was that of preventive penology, not child protection. This crusade registered no real reaction against child beating. The virtue of removing children from their homes was not to point up abuse or neglect and protect its victims, it was to decrease the likelihood that parental inadequacies, the "cause of poverty," would transfer themselves to the child and hence to the next generation of society (Giovannoni, 1971:652). Thus, as indicated by Zalba (1966), the whole nineteenth century movement toward institutionalization actually failed to differentiate between abuse and poverty and therefore registered no social reaction against beating as a form of deviance.

Mary Ellen, the SPCC, and a Short-Lived Social Reaction

The first period when public interest focussed on child abuse occurred in the last quarter of the nineteenth century. In 1875, the Society for the Prevention of Cruelty to Animals intervened in the abuse case of a nine-year-old girl named Mary Ellen who had been treated viciously by foster parents. The case of Mary Ellen was splashed across the front pages of the nation's papers with dramatic results. As an outgrowth of the journalistic clamor, the New York Society for the Prevention of Cruelty to Children was formed. Soon incorporated under legislation that required law enforcement and court officials to aid agents of authorized cruelty societies, the NYSPCC and other societies modeled after it undertook to prevent abuse.

Though the police functions of the anti-cruelty societies represented a new reaction to abuse, their activities did not signify a total break with the society-saving emphasis of the House of Refuge Movement. In fact, three lines of evidence suggest that the SPCC enforcement efforts actually withheld a fixed label of deviancy from the perpetrators of abuse, in much the same manner as had the House of Refuge reforms. First, the "saving" of the child actually boosted the number of children placed in institutions, consequently supporting House of Refuge activities (Thomas, 1972:311). Second, according to Falks (1970:176), interorganizational dependency grew between the two reform movements, best evidenced by the success of SPCC efforts in increasing public support to child-care institutions under the auspices of House of

Refuge groups. Finally, and perhaps most convincingly, natural parents were not classified as abusers of the great majority of the so-called "rescued children." In fact, the targets of these savings missions were cruel employers and foster or adopted parents (Giovannoni, 1971:653). Rarely did an SPCC intervene against the "natural" balance of power between parents and children. The firmness of the SPCC's alleged social action against abuse appears significantly dampened by its reluctance to shed identification with the refuge house emphasis on the "industrial sins of the city" and to replace it with a reaction against individuals.

The decline of the SPCC movement is often attributed to lack of public interest, funding problems, mergers with other organizations and the assumption of protection services by public agencies (Felder, 1971:187). Its identification with the House of Refuge Movement also contributed to its eventual demise. More specifically, the House of Refuge emphasis on the separation of child from family, a position adopted and reinforced by the SPCC's activities, came into conflict with perspectives advocated by the newly-emerging professions of social work and child psychology (Kadushen, 1957:202f). Instead of removing the child from the home, these new interests emphasized efforts to unite the family (Thomas, 1972). This latter position, backed by the power of professional expertise, eventually undercut the SPCC's policy of preventive policing by emphasizing the protection of the home.

The erosion of the SPCC position was foreshadowed by the 1909 White House Conference on Children. This Conference proclaimed that a child should not be removed from his or her home for reasons of poverty alone, and called for service programs and financial aid to protect the home environment. Yet, the practice of preventive policing and institutionalization did not vanish, due, in part, to the development of the juvenile court system. The philosophy and practice of this system continued to identify abuse and neglect with poverty and social disorganization.

The Juvenile Court and the Continued Shadow of Abuse

The founding of the first juvenile court in Illinois in 1899 was originally heralded as a major landmark in the legal protection of juveniles. By 1920, courts were established in all but three states. Nonetheless, it is debatable that much reform was accomplished by juvenile court legislation. Coalitions of would-be reformers (headed by various female crusaders and the commissioners of several large public reformatories) argued for the removal of youthful offenders from adult institutions and advocated alteration of the punitive, entrepreneurial and sectarian "House of Refuge" institutions (Fox, 1970:1225–29). More institutions and improved conditions were demanded (Thomas, 1972:323). An analysis of the politics of juvenile court legislation suggests, however, that successful maneuvering by influential sectarian entrepreneurs resulted in only a partial achievement of reformist goals (Fox, 1970:1225–26). Legislation did remove juveniles from adult institutions. It did not reduce the House of Refuge Movement's control of juvenile institutions. Instead, legislation philosophically supported and financially reinforced the Movement's "society-saving" operation of sectarian industrial schools (Fox, 1970:1226–27).

The channeling of juvenile court legislation into the "society-saving" mold of the House of Refuge Movement actually withheld a deviant label from abusive parents. Even the reformers, who envisioned it as a revolution in child protection, did not see the court as protection from unfit parents. It was meant instead to prevent the development of "lower class" delinquency (Platt, 1969) and to rescue "those less fortunate in the social order" (Thomas, 1972:326). Again, the victims of child battering were characterized as pre-delinquents, as part of the general "problem" of poverty. These children, not their guardians, were the targets of court action and preventive policies. The courts, like the House of Refuge and SPCC movements before them, constrained any social reaction which would apply the label of deviant to parents who abused their children.

SOCIAL REACTION AT MID-CENTURY: THE CULTURAL SETTING FOR THE "DISCOVERY" OF ABUSE

The Decline of Preventative Penology

As noted, preventative penology represented the philosophical basis for various voluntary associations and legislative reform efforts resulting in the institutionalization of neglected or abused children. Its primary emphasis was on the protection of society. The decline of preventive penology is partially attributed to three variables: the perceived failure of "institutionalization," the impact of the "Great Depression" of the 1930s, and a change in the cultural meaning of "adult vices."

In the several decades prior to the discovery of abuse, the failure of institutionalization to "reorder" individuals became increasingly apparent. This realization undermined the juvenile courts' role in administering a pre-delinquency system of crime prevention. Since the rise of juvenile courts historically represented a major structural support for the notion of preventative penology, the lessening of its role removed a significant barrier to concern with abuse as an act of individual victimization. Similarly, the widespread experience of poverty during the Great Depression weakened other beliefs in preventive penology. As impersonal economic factors impoverished a great number of citizens of good moral credentials, the link between poverty and immorality began to weaken.

Another characteristic of the period immediately prior to the discovery of abuse was a changing cultural awareness of the meaning of adult vice as indices of the future character of children. "Parental immoralities that used to be seen as warnings of oncoming criminality in children [became] acceptable factors in a child's homelife" (Fox, 1970:1234). Parental behavior such as drinking, failing to provide a Christian education, and refusing to keep a child busy with useful labor, were no longer classified as nonacceptable nor deemed symptoms of immorality transmitted to the young. Hence, the saving of society from the tainted young became less of a mandate, aiding the perception of social harm against children as "beings" in themselves.

Advance of Child Protection

Concurrent with the demise of "society-saving" in the legal sphere, developments in the fields of child welfare and public policy heightened interest in the problems of the child as an individual. The 1909 White House Conference on Children spawned both the "Mother's Aid" Movement and the American Association for the Study and Prevention of Infant Mortality. The former group, from 1910 to 1930, drew attention to the benefits of keeping children in the family while pointing out the detrimental effects of dehumanizing institutions. The latter group then, as now, registered concern over the rate of infant deaths.

During the first half of the twentieth century, the Federal Government also met the issue of child protection with legislation that regulated child labor, called for the removal of delinquent youths from adult institutions, and established, in 1930, a bureaucratic structure whose purpose revolved around child protection. The Children's Bureau of HEW immediately adopted a "Children's Charter" promising every child a home with love and security plus full-time public services for protection from abuse, neglect, exploitation or moral hazard (Radbill, 1968:15).

Despite the growth of cultural and structural dispositions favoring the protection and increased rights of children, there was still no significant attention given to perpetrators of abuse, in the courts (Paulsen, 1966:710), in the legislature (DeFrancis, 1967:3), or by child welfare agencies (Zalba, 1966). While this inactivity may have been partly caused by the lack of effective mechanisms for obtaining data on abuse (Paulsen, 1966:910), these agencies had little social incentive for interfering with an established power set—the parent over the child. As a minority group possessing neither the collective awareness nor the elementary organizational skills necessary to address their grievances to either the courts or to the legislators, abused and neglected children awaited the advocacy of some other organized interest. This outside intervention would not, however, be generated by that sector of "organized helping" most closely associated with the protective needs of children—the growing web of child welfare bureaucracies at State and Federal levels. Social work had identified its professional advance with the adoption of the psychoanalytic model of casework (Zalba, 1966). This perspective, rather than generating a concern with political inequities internal to the family, focused instead on psychic disturbances internal to its members. Rather than challenging the strength of parents, this served to reinforce the role of powerful guardians in the rearing of young.

Nor would advocacy come from the public at large. Without organized labeling interests at mid-century, child abuse had not become an issue publicly regarded as a major social problem. In fact, a fairly general tolerance for abuse appeared to exist. This contention is supported by the findings of a nationwide study conducted by NORC during the period in which laws against abuse were actually being adopted (Gil & Nobel, 1969). Despite the wide-scale publicizing of abuse in this "post-discovery" period, public attitudes remained lenient. Data revealed a high degree of empathy with convicted or suspected perpetrators (Gil, 1970:63–67). These findings are understandable in light of cultural views accepting physical force against children as a nearly universally applied precept of intrafamilial organization (Goode, 1971). According to the coordinator of the national survey, "Culturally determined permissive attitudes toward the use of physical force in child rearing seem to constitute the common core of all physical abuse of children in American society" (Gil, 1970:141).

While the first half of the twentieth century is characterized by an increasing concern for child welfare, it developed with neither an organizational nor attitudinal reaction against child battering as a specific form of deviance. The "discovery" of abuse, its definition as a social problem and the socio-legal reaction against it, awaited the coalition of organized interests.

THE ORGANIZATION OF SOCIAL REACTION
AGAINST THE "BATTERED CHILD SYNDROME"

What organization of social forces gave rise to the discovery of abuse as deviance? The discovery is not attributable to any escalation of abuse itself. Although some authors have recently suggested that the increasing nuclearization of the family may increase the victimization of its offspring (Skolnick & Skolnick, 1971), there has never been any evidence that, aside from reporting inflation due to the impact of new laws, battering behavior was actually increasing (Eads, 1972). The attention here is on the organizational matrix encouraging a recognition of abuse as a social problem. In addressing this issue I will examine factors associated with the organizational structure of the medical profession leading to the discovery of abuse by pediatric radiologists rather than by other medical practitioners.

The "discovery" of abuse by pediatric radiology has often been described chronologically (Radbill, 1968:15; McCoid, 1965:2–5; Thomas, 1972:330). John Caffey (1946) first linked observed series of long bone fractures in children with what he termed some "unspecific origin." Although his assumption was that some physical disturbance would be discovered as the cause of this pattern of "subdural hematoma," Coffey's work prompted a series of further investigations into various bone injuries, skeletal trauma, and multiple fractures in young children. These research efforts lead pediatric radiology gradually to shift its diagnosis away from an internal medical explication toward the ascription of social cause.

In subsequent years it was suggested that what was showing up on x-rays might be the results of various childhood accidents (Barmeyer, *et al.,* 1951), of "parental carelessness" (Silverman, 1953), of "parental conduct" (Bakwin, 1956), and most dramatically, of the "indifference, immaturity and irresponsibility of parents" (Wooley & Evans, 1955). Surveying the progression of this research and reviewing his own investigations, Coffey (1957) later specified "misconduct and deliberate injury" as the primary etiological factors associated with what he had previously labelled "unspecific trauma." The discovery of abuse was on its way. Both in scholarly research (McCoid, 1966:7) and journalist outcry (Radbill, 1968:16), the last years of the fifties showed dramatically increased concern for the beaten child.

Why did pediatric radiologists and not some other group "see" abuse first? Legal and social welfare agents were either outside the scene of abusive behavior or inside the constraining vision of psychoanalytically committed casework. But clinicians, particularly hospital physicians and pediatricians, who encountered abused children more immediately, should have discovered "abuse" before the radiologists.

Four factors impeded the recognition of abuse (as it was later labelled). First, some early research maintained that doctors in emergency room settings were simply unaware of the possibilities of "abuse" as a diagnosis (Bain, 1963; Boardman, 1962). While this may be true, the massive symptoms (blood, burns, bruises) emergency room doctors faced far outweighed the lines appearing on the x-ray screens of radiologic specialists. A second line of evidence contends that many doctors were simply psychologically unwilling to believe that parents would inflict such atrocities on their own children (Elmer, 1960; Fontana, Donovan, Wong, 1963; Kempe *et al.,* 1963). This position is consistent with the existing cultural assumptions pairing parental power with parental wisdom and benevolence. Nonetheless, certain normative and structural elements within professional medicine appear of greater significance in re-inforcing the physician's reluctance to get involved, even diagnostically. These factors are the "norm of confidentiality between doctor and client" and the goal of profes-sional autonomy.

The "norm of confidentiality" gives rise to the third obstacle to a diagnosis of abuse: the possibility of legal liability for violating the confidentiality of the physician-patient relationship (Boardman, 1962). Interestingly, although some re-search connotes doctors' concern over erroneous diagnosis (Braun, Braun & Si-monds, 1953), physicians primarily view the parent, rather than the child, as their real patient. On a strictly monetary level, of course, it is the parent who contracts with that doctor. Additional research has indicated that, particularly in the case of pediatricians, the whole family is viewed as one's clinical domain (Bucher & Strauss, 1961:329). It is from this vantage point that the impact of possible liability for a di-agnostic disclosure is experienced. Although legal liability for a diagnosis of abuse may or may not have been the risk (Paulsen, 1967b:32), the belief in such liability could itself have contributed to the narrowness of a doctor's diagnostic perceptions (McCoid, 1966:37).

A final deterrent to the physician's "seeing" abuse is the reluctance of doctors to become involved in a criminal justice process that would take both their time (Bain, 1963:895) and ability to guide the consequences of a particular diagnosis (Board-man, 1962:46). This deterrent is particularly related to the traditional success of or-ganized medicine in politically controlling the consequences of its own performance, not just for medical practitioners but for all who come in contact with a medical prob-lem (Freidson, 1969:106; Hyde, *et al.,* 1954).

The political control over the consequences of one's profession would be jeopar-dized by the medical diagnosis of child abuse. Doctors would be drawn into judicial proceedings and subordinated to a role as witnesses. The outcome of this process would be decided by criminal justice standards rather than those set forth by the medical profession. Combining this relatively unattractive alternative with the obvi-ous and unavoidable drain on a doctor's financial earning time, this fourth obstacle to the clinician's discovery of abuse is substantial.

Factors Conducive to the Discovery of Abuse by Pediatric Radiology

Why didn't the above factors inhibit the discovery of abuse by pediatric radiologists as well as by clinicians? First it must be recognized that the radiologists in question (Caffey, Barmeyer, Silverman, Wooley and Evans) were all researchers of children's x-rays. As such, the initial barrier becomes irrelevant. The development of diagnos-tic categories was a consequence rather than a pre-condition of the medical mission. Regarding the psychological denial of parental responsibility for atrocities, it must be remembered that the dramatic character of a beating is greatly reduced by the time it reaches an x-ray laboratory. Taken by technicians and developed as black and white prints, the radiologic remnants of abuse carry with them little of the horror of the bloody assault.

With a considerable distance from the patient and his or her family, radiologists are removed from the third obstacle concerning legal liabilities entailed in violating the doctor-patient relationship. Unlike pediatricians, radiologists do not routinely regard the whole family as one's clinical domain. Of primary importance is the individual whose name or number is imprinted on the x-ray frames. As such, fears about legal sanctions instigated by a parent whom one has never seen are less likely to deter the recognition of abuse.

Given the irrelevance of the first three obstacles, what about the last? Pediatric radiologists are physicians, and as such would be expected to participate in the "professional control of consequences" ethos. How is it that they negotiate this obstacle in favor of public recognition and labelling of abuse?

The Discovery: An Opportunity for Advancement Within the Medical Community

To ask why the general norm of "professional control of consequences" does not apply equally to radiologists as to their clinical counterparts is to confuse the reality of organized medicine with its image. Although the medical profession often appears to outsiders as a separate and unified community within a community (Goode, 1957), and although medical professionals generally favor the maintenance of this image (Glaser, 1960), it is nonetheless more adequately described as an organization of internally competing segments, each striving to advance its own historically derived mission and future importance (Bucher & Strauss, 1961). In analyzing pediatric radiology as one such segment, several key variables facilitated its temporary parting with the dominant norms of the larger medical community. This parting promoted the elevation of its overall status within that community.

The first crucial element is that pediatric radiology was a marginal specialty within organized medicine. It was a research-oriented subfield in a profession that emphasized face-to-face clinical interaction. It was a safe intellectual endeavor within an overall organization which placed a premium on risky pragmatic enterprise. Studies of value orientations among medical students at the time of the "discovery" of abuse have suggested that those specialties which stress "helping others," "being of service," "being useful," and "working with people" were ranked above those which work "at medical problems that do not require frequent contact with patients" (Cahalan, 1957). On the other hand, intellectual stimulation afforded very little prestige. Supporting this conclusion was research indicating that although forty-three percent of practicing physicians selected "close patient relations" as a mandate of their profession, only twenty-four percent chose "research" as worthy of such an evaluation (Philips, 1964). Pairing this ranking system with the profession's close-knit, "fraternity-like" communication network (Hall, 1946), one would expect research-oriented radiologists to be quite sensitive about their marginal evaluation by colleagues.

Intramedical organizational rankings extend along the lines of risk-taking as well as patient-encounters. Here, too, pediatric radiologists have traditionally ranked lower than other medical specialties. Becker's (1961) study of medical student culture suggests that the most valued specialties are those which combine wide experiences with risk and responsibility. These are most readily "symbolized by the possibility of killing or disabling patients in the course of making a mistake" (Freidson, 1969:

107). From this perspective, it is easy to understand why surgery and internal medicine head the list of the most esteemed specialties. Other research has similarly noted the predominance of surgeons among high elected officials of the American Medical Association (Hall, 1946). Devoid of most risk taking, little involved in life or death decisions, pediatric radiologists are again marginal to this ethos of medical culture.

The "discovery" of child abuse offered pediatric radiologists an alternative to their marginal medical status. By linking themselves to the problem of abuse, radiologists became indirectly tied into the crucial clinical task of patient diagnosis. In addition, they became a direct source of input concerning the risky "life or death" consequences of child beating. This could represent an advance in status, a new basis for recognition within the medical profession. Indeed, after initial documentation of abuse, literature in various journals of radiology, roentgenology and pediatrics, articles on this topic by Wooley and Evans (1955) and Gwinn, deWin and Peterson (1961) appeared in the *Journal of the American Medical Association*. These were among the very few radiologic research reports published by that prestigious journal during the time period. Hence, the first factor conducive to the radiological discovery of abuse was a potential for intraorganizational advance in prestige.

The Discovery: An Opportunity for Coalition Within the Medical Community

A second factor encouraging the discovery of abuse by relatively low-status pediatric radiologists concerns the opportunity for a coalition of interests with other more prestigious segments within organized medicine. The two other segments radiologists joined in alliance were pediatrics and psychodynamically oriented psychiatry. By virtue of face-to-face clinical involvements, these specialties were higher ranking than pediatric radiology. Nevertheless each contained a dimension of marginality. Pediatrics had attained valued organizational status several decades prior to the discovery of abuse. Yet, in an age characterized by preventive drugs and treatments for previously dangerous or deadly infant diseases, it was again sliding toward the margins of the profession (Bucher & Strauss, 1961). Psychodynamic psychiatry (as opposed to its psychosomatic cousin) experienced marginality in dealing with non-physical problems.

For both pediatrics and psychodynamic psychiatry, links with the problem of abuse could partially dissipate the respective marginality of each. Assuming a role in combating the "deadly" forces of abuse could enlarge the "risky" part of the pediatric mission. A symbolic alliance of psychodynamic psychiatry with other bodily diagnostic and treatment specialties could also function to advance its status. Neither of these specialties was in a position to "see" abuse before the radiologists. Pediatricians were impeded by the obstacles discussed above. Psychiatrists were blocked by the reluctance of abusive parents to admit their behavior as problematic (Steele & Pollock, 1968). Nonetheless, the interests of both could perceivably be advanced by a coalition with the efforts of pediatric radiologists. As such, each represented a source of potential support for pediatric radiologists in their discovery of abuse. This potential for coalition served to reinforce pediatric radiology in its movement toward the discovery of abuse.

The Discovery: An Opportunity for the Application of an Acceptable Label

A crucial impediment to the discovery of abuse by the predominant interests in organized medicine was the norm of controlling the consequences of a particular diagnosis. To diagnose abuse as social deviance might curtail the power of organized medicine. The management of its consequences would fall to the extramedical interests of formal agents of social control. How is it then, that such a diagnosis by pediatric radiology and its endorsement by pediatric and psychiatric specialties, is said to have advanced these specialties within the organization of medicine? Wasn't it more likely that they should have received criticism rather than acclaim from the medical profession?

By employing a rather unique labelling process the coalition of discovery interests was able to convert the possible liability into a discernible advantage. The opportunity of generating a medical, rather than socio-legal label for abuse provided the radiologists and their allies with a situation in which they could both reap the rewards associated with the diagnosis and avoid the infringement of extra-medical controls. What was discovered was no ordinary behavior form but a "syndrome." Instead of departing from the tradition of organized medicine, they were able to idealize its most profound mission. Possessing a repertoire of scientific credibility, they were presented with the opportunity "to label as illness what was not previously labeled at all or what was labeled in some other fashion, under some other institutional jurisdiction" (Freidson, 1971:261).

The symbolic focal point for the acceptable labeling of abuse was the 1962 publication of an article entitled "The Battered-Child Syndrome" in the *Journal of the American Medical Association* (Kempe *et al.*, 1962). This report, representing the joint research efforts of a group of radiologic, pediatric, and psychiatric specialists, labelled abuse as a "clinical condition" existing as an "unrecognized trauma" (Kempe, 1962:17). It defined the deviance of its "psychopathic" perpetrators as a product of "psychiatric factors" representing "some defect in character structure" (Kempe, 1962:24). As an indicator of prestige within organized medicine, it is interesting to note that the position articulated by those labellers was endorsed by the editorial board of the AMA in that same issue of *JAMA*.

As evidenced by the AMA editorial, the discovery of abuse as a new "illness" reduced drastically the intra-organizational constraints on doctors "seeing" abuse. A diagnostic category had been invented and publicized. Psychological obstacles in recognizing parents as capable of abuse were eased by the separation of normatively powerful parents from non-normatively pathological individuals. Problems associated with perceiving parents as patients whose confidentiality must be protected were reconstructed by typifying them as patients who needed help. Moreover, the maintenance of professional autonomy was assured by pairing deviance with sickness. This last statement is testimony to the power of medical nomenclature. It was evidenced by the fact that (prior to its publication) the report which coined the label "battered child syndrome" was endorsed by a Children's Bureau conference which included social workers and law enforcement officials as well as doctors (McCoid, 1965:12).

The Generation of the Reporting Movement

The discovery of the "battered child syndrome" was facilitated by the opportunities for various pediatric radiologists to advance in medical prestige, form coalitions with other interests, and invent a professionally acceptable deviant label. The application of this label has been called the child abuse reporting movement. This movement was well underway by the time the 1962 Children's Bureau Conference confirmed the radiological diagnosis of abuse. Besides foreshadowing the acceptance of the sickness label, this meeting was also the basis for a series of articles to be published in *Pediatrics* which would further substantiate the diagnosis of abuse. Soon, however, the reporting movement spread beyond intra-organizational medical maneuvering to incorporate contributions from various voluntary associations, governmental agencies, as well as the media.

Extramedical responses to the newly discovered deviance confirmed the recognition of abuse as an illness. These included reports by various social welfare agencies which underscored the medical roots of the problem. For instance, the earliest investigations of the problem by social service agents resulted in a call for cooperation with the findings of radiologists in deciding the fate of abusers (Elmer, 1960: 100). Other studies called for "more comprehensive radiological examinations" (Boardman, 1962:43). That the problem was medical in its roots as well as consequences was reinforced by the frequent referral of caseworkers to themselves as "battered child therapists" whose mission was the "curing" of "patients" (Davoren, 1968). Social welfare organizations, including the Children's Division of the American Humane Association, the Public Welfare Association, and the Child Welfare League, echoed similar concerns in sponsoring research (Children's Division, 1963; De Francis, 1963) and lobbying for "treatment based" legislative provisions (McCoid, 1965).

Not all extramedical interests concurred with treatment of abusers as "sick." Various law enforcement voices argued that the abuse of children was a crime and should be prosecuted. On the other hand, a survey of thirty-one publications in major law journals between 1962–1972 revealed that nearly all legal scholars endorsed treatment rather than punishment to manage abusers. Lawyers disagreed, however, as to whether reports should be mandatory and registered concern over who should report to whom. Yet, all concurred that various forms of immunity should be granted reporters (Paulsen, 1967a; De Francis, 1967). These are all procedural issues. Neither law enforcers nor legal scholars parted from labelling abuse as a problem to be managed. The impact of the acceptable discovery of abuse by a respected knowledge sector (the medical profession) had generated a stigmatizing scrutiny bypassed in previous eras.

The proliferation of the idea of abuse by the media cannot be underestimated. Though its stories were sensational, its credibility went unchallenged. What was publicized was not some amorphous set of muggings but a "syndrome." Titles such as "Cry rises from beaten babies" (*Life,* June 1963), "Parents who beat children" (*Saturday Evening Post,* October 1962), "The shocking price of parental anger" (*Good Housekeeping,* March 1964), and "Terror struck children" (*New Republic,* May 1964) were all buttressed by an awe of scientific objectivity. The problem had become "real" in the imaginations of professionals and laymen alike. It was rediscovered visually by ABC's "Ben Casey," NBC's "Dr. Kildare," and CBS's "The Nurses," as well as in several other television scripts and documentaries (Paulsen, 1967b:488–489).

Discovered by the radiologists, substantiated by their colleagues, and distributed by the media, the label was becoming widespread. Despite this fact, actual reporting laws were said to be the cooperative accomplishments of zealous individuals and voluntary associations (Paulsen, 1967b:491). Who exactly were these "zealous individuals"?

Data on legislative lobbyists reveal that, in almost every state, the civic committee concerned with abuse legislation was chaired by a doctor who "just happened" to be a pediatrician (Paulsen, 1967b:491). Moreover, "the medical doctors who most influenced the legislation frequently were associated with academic medicine" (Paulsen, 1967b:491). This information provides additional evidence of the collaborative role of pediatricians in guiding social reaction to the deviance discovered by their radiological colleagues.

Lack of Resistance to the Label

In addition to the medical interests discussed above, numerous voluntary associations provided support for the movement against child abuse. These included the League of Women Voters, Veterans of Foreign Wars, the Daughters of the American Republic, the District Attorneys Association, Council of Jewish Women, State Federation of Women's Clubs, Public Health Associations, plus various national chapters of social workers (Paulsen, 1967b:495). Two characteristics emerge from an examination of these interests. They either have a professional stake in the problem or represent the civic concerns of certain upper-middle class factions. In either case the labelers were socially and politically removed from the abusers, who in all but one early study (Steele and Pollock), were characterized as lower class and minority group members.

The existence of a wide social distance between those who abuse and those who label facilitates not only the likelihood of labelling but nullifies any organized resistance to the label by the "deviant" group itself. Research findings which describe abusers as belonging to no outside-the-family associations or clubs (Young, 1964) or which portray them as isolates in the community (Giovannoni, 1971) reinforce the conclusion. Labelling was generated by powerful medical interests and perpetuated by organized media, professional and upper-middle class concerns. Its success was enlarged by the relative powerlessness and isolation of abusers, which prevented the possibility of organized resistance to the labelling.

THE SHAPE OF SOCIAL REACTION

I have argued that the organizational advantages surrounding the discovery of abuse by pediatric radiology set in motion a process of labelling abuse as deviance and legislating against it. The actual shape of legislative enactments has been discussed elsewhere (De Francis, 1967; Paulsen, 1967a). The passage of the reporting laws encountered virtually no opposition. In Kentucky, for example, no one even appeared to testify for or against the measure (Paulsen, 1967b:502). Any potential opposition from the American Medical Association, whose interests in autonomous control of the consequences of a medical diagnosis might have been threatened, had been undercut by the radiologists' success in defining abuse as a new medical problem. The AMA, unlikely to argue against conquering illness, shifted to support reporting legislation which would maximize a physician's diagnostic options.

The consequences of adopting a "sick" label for abusers is mirrored in two findings: the low rate of prosecution afforded offenders and the modification of reporting statutes so as exclusively to channel reporting toward "helping services." Regarding the first factor, Grumet (1970:306) suggests that despite existing laws and reporting statutes, actual prosecution has not increased since the time of abuse's "discovery." In support is Thomas (1972) who contends that the actual percentage of cases processed by family courts has remained constant during the same period. Even when prosecution does occur, convictions are obtained in only five to ten percent of the cases (Paulsen, 1966b). And even in these cases, sentences are shorter for abusers than for other offenders convicted under the same law of aggravated assault (Grumer, 1970:307).

State statutes have shifted on reporting from an initial adoption of the Children's Bureau model of reporting to law enforcement agents, toward one geared at reporting to child welfare or child protection agencies (De Francis, 1970). In fact, the attention to abuse in the early sixties has been attributed as a factor in the development of specialized "protective interests" in states which had none since the days of the SPCC crusades (Eads, 1969). This event, like the emphasis on abuser treatment, is evidence of the impact of labelling of abuse as an "illness."

REFERENCES

Bain, Katherine. 1963. "The physically abused child." *Pediatrics* 31 (June): 895–897.

Bakwin, Harry. 1956. "Multiple skeletal lesions in young children due to trauma." *Journal of Pediatrics* 49 (July): 7–15.

Barmeyer, G. H., L. R. Anderson and W. B. Cox. 1951. "Traumatic periostitis in young children." *Journal of Pediatrics* 38 (Feb): 184–190.

Becker, Howard S. 1963. *The Outsiders.* New York: The Free Press.

Becker, Howard S. et al. 1961. *Boys in White.* Chicago: University of Chicago Press.

Boardman, Helen. 1962. "A project to rescue children from inflicted injuries." *Journal of Social Work* 7 (January): 43–51.

Braun, Ida G., Edgar J. Braun and Charlotte Simonds. 1963. "The mistreated child." *California Medicine* 99 (August): 98–103.

Bremner, R. 1970. *Children and Youth in America: A Documentary History, Vol. 1.* Cambridge, Mass.: Harvard University Press.

Bucher, Rue and Anselm Strauss. 1961. "Professions in process." *American Journal of Sociology* 66 (January): 325–334.

Caffey, John. 1946. "Multiple fractures in the long bones of infants suffering from chronic subdural hematoma." *American Journal of Roentgenology* 56(August): 163–173.

———. 1957. "Traumatic lesions in growing bones other than fractures and lesions: clinical and radiological features." *British Journal of Radiology* 30(May): 225–238.

Cahalan, Don. 1957. "Career interests and expectations of U.S. medical students." 32: 557–563.

Chambliss, William J. 1964. "A sociological analysis of the law of vagrancy." *Social Problems* 12(Summer): 67–77.

Children's Division. 1963. *Child Abuse—Preview of a Nationwide Survey.* Denver: American Humane Association (Children's Division).

Davoren, Elizabeth. 1968. "The role of the social worker." Pp. 153–168 in Ray E. Helfer and Henry C. Kempe (eds.), *The Battered Child.* Chicago: University of Chicago Press.

De Francis, Vincent. 1963. "Parents who abuse children." *PTA Magazine* 58(November): 16–18.

———. 1967. "Child abuse—the legislative response." *Denver Law Journal* 44(Winter): 3–41.

———. 1970. *Child Abuse Legislation in the 1970's.* Denver: American Humane Association.

Eads, William E. 1969. "Observations on the establishment of child protection services in California." *Stanford Law Review* 21(May): 1129–1155.

Earle, Alice Morse. 1926. *Child Life in Colonial Days.* New York: Macmillan.

Elmer, Elizabeth. 1960. "Abused young children seen in hospitals." *Journal of Social Work* 3(October): 98–102.

Felder, Samuel. 1971. "A lawyer's view of child abuse." *Public Welfare* 29: 181–188.

Folks, Homer. 1902. *The Case of the Destitute, Neglected and Delinquent Children.* New York: Macmillan.

Fontana, V., D. Donovan and R. Wong. 1963. "The maltreatment syndrome in children." *New England Journal of Medicine* 269(December): 1389–1394.

Fox, Sanford J. 1970. "Juvenile justice reform: an historical perspective." *Stanford Law Review* 22(June): 1187–1239.

Freidson, Eliot J. 1968. "Medical personnel: physicians." Pp. 105–114 in David L. Sills (ed.), *International Encyclopedia of the Social Sciences.* Vol. 10. New York: Macmillan.

———. 1971. The Profession of Medicine: *A Study in the Sociology of Applied Knowledge.* New York: Dodd, Mead and Co.

Gil, David. 1970. *Violence Against Children.* Cambridge, Mass.: Harvard University Press.

Gil, David and John H. Noble. 1969. "Public knowledge, attitudes and opinions about physical child abuse." *Child welfare* 49(July): 395–401.

Giovannoni, Jeanne. 1971. "Parental mistreatment." *Journal of Marriage and the Family* 33(November): 649–657.

Glaser, William A. 1960. "Doctors and politics." *American Journal of Sociology* 66(November): 230–245.

Goode, William J. 1957. "Community within a community: the profession." *American Sociological Review* 22(April): 194–200.

———. 1971. "Force and violence in the family." *Journal of Marriage and the Family* 33(November): 424–436.

Grumet, Barbara R. 1970. "The plaintive plaintiffs: victims of the battered child syndrome." *Family Law Quarterly* 4(September): 296–317.

Gusfield, Joseph R. 1963. *Symbolic Crusades.* Urbana, Ill.: University of Illinois Press.

Gwinn, J. J., K. W. Lewin and H. G. Peterson. 1961. "Roentgenographic manifestations of unsuspected trauma in infancy." *Journal of the American Medical Association* 181(June): 17–24.

Hall, Jerome. 1952. *Theft, Law and Society.* Indianapolis: Bobbs-Merrill Co.

Hall, Oswald. 1946. "The informal organization of medicine." *Canadian Journal of Economics and Political Science* 12(February): 30–41.

Hyde, D. R., P. Wolff, A. Gross and E. L. Hoffman. 1954. "The American Medical Association: power, purpose and politics in organized medicine." *Yale Law Journal* 63(May): 938–1022.

Kadushin, Alfred. 1967. *Child Welfare Services.* New York: Macmillan.

Kempe, C. H., F. N. Silverman, B. F. Steele, W. Droegemuller and H. K. Silver. 1962. "The battered-child syndrome." *Journal of the American Medical Association* 181(July): 17–24.

Lemert, Edwin M. 1974. "Beyond Mead: the societal reaction to deviance." *Social Problems* 21(April): 457–467.

McCoid, A. H. 1965. "The battered child syndrome and other assaults upon the family." *Minnesota Law Review* 50(November): 1–58.

Paulsen, Monrad G. 1966. "The legal framework for child protection." *Columbia Law Review* 66(April): 679–717.

———. 1967. "Child abuse reporting laws: the shape of the legislation." *Columbia Law Review* 67(January): 1–49.

Philips, Bernard S. 1964. "Expected value deprivation and occupational preference." *Sociometry* 27(June): 15–160.

Platt, Anthony M. 1969. *The Child Savers: The Invention of Juvenile Delinquency*. Chicago: University of Chicago Press.

Quinney, Richard, 1970. *The Social Reality of Crime*. Boston: Little, Brown.

Radbill, Samuel X. 1968. "A history of child abuse and infanticide." Pp. 3–17 in Ray E. Helfer and Henry C. Kempe (eds.), *The Battered Child*. Chicago: University of Chicago Press.

Rothman, David J. 1971. *The Discovery of the Asylum: Social Order and Disorder in the New Republic*. Boston: Little, Brown.

Shepard, Robert E. 1965. "The abused child and the law." *Washington and Lee Law Review* 22(Spring): 182–195.

Silverman, F. N. 1965. "The roentgen manifestations of unrecognized skeletal trauma in infants." *American Journal of Roentgenology, Radium and Nuclear Medicine* 69(March): 413–426.

Skolnick, Arlene and Jerome H. Skolnick. 1971. *The Family in Transition*. Boston: Little, Brown.

Steele, Brandt and Carl F. Pollock. 1968. "A psychiatric study of parents who abuse infants and small children." Pp. 103–147 in Ray E. Helfer and Henry C. Kempe (eds.), *The Battered Child*. Chicago: University of Chicago Press.

Sutherland, Edwin H. 1950. "The diffusion of sexual psychopath laws." *American Journal of Sociology* 56(September): 142–148.

Thomas, Mason P. 1972. "Child abuse and neglect: historical overview, legal matrix and social perspectives." *North Carolina Law Review* 50(February): 239–249.

Woolley, P V. and W. A. Evans Jr. 1955. "Significance of skeletal lesions in infants resembling those of traumatic origin." *Journal of the American Medical Association* 158(June): 539–543.

Young, Leontine. 1964. *Wednesday's Children: A Study of Child Neglect and Abuse*. New York: McGraw-Hill.

Zalba, Serapio R. 1966. "The abused child. I. A survey of the problems." *Social Work* 11(October): 3–16.

QUESTIONS FOR DISCUSSION

1. How do the organizations of social forces give rise to the deviant labeling of child abuse?

2. How did the reform movements prior to the 1960s help and hurt parent-child relations?

3. Why are pediatric radiologists and not some other group (social welfare agents, law enforcement, hospital physicians) credited for the discovery of child abuse and setting in motion a process of labeling child abuse as deviance?

Cult Update

Justine Janette

Scientology

If you communicated with CAN [Cult Awareness Network] in any way over the years—if you ever called, sent a letter or a check, or received their newsletter—Scientology may now have your name or other personal information. Judge Thomas Quinn recently ordered that all CAN records be turned over to the Cook County Sheriff for public sale. Instead Scientology attorney Kendrick Moxon accepted the records as settlement in full of the near two million dollar Jason Scott Judgment against CAN. (The auction was expected to produce only a small fraction of the Jason Scott judgment since Scientology was expected to be the only major bidder.) They now belong to Gerald Beenie, a Scientologist who bought the Jason Scott judgment for $25,000. In over 50 lawsuits all of which but one for near two million dollars were either won by CAN or dismissed, Scientology drove CAN out of business and subsequently into bankruptcy. The CAN service mark and logo were sold to a Scientology attorney for $20,000 several years ago. Scientology also got the phone number. Scientologists now answer the phones "Cult Awareness Network." The records include all confidential correspondence, financial records, and membership and subscriber lists including more than 10,000 names. Any questions may be sent to Edward Lottick, M.D., acting chairman of the board of the (original) Cult Awareness Network and president of the Foundation for Human Rights: 41 Gershom Pl, Kingston, PA 18704; 570-287-1377; elottick@aol.com.

Aum Shin Rikyo

Despite Aum Shin Rikyo's horrible terrorist gassing of the Tokyo subways in 1995, and despite prison terms for many of Aum's leaders and the loss of its tax privileges as a religious organization, the Japanese-based cult is not only still in existence, it is undergoing a revival. Membership is up, businesses are flourishing, and property purchases are soaring. Japan's Public Security Investigation Agency (PSIA) has published two reports warning of the resurgence of the cult. "As far as we can see," says the head of the PSIA's Aum unit, "the potential danger represented by the cult hasn't diminished at all."

Branch Davidians

While he was a member of the Branch Davidians, George Roden, 60, was part of the cult's leadership under guru David Koresh. Roden left the group in 1987 after a gun

"Cult Update" by Justine Janette from *Skeptic*, Spring 1999, v7 i2 p18. Reprinted by permission of Skeptic Magazine (www.skeptic.com).

battle with Koresh. In December, Roden was found dead on the grounds outside a psychiatric hospital by an employee of Big Spring State Hospital in Texas. Roden was apparently trying to escape from the hospital. A preliminary autopsy indicated he died of a heart attack.

Concerned Christians

Twenty members of the Denver-based group Concerned Christians may have taken up residence in Rafina, a small town 15 miles west of Athens, Greece. Recently deported from Israel for allegedly plotting to instigate a shoot-out with Jerusalem police in an attempt to hasten the return of Christ, they are thought to have joined other Concerned Christians members in Greece. Greek security officials are investigating reports from the media that the 20 cult members joined other members of their group in rented apartments and villas in the nearby hillside community of Neo Voutza. Authorities ordered surveillance of the two sites and together with immigration officials are attempting to determine if these individuals are in fact members of the Denver-based group. Concerned Christians leader and former Denver resident Monte Kim Miller disappeared with his followers in late September, 1998. Some members resurfaced in Jerusalem only to be deported by Israel. Miller, who founded the group, prophesied an apocalypse would strike Denver last fall, and intends to die on the streets of Jerusalem in December, 1999, only to rise again in three days. He has made other doomsday predictions and claims to be the voice of God.

Genesis Associates

Known as "the two Pats," Patricia Mansmann, psychologist, and Patricia Neuhusel, licensed social worker, co-founded Genesis Associates in Pennsylvania in the late 1980s. Genesis was conceived as a professional practice specializing in drug and alcohol counseling, but now both founders may have their licenses revoked. Apart from several active lawsuits filed against Genesis, Prosecuting Attorney Bernadette Paul and the state Bureau of Professional and Occupational Affairs filed formal charges in February contending "the two Pats" used dangerous and unprofessional counseling techniques. Genesis has been under scrutiny for years because of the Pats' use of techniques they call "rage work" and "detachment" therapies. "Rage work" involves members beating each other with pillows and plastic bats while others scream obscenities at them. "Detachment" therapy requires that couples split up and that parents cut off all contact with children, associating only with those in the Genesis "Network." Neither of these practices is accepted as standard, and may be harmful to clients. Mansmann, whose psychologist's license was suspended for two years, violated her suspension agreement by continuing to regularly treat clients. Like the complaint filed in 1996, the state again seeks to revoke the pair's licenses, but permanently this time, accompanied by fines of $10,000 for each count of wrongdoing.

Heaven's Gate

In March, 1997, 39 Heaven's Gate cult members committed mass suicide in a rented southern California mansion. On February 22, after a two-year court battle for the rights to property of deceased cult members, surviving ex-cult members Mark and Sarah King of Phoenix were denied rights to the belongings. The property will go to the San Diego Public Administrator's Office to auction off, with proceeds going to families of the deceased members to offset expenses for funerals and burials. "We're pleased that the court validated what we believed to be all along, that this is a simple case of California law," said Public Administrator Don Billings, according to the Associated Press [February 23, 1999].

International Church of Christ (ICC)

"I have known people who come out, and they're destroyed. Our goal is to limit their effectiveness on the campus," said Adam Looney of Christ in Action Student Ministries when asked about the International Church of Christ (ICC), which is recruiting student "disciples" at Texas Tech University campus. According to Grear Howard of Baptist Student Ministries, ICC encourages students to cut off communication with their parents, sever relationships with anyone not involved with ICC, and donate large amounts of money. A dozen Texas Tech University ministers, citing concerns about ICC at 30 other American universities, are uniting to urge students on their campus to avoid the religious group, calling it America's most dangerous cult.

Landmark Education Corporation

File magazine writer Rosemary Mahoney, on an assignment for the publication, took The Landmark Forum course, Landmark's primary educational program. She then wrote an article for the September 1998 issue of *File,* where she asserted that The Landmark Forum was an elaborate pyramid scheme and implied that hypnosis was used to convince Landmark students that they had experienced actual results. In response to the article, Landmark Education Corporation filed a complaint against *File* and Mahoney stating the article damaged their business reputation and defamed the reputation of Landmark course leader Beth Handel. Landmark, based in San Francisco with offices in 35 cities in the United States, is seeking $10,000,000 in actual and punitive damages.

Millennium Cults

A number of law enforcement agencies are taking precautions because of the possibility of apocalyptic cults creating violence to help bring about world destruction on the millennium (seen as either December 31, 1999 or December 31, 2000). "With the coming of the next millennium, some religious apocalyptic groups or individuals may turn to violence as they seek to achieve dramatic effects to fulfill their prophecies; FBI Director Louis Freeh warned at a February congressional hearing on counter-terrorism. Freeh went on to cite "rogue terrorists," such as Saudi dissident Osama bin Laden, as potentially the most urgent risk to U.S. interests worldwide, while cautioning that the domestic threat must not be ignored. "The possibility of an indigenous group like Aum Supreme Truth [Aum Shin Rikyo] cannot be excluded," he stated, referring to the cult responsible for the 1995 nerve gas attack in the Tokyo subway system which injured thousands and killed twelve.

Movement of Spiritual Inner Awareness (MSIA)

John-Roger, or J-R, leader of the MSIA cult, prevailed in a lawsuit against University of California, San Diego, religion professor David Lane. MSIA sued Lane for providing on his web page a copy of the book *Life 102: What to Do When Your Guru Sues You*. Written by Peter McWilliams, former 15-year devotee of John-Roger, the book exposes John-Roger and the harmful cult practices McWilliams encountered. As part of a legal settlement, McWilliams gave MSIA rights to the book; but before the lawsuit concluded McWilliams gave universal permission to mirror the book on web sites. Lane hosted the book on his site even though J-R was threatening to sue, because Lane thought that once material was on the web with permission from the author and publisher, it could remain there. And he believed the material was too important to take away from the public. In addition, he was making no money from having it on his site. A judge recently ordered Lane to remove the book from his site, and even made him responsible for J-R's court costs and legal fees. Lane may go bankrupt and is considering an appeal. He may be reached at dlane@weber.ucsd.edu.

Neo-Nazism

In February 1999 hundreds of Neo-Nazis clashed with police in Budapest, Hungary. Thirty-four were arrested and eight officers were injured. Skinheads from Croatia, Slovenia, Germany and England met to mark the anniversary of a retreat by elite Nazi troops in February 1945, from the advancing Russian army. Carrying Nazi flags and dressed in black, the neo-Nazis marched in formation and gave speeches, insisting as one attendee told RTL Klub TV news that "We are here in peace." Later that evening while celebrating at the Viking beer hall they clashed with police on routine patrol.

Satanism

On February 19, satanic cult members in Russia were convicted on charges of murder with aggravating circumstances. The cult members, ranging in age from under 20 to 80, stood trial in the city of Donskoi for two ritual murders in 1998. The cult's leader, Yelena Kuzina, 80, was sentenced to five years in prison. Her oldest son, second in command in the cult, was sentenced to ten years. Defense lawyers for the Satanists said they would protest the sentences.

Wiccan Cult

Randall James, a member of the so-called Wiccan Cult, committed suicide in Texas at the age of 16. His parents said they were unaware of their son's inner torment until his suicide. His journals revealed a confused youngster torn between a dangerous cult and doing what was right. The group is to be distinguished from the traditional Wiccan movement, whose members claim to practice white magic which attempts to create harmony with nature.

Yahweh Ben Yahweh

In 1992, former Oakland Raiders defensive player Robert Rozier admitted to killing seven men as part of his cult activity with leader Yahweh ben Yahweh. For his testimony against Yahweh, Rozier was given a reduced prison sentence and put in the Federal witness protection program in California. He was recently arrested for violating his program agreement by writing bad checks totaling $125.24, which due to the "three strikes" California law might send him to prison for life. In his 1992 trial, Rozier recounted how he and other Rahweh followers went in search of killing white people, whom Yahweh called "white devils," to be part of Yahweh's religious sect. Following this testimony jurors convicted ben Yahweh and six followers of conspiring to commit murder to maintain their religious empire. Yahweh's cult considers itself a lost black tribe of Israel.

QUESTIONS FOR DISCUSSION

1. What is the author's overall theme in discussing the fourteen cults?

2. Which cults advocate severing family relationships while in the group? Why would severing family relationships be implemented in the groups?

3. Which group was labeled "America's most dangerous cult"? Why did the group get this label? What does this say about America's value system?

C H A P T E R 2

DEVIANT EVENTS
AND SOCIAL CONTROL

Cheating at Small Colleges: An Examination
of Student and Faculty Attitudes and Behaviors

*Melody A. Graham, Jennifer Monday,
Kimberly O'Brien, and Stacey Steffen*

It has been well documented that academic dishonesty is a problem on college campuses and universities (Barnett & Dalton, 1981; Hale, 1987; Stevens & Stevens, 1987). Even so, academic integrity is a desirable characteristic for college students and one that institutions strive to model.

Research has identified a number of reasons why students cheat; stress, opportunity, personality characteristics, and low academic achievement are often cited (Barnett & Dalton, 1981). Some studies have suggested that women are less likely than men to be dishonest (Ward & Beck, 1989) and that younger students (i.e., junior high and high school age) cheat more than college students (Evans & Craig, 1990). Institutional influences such as honor codes, faculty involvement, and penalties for cheating have also been investigated in relationship to cheating (May & Loyd, 1993) and have been found to be a deterrent.

"Cheating at Small Colleges: An Examination of Student and Faculty Attitudes and Behaviors" by Melody A. Graham, Jennifer Monday, Kimberly O'Brien, and Stacey Steffen from *Journal of College Student Development,* July 1994, vol. 35, pp. 255–260. Reprinted with permission of American College Personnel Association.

A few studies have examined faculty perceptions of cheating and have found that faculty are aware that cheating is a problem (Evans & Craig, 1990). However, there is some indication that faculty members do not agree on a definition of cheating (Barnett & Dalton, 1981), and when they do agree there is a wide range of ways faculty deal with cheating.

The majority of previous work done in this area has focused on academic dishonesty at large universities or state schools that have large class sizes and environmental conditions that may be more conducive to cheating. This study was designed to examine both student and faculty perceptions of cheating at small liberal arts colleges. Specifically, this study was designed to investigate four different research questions: (a) What behaviors do faculty and students perceive as cheating and how severe an offense is cheating considered to be?; (b) How are attitudes toward cheating related to self-reported behavior?; (c) What is the relationship between faculty attitudes about cheating, their behavior during tests, and written policy statements on syllabi?; (d) How much do faculty think students cheat?; and (e) What is the relationship between student cheating and background variables such as religiosity? It was hoped that a definition of cheating would emerge from these results that could be used to develop appropriate faculty policies.

METHOD

Participants

Four hundred and eight students were sampled from two colleges in the midwest. Approximately 70% of the subjects were enrolled in a private Catholic college, whereas the other 30% attended a community college. Three-fourths of the respondents were female (which corresponds to the population at the schools). Ages ranged from 17 to 62, with the median age at 18.5 years and the mean age at 33.7. Students were equally sampled by their year in school, with slightly fewer seniors represented (18.9%). Ninety-six percent of all subjects were White, which again corresponds to the population at the schools.

A second survey was completed by 48 faculty members from the private Catholic college, with a response rate of 45%. Ninety percent of the faculty responding were full-time at the school. Slightly more females responded (48.3%) than males (37.5%). Almost all of the faculty were White (96%), and they reported teaching on average 12.3 years.

Procedure

Classes at the private college were selected for inclusion in the sample in order to adequately represent all majors and years in school. Business and nursing classes were oversampled, given the large number of students enrolled in these majors at the college. A convenience sample was obtained at the community college. At both schools, faculty members were approached and asked if they would distribute a survey during the selected class period. All but one faculty member agreed to participate. Surveys were then hand delivered to the appropriate faculty member. The instructor read out loud to the class a prepared set of instructions. The students were informed that the survey was both voluntary and confidential and any questions they did not understand or wished not to answer should be left blank. Faculty surveys were sent to all full- and part-time faculty members through campus mail.

Measurement

Both questionnaires (faculty and student) asked for some background information, such as age, sex, race, and religiosity. In addition, students were asked their current housing arrangement, GPA, primary source of financial support, and number of hours worked. Attitudes toward cheating were measured by a 14 item scale developed by Gardner and Melvin (1988). The internal consistency reliability for the scale as measured by coefficient alpha was found to be .83 for faculty and .82 for students. High scores on this scale indicate less tolerance or a more condemnatory attitude toward cheating. The scale used a 5 point rating ranging from strongly agree to strongly disagree.

Three additional scales were used that listed 17 behaviors that could be thought of as cheating. Students and faculty were asked to rate the severity of the cheating behavior on a 4 point scale; 1 = not cheating, 2 = not very severe form of cheating, 3 = severe form of cheating and 4 = very severe form of cheating. Faculty and students were also asked to fill out the scale from the opposite perspective (i.e., students rated how faculty felt about each behavior and faculty rated how students felt about each behavior). Students were also asked to indicate how often they had engaged in each of the behaviors listed, and how often they thought the behaviors occurred on their campus (both 4 point scales ranging from not at all to a lot). Faculty were asked how often they thought the behaviors occurred on their campus. Finally, both students and faculty were asked to rank reasons why a student would cheat and why they would not.

RESULTS

Differences in attitudes and behaviors between the Catholic school and community college were analyzed using *t-tests*. There were no statistically significant differences in attitude toward cheating between the two groups or in the amount of cheating that was perceived on each campus. Differences in the amount of cheating in which students engaged were only found in one of the 17 cheating behaviors: the Catholic college students were more likely to study notes taken by someone else, $t(475) = 2.6p < .05$. Therefore, for the remaining analyses described, the community college and Catholic college students are combined.

Definitions of Cheating

Table 1 lists the 17 cheating behaviors from the questionnaire and reports and percentage of students and faculty who consider each behavior cheating.

The first 11 behaviors listed show 100% agreement among the faculty about what can be defined as cheating. There were no behaviors in this group upon which 100% of the students agreed, however more than 93% of the students agreed these same 11 items were cheating.

TABLE 1   Percent of Students and Faculty Who View Behavior as Cheating
and Percent of Students Who Report Having Engaged in Each Behavior

Behavior	Percent of Students View as Cheating	Percent of Faculty View as Cheating	Percent of Students Have Done Behavior
Looking at notes during a test	99.6	100.0	25.8
Arranging to give or receive answers by signal	98.9	100.0	4.5
Copying during an exam	98.9	100.0	26.0
Taking test for someone else	93.5	100.0	2.7
Asking for an answer during an exam	98.2	100.0	19.7
Giving answers during an exam	97.9	100.0	20.6
Copying someone else's term paper	97.2	100.0	13.7
Allowing a student to copy on a test	96.0	100.0	23.5
Having someone write a term paper for you	95.9	100.0	7.4
Finding a copy of an exam and memorizing the answers	95.1	100.0	17.1
Writing a paper for some-one else	93.6	100.0	9.5
Giving test questions to a student in a later session	86.8	97.9	46.2
Getting answers from a student in an earlier session	92.9	91.7	49.6
Not contributing a fair share in a group project	79.4	79.6	36.4
Allowing someone to copy homework	74.6	83.0	63.1
Using an old test to study without the teacher's knowledge	66.0	83.3	37.5
Using a paper for more than one class	45.9	77.1	53.6

Note: Percent who responded that the behavior was not very severe, severe, or very severe form of cheating.

Table 2 gives the mean severity rating by students and faculty for each of the behaviors. In every case, faculty rated the behaviors as significantly more severe than the students. Students and faculty agreed on the top three most serious forms of cheating: "taking a test for someone else," "copying someone else's term paper," and "having someone write a term paper for you." In general, the severity ratings followed the same order for faculty and students, except that students did not perceive writing a term paper for someone as severe a form of cheating as other cheating behaviors.

Faculty were also asked to rate how severe they thought students felt each behavior was, and students gave estimates of the faculty's ratings of severity. By comparing the student's severity ratings with faculty's perceptions of students' rating (columns 2

**Student and Faculty's Mean Severity Rating and Students' Perception of Faculty's   TABLE 2
Rating of Severity and Faculty's Perception of Students' Rating of Severity**

Behavior	Faculty's Severity Rating	Students' Severity Rating	Faculty's Perception of Students	Students' Perception of Faculty
Taking a test for someone else	3.91	3.76	3.56	3.86
Copying someone else's term paper	3.80	3.35	3.31	3.68
Having someone write a term paper for you	3.80	3.35	2.95	3.66
Copying during an exam	3.73	3.28	2.91	3.77
Arranging to give or receive answers by signal	3.68	3.25	3.02	3.65
Writing a paper for someone else	3.64	2.91	2.52	3.50
Giving answers during an exam	3.60	3.30	2.87	3.66
Asking for an answer during an exam	3.60	3.29	3.04	3.66
Allowing a student to copy on a test	3.55	3.21	2.75	3.62
Looking at notes during a test	3.53	3.25	2.98	3.66
Finding a copy of an exam and memorizing the answers	3.39	2.97	2.48	3.57
Getting answers from a student in an earlier session	3.36	2.73	2.18	3.32
Giving test questions to a student in a later session	3.20	2.65	2.30	3.18
Not contributing a fair share in a group project	2.71	2.41	1.91	2.71
Allowing someone to copy homework	2.52	2.23	1.80	2.67
Using a paper for more than one class	2.40	1.70	1.69	2.24
Using an old test to study without the teacher's knowledge	2.23	2.02	1.60	2.59

Note: Severity was rated on a 4-point scale with 1 = not cheating and 4 = very severe.

& 3 in Table 2) it was found, in every case but one, faculty underestimate how severe students think cheating actually is. The one exception is using a paper in more than one class. On the other hand, when faculty severity ratings were compared with students' perception of faculty (columns 1 & 4 in Table 2) it was found that students are accurate in their estimates of how severe faculty think cheating is.

Amount of Cheating

Table 1 also indicates the percentage of students who have engaged in each of the behaviors defined as cheating. The most severe behaviors (taking a test for someone else, copying someone else's term paper, and having someone write a term paper for you) seldom were cited by students, although 14% did report copying someone else's term paper. Table 1 also shows there was a dramatic increase in the rate of cheating in the last five behaviors listed. These behaviors are also the ones that both students and faculty have a harder time agreeing on as cheating. The most common forms of cheating are allowing someone to copy homework and using a paper for more than one class. About one-fourth of the population sampled had copied during a test. The 17 cheating behaviors were combined to form a scale ranging from 17 to 60. Low scores indicate little to no cheating and high scores indicate a great amount of cheating. The mean for the scale was 23.3 with a standard deviation of 6.27. The alpha reliability of the scale was .89. When looking at all the cheating behaviors combined, 89.9% of the students surveyed said they had engaged in some form of cheating at least once.

Correlates to Cheating

Pearson product-moment correlations and one-way analysis of variance (ANOVA) were calculated between the cheating scale and the variables of interest. It was found that students who cheat a lot have a lower GPA, $r = -.22$, $p < .01$; are traditional age students, $r = -.266$, $p < .01$; live in campus housing, $F(3, 447) = 12.91$, $p < .001$; have parents who pay for their education, $F(3, 447) = 11.25$, $p < .01$; and were female, $t(477) = 2.0$, $p < .05$.

The 13 attitude questions were also combined to form a scale with a range of 13 to 70. Low scores indicate lenient attitudes toward cheating. The scale had a mean of 26 and a standard deviation of 5.7. This scale was found to be related to cheating ($r = .394, p < .01$) in that those with lenient attitudes toward cheating reported having engaged in more forms of cheating. Two other attitudinal variables were also related to the cheating scale: believing the various cheating behaviors were less severe ($r = -.414, p < .01$), and feeling that other students cheat a lot ($r = .148, p < .05$). Variables that were not related to cheating were: number of activities in which the student was engaged, year in school, academic major, and religiosity.

Although the above variables had a univariate relationship with cheating, when they all were entered into a multiple regression equation, a different picture emerged. A stepwise multiple regression was performed using 14 different predictor variables. The equation was successful at explaining 22% of the variance in cheating ($R^Z = .474$), $F(3, 253) = 24.37$, $p < .01$. However, only three of the variables made a significant independent contribution to the equation. These variables were: attitudes toward cheating, ratings of how severe of an offense cheating is considered to be, and beliefs about how much others cheat. Students with lenient attitudes toward cheating, who believe cheating is not that severe of an offense, and who think that a large number of other students at their school cheat are more likely to have engaged in various types of cheating.

Although the background variables (e.g., age, living arrangement, source of payment for college) do not explain cheating behavior as well as the attitudinal variables, they are significantly related to attitudes toward cheating. Students who had lenient attitudes toward cheating were younger ($r = -.306, p < .01$), had lower GPA's ($r = -.294, p < .01$), and were less religious ($r = .145, p < .05$).

Reasons for Cheating

Faculty and students agree on the top three reasons why a student *would* cheat. The reasons why students report that they cheat are that: (a) they need a better grade (72.5% of students and 84.9% of faculty), (b) they didn't have time to study (60.4% of students and 69.9% of faculty), and (c) they saw an opportunity and just took it (33.5% of students and 61.5% of faculty). It was also found that 25% of students reported that they would cheat if they thought the teacher was unfair, whereas no faculty members reported this as a reason a student would cheat.

Faculty and students also agreed on the top reasons why a student *would not* cheat. The top reason reported for why a student would not cheat was because they believe it is wrong (52.5% of students and 80.5% of faculty). The second reason was the high cost of getting caught (51.2% of students and 59.0% of faculty), followed by the high likelihood of getting caught (43.8% of students and 46.1% of faculty). Faculty were more likely than students to think that fewer students would cheat because they didn't want the faculty to think less of them (29.3% of students and 46.3% of faculty). Students were more likely than faculty members to respond that not cheating is the best way to get ahead (44.9% of students and 26.3% of faculty).

Faculty Policies

Although students believe that getting caught is a disincentive to cheating, faculty members do not have very stringent policies against cheating. Only 64.3% of the faculty surveyed had a statement on their syllabi about cheating, and 20% reported that they do not watch students while they are taking tests. Though the percentage of faculty that have caught a student cheating is high (78.7%), only 9% penalized the student for cheating (by failing the assignment, deducting points, or failing the course). Faculty attitudes toward cheating (i.e., believe cheating is morally wrong or that under no circumstances is cheating justified) were more likely to have a statement on their syllabi about cheating ($r = -.316$, $p < .05$), were more likely to think that more cheating is occurring on their campus than those with less conservative attitudes ($r = -.323$, $p < .05$), and were more likely to be female ($r = -.385$, $p < .01$). Nothing was significantly related to what faculty do while students are taking a test.

DISCUSSION

From these findings, it is clear that cheating is a problem. Close to 90% of all the students surveyed admitted they had engaged in some form of cheating at least once. However, defining cheating is not as clear cut. There were no forms of behavior that all students defined as cheating, however there were 9 behaviors that 100% of the faculty and more than 93% of the students agreed upon. Students also seemed to make a distinction between cheating that they engaged in themselves and cheating that they allowed to happen. The top five behaviors that students considered cheating are things that a student can do themselves, as opposed to allowing someone else to do. It appears that students feel that it is worse to cheat by oneself rather than allow someone to cheat from their work. Faculty do not make this distinction.

In every case, faculty viewed cheating as a significantly more severe offense than did the students. It does appear that students have a relatively good understanding of how faculty view cheating, whereas faculty underestimate how severe cheating is to students. It is easy to understand why faculty underestimate how severe students think cheating is when 90% of them have cheated. Therefore, viewing cheating as severe is not in and of itself a deterrent to cheating.

Attitudes toward cheating and norms about cheating were the best predictors of who will cheat. Students who cheat have lenient attitudes toward cheating, think that a large number of other students at their school cheat, and believe cheating is not that severe of an offense. These attitudes toward cheating were more important than any of the background variables that were measured. However, the background variables such as age, religiosity, and GPA were good predictors of attitudes. Younger, less religious students with low GPAs are more likely to hold lenient attitudes toward cheating.

Students report that a good reason not to cheat is the high cost of getting caught. However, after reviewing faculty policies it does not appear that is the case. Only 9% of the faculty who had caught a student cheating penalized the student by deducting points, failing the class, etc. It appears that although faculty agree that cheating is severe, there is not much consensus on what action is appropriate after a cheater has been caught. Students are also less likely to cheat when all the faculty agree that the act is cheating. For behaviors that faculty do not agree upon (i.e., allowing someone to copy homework), there is a far greater amount of student cheating.

After examining reasons students do and do not cheat, it is clear that students cheat for situational reasons, and they don't cheat because they hold values that are inconsistent with cheating. Students also view the classroom as a reciprocal process—when faculty are unfair students see this as a violation of the rules and, thus, feel freer to cheat. It is not clear from this study how students define "unfair" or exactly what behaviors students consider fair.

Implications for Professional Practice

This research suggests a number of implications for faculty and colleges. First, it appears that if institutions want to decrease cheating, they need to be clear about how cheating is defined. When faculty agree that a behavior is cheating, students are less likely to engage in that behavior. To avoid mixed messages being sent to students from different faculty members and the institution, it would be helpful to have a handbook that clearly defines cheating and the various forms that cheating can take. Faculty should also clearly define what is acceptable behavior in their class and what is not (i.e., it is okay to work together on homework in this class, but using a paper from a previous class is not acceptable).

Students are less likely to engage in cheating when they think the cost of getting caught is high, and they believe that there is a good chance they will get caught. Therefore, the penalties for engaging in cheating need to be clearly defined and communicated to students. This has been supported in other research that has found that institutions with honor codes and clear penalties for violating the code have less cheating than institutions without honor codes (May & Loyd, 1993). The handbook could specify a clear procedure for dealing with cheating.

It is also clear that there are situational influences on cheating; if there is an opportunity, even students with attitudes against cheating may be tempted to take the opportunity. Faculty should be diligent in decreasing situations where cheating can occur. Faculty should also be aware that behavior that is perceived as unfair by the students may induce cheating. More research is needed to discover what behaviors students perceive as unfair and how faculty can avoid being labeled unfair.

Limitations of Study

There were a number of limitations of the study. First, there was a less than 50% response rate by the faculty. It could be faculty that are more concerned with academic cheating were more likely to respond, and thus the faculty attitudes reported may be skewed. There are also problems with asking people to recall behavior that they may feel is socially unacceptable. Confidentiality was stressed in an effort to reduce students' apprehension about cheating. The sample was predominately White and female. Although this represents the population sampled, it does limit generalizability of the findings.

Given these limitations, cheating does not appear to be as much a problem on small campuses as on large campuses. However, it may be easier to control situations on smaller campuses with smaller class sizes. The strongest influence on cheating appears to be attitudes and norms about cheating. Small colleges should work on developing an environment in which cheating is not tolerated. This can be done through making the schools' views about cheating clear. Institutions should also work on encouraging students to explore their own attitudes toward cheating in an effort to help students clarify how they feel. Future research should explore the best way to foster attitudes against cheating. This study suggests that both faculty and student behavior in the classroom should be explored when investigating cheating.

REFERENCES

Barnett, D. C., & Dalton, J. C. (1981). Why college students cheat. *Journal of College Student Personnel. 22,* 545–551.

Evans, E. D., & Craig, D. (1990). Teacher and student perceptions of academic cheating in middle and senior high schools. *Journal of Educational Research, 84* (1), 44–52.

Gardner, W. M., & Melvin, K. B. (1988). A scale for measuring attitude toward cheating. *Bulletin of the Psychonomic Society. 26,* 429–432.

Hale, J. L. (1987). Plagiarism in classroom settings. *Communication Research Reports. 4* (2), 66–70.

May, K. M. & Loyd, B. H. (1993). Academic dishonesty: The honor system and students' attitudes. *Journal of College Student Development. 34,* 125–129.

Stevens, G. E., & Stevens, F. W. (1987). Ethical inclinations of tomorrow's managers revisited: How and why students cheat. *Journal of Education for Business, 63* (1), 24–29.

Ward, D. A., & Beck, W. L. (1989). Gender and dishonesty. *The Journal of Social Psychology,* 130, 333–339.

QUESTIONS FOR DISCUSSION

1. Discuss the authors' views about how attitudes toward cheating are related to self-reporting behavior and cheating.

2. Discuss the reasons the authors give about why some students cheat while others do not.

3 Discuss the article's implications for professional practice and ethics at your college.

Deviant Places: A Theory of the Ecology of Crime
Rodney Stark

Norman Hayner, a stalwart of the old Chicago school of human ecology, noted that in the area of Seattle having by far the highest delinquency rate in 1934, "half the children are Italian." In vivid language, Hayner described the social and cultural shortcomings of these residents: "largely illiterate, unskilled workers of Sicilian origin. Fiestas, wine-drinking, raising of goats and gardens . . . are characteristic traits." He also noted that the businesses in this neighborhood were run down and on the wane and that "a number of dilapidated vacant business buildings and frame apartment houses dot the main street," while the area has "the smallest percentage of homeowners and the greatest aggregation of dilapidated dwellings and run-down tenements in the city" (Hayner, 1942:361–363). Today this district, which makes up the neighborhood surrounding Garfield High School, remains the prime delinquency area. But there are virtually no Italians living there. Instead, this neighborhood is the heart of the Seattle black community.

Thus we come to the point. How is it that neighborhoods can remain the site of high crime and deviance rates *despite a complete turnover in their populations?* If the Garfield district was tough *because* Italians lived there, why did it stay tough after they left? Indeed, why didn't the neighborhoods the Italians departed to become tough? Questions such as these force the perception that the composition of neighborhoods, in terms of characteristics of their populations, cannot provide an adequate explanation of variations in deviance rates. Instead, *there must be something about places as such* that sustains crime.[1]

This paper attempts to fashion an integrated set of propositions to summarize and extend our understanding of ecological sources of deviant behavior. In so doing, the aim is to revive a *sociology* of deviance as an alternative to the social psychological approaches that have dominated for 30 years. That is, the focus is on traits of places and groups rather than on traits of individuals. Indeed, I shall attempt to show that by adopting survey research as the *preferred* method of research, social scientists lost touch with significant aspects of crime and delinquency. Poor neighborhoods disappeared to be replaced by individual kids with various levels of family income, but no detectable environment at all. Moreover, the phenomena themselves became bloodless, sterile, and almost harmless, for questionnaire studies cannot tap homicide, rape, assault, armed robbery, or even significant burglary and fraud—too few people are involved in these activities to turn up in significant numbers in feasible samples, assuming that such people turn up in samples at all. So delinquency, for example, which once had meant offenses serious enough for court referrals, soon meant taking $2 out of mom's purse, having "banged up something that did not belong to you," and having a fist fight. This transformation soon led repeatedly to the "discovery" that poverty is unrelated to delinquency (Tittle, Villemez, and Smith, 1978).

Yet, through it all, social scientists somehow still knew better than to stroll the streets at night in certain parts of town or even to park there. And despite the fact that countless surveys showed that kids from upper- and lower-income families scored the same on delinquency batteries, even social scientists knew that the parts of town that scared them were not upper-income neighborhoods. In fact, when the literature was examined with sufficient finesse, it was clear that class *does* matter—that serious offenses are very disproportionately committed by a virtual under class (Hindelang, Hirschi, and Weis, 1981).

So, against this backdrop, let us reconsider the human ecology approach to deviance. To begin, there are five aspects of urban neighborhoods which characterize high deviance areas of cities. To my knowledge, no member of the Chicago school ever listed this particular set, but these concepts permeate their whole literature starting with Park, Burgess, and McKenzie's classic, *The City* (1925). And they are especially prominent in the empirical work of the Chicago school. (Faris and Dunham, 1939; Shaw and McKay, 1942). Indeed, most of these factors were prominent in the work of 19th-century moral statisticians such as the Englishmen Mayhew and Buchanan, who were doing ecological sociology decades before any member of the Chicago school was born. These essential factors are (1) density; (2) poverty; (3) mixed use; (4) transience; and (5) dilapidation.

Each of the five will be used in specific propositions. However, in addition to these characteristics of places, the theory also will incorporate some specific *impacts* of the five on the moral order as *people respond to them.* Four responses will be assessed: (1) moral cynicism among residents; (2) increased opportunities for crime and deviance; (3) increased motivation to deviate; and (4) diminished social control.

Finally, the theory will sketch how these responses further *amplify* the volume of deviance through the following consequences: (1) by attracting deviant and crime-prone people and deviant and criminal activities to a neighborhood; (2) by driving out the least deviant; and (3) by further reductions in social control.

The remainder of the paper weaves these elements into a set of integrated propositions, clarifying and documenting each as it proceeds. Citations will not be limited to recent work, or even to that of the old Chicago school, but will include samples of the massive 19th-century literature produced by the moral statisticians. The aim is to help contemporary students of crime and deviance rediscover the past and to note the power and realism of its methods, data, and analysis. In Mayhew's (1851) immense volumes, for example, he combines lengthy, first-person narratives of professional criminals with a blizzard of superb statistics on crime and deviance.

Before stating any propositions, one should note the relationship between this essay and ongoing theoretical work, especially my deductive theory of religion (Stark and Bainbridge, 1987). A major impediment to the growth of more formal and fully deductive theories in the social sciences is that usually one lacks the space necessary to work out the links between an initial set of axioms and definitions and the relevant set of propositions (statements deduced from the axioms and definitions). In consequence, it is not shown that the propositions outlined here follow logically from my axiomatic system, but they can be derived. For those interested in these matters, one can refer to the more complete formulation of control theory that was derived in *A Theory of Religion* (Stark and Bainbridge, 1987) to explain the conditions under which people are recruited by deviant religious movements. In any event, logical steps from one proposition to another will be clear in what follows, but the set as a whole must be left without obvious axiomatic ancestry.

Proposition 1: *The greater the density of a neighborhood, the more association between those most and least predisposed to deviance.*

At issue here is not simply that there will be a higher proportion of deviance-prone persons in dense neighborhoods (although, as will be shown, that is true, too), rather it is proposed that there is a higher average level of interpersonal interactions in such neighborhoods and that individual traits will have less influence on patterns of contact. Consider kids. In low-density neighborhoods—wealthy suburbs, for example—some active effort is required for one 12-year-old to see another (a ride from a parent often is required). In these settings, kids and their parents can easily limit contact with bullies and those in disrepute. Not so in dense urban neighborhoods—the "bad" kids often live in the same building as the "good" ones, hang out close by, dominate the nearby playground, and are nearly unavoidable. Hence, peer groups in dense neighborhoods will tend to be inclusive, and all young people living there will face maximum peer pressure to deviate—as differential association theorists have stressed for so long.

Proposition 2: *The greater the density of a neighborhood, the higher the level of moral cynicism.*

Moral cynicism is the belief that people are much worse than they pretend to be. Indeed, Goffman's use of the dramaturgical model in his social psychology was rooted in the fact that we require ourselves and others to keep up appearances in public. We all, to varying degrees, have secrets, the public airing of which we would find undesirable. So long as our front-stage performances are credible and creditable, and we shield our backstage actions, we serve as good role models (Goffman, 1959, 1963). The trouble is that in dense neighborhoods it is much harder to keep up appearances—whatever morally discreditable information exists about us is likely to leak.

Survey data suggest that upper-income couples may be about as likely as lower-income couples to have physical fights (Stark and McEvoy, 1970). Whether that is true, it surely is the case that upper-income couples are much less likely to be *overheard* by the neighbors when they have such a fight. In dense neighborhoods, where people live in crowded, thin-walled apartments, the neighbors do hear. In these areas teenage peers, for example, will be much more likely to know embarrassing things about one another's parents. This will color their perceptions about what is normal, and their respect for the conventional moral standards will be reduced. Put another way, people in dense neighborhoods will serve as inferior role models for one another—the same people would *appear* to be more respectable in less dense neighborhoods.

Proposition 3: *To the extent that neighborhoods are dense and poor, homes will be crowded.*

The proposition is obvious, but serves as a necessary step to the next propositions on the effects of crowding, which draw heavily on the fine paper by Gove, Hughes, and Galle (1979).

Proposition 4: *Where homes are more crowded, there will be a greater tendency to congregate outside the home in places and circumstances that raise levels of temptation and opportunity to deviate.*

Gove and his associates reported that crowded homes caused family members, especially teenagers, to stay away. Since crowded homes will also tend to be located in mixed-use neighborhoods (see Proposition 9), when people stay away from home they will tend to congregate in places conducive to deviance (stores, pool halls, street corners, cafes, taverns, and the like).

Proposition 5: *Where homes are more crowded, there will be lower levels of supervision of children.*

This follows from the fact that children from crowded homes tend to stay out of the home and that their parents are glad to let them. Moreover, Gove and his associates found strong empirical support for the link between crowding and less supervision of children.

Proposition 6: *Reduced levels of child supervision will result in poor school achievement, with a consequent reduction in stakes in conformity and an increase in deviant behavior.*

This is one of the most cited and strongly verified causal chains in the literature on delinquency (Thrasher, 1927; Toby and Toby, 1961; Hirschi, 1969; Gold, 1970; Hindelang, 1973). Indeed, Hirschi and Hindelang (1977:583) claim that the "school variables" are among the most powerful predictors of delinquency to be found in survey studies: "Their significance for delinquency is nowhere in dispute and is, in fact, one of the oldest and most consistent findings of delinquency research."

Here Toby's (1957) vital concept of "stakes in conformity" enters the propositions. Stakes in conformity are those things that people risk losing by being detected in deviant actions. These may be things we already possess as well as things we can reasonably count on gaining in the future. An important aspect of the school variables is their potential for future rewards, rewards that may be sacrificed by deviance, but only for those whose school performance is promising.

Proposition 7: *Where homes are more crowded, there will be higher levels of conflict within families, weakening attachment and thereby stakes in conformity.*

Gove and his associates found a strong link between crowding and family conflict, confirming Frazier's (1932:636) observations:

> So far as children are concerned, the house becomes a veritable prison for them.
> There is no way of knowing how many conflicts in Negro families are set off by

the irritations caused by overcrowding people, who come home after a day of frustration and fatigue, to dingy and unhealthy living quarters.

Here we also recognize that stakes in conformity are not merely material. Indeed, given the effort humans will expend to protect them, our attachments to others are among the most potent stakes in conformity. We risk our closest and most intimate relationships by behavior that violates what others expect of us. People lacking such relationships, of course, do not risk their loss.

Proposition 8: *Where homes are crowded, members will be much less able to shield discreditable acts and information from one another, further increasing moral cynicism.*

As neighborhood density causes people to be less satisfactory role models for the neighbors, density in the home causes moral cynicism. Crowding makes privacy more difficult. Kids will observe or overhear parental fights, sexual relations, and the like. This is precisely what Buchanan noted about the dense and crowded London slums in 1846 (in Levin and Lindesmith, 1937:15):

> In the densely crowded lanes and alleys of these areas, wretched tenements are found containing in every cellar and on every floor, men and women, children both male and female, all huddled together, sometimes with strangers, and too frequently standing in very doubtful consanguinity to each other. In these abodes decency and shame have fled; depravity reigns in all its horrors.

Granted that conditions have changed since then and that dense, poor, crowded areas in the center cities of North America are not nearly so wretched. But the essential point linking "decency" and "shame" to lack of privacy retains its force.

Proposition 9: *Poor, dense neighborhoods tend to be mixed-use neighborhoods.*

Mixed use refers to urban areas where residential and commercial land use co-exist, where homes, apartments, retail shops, and even light industry are mixed together. Since much of the residential property in such areas is rental, typically there is much less resistance to commercial use (landlords often welcome it because of the prospects of increased land values). Moreover, the poorest, most dense urban neighborhoods often are adjacent to the commercial sections of cities, forming what the Chicago school called the "zone of transition" to note the progressive encroachments of commercial uses into a previously residential area. Shaw and McKay (1942:20) describe the process as follows:

> As the city grows, the areas of commerce and light industry near the center encroach upon areas used for residential purposes. The dwellings in such areas, often already undesirable because of age, are allowed to deteriorate when such invasion threatens or actually occurs, as further investment in them is unprofitable. These residences are permitted to yield whatever return can be secured in their dilapidated condition, often in total disregard for the house laws. . . .

Shaw and McKay were proponents of the outmoded concentric zonal model of cities, hence their assumption that encroachment radiates from the city center. No matter, the important point is that the process of encroachment occurs whatever the underlying shape of cities.

Proposition 10: *Mixed use increases familiarity with and easy access to places offering the opportunity for deviance.*

A colleague told me he first shoplifted at age eight, but that he had been "casing the joint for four years." This particular "joint" was the small grocery store at the corner of the block where he lived, so he didn't even have to cross a street to get there. In contrast, consider kids in many suburbs. If they wanted to take up shoplifting they would have to ask mom or dad for a ride. In purely residential neighborhoods there simply are far fewer conventional opportunities (such as shops) for deviant behavior.

Proposition 11: *Mixed-use neighborhoods offer increased opportunity for congregating outside the home in places conducive to deviance.*

It isn't just stores to steal from that the suburbs lack, they also don't abound in places of potential moral marginality where people can congregate. But in dense, poor, mixed-use neighborhoods, when people leave the house they have all sorts of places to go, including the street corner. A frequent activity in such neighborhoods is leaning. A bunch of guys will lean against the front of the corner store, the side of the pool hall, or up against the barber shop. In contrast, out in the suburbs young guys don't gather to lean against one another's houses, and since there is nowhere else for them to lean, whatever deviant leanings they might have go unexpressed. By the same token, in the suburbs, come winter, there is no close, *public* place to congregate indoors.

Thus, we can more easily appreciate some fixtures of the crime and delinquency research literature. When people, especially young males, congregate and have nothing special to do, the incidence of their deviance is increased greatly (Hirschi, 1969). Most delinquency, and a lot of crime, is a social rather than a solitary act (Erickson, 1971).

Proposition 12: *Poor, dense, mixed-use neighborhoods have high transience rates.*

This aspect of the urban scene has long attracted sociological attention. Thus, McKenzie wrote in 1926 (p. 145): "Slums are the most mobile . . . sections of a city. Their inhabitants come and go in continuous succession."

Proposition 13: *Transience weakens extra-familial attachments.*

This is self-evident. The greater the amount of local population turnover, the more difficult it will be for individuals or families to form and retain attachments.

Proposition 14: *Transience weakens voluntary organizations, thereby directly reducing both informal and formal sources of social control* (see Proposition 25).

Recent studies of population turnover and church membership rates strongly sustain the conclusion that such membership is dependent upon attachments, and hence suffers where transience rates reduce attachments (Wuthnow and Christiano, 1979; Stark, Doyle, and Rushing, 1983; Welch, 1983; Stark and Bainbridge, 1985). In similar fashion, organizations such as PTA or even fraternal organizations must suffer where transience is high. Where these organizations are weak, there will be reduced community resources to launch local, self-help efforts to confront problems such as truancy or burglary. Moreover, neighborhoods deficient in voluntary organizations also will be less able to influence how external forces such as police, zoning boards, and the like act vis-à-vis the community, a point often made by Park (1952) in his discussions of natural areas and by more recent urban sociologists (Suttles, 1972; Lee, Oropesa, Metch, and Guest, 1984; Guest, 1984).

In their important recent study, Simcha-Fagan and Schwartz (1986) found that the association between transience and delinquency disappeared under controls for organizational participation. This is not an example of spuriousness, but of what Lazarsfeld called "interpretation" (Lazarsfeld, Pasanella, and Rosenberg, 1972). Transience *causes* low levels of participation, which in turn *cause* an increased rate of delinquency. That is, participation is an *intervening variable* or *linking mechanism* between transience and delinquency. When an intervening variable is controlled, the association between X and Y is reduced or vanishes.

Proposition 15: *Transience reduces levels of community surveillance.*

In areas abounding in newcomers, it will be difficult to know when someone doesn't live in a building he or she is entering. In stable neighborhoods, on the other hand, strangers are easily noticed and remembered.

Proposition 16: *Dense, poor, mixed-used, transient neighborhoods will also tend to be dilapidated.*

This is evident to anyone who visits these parts of cities. Housing is old and not maintained. Often these neighborhoods are very dirty and littered as a result of density, the predominance of renters, inferior public services, and a demoralized population (see Proposition 22).

Proposition 17: *Dilapidation is a social stigma for residents.*

It hardly takes a real estate tour of a city to recognize that neighborhoods not only reflect the status of their residents, but confer status upon them. In Chicago, for example, strangers draw favorable inferences about someone who claims to reside in Forest Glen, Beverly, or Norwood Park. But they will be leery of those who admit to living on the Near South Side. Granted, knowledge of other aspects of communities enters into these differential reactions, but simply driving through a neighborhood such as the South Bronx is vivid evidence that very few people would actually *want* to live there. During my days as a newspaper reporter, I discovered that to move just a block North, from West Oakland to Berkeley, greatly increased social assessments of individuals. This was underscored by the frequent number of times people told me they lived in Berkeley although the phone book showed them with an Oakland address. As Goffman (1963) discussed at length, stigmatized people will try to pass when they can.

Proposition 18: *High rates of neighborhood deviance are a social stigma for residents.*

Beyond dilapidation, neighborhoods abounding in crime and deviance stigmatize the moral standing of all residents. To discover that you are interacting with a person through whose neighborhood you would not drive is apt to influence the subsequent interaction in noticeable ways. Here is a person who lives where homicide, rape, and assault are common, where drug dealers are easy to find, where prostitutes stroll the sidewalks waving to passing cars, where people sell TVs, VCRs, cameras, and other such items out of the trunks of their cars. In this sense, place of residence can be a dirty, discreditable secret.

Proposition 19: *Living in stigmatized neighborhoods causes a reduction in an individual's stake in conformity.*

This is simply to note that people living in slums will see themselves as having less risk by being detected in acts of deviance. Moreover, as suggested below in Propositions 25–28, the risks of being detected also are lower in stigmatized neighborhoods.

Proposition 20: *The more successful and potentially best role models will flee stigmatized neighborhoods whenever possible.*

Goffman (1963) has noted that in the case of physical stigmas, people will exhaust efforts to correct or at least minimize them—from plastic surgery to years of therapy. Presumably it is easier for persons to correct a stigma attached to their neighborhood than one attached to their bodies. Since moving is widely perceived as easy, the stigma of living in particular neighborhoods is magnified. Indeed, as we see below, some people do live in such places because of their involvement in crime and deviance. But, even in the most disorderly neighborhoods, *most* residents observe the laws and norms. Usually they continue to live there simply because they can't afford better. Hence, as people become able to afford to escape, they do. The result is a process of selection whereby the worst role models predominate.

Proposition 21: *More successful and conventional people will resist moving into a stigmatized neighborhood.*

The same factors that *pull* the more successful and conventional out of stigmatized neighborhoods *push* against the probability that conventional people will move into these neighborhoods. This means that only less successful and less conventional people *will* move there.

Proposition 22: *Stigmatized neighborhoods will tend to be overpopulated by the most demoralized kinds of people.*

This does not mean the poor or even those engaged in crime or delinquency. The concern is with persons unable to function in reasonably adequate ways. For here will congregate the mentally ill (especially since the closure of mental hospitals), the chronic alcoholics, the retarded, and others with limited capacities to cope (Faris and Dunham, 1939; Jones, 1934).

Proposition 23: *The larger the relative number of demoralized residents, the greater the number of available "victims."*

As mixed use provides targets of opportunity by placing commercial firms within easy reach of neighborhood residents, the demoralized serve as human targets of opportunity. Many muggers begin simply by searching the pockets of drunks passed out in doorways and alleys near their residence.

Proposition 24: *The larger the relative number of demoralized residents, the lower will be residents' perception of chances for success, and hence they will have lower perceived stakes in conformity.*

Bag ladies on the corner, drunks sitting on the curbs, and schizophrenics muttering in the doorways are not advertisements for the American Dream. Rather, they testify that people in this part of town are losers, going nowhere in the system.

Proposition 25: *Stigmatized neighborhoods will suffer from more lenient law enforcement.*

This is one of those things that "everyone knows," but for which there is no firm evidence. However, evidence may not be needed, given the many obvious reasons why the police would let things pass in these neighborhoods that they would act on in better neighborhoods. First, the police tend to be reactive, to act upon complaints rather than seek out violations. People in stigmatized neighborhoods complain less often. Moreover, people in these neighborhoods frequently are much less willing to testify when the police do act—and the police soon lose interest in futile efforts to find evidence. In addition, it is primarily vice that the police tolerate in these neighborhoods, and the police tend to accept the premise that vice will exist *somewhere.* Therefore, they tend to condone vice in neighborhoods from which they do not receive effective pressures to act against it (see Proposition 14). They may even believe that by having vice limited to a specific area they are better able to regulate it. Finally, the police frequently come to share the outside community's view of stigmatized neighborhoods—as filled with morally disreputable people, who deserve what they get.

Proposition 26: *More lenient law enforcement increases moral cynicism.*

Where people see the laws being violated with apparent impunity, they will tend to lose their respect for conventional moral standards.

Proposition 27: *More lenient law enforcement increases the incidence of crime and deviance.*

This is a simple application of deterrence theory. Where the probabilities of being arrested and prosecuted for a crime are lower, the incidence of such crimes will be higher (Gibbs, 1975).

Proposition 28: *More lenient law enforcement draws people to a neighborhood on the basis of their involvement in crime and deviance.*

Reckless (1926:165) noted that areas of the city with "wholesome family and neighborhood life" will not tolerate "vice," but that "the decaying neighborhoods have very little resistance to the invasions of vice." Thus, stigmatized neighborhoods become the "soft spot" for drugs, prostitution, gambling, and the like. These are activities that require public awareness of where to find them, for they depend on customers rather than victims. Vice can function only where it is condoned, at least to some degree. In this manner, McKenzie (1926:146) wrote, the slum "becomes the hiding-place for many services that are forbidden by the mores but which cater to the wishes of residents scattered throughout the community."

Proposition 29: *When people are drawn to a neighborhood on the basis of their participation in crime and deviance, the visibility of such activities and the opportunity to engage in them increases.*

It has already been noted that vice must be relatively visible to outsiders in order to exist. Hence, to residents, it will be obvious. Even children not only will know *about* whores, pimps, drug dealers, and the like, they will *recognize* them. Back in 1840, Allison wrote of the plight of poor rural families migrating to rapidly growing English cities (p. 76):

> The extravagant price of lodgings compels them to take refuge in one of the
> crowded districts of the town, in the midst of thousands in similar necessitous
> circumstances with themselves. Under the same roof they probably find a nest
> of prostitutes, in the next door a den of thieves. In the room which they occupy
> they hear incessantly the level of intoxication or are compelled to witness the
> riot of licentiousness.

In fact, Allison suggested that the higher social classes owed their "exemption from atrocious crime" primarily to the fact that they were not confronted by the temptations and seductions to vice that assail the poor. For it is the "impossibility of concealing the attractions of vice from the younger part of the poor in the great cities which exposes them to so many causes of demoralization."

Proposition 30: *The higher the visibility of crime and deviance, the more it will appear to others that these activities are safe and rewarding.*

There is nothing like having a bunch of pimps and bookies flashing big wads of money and driving expensive cars to convince people in a neighborhood that crime pays. If young girls ask the hookers on the corner why they are doing it, they will reply with tales of expensive clothes and jewelry. Hence, in some neighborhoods, deviants serve as role models that encourage residents to become "street wise." This is a form of "wisdom" about the relative costs and benefits of crime that increases the likelihood that a person will spend time in jail. The extensive recent literature on perceptions of risk and deterrence is pertinent here (Anderson, 1979; Jenson, Erickson, and Gibbs, 1978; Parker and Grasmick, 1979).

CONCLUSION

A common criticism of the ecological approach to deviance has been that although many people live in bad slums, most do not become delinquents, criminals, alcoholics, or addicts. Of course not. For one thing, as Gans (1962), Suttles (1968), and others have recognized, bonds among human beings can endure amazing levels of stress and thus continue to sustain commitment to the moral order even in the slums. Indeed, the larger culture seems able to instill high levels of aspiration in people even in the worst ecological settings. However, the fact that most slum residents aren't criminals is beside the point to claims by human ecologists that aspects of neighborhood structure can sustain high rates of crime and deviance. Such propositions do not imply that residence in such a neighborhood is either a necessary or a sufficient condition for deviant behavior. There is conformity in the slums and deviance in affluent suburbs. All the ecological propositions imply is a substantial correlation between variations in neighborhood character and variations in crime and deviance rates. What an ecological theory of crime is meant to achieve is an explanation of why crime and deviance are so heavily concentrated in certain areas, and to pose this explanation in terms that do not depend entirely (or even primarily) on *compositional* effects—that is, on answers in terms of "kinds of people."

To say that neighborhoods are high in crime because their residents are poor suggests that controls for poverty would expose the spuriousness of the ecological effects. In contrast, the ecological theory would predict that the deviant behavior of the poor would vary as their ecology varied. For example, the theory would predict less deviance in poor families in situations where their neighborhood is less dense and more heterogeneous in terms of income, where their homes are less crowded and dilapidated, where the neighborhood is more fully residential, where the police are not permissive of vice, and where there is no undue concentration of the demoralized.

As reaffirmed in the last paragraphs of this essay, the aim here is not to dismiss "kinds of people" or compositional factors, but to restore the theoretical power that was lost when the field abandoned human ecology. As a demonstration of what can be regained, let us examine briefly the most serious and painful issue confronting contemporary American criminology—black crime.

It is important to recognize that, for all the pseudo-biological trappings of the Chicago school (especially in Park's work), their primary motivation was to refute "kinds of people" explanations of slum deviance based on Social Darwinism. They regarded it as their major achievement to have demonstrated that the real cause of slum deviance was social disorganization, not inferior genetic quality (Faris, 1967).

Today Social Darwinism has faded into insignificance, but the questions it addressed remain—especially with the decline of human ecology. For example, like the public at large, when American social scientists talk about poor central city neighborhoods, they mainly mean black neighborhoods. And, since they are not comfortable with racist explanations, social scientists have been almost unwilling to discuss the question of why black crime rates are so high. Nearly everybody knows that in and of itself, poverty offers only a modest part of the answer. So, what else can safely be said about blacks that can add to the explanation? Not much, *if* one's taste is for answers based on characteristics of persons. A lot, if one turns to ecology.

Briefly, my answer is that high black crime rates are, in large measure, the result of *where* they live.

For several years there has been comment on the strange fact that racial patterns in arrest and imprisonment seem far more equitable in the South than in the North and West. For example, the ratio of black prison inmates per 100,000 to white prison inmates per 100,000 reveals that South Carolina is the most equitable state (with a ratio of 3.2 blacks to 1 white), closely followed by Tennessee, Georgia, North Carolina, Mississippi, and Alabama, while Minnesota (22 blacks to 1 white) is the least equitable, followed by Nebraska, Wisconsin, and Iowa. Black/white arrest ratios, calculated the same way, also show greater equity in the South while Minnesota, Utah, Missouri, Illinois, and Nebraska appear to be least equitable (Stark, 1986). It would be absurd to attribute these variations to racism. Although the South has changed immensely, it is not credible that cops and courts in Minnesota are far more prejudiced than those in South Carolina.

But what *is* true about the circumstance of Southern blacks is that they have a much more normal ecological distribution than do blacks outside the South. For example, only 9% of blacks in South Carolina and 14% in Mississippi live in the central core of cities larger than 100,000, but 80% of blacks in Minnesota live in large center cities and 85% of blacks in Nebraska live in the heart of Omaha. What this means is that large proportions of Southern blacks live in suburbs, small towns, and rural areas where they benefit from factors conducive to low crime rates. Conversely, blacks outside the South are heavily concentrated in precisely the kinds of places explored in this essay—areas where the probabilities of *anyone* committing a crime are high. Indeed, a measure of black center city concentration is correlated .49 with the black/white arrest ratio and accounts for much of the variation between the South and the rest of the nation (Stark, 1986).

"Kinds of people" explanations could not easily have led to this finding, although one might have conceived of "center city resident" as an individual trait. Even so, it is hard to see how such an individual trait would lead to explanations of why place of residence mattered. Surely it is more efficient and pertinent to see dilapidation, for example, as a trait of a building rather than as a trait of those who live in the building.

Is there any reason why social scientists must cling to individual traits as the *only* variables that count? Do I hear the phrase "ecological fallacy"? What fallacy? It turns out that examples of this dreaded problem are very hard to find and usually turn out to be transparent examples of spuriousness—a problem to which *all* forms of non-experimental research are vulnerable (Gove and Hughes, 1980; Stark, 1986; Lieberson, 1985).

Finally, it is not being suggested that we stop seeking and formulating "kinds of people" explanations. Age and sex, for example, have powerful effects on deviant behavior that are not rooted in ecology (Gove, 1985). What is suggested is that, although males will exceed females in terms of rates of crime and delinquency in all neighborhoods, males in certain neighborhoods will have much higher rates than will males in some other neighborhoods, and female behavior will fluctuate by neighborhood too. Or, to return to the insights on which sociology was founded, social structures are real and cannot be reduced to purely psychological phenomena. Thus, for example, we can be sure that an adult, human male will behave somewhat differently if he is in an all-male group than if he is the only male in a group—and no sex change surgery is required to produce this variation.

REFERENCES

Allison, Archibald. 1840. The Principles of Population and the Connection With Human Happiness. Edinburgh: Blackwood.

Anderson, L. S. 1979. The deterrent effect of criminal sanctions: Reviewing the evidence. In Paul J. Brantingham and Jack M. Kress (eds.), Structure, Law and Power. Beverly Hills: Sage.

Bursik, Robert J., Jr., and Jim Webb. 1982. Community change and patterns of delinquency. American Journal of Sociology 88:24–42.

Erickson, Maynard L. 1971. The group context of delinquent behavior. Social Problems 19:114–129.

Faris, Robert E. L. 1967. Chicago Sociology, 1920–1932. San Francisco: Chandler.

Faris, Robert E. L. and Warren Dunham. 1939. Mental Disorder in Urban Areas. Chicago: University of Chicago Press.

Frazier, E. Franklin. 1932. The Negro in the United States. New York: Macmillan.

Gans, Herbert J. 1962. The Urban Villagers. New York: Free Press.

Gibbs, Jack P. 1975. Crime, Punishment, and Deterrence. New York: Elsevier.

Goffman, Erving. 1959. Presentation of Self in Everyday Life. New York: Doubleday.

———. 1963. Stigma. Englewood Cliffs, NJ: Prentice-Hall.

Gold, Martin. 1970. Delinquent Behavior in an American City. Belmont, CA: Brooks/Cole.

Gove, Walter R. 1985. The effect of age and gender on deviant behavior: A biopsychological perspective. In Alice Rossi (ed.), Gender and the Life Course. New York: Aldine.

Gove, Walter R. and Michael L. Hughes. 1980. Reexamining the ecological fallacy: A study in which aggregate data are critical in investigating the pathological effects of living alone. Social Forces 58:1,157–1,177.

Gove, Walter R., Michael L. Hughes, and Omer R. Galle, 1979. Overcrowding in the home. American Sociological Review 44:59–80.

Guest, Avery M. 1984. Robert Park and the natural area: A sentimental review. Sociology and Social Research 68:1–21.

Hayner, Norman S. 1942. Five cities of the Pacific Northwest. In Clifford Shaw and Henry McKay (eds.), Juvenile Delinquency and Urban Areas. Chicago: University of Chicago Press.

Hindelang, Michael J. 1973. Causes of delinquency: A partial replication and extension. Social Problems 20:471–478.

Hindelang, Michael J., Travis Hirschi, and Joseph G. Weis. 1981. Measuring Delinquency. Beverly Hills: Sage.

Hirschi, Travis. 1969. Causes of Delinquency. Berkeley: University of California Press.

Hirschi, Travis and Michael J. Hindelang. 1977. Intelligence and delinquency: A revisionist view. American Sociological Review 42:571–587.

Jensen, Gary F., Maynard L. Erickson, and Jack Gibbs. 1978. Perceived risk of punishment and self-reported delinquency. Social Forces 57:57–58.

Jones, D. Caradog. 1934. The Social Survey of Merseyside, Vol. III. Liverpool: University Press of Liverpool.

Lazarsfeld, Paul F., Ann K. Pasanella, and Morris Rosenberg. 1972. Continuities in the Language of Social Research. New York: Free Press.

Lee, Barrett A., Ralph S. Oropesa, Barbara J. Metch, and Avery M. Guest. 1984. Testing the decline-of-community thesis: Neighborhood organizations in Seattle, 1929 and 1979. American Journal of Sociology 89:1161–1188.

Levin, Yale and Alfred Lindesmith. 1937. English Ecology and Criminology of the Past Century. Journal of Criminal Law and Criminology 27:801–816.

Lieberson, Stanley. 1985. Making It Count: The Impoverishment of Social Research and Theory. Berkeley: University of California Press.

Mayhew, Henry. 1851. London Labor and the London Poor. London: Griffin.

McKenzie, Roderick. 1926. The scope of human ecology. Publications of the American Sociological Society 20:141–154.

Minor, W. William and Joseph Harry. 1982. Deterrent and experimental effects in perceptual deterrence research. Journal of Research in Crime and Delinquency 18:190–203.

Park, Robert E. 1952. Human Communities: The City and Human Ecology. New York: The Free Press.

Park, Robert E., Ernest W. Burgess, and Roderick McKenzie. 1925. The City. Chicago: University of Chicago Press.

Parker, J. and Harol G. Grasmick. 1979. Linking actual and perceived certainty of punishment: An exploratory study of an untested proposition in deterrence theory. Criminology 17:366–379.

Reckless, Walter C. 1926. Publications of the American Sociological Society 20:164–176.

Shaw, Clifford R. and Henry D. McKay. 1942. Juvenile Delinquency and Urban Areas. Chicago: University of Chicago Press.

Simcha-Fagan, Ora and Joseph E. Schwartz. 1986. Neighborhood and delinquency: An assessment of contextual effects. Criminology 24:667–699.

Stark, Rodney. 1986. Crime and Deviance in North America: ShowCase. Seattle: Cognitive Development Company.

Stark, Rodney and William Sims Bainbridge. 1985. The Future of Religion. Berkeley: University of California Press.

———. 1987. A Theory of Religion. Bern and New York: Lang.

Stark, Rodney and James McEvoy. 1970. Middle class violence. Psychology Today 4:52–54, 110–112.

Stark, Rodney, Daniel P. Doyle, and Jesse Lynn Rushing. 1983. Beyond Durkheim: Religion and suicide. Journal for the Scientific Study of Religion 22:120–131.

Suttles, Gerald. 1968. The Social Order of the Slum. Chicago: University of Chicago Press.

———. 1972. The Social Construction of Communities. Chicago: University of Chicago Press.

Thrasher, Frederick M. 1927. The Gang. Chicago: University of Chicago Press.

Tittle, Charles R., Wayne J. Villemez, and Douglas A. Smith. 1978. The myth of social class and criminality: An empirical assessment of the empirical evidence. American Sociological Review 43:643–656.

Toby, Jackson. 1957. Social disorganization and stake in conformity: Complementary factors in the predatory behavior of hoodlums. Journal of Criminal Law, Criminology and Police Science 48:12–17.

Toby, Jackson and Marcia L. Toby. 1961. Law School Status as a Predisposing Factor in Subcultural Delinquency. New Brunswick: Rutgers University Press.

Welch, Kevin. 1983. Community development and metropolitan religious commitment: A test of two competing models. Journal for the Scientific Study of Religion 22:167–181.

Wuthnow, Robert and Kevin Christiano. 1979. The effects of residental migration on church attendance. In Robert Wuthnow (ed.), The Religious Dimension. New York: Academic Press.

NOTE

1. This is *not* to claim that neighborhoods do not change in terms of their levels of crime and deviance. Of course they do, even in Chicago (Bursik and Webb, 1982). It also is clear that such changes in deviance levels often are accompanied by changes in the kinds of people who live there. The so-called gentrification of a former slum area would be expected to reduce crime and deviance there as the decline of a once nicer neighborhood into a slum would be expected to increase it. However, such changes involve much more than changes in the composition of the population. Great physical changes are involved too, and my argument is that these have effects of their own.

QUESTIONS FOR DISCUSSION

1. Discuss how three of the five essential factors (or aspects) of urban neighborhoods, which are characterized as having high deviance, are in your geographical location.

2. Relate five of the thirty propositions to your geographical location.

3. Explain how the ecological theory can explain deviance.

Parade Strippers: Being Naked in Public

Craig J. Forsyth

BEING NAKED IN PUBLIC

As a topic for research, being naked in public can be discussed under the broad umbrella of exhibitionism or within the narrow frame of fads or nudity (Bryant 1977). In general, exhibitionism involves flaunting oneself in order to draw attention. In the field of deviance the term exhibitionism may also refer to behavior involving nudity for which the public shows little tolerance (Bryant 1977, p. 100; Bartol 1991, p. 280). This research, however, focuses on a form of public nudity that has a degree of social acceptance.

An extensive sociological study of public nudity was *The Nude Beach* (Douglas et al. 1977). Weinberg's (1981a, 1981b) study of nudists represents another type and degree of public nakedness. Other research has addressed the topics of streaking (running nude in a public area) (Toolan et al. 1974; Anderson 1977; Bryant 1982) and mooning (the practice of baring one's buttocks and prominently displaying the naked buttocks out of an automobile or a building window or at a public event) (Bryant 1977, 1982). Both streaking and mooning were considered fads. One question considered by sociological research on nakedness is when and why it is permissible, appropriate, or acceptable to be naked in public (Aday 1990). Researchers have also addressed some possible motivations or rationales for public nudity. Toolan et al. (1974, p. 157), for example, explain motivations for streaking as follows:

> While streaking is not in itself a sex act, it is at least a more-than-subtle assault upon social values. Its defiance serves as a clarion call for others to follow suit, to show "the squares" that their "old hat" conventions, like love, marriage, and the family, are antiquated.

Both Bryant (1982) and Anderson (1977) say that streaking began as a college prank that spread quickly to many campuses. As a fad, it still retained parameters of time and place. Bryant (1982, p. 136) contended that it was one generation flaunting their liberated values in the faces of the older, more conservative generation. Anderson (1977, p. 232) said that it embodied the new morality and thus was "perceived by many to be a challenge to traditional values and laws."

Mooning, like streaking, was considered a prank and an insult to conformity and normative standards of behavior. Neither streaking nor mooning had any erotic

---

value (Bryant 1982). Unlike streaking, mooning is still relatively common on college campuses.

Nudism in nudist camps has had little erotic value. Indeed, nudity at nudist camps has been purposively antierotic. Weinberg (1981b, p. 337) believes that the nudist camp would "anesthetize any relationship between nudity and sexuality." One strategy used by nudist camps to ensure this was to exclude unmarried people.

> Most camps, for example, regard unmarried people, especially single men, as a threat to the nudist morality. They suspect that singles may indeed see nudity as something sexual. Thus, most camps either exclude unmarried people (especially men), or allow only a small quota of them (Weinberg 1981b, p. 337). . . .

Nude sunbathing incorporates many rationales from voyeurism to lifestyle and in many cases has a degree of erotic value. The sexuality of the nude beach has been evaluated as situational.

> Voyeurism . . . poses a dilemma for the nude beach naturalists, those who share in some vague way the hip or casual vision of the nude beach. . . . voyeurs have became the plague of the nude scene. . . . The abstract casual vision of the beach does not see it as in any way a sex trip, but the casual vision of life in general certainly does not exclude or downgrade sex (Douglas et al. 1977, pp. 126–27).

Similar to the nudist in the nudist camp, nude beachers expressed contempt for the "straight" voyeur.

> Sometimes I really feel hostile to the lookers. Obviously you can't look at people that way even if they are dressed . . . it really depends on your attitude in looking. I've even told a couple of people to fuck off . . . and some people to leave. I was thinking this would be the last time I would come down here . . . there were too many sightseers . . . it sort of wrecks your time to have someone staring at you (Douglas et al. 1977, p. 130). . . .

[What about parade stripping, the most recent phenomenon of being naked in public, as practiced on Mardi Gras day in New Orleans?]

MARDI GRAS: DEVIANCE BECOMES NORMAL

On Mardi Gras day in New Orleans many things normally forbidden are permitted. People walk around virtually nude, women expose themselves from balconies, and the gay community gives new meaning to the term outrageous. Laws that attempt to legislate morality are informally suspended. It is a sheer numbers game for the police; they do not have the resources to enforce such laws. . . .

The celebration of carnival or Mardi Gras as it occurs in New Orleans and surrounding areas primarily involves balls and parades. These balls and parades are produced by carnival clubs called "krewes." Parades consist of several floats, usually between fifteen and twenty-five, and several marching bands that follow each float. There are riders on the floats. Depending on the size of the float, the number of riders can vary from four to fifteen. The floats roll through the streets of New Orleans on predetermined routes. People line up on both sides of the street on the routes. The float riders and the viewers on the street engage in a sort of game. The riders have bags full of beads or other trinkets that they throw out to the viewers along the route. The crowds scream at the riders to throw them something. Traditionally, the scream has been "throw me something mister." Parents put their children on their shoulders or have ladders with seats constructed on the top in order to gain some advantage in catching some of these throws. These "advantages" have become fixtures, and Mardi Gras ladders are sold at most local hardware stores. It is also advantageous if the viewer knows someone on the float or is physically closer to the float. Another technique is to be located in temporary stands constructed along the parade route that "seat" members of the other carnival krewes in the city or other members of the parading krewe.

In recent years another technique has emerged. Women have started to expose their breasts in exchange for throws. The practice has added another permanent slogan to the parade route. Many float riders carry signs that say "show me your tits"; others merely motion to the women to expose themselves. In some cases, women initiate the encounter by exposing their breasts without any prompting on the part of the float rider.

The author became aware of the term "beadwhore" while viewing a Mardi Gras parade. There were several women exposing their breasts to float riders. I had my 3-year-old son on my shoulders and I was standing in front of the crowd next to the floats. I am also a tall person. All of these factors usually meant that we caught a lot of throws from the float riders, but we caught nothing. Instead, the float riders were rewarding the parade strippers. As we moved away to find a better location, a well-dressed older woman, who had been standing behind the crowd, said to me:

> You can't catch anything with those beadwhores around. Even cute kids on the shoulders of their fathers can't compete with boobs. When the beadwhores are here, you just need to find another spot.

The term was also used by some of the interviewees [in this research].

METHODOLOGY

Data for this research were obtained in two ways: interviews and observations in the field. Interview data were gotten from an available sample of men who ride parade floats (N = 54) and from women who expose themselves (N = 51). These interviews ranged in length from 15 to 45 minutes. In the interviews with both float riders and parade strippers an interview guide was used to direct the dialogue. The guide was intended to be used as a probing mechanism rather than as a generator of specific responses. Respondents were located first through friendship networks and then by snowballing. Snowball sampling is a method through which the researcher develops an ever-increasing set of observations (Babbie 1992). Respondents in the study were asked to recommend others for interviewing, and each of the subsequently interviewed participants was asked for further recommendations. Additional informal interviews were carried out with other viewers of Mardi Gras.

Observations were made at Mardi Gras parades in the city of New Orleans over two carnival seasons: 1990 and 1991. Altogether, 42 parades were observed. The author assumed the role of "complete observer" for this part of the project (Babbie

1992, p. 289). This strategy allows the researcher to be unobtrusive and not affect what is going on. The author has lived a total of 24 years in New Orleans and has been a complete participant in Mardi Gras many times. Observations were made at several different locations within the city.

FINDINGS

The practice of parade stripping began in the late 1970s but its occurrence sharply increased from 1987 to 1991. During this study, no stripping occurred in the day-time. It always occurred in the dark, at night parades. Strippers were always with males. Those interviewed ranged in age from 21 to 48; the median age was 22. Most of them were college students. Many began stripping during their senior year in high school, particularly if they were from the New Orleans area. If from another area, they usually began in college. All of the strippers interviewed were in one location, a middle-class white area near two universities. Both riders and strippers said it was a New Orleans activity not found in the suburbs, and they said it was restricted to only certain areas of the city. One float rider said:

> In Metairie [the suburbs] they do it rarely if at all, but in New Orleans they
> have been doing it for the last ten years. Mostly I see it in the university section
> of the city during the night parades.

Parade strippers often attributed their first performances to alcohol, to the coaxing of the float riders, to other strippers in the group, or to a boyfriend. This is consistent with the opinion of Bryant (1982, pp. 141–42), who contended that when females expose themselves it is usually while drinking. Alcohol also seemed to be involved with the float riders' requests for women to expose themselves. One rider stated:

> Depending on how much I have had to drink, yes I will provoke women to ex-
> pose themselves. Sometimes I use hand signals. Sometimes I carry a sign which
> says "show me your tits." If I am real drunk I will either stick the sign in their
> face or just scream at them "show me your tits."

Data gained through both interviews and observation indicated that parade stripping is usually initiated by the float riders. But many of the women indicated that they were always aware of the possibility of stripping at a night parade. Indeed, some females came well prepared for the events. An experienced stripper said:

> I wear an elastic top. I practice before I go to the parade. Sometimes I practice
> between floats at parades. I always try to convince other girls with us to show
> 'em their tits. I pull up my top with my left hand and catch beads with my right
> hand. I get on my boyfriend's shoulder. I do it for every float . . . I'll show my
> breasts longer for more stuff and I'll show both breasts for more stuff.

Other parade strippers gave the following responses when asked, "Why do you expose yourself at parades?"

> I'm just a beadwhore. What else can I say?

> I expose myself because I'm drunk and I'm encouraged by friends and strangers
> on the floats.

I get drunk and like to show off my breasts. And yes they are real.

Basically for beads. I do not get any sexual gratification from it.

I only did it once. I did it because a float rider was promising a pair of glass beads.

When I drink too much at a night parade, I turn into a beadwhore.

It's fun.

I exposed myself on a dare. Once I did it, I was embarrassed.

Only one woman admitted that she did it for sexual reasons. At 48, she was the oldest respondent. When asked why she exposed her breasts at parades, she said:

Sexual satisfaction. Makes me feel young and seductive. My breasts are the best feature I have.

One woman who had never exposed herself at parades commented on her husband's efforts to have her participate during the excitement of a parade.

We were watching a parade one night and there were several women exposing their breasts. They were catching a lot of stuff. My husband asked me to show the people on the float my breasts so that we could catch something. He asked me several times. I never did it and we got into an argument. It seemed so unlike him, asking me to do that.

Float riders often look on bead tossing as a reward for a good pair of breasts, as the following comments show:

The best boobs get the best rewards.

Ugly women get nothing.

Large boobs get large rewards.

When parade strippers exposed themselves they were not as visible to people not on the float as one would think. Strippers were usually on the shoulders of their companions and very close to the float. For a bystander to get a "good look" at the breasts of the stripper was not a casual act. A person had to commit a very deliberate act in order to view the event. Those who tried to catch a peek but were either not riding the floats or not among the group of friends at the parade were shown both pity and contempt.

I hate those fuckers [on the ground] who try to see my boobs. If I'm with some people they can look. That's ok. But those guys who seek a look they are disgusting. I bet they can't get any. They probably go home and jerk off. I guess I feel sorry for them too. But I still don't like them. You know it's so obvious, they get right next to the float and then turn around. Their back is to the float. They are not watching the parade. We tell them to "get the fuck out of here asshole" and they leave.

Like a small minority of nude sunbathers who like to be peeped at (Douglas et al. 1977, p. 128), there are strippers who like the leering of bystanders. Our oldest respondent, mentioned earlier, said she enjoyed it. "I love it when they look. The more they look the more I show them," she remarked.

Parade strippers most often perform in the same areas. Although parade stripping usually involves only exposing breasts, three of the interviewees said they had exposed other parts of their bodies in other public situations.

Strippers and their male companions tried to separate themselves from the crowds; they developed a sense of privacy needed to perform undisturbed (Sommer 1969; Palmer 1977). Uninvited "peepers" disturbed the scene and were usually removed through verbal confrontation.

Most strippers and others in attendance apparently compartmentalized their behavior (Schur 1979, p. 319; Forsyth and Fournet 1987). It seemed to inflict no disfavor on the participants, or if it did they seemed to manage the stigma successfully (Gramling and Forsyth 1987). . . .

CONCLUSION

Parade stripping seemed to exist because trinkets and beads were given; for those interviewed, there was no apparent sexuality attached except in one case.

Parade stripping is probably best understood as "creative deviance" (Douglas et al. 1977, p. 238), deviance that functions to solve problems or to create pleasure for the individual. Many forms of deviance, however, do not work in such simplistic ways.

> Most people who go to a nude beach, or commit any other serious rule violation, do not find that it *works* [emphasis added] for them. They discover they are too ashamed of themselves or that the risk of shaming by others is too great, so they do not continue. Other people find it hurts them more (or threatens them) or, at the very least, does not do anything good for them. So most forms of deviance do not spread (Douglas et al. 1977, p. 239).

Some forms of deviance apparently do "work," and parade stripping is one of them. The beadwhore engages in a playful form of exhibitionism. She and the float rider both flirt with norm violation. The stripper gets beads and trinkets and the float rider gets to see naked breasts. Both receive pleasure in the party atmosphere of Mardi Gras, and neither suffers the condemnation of less creative and less esoteric deviants.

REFERENCES

Aday, David P. 1990. *Social Control at the Margins*. Belmont, CA: Wadsworth.

Anderson, William A. 1977. "The Social Organizations and Social Control of a Fad." *Urban Life* 6:221–40.

Babbie, Earl. 1992. *The Practice of Social Research*. Belmont, CA: Wadsworth.

Bartol, Curt R. 1991. *Criminal Behavior: A Psychosocial Approach*. Englewood Cliffs, NJ: Prentice-Hall.

Bryant, Clifton D. 1977. *Sexual Deviancy in Social Context*. New York: New Viewpoints.

Bryant, Clifton D. 1982. *Sexual Deviancy and Social Proscription: The Social Context of Carnal Behavior*. New York: Human Sciences Press.

Douglas, Jack D., Paul K. Rasmussen, and Carol A. Flanagan. 1977. *The Nude Beach*. Beverly Hills, CA: Sage.

Forsyth, Craig J., and Lee Fournet. 1987. "A Typology of Office Harlots: Party Girls, Mistresses and Career Climbers." *Deviant Behavior* 8:319–328.

Gramling, Robert, and Craig J. Forsyth. 1987. "Exploiting Stigma." *Sociological Forum* 2: 401–415.

Palmer, C. Eddie. 1977. "Microecology and Labeling Theory: A Proposed Merger," pp. 12–17 in *Sociological Stuff*, edited by H. Paul Chalfant, Evans W. Curry, and C. Eddie Palmer. Dubuque, IA: Kendall/Hunt.

Schur, Edwin M. 1979. *Interpreting Deviance*. New York: Harper & Row.

Sommer, Robert. 1969. *Personal Space*. Englewood Cliffs, NJ: Prentice-Hall.

Toolan, James M., Murray Elkins, and Paul D'Encarnacao. 1974. "The Significance of Streaking." *Medical Aspects of Human Sexuality* 8:152–165.

Weinberg, Martin S. 1981a. "Becoming a Nudist," pp. 291–304 in *Deviance: An Interactionist Perspective*, edited by Earl Rubington and Martin S. Weinberg. New York: Macmillan.

Weinberg, Martin S. 1981b. "The Nudist Management of Respectability," pp. 336–345 in *Deviance: An Interactionist Perspective*, edited by Earl Rubington and Martin S. Weinberg. New York: Macmillan.

Questions for Discussion

1. What definition of deviance best describes parade strippers and beadwhores? Why are parade strippers and beadwhores deviant if law enforcement officials know about their law-violating behaviors but do not arrest them?

2. According to the author, why do parade strippers and beadwhores expose their bodies?

3. According to the author, how does deviance become "normal" at Mardi Gras?

C H A P T E R 3

BECOMING DEVIANT

Deviance as Fun

Jeffrey W. Riemer

Deviant behavior as a fun activity has been consistently ignored in the work of social scientists. Rarely has deviance been considered a spontaneous, "just for the hell of it" activity, in which the participants engage simply for the pleasure it provides. Rather than incorporating this simple, yet potentially useful, dimension into our understanding of deviance phenomena, sociologists have continued to be preoccupied with elaborating the more sober social, psychological, and sociological explanations for this activity.

Deviance is seldom treated as a frivolous, flippant activity. Yet, this hedonistic "pleasure of the moment" explanation squares well with a common sense interpretation for some of the behavior usually recognized as deviant. In an era of multicausal explanations we may be inadvertently ignoring an important dimension for better understanding deviant behavior.

Vandalism could be partially explained this way, as could some instances of premarital and extra-marital sex, homosexuality (or bisexuality), drug use (or experimentation), drunkenness, shoplifting, auto theft, gambling, fist fights, reckless driving, profanity and prostitution, at least from the client's point of view—to mention a few.

Gustav Ichheiser (1970) has suggested that we are often "blind" to the obvious. We overlook or ignore what we believe everyone already knows. He suggests that what

"Deviance as Fun" by Jeffrey W. Riemer, from *Adolescence*, vol. 16, no. 61 (Spring 1981). Reprinted with permission from Libra Publishers, Inc.

is "taken for granted" is usually neglected or treated as unimportant. This seems to be what has happened with the fun dimension of deviance. When something squares well with common sense knowledge it does not nullify its theoretical importance.

THE LITERATURE

Some social scientists have treated this dimension of deviance in an ancillary way in their writings, while most have totally ignored it. In fact, at least one has been literally forced to recognize this alternative by a respondent who was unwilling to accept a more sophisticated interpretation of her behavior constructed by serious-minded social scientists. Simmons relates the following:

> Several years ago a lesbian interrupted my interview with her and said, "We've kept talking about emotional turmoil and male avoidance. But the truth is I go to bed with her because it's fun." (Simmons, 1969:63)

Garfinkel (1964) has actually created deviant behavior by purposely having his students violate "taken for granted" background expectancies in routine social situations in an attempt to understand the common sense world and how social order emerges. Many of his quasi-experimental researches are quite humorous and, in addition, demonstrate subject reactions to folkway violations as well as experimenter discomfort and pleasure at being "deviant." But Garfinkel fails to interpret any of these research designs as exercises in "fun" or "deviance."

Deviant behavior texts through the years have routinely ignored this dimension with no mention of "fun," "pleasure," "hedonism," or the like, in reference to the causal or contributory factors that influence deviant behavior. Most theorists have continued to delve into the more serious publicly recognized forms of objectionable behavior, often including what is usually recognized as criminal behavior (Thio, 1978; Goode, 1978; Akers, 1977; Sagarin and Montanino, 1977). Certainly, all crime is deviance but not all deviance is crime (Glaser, 1971).

Mundane deviance has been typically neglected. Denzin has argued that sociologists should pay more attention to the "mundane, routine, ephemeral, or normal" forms of deviance that surround all of our everyday life experiences (Denzin, 1970: 121). However, his suggestion has fallen on deaf ears or met objection. Most current students of deviant behavior continue to agree with Gibbons and Jones (1971) who argue that "omnibus" definitions of deviance are unfruitful and loaded with triviality. Mundane deviance is seen as inconsequential and not worth consideration.

The major exception is Loftland (1969) who, in passing, recognized the "deviance as fun" alternative.

> At least some kinds of prohibited activities are claimed by some parts of the population to be *in themselves* fun, exciting and adventurous. More than simply deriving pleasant fearfulness from violating the prohibition per se, there can exist claims that the prohibited activity *itself* produces a pleasant level of excitation. (Loftland, 1969:109)

Even so, after eight years we find little incorporation of this argument.

Research studies have also consistently ignored the deviance as fun dimension. Research into deviant behavior regularly employs the more accepted sophisticated conceptual frameworks for describing and explaining the behavior in question. Again, a somber and serious approach is taken.

One exception, Riemer (1978), has looked at deviance in the workplace as a fun activity. Four activities that building construction workers routinely engage in while being paid to work are discussed. These include: drinking alcoholic beverages, "girl watching" and other sexually related activities, stealing, and loafing. Each of these activities is engaged in periodically by some workers as a pleasant change of pace from the routine and boredom of work when the opportunities arise. These activities typically arise spontaneously (given the opportunity) with the workers giving a "just for the hell of it" rationale for their participation. This fun explanation is not meant to be the sole reason for this worker behavior but it cannot be ignored as contributory.

Unfortunately, it is the juvenile delinquency theorists who have taken a more assimilating view of deviance as a fun activity. Ferdinand (1966) in developing a typology of delinquency, included the "mischievous-indulgent" and the "disorganized acting-out" types which are both motivated in their actions by some degree of hedonism and spontaneity. Along this same line, Gibbons (1965) developed a series of delinquent categories in which he included the "casual gang delinquent" (who regards himself as a non-delinquent but likes to have fun) and the "auto-thief joyrider."

Similarly, Briar and Piliavin have argued for the influence of situationally induced motives to deviate. They suggest that some delinquency may be

> prompted by short-term situationally induced desires experienced by all boys to obtain valued goods, to portray courage in the presence of, or be loyal to peers, to strike out at someone who is disliked, or simply to "get kicks." (Briar and Piliavin, 1965:36)

Matza (1961) and Cohen (1955) have also built an "adventure," "play," and "fun" dimension into their explanation of delinquent behavior.

To offer a twist on the Matza and Sykes (1961) thesis that much juvenile behavior could be analyzed as an extension of adult behavior, it is argued here that some adult deviance could be analyzed as an extension of juvenile delinquency. Some deviant behavior may arise as a simple pleasure-seeking activity.

IMPLICATIONS

The argument presented here is not meant to dispute any particular theoretical orientation for deviant behavior. Rather, it calls attention to them all for neglecting a seemingly fruitful area of pursuit for helping to explain this behavior. Certainly, the fun dimension should not be construed as a unidimensional explanation for deviance phenomena. It is simply another perspective that should be incorporated into our existing approaches.

A deviance as fun argument suggests that all persons are deviant at least some of the time. Each of us departs from prevailing normative standards on occasion, weighing the possibility of being publicly designated as deviant and sanctioned accordingly, and choosing a seemingly pleasant diversion from our normative routines. Accordingly, this fun dimension implies a "normal" deviance that exists in society (Durkheim, 1964; Goffman, 1963).

For most of us, most of the time, these pleasant diversions remain non-problematic. But occasionally an act that begins innocently evolves into a serious legal infraction, e.g., the party-goer who drinks to excess as "the life of the party" and as a result of the intoxication becomes involved in a serious traffic accident on the journey home. What began as fun turned into tragedy.

What is pleasurable, enjoyable, or fun lies in the eyes of the beholder. The current labeling and political theorists might well address what is considered acceptable fun. "One man's pleasure is another man's pain."

The choice to violate normative standards is a choice we all have. For some, this choice represents an enjoyable and exciting alternative. It is within this context that we may learn more about deviant behavior and social control.

REFERENCES

Akers, Ronald L. 1977. *Deviant Behavior.* Belmont, Calif.: Wadsworth.

Becker, Howard S. 1963. *Outsiders.* New York: Free Press.

Birenbaum, Arnold, and Edward Sagarin. 1976. *Norms and Human Behavior.* New York: Praeger.

Briar, Scott, and Irving Piliavin. 1965. "Delinquency, Situational Inducements, and Commitment to Conformity," *Social Problems,* 13 (Summer): 35–45.

Cohen, Albert K. 1955. *Delinquent Boys: The Culture of the Gang.* Glencoe, Ill.: Free Press.

Davis, Nanette J. 1975. *Sociological Constructions of Deviance.* Dubuque, Iowa: Wm. C. Brown Company.

Denzin, Norman K. 1970. "Rules of Conduct and the Study of Deviant Behavior: Some Notes on the Social Relationship," in *Deviance and Respectability,* Jack D. Douglas (Ed.). New York: Basic Books, pp. 120–159.

Durkheim, Emile. 1964. *The Rules of Sociological Method.* New York: Free Press.

Feldman, Saul D. 1978. *Deciphering Deviance.* Boston: Little, Brown.

Ferdinand, Theodore N. 1966. *Typologies of Delinquency: A Critical Analysis.* New York: Random House.

Finestone, Harold. 1957. "Cats, Kicks and Color," *Social Problems,* 5 (July): 3–13.

Garfinkel, Harold. 1964. "Studies of the Routine Grounds of Everyday Activities," *Social Problems,* 11 (Winter): 225–250.

Gibbons, Don C. 1965. *Changing the Lawbreaker: The Treatment of Delinquents and Criminals.* Englewood Cliffs, N.J.: Prentice-Hall.

Gibbons, Don C., and Joseph F. Jones. 1971. "Some Critical Notes on Current Definitions of Deviance," *Pacific Sociological Review,* 14 (January): 20–37.

Glaser, Daniel. 1971. *Social Deviance.* Chicago: Markham.

Goffman, Erving. 1963. *Stigma.* Englewood Cliffs, N.J.: Prentice-Hall.

Goode, Erich. 1978. *Deviant Behavior.* Englewood Cliffs, N.J.: Prentice-Hall.

Ichheiser, Gustav. 1970. *Appearances and Realities.* San Francisco: Jossey-Bass.

Loftland, John. 1969. *Deviance and Identity.* Englewood Cliffs, N.J.: Prentice-Hall.

Matza, David. 1961. "Subterranean Traditions of Youth," *Annals of the American Academy of Political and Social Science,* 338 (November).

Matza, David, and Gresham M. Sykes. 1961. "Juvenile Delinquency and Subterranean Values," *American Sociological Review,* 26: 712–719.

Riemer, Jeffrey W. 1978. "'Deviance' as Fun—A Case of Building Construction Workers at Work," in *Social Problems—Institutional and Interpersonal Perspectives,* K. Henry (Ed.), Glenview, Ill.: Scott, Foresman, pp. 322–332.

Sagarin, Edward, and Fred Montanino (Eds.). 1977. *Deviants: Voluntary Actors in a Hostile World*. Morristown, N.J.: General Learning Press.

Simmons, J. L. 1978. *Deviants*. Berkeley: Glendessary Press.

Thio, Alex. 1978. *Deviant Behavior*. Boston: Houghton Mifflin.

QUESTIONS FOR DISCUSSION

1. What does the author mean by deviance as fun? Discuss a "fun," "pleasurable," "hedonistic" example.

2. Discuss the author's views on deviance in the workplace as a fun activity.

3. Can deviance be fun when students violate "taken for granted" expectations in routine social situations? Explain your response.

Learning to Strip: The Socialization Experiences of Exotic Dancers

Jacqueline Lewis

INTRODUCTION

Entering any new job or social role requires a process of socialization where the individual acquires the necessary values, attitudes, interests, skills and knowledge in order to be competent at her/his job. As with any new job or social role, becoming an exotic dancer requires a process of socialization. For exotic dancers, achieving job competence involves getting accustomed to working in a sex-related occupation, and the practice of taking their clothes off in public for money. In addition, in order to be a successful exotic dancer, women must also learn how to manipulate clientele and to rationalize such behaviour and their involvement in a deviant occupation.[1] For some dancers, the socialization process is partially anticipatory in nature, although, dancers reported that most of their socialization occurred once they had made their decision to dance and found themselves actually working in the strip club environment. In this paper, I explore the factors influencing entry into exotic dancing, the socialization experiences of exotic dancers and the process of obtaining job competence.

Background

Since the late 1960s, exotic dancing and the experiences of exotic dancers have been the focus of academic inquiry (Boles & Garbin, 1974a, 1974b, 1974c; Carey,

"Learning to Strip: The Socialization Experiences of Exotic Dancers" by Jacqueline Lewis. Published in *The Canadian Journal of Human Sexuality*, 1999, vol. 7, pp. 51–66. Reprinted with permission.

Peterson & Sharpe, 1974; Dressel & Petersen, 1982a, 1982b; Enck & Preston, 1988; Forsyth & Deshotels, 1997; McCaghy & Skipper, 1969, 1972; Petersen & Dressel, 1982; Prus, 1980; Reid, Epstein & Benson, 1994; Ronai & Ellis, 1989, Ronai, 1992; Skipper & McCaghy, 1971; Thompson & Harred, 1992). The relevance of some of the available literature to the present study is, however, limited by the focus of the articles. Within this literature on exotic dancers, only the articles by Boles and Garbin (1974b, 1974c), Carey et al. (1974), Dressel and Petersen (1982b), McCaghy and Skipper (1971), Prus (1980), Skipper and McCaghy (1972), and Thompson and Harred (1992) address the socialization experiences of dancers in any detail. Dressel and Petersen's (1982b) focus on the socialization of male exotic dancers makes their work of limited applicability to the present study.

Although much of this research was conducted over 15 to 20 years ago, some of it remains relevant to the work reported here. For example, the findings of Boles and Garbin (1974b, 1974c), Carey et al. (1974), McCaghy and Skipper (1971), Prus (1980) and Skipper and McCaghy (1972) provide an historical point of comparison that indicates some consistency between past and current research findings on the occupational socialization of exotic dancers.

The literature on occupational socialization of exotic dancers emphasizes two basic themes: (1) the factors that influence entry into dancing; and (2) anticipatory and on-the-job socialization experiences. Two types of models have been advanced to explain entry into exotic dancing: (1) career contingency models (Skipper & McCaghy, 1972; Carey et al., 1974; Thompson & Harred, 1992); and (2) conversion models (Boles & Garbin, 1974b; Carey et al., 1974; Thompson & Harred, 1992). In some research reports, these models are used on their own (e.g., Skipper & McCaghy, 1972; Boles & Garbin, 1974b), and in others they are used in combination (Carey et al., 1974; Thompson & Harred, 1992). Although a variety of singular and combined models have been used to explain entry into exotic dancing, there are several common factors that are identified across the studies: (1) knowledge and accessibility of an opportunity structure that makes exotic dancing an occupational alternative (Carey et al., 1974; Skipper & McCaghy, 1972; Prus, 1980; Thompson & Harred, 1992); (2) an awareness of the economic rewards associated with being an exotic dancer (Boles & Garbin, 1974b; Carey et al., 1974; Dressel & Petersen, 1982b; Skipper & McCaghy, 1972; Prus, 1980; Thompson & Harred, 1992); (3) a recruitment process involving personal networks (Boles & Garbin, 1974b; Dressel & Petersen, 1982b; Thompson & Harred, 1992); and (4) financial need or a need for employment (Boles & Garbin, 1974b, 1974c; Carey et al., 1974; Prus, 1980; Thompson & Harred, 1992).

With respect to the anticipatory and on-the-job socialization experiences of dancers (Boles & Garbin, 1974c; Dressel & Petersen, 1982b; Thompson & Harred, 1992), early research found that most female dancers had either professional training in dance, music or theatre, had been previously employed in the entertainment industry, or received extensive training in stripping prior to dancing before an audience (Boles & Garbin, 1974c; McCaghy & Skipper, 1972; Prus, 1980). However, despite their advanced (anticipatory) preparation, a large part of the occupational socialization dancers experienced occurred through informal channels after they had entered the occupation. Through observing and interacting with other subcultural members, dancers learned the tricks of the trade, such as how to: interact with customers for profit; manage their deviant lifestyle; and be successful at their job (Boles & Garbin, 1974c; Dressel & Petersen, 1982; McCaghy & Skipper, 1972; Thompson & Harred, 1992).

Method

This study used a combination of field observations inside strip clubs, and interviews with exotic dancers and other club staff to identify issues associated with the work and careers of exotic dancers. Observations were conducted at clubs in several cities in southern Ontario. Observational data were collected primarily to supplement interview data and to assist us in describing the work environment of exotic dancers including: physical setting; contacts between those present in the club (employees and clients); and the atmosphere of different clubs.

Thirty semi-structured, in-depth interviews were conducted with female exotic dancers, club staff and key informants. Participants were recruited either by the research team during field trips to the clubs or by dancers who had participated in the study. Each interview was audiotaped and took place in a location chosen by the respondent (e.g., respondent's home, a research team member's office, a private space at a strip club, a local coffee shop). Interviews lasted anywhere from one to three hours, with the majority taking approximately one and a half hours. All interviews were conducted informally to allow participants to freely express themselves, and to allow for exploration of new or unanticipated topics that arose in the interview.

The interviews explored each woman's work history, her perception of her future in the occupation, a description of her work, the various forms of interaction engaged in with clients, use of drugs and alcohol, current sexual practices, perception of risk for HIV and other STDs associated with dancing, sexual health-maintaining strategies, factors influencing risk and ability to maintain sexual health, and the presence and/or possibility of a community among exotic dancers. Interviews with other club employees were designed to tap their experiences in, and impressions of, club-related activities.

As interviews were collected and transcribed, it became increasingly apparent that there was a variety of recurrent themes that ran throughout the interviews (e.g., motivations for entry, socialization process, health and safety concerns, relationships between club employees, impact of dancing on dancers' lives, etc.). Coding categories were developed to fit with these emerging themes. All interviews were then coded in Nud*ist, a qualitative analysis software package, by members of the research team. As noted by Glaser and Strauss (1967), "[. . .] in discovering theory, one generates conceptual categories or their properties from evidence; then the evidence from which the category emerged is used to illustrate the concept" (p. 23). The quotes that appear in this paper were selected as examples of the responses provided by the women interviewed that fit the various conceptual categories that emerged during data analysis.

BECOMING AN EXOTIC DANCER

Unlike other more conventional occupations with formally structured socialization programs, the socialization experiences of the women we spoke with were informal in nature. Dancers reported that they acquired the requisite skills for the job through informal socialization processes that were either: (1) anticipatory in nature, occurring prior to dancing; and/or (2) that occurred on-the-job, once they were employed to dance in a strip club.

Anticipatory Socialization

Early studies of female exotic dancers (see Boles & Garbin, 1974; McCaghy & Skipper, 1972) found that most dancers had fairly broad anticipatory socialization experiences, having been previously employed in an entertainment-related job, having some type of professional training in dance, music or theatre, or having an agent who helped prepare them for the career of exotic dancing. In this study, we, however, found little indication of the latter two types of anticipatory socialization experiences.

Although one woman had a background in drama, she talked about how it actually did little to prepare her for the job:

> I thought you know, O.K. being in Drama, ya, I'm kind of a freer person, whatever. But, like, actually taking off your clothes—nothing, nothing prepares you for it. Nothing. Seconds before I went up to go dance [for the first time], I'm thinking, oh my God, I can't do this, I can't do this. I can't do this. Then my music started playing and I'm like, I guess I have to now. And you know, your stomach's all in knots and you just do it. There's no way to describe it. You just do it.

Even the few women who indicated that they began dancing with the help of an agent talked about how they received little job preparation. For example, one woman said:

> I responded to an ad in the local paper and there was a number and you phoned the number and then you met with this guy and he made you sign a contract and then he kind of talked to you about what goes on. There was no training. Then he just took me to the bar later that evening and that was it.

Although the experiences dancers reported during their interviews varied, the women we spoke with who reported engaging in anticipatory socialization, talked about spending time in strip clubs before deciding to dance. In recalling their entry into exotic dancing, some of the dancers we interviewed spoke of being curious about dancing, and wanting to find out if it was something they could do. These women reported that they sussed out and gained familiarity with dancing by going out to strip clubs on their own and talking to dancers or by going out to the clubs with friends who hung out at or worked in strip clubs.

> So, I read some more about it. I read a couple of books on the sex industry and strippers in particular and burlesque dancers. Um, and then I visited a lot of the clubs and tried to talk to the dancers about how they got interested in it and how they get paid and what the job entails. They were pretty open to talking to me about it.

> I had a girlfriend who was a pretty promiscuous person, you know . . . We used to get together and like just hang out in dance clubs. Not strip clubs, but just normal clubs . . . She's a pretty cool person, you know, she's my best friend and my daughter's Godmother and she told me that she was working in a club as a

waitress. So I said, great, you know, "What kind of club is it?" Then she said, "It's a strip bar." And, I'm kind of like, "What? You're working in a strip bar?" And, she's like, "Yeah, I make really good money being a waitress there." I'm like, "O.K. whatever," you know. Then eventually she came out with it, and she's like, I'm not really just waitressing, I'm dancing too." And I'm like, "Wow, oh, how much money are you making doing that?" And she's like, "Well, really good and you know if you want to help make your daughter's life better, why don't you come with me one night?" And I'm like, "Oh, man, I don't know if I could do that, you know." I got real scared and everything, but we set a date for the next Friday.

The other women who had anticipatory socialization experiences reported experiencing a more gradual drift into dancing (Matza, 1992).[2] Instead of purposefully going to strip clubs and talking to people in the industry with the intention of sussing it out, these women drifted into dancing through associations they had with people in the industry or by working as a waitress in one of the clubs.

I didn't start out dancing. First I was a waitress. Eventually, I quit waitressing and I went and started dancing at a strip club.

I waitressed for about a year at the Maverick, and then I started to dance. I've been dancing for 7 years, just over 7 years.

I used to date this guy and some of his friends worked in the clubs, so we could go and hang out. He used to try to get me to try it [dancing], but I wouldn't. But, once we broke up, I decided to try it.

This one woman I sort of knew, she had danced a few years before and it just came up in discussion that she used to be a dancer. So she kind of gave me a little information and where to go and what to do. So that is how I got started.

According to Ritzer and Walczak (1986), "[. . .] deviant occupational skills may be learned through involvement in different but related occupations or through non-occupational activities" (p. 144). Through hanging out with people associated with the industry or by working in a strip club in some other capacity, these women experienced a form of anticipatory socialization that enabled them to view dancing as a viable job option. As noted by Matza (1992), "some learning is truly a discovery [for the individual], for until they have experimented with the forbidden, [. . . they] are largely unaware that infraction is feasible behavior" (p. 184).

A lot of my friends and a lot of the group that I used to hang around with while I was waitressing were uh, we were all in the same circles with the guys from a strip club for women and uh, the two clubs were connected, and so they kept saying "try it" and, you know, "go to this bar, start there" and that's just how I ended up there.

I lived with a guy when I was at [high] school . . . I was still a virgin. I slept with him on my seventeenth birthday and he blew my mind. The first thing he asked me was if I masturbated. And, like I'm a hick town girl, naive as shit and it was like wow. This guy's cool. And, I moved in with him the next day . . . We moved right into the city, downtown Toronto and he was hanging out with strippers . . . And I used to threaten him, you know, if you keep hanging out with these girls, I'm gonna become one. And I did.

> I used to waitress at a saloon and then I was hanging out with some of the girls
> and then dating guys from a dance club for women, heaven forbid, and it just
> went from there, I guess that's how I got into dancing.

> I started out waitressing in one of the local clubs. Watching the girls and
> the money they were making back then, I thought well, I'll try it [referring
> to dancing].

The experiences of the women who drifted into dancing can be viewed as a form of recruitment or conversion process whereby the individual is gradually introduced/exposed to the inner world of a new role or career and gives up one view of that role, or one world view, for another (see Becker, 1964; Lofland & Stark, 1965; Prus, 1977). According to Lofland and Stark (1965), the reinforcement and encouragement made available through intensive interaction with subgroup members is necessary if the recruit is to experience a complete conversion process.

Regardless of how they began their process of occupational socialization, in providing themselves with time to think things through, and to learn to identify with the norms, values and beliefs of the dancing subculture prior to entering it, these women were engaging in a form of role rehearsal and anticipatory socialization. Such efforts provided them with the opportunity to prepare themselves for the eventual reality of their new status, thereby easing the difficulties associated with the transition. Through engaging in anticipatory socialization, the women interviewed became accustomed to the strip club environment and the idea of taking their clothes off in public for money, thereby facilitating their entry into dancing.

On-the-Job Socialization

Similar to the socialization experiences of individuals in other occupations, novice dancers learn through interaction and observation while on-the-job. Since exotic dancers, however, have little, if any, formal training, learning through observation and interaction is crucial for attaining job competence (see Sanders, 1974). Although some of this learning may be anticipatory in nature and occur prior to the initial dancing experience, it takes some time and experience to move from being a novice dancer to a seasoned pro. Since there is no formal certification structure, peers play an important role in this transformation process. During this period, novices can continue to acquire knowledge from those around them about how to be successful at their job. Experienced strip club staff can therefore play an important role in the socialization process of the novice dancer. As one woman noted:

> You learn as you go. Other people in the club give you advice. And, you know,
> you gradually learn about how to make more money and who to talk to and
> that kind of stuff as you go.

Through talking to and receiving advice about the job from other staff members, novice dancers learn how to handle situations that may arise while working in the club, and how to dance for profit.

> The DJ at the first club I danced at was very good. On my first night he was
> like, "don't worry about it . . . You know, just go up there and do your thing

and you know, don't worry about it." And the other girls were kind of support-ive, like, "Oh, you'll get used to it, it's not that bad after a while." You know, some of them kind of take you under their wing and sort of show you the ropes so to speak.

I had just gone to the DJ booth and given him my music for when I was going to go on stage and he said it would be after a few girls, because he already had a few on the list . . . I had no idea what to do with myself after I had given the music to him. He said, "Just hang out and, you know, talk to people, be friendly, you know." So I was just walking back to put my CDs back in my bag, the ones I wasn't going to use, and three guys called me over to their table.

I learned a lot just watching the other women. Some of them had been dancing for a while and they were really good at handling customers when they tried to break the rules.

Other dancers play a particularly important role in the socialization process. As the following quotes illustrate, novices can learn how to dress, dance and interact with customers for profit, through observing and interacting with dancers more ex-perienced than themselves.

Most of the dancers are really nice, like, they're really understanding. They knew, you know, I hadn't danced for very long. Everybody was offering me ad-vice. There were a few that were kind of like, stay away from me and I'll stay away from you sort of thing.

I get ideas for my show from watching, you know, the ones that have been do-ing this [dancing] for a while. There were these three other dancers that were there [at the club she had begun working at]. Normally they have six on at a time. But, there were these three other dancers there that were amazing. Like, they couldn't have helped me more. And you know, they knew, like, at that time I'd only been dancing for about a week and you know, they were offering me advice left, right and centre. And you know, they were just so nice. They couldn't be more helpful.

One woman explained how a friend of hers, who was an experienced dancer, helped teach her how to table dance.

[Talking about her first table dance] So, she comes up beside me and the next thing I know, both our tops are off and she's like all rubbing close to me and I'm like going, "Oh my God." I never thought of her that way before, you know. Cuz we've always just been friends, you know. So, it was kind of a funny experi-ence. But, he ended up spending like a hundred dollars on songs. So, I'm think-ing, hey, this is great, you know. I mean, this is awesome. I've got money to come home with, you know, it wasn't a wasted night. I thought, O.K., I can deal with this a couple nights a week.

She went on to describe how her friend, along with a few other dancers, also helped her with her first stage performance:

The only thing that was really scary after my first night in the club was the stage, because I had never been on a stage before, and I'm thinking, "Oh my God, I don't have big breasts, I'm not like toned and tanned and blonde" or whatever. So, my friend's like, "Well, we can do like the dance that we did with that guy. We should do a dance on stage together." And, I'm like, "But I'm not gay and I'm not going to be able to make them think that I am." She's like, "Well, don't worry, just follow my lead [. . .]" The stage show actually went well because two other girls came up, so there was four girls on stage. So it made everybody kind of, you know, sort of stare at the stage and everyone was happy, so I was like, "O.K., this isn't so bad", you know? And then after doing that a few times I decided that, you know, I wanted to try it on my own. So I did, and I didn't like it as much because, you know, you sort of feel really centred out. But eventually I got used to it and I was able to do it, you know. Now I've got the hang of it.

RATIONALIZING PARTICIPATION IN A DEVIANT OCCUPATION

Since exotic dancing is viewed as a deviant occupation in our society, if novice dancers are to retain a valued sense of self, they must learn ways to justify their involvement in the strip club subculture. According to Sykes and Matza (1957), in order to deviate people must have access to a set of rationalizations or neutralizations that allow them to reduce the guilt they feel about violating social norms. Neutralization makes norm violations "morally feasible since it serves to obliterate, or put out of mind, the dereliction implicit in it" (Matza, 1992, p. 182).

During interviews with dancers, it became apparent that dancers typically rely on several "techniques of neutralization" (Sykes & Matza, 1957) to justify their involvement in deviant behaviour. Similar to Thompson and Harred's (1992) research on topless dancers, we found that the dancers we interviewed tended to rely primarily on three of Sykes and Matza's (1957) techniques of neutralization.

They denied injury or harm:

> Ya well, we pretend [that they like the customers], but what do they really expect. Do they really think we are there because we like them, that we like to dance for men—no. And really, who are we hurting? We may take their money, and although sometimes it may be a lot, but, they are adults, they should know better. And besides, it's just money.

They condemned the condemners:

> People may judge us and say that dancing is bad, but they seem to forget who it is we are dancing for—doctors, lawyers, sports figures. If it wasn't for them there would be no dancing—so maybe the focus is on the wrong people [the dancers rather than the customers].

> As soon as you tell people you dance, it's "Oh." It's a totally different idea of what kind of person you are, or however you are is a put on. I just think what is the big deal. We are all the same here.

> So I take my clothes off for a living. Doesn't make you any different. You all go there smoking dope and drinking beer anyway so. I mean I don't know how many people I know that work at a car plant and say, "I go to work have a couple

beers, smoke a doobie and go to work." And I said, "Where'd you guys get the beer." "Aw the guys in the parking lot sell it." And I'm laughing my guts out thinking are you guys serious? I think Jesus, these are people that build our motor vehicles and people are driving around in these things.

A guy friend of mine, known him for years, government employee, a very high job, used to smoke crack in the parking lot before he went to work. He has this home in the city, a historical home, I don't know a four or five hundred thousand dollar home . . . perfect job, normal job like everybody else, a wife and three kids at home, but smokes crack in the parking lot before he goes to work in the morning. Like how is he any better than me? I would just think you guys have the nerve to judge me and think you are a bit of a weirdo cuz you take your clothes off, wow how smart could you be? At least I'm not hooked on crack or anything. That's what I'm saying, some people that I know really well, you'd be shocked to hear what they do. So like I think what I do is just a drop in the bucket. At least I have an excuse, I'm a stripper. Like what is your excuse? You are a government employee, got a good job and you're the head of what? So I'm kinda laughing who are you to judge me.

And they appealed to higher loyalties:[3]

Well, they say that you're not supposed to show your body to lustful men and that that's a sin. So I assume that like, obviously God wasn't gonna be very happy that I was doing something like this. But, the other way I looked at it was, I have a daughter who is two years old and the government really doesn't give you enough to survive, so I had to do something. And I figured that if it's a sin to take off your clothes and it's a sin to let your child starve, definitely, I would take care of the second one, and it's probably more normal.

I had all these bills and I needed to feed my kids and well, what was I going to do. I do what I have to do to get by.

If you need to feed your kids, what are you going to do?

In addition to using some of Sykes and Matza's (1957) techniques of neutralization to justify their involvement in exotic dancing, we found that dancers used the technique of normalization. As the following quote illustrates, some women attempted to justify or neutralize their involvement in exotic dancing by refuting the deviancy associated with it.

And I looked at the salaries these people were making and it was, you know, a thousand dollars a night, some nights, and it was really, really substantially helping with their tuition. And these were people working on Master's degrees and Doctorates and all kinds of things and I thought, "Wow, if they can do this, hey, maybe I can."

Despite the deviancy associated with being an exotic dancer and the negative aspects of the job, most of the women we spoke with seemed to be able to rationalize or justify their involvement in exotic dancing. In summarizing the use of justifications by exotic dancers, one woman said:

You can justify it because you bring home money and at the end of the night that feels great. You don't reflect on, you know, how you were degraded, the leering and the other bad stuff. You know, you don't think about it because you've got a big wad of money in your hand.

In other words, the major incentive for entering dancing, money,[4] is also used as the main justification or rationale for continuing to do it. As with Hong and Duff's (1977) study of taxi-dancers, the neutralization techniques or rationalizations used by exotic dancers to downplay the norm-violating nature of their behaviour, soothe guilt feelings, and cope with the unpleasant aspects of their jobs, were learned during the informal socialization processes that occur on-the-job.

PUTTING ON A SHOW

Beyond acquiring the courage to take off one's clothes in public and learning how to justify one's actions, obtaining competence as an exotic dancer also requires learning to be good at the job. In order to become a successful exotic dancer, the novice dancer must learn how to put on a good show or performance. As with any successful performer, dancers need to learn how to use impression management skills to create an illusion that will allow them to control/manipulate their audience in order to achieve some specified goal, in this case the acquisition of money. In their interviews, the women talked about how their job required they put on a skilful performance that would lure men in and get them to spend their money on dancers.

> A dance is not just dancing, it is the way you present yourself, the way you talk to the customer, the way you introduce yourself. If you gonna have a smile, right away it's gonna be easy [to make money].

> Just turn the guys on, make them think that we are like, you know, licking each other [when performing with another woman]. But we weren't, it's all show. I mean, you don't have to do anything, you know, that's real. You just have to make it look real. So, you know, you would lift up the girl's leg, put your head down, you know, pretend that you're like, oh, you know, that kind of stuff. You know, like men are kind of stupid, so they buy it, right.

> Sometimes you just look at a customer, the way he reacts . . . I can tell what they like. I'm always doing things that flatter my body. I touch my boobs all the time. I touch myself all the time. It's kind of masturbation but in front of people . . . It gets the men going and keeps them coming back.

As dancers reported in their interviews, learning how to control or manipulate an audience is acquired through observation and interaction with subcultural members within the club setting.

> I was really glad I waitressed before dancing. I got to overhear a lot of the conversations between the dancers and the customers. It was that way that I figured out how to operate and ways to play the men for their money.

> Some of the girls that have been dancing a while here were really nice to me. They gave me advice on how to keep the guys interested so they will buy several table dances in a row.

Skill development, improvement and job competence more generally were affirmed by coworkers through praise, and by customers through applause, requests for table dances, the development of a regular clientele, and increased take-home pay.

TYPOLOGY OF DANCERS

Although the women interviewed reported that they experienced a process of adjustment in becoming a dancer, this process differed somewhat according to the type of dancer each woman could be classified as. Based on the interview data collected, there appear to be two types of dancers: the career dancer and the goal-oriented dancer. Both types of dancers report money as the primary motivating factor for entry into dancing; however, they differ in the types of future they envision for themselves. Despite the fact that most of the women we spoke with told us that they never intended on making dancing a career, some ended up staying in the industry for many years, essentially making it one. Other women reported that they entered the world of dancing with the expectation that dancing would be their career for a while. Whether they intended on making dancing a career or not, the career dancers we spoke with tended to possess limited skill training and education. As a result, they saw dancing as an employment opportunity that enabled them to make a decent living that would otherwise be unavailable to them through other channels.

> This is a career for me, it's seventeen years. I don't want to stop this now. And besides, what other job could I get where I can earn this kind of money.

> There really are no jobs for women like me who have little education. At least none where I could make this much money.

> You know I'm not educated . . . uh, it's hard to, hard to get back out into the real world once you're in there, it's like I feel like, what else can I do?

> I've been a stripper for 7 years, what else am I gonna be able to do? You know, even if I try I'm always gonna be a dancer, I'm always gonna be labelled. I make good money, so why go work for minimum wage.

In contrast with the career dancer, the goal-oriented dancer enters dancing with a specific goal in mind.

> I don't look at it like a career so it's kind of like a means to an end. You know how you put yourself on a program, like a five-year program. Get in there and make a whack of cash and then go on to something else. Like that can't be the only thing that I want to do for the rest of my life.

> There's aspects of the job I like, I mean, I do like some of the girls that work there, some of the bar staff. You know, they're fun to be around at work. And the guys, if they're nice, I can, you know, have had some good conversations. But, I do not like taking my clothes off you know? And I don't want to make this a career. It is a means to an end.

Some dancers report being motivated to enter dancing in order to make the money they needed to get or stay out of debt.

I'm getting my Honours Bachelor of Arts in Drama and I want to eventually open my own Drama Therapy Clinic. So, this is just a means of getting there because the money is really good and I'd like to start saving. You know, I've spent all my money on my education and I haven't put any aside for my future, so this would be a quick way to do it, cuz the money's really good and it's really fast.

The bills kept coming in and coming in and I couldn't keep my head above water and everybody was threatening to take me to court and I had all these debts and I just I needed money fast. So, I thought I could dance for a bit until I got on top of things.

I don't want to do it, but you have to, I have to do it, I don't have a choice. I have a car payment, I have to pay my rent, I can't not do it.

Nothing else will pay my bills. So that's it.

One specific group of goal-oriented dancers are the students. These women report that for them dancing is a short-term job that pays well and that can fit in with their class schedule.

It's ideal when you're going to school because you just make your own schedules. When I have exam week I don't go at all. So, it fits in with school. So, I guess, I mean, I don't think I would work [as a dancer] once I finish school, unless I couldn't find a job or something.

I have two hundred from my Child Tax and a hundred from my support. And then I was given eleven thousand from OSAP [Ontario Student Assistance Program] to do me 'til January. But then I have tuition and daycare costs and they have gone up, and prescriptions that I have to cover. I'm really down and out right now. So, I really have no choice except to dance—there really is no other way to pay the bills and keep from getting farther in debt while I'm in school.

The commonality among goal-oriented dancers is that dancing is seen as a short-term thing, a means to an end, once the end is achieved (e.g., they graduate from university, pay off their debts, etc.), the plan is to leave dancing. It is important to note though, that although many goal-oriented dancers reported planning on leaving, some spoke of difficulties exiting once they got used to the money they could earn.

It's kinda hard once you get used to the money to leave [dancing]. I mean, like, I always said I would leave when I got out of debt, but the money draws you back.

I've wanted out for so many years now and just didn't know how. You get so trapped in there and I didn't know what to do or what I could do.

I started dancing to help pay off the mortgage on my house and get rid of some debts. I thought it was a one shot deal, but I seem to fall back on it whenever I need money.

The type of dancer one identifies as has implications for the socialization experiences of dancers. Women who see dancing as a career, rather than as a temporary job, tend to be more inclined to get involved in the "dancer life," develop relationships with other dancers and club employees, and become immersed in the strip club subculture. As a result, they are likely to experience a more complete socialization process than goal-oriented dancers. Goal-oriented dancers, in contrast, tend to limit their ties to others in the business. As the following quotes illustrate, they try to keep dancing and their private lives separate.

> I don't hang out with other dancers. When I leave here I go back to my other life.

> Although I try to be friendly to everyone here [at the club] I stick to myself as much as possible and when I leave [work], I try to leave it and everybody associated with it behind.

The implication of keeping the two aspects of their lives separate is that goal-oriented dancers have to contend with the stigma associated with dancing on their own and, as a result, often live very closeted/secretive lives.

> I work really hard at keeping this [dancing] a secret from my family. It is hard cuz I still live at home with my parents. So, I keep my costumes in the trunk of my car and I make sure I am the only one with a key.

> And I went home to visit and my mom's like, "So, how's your summer going?" And she's asking me all these questions and I had to lie and say that I was working for a security company. I hate it [lying], cuz my mom and I just started to get really close again and here I was suddenly back to the way I was when I was a teenager, the lying and you know, staying out all hours of the night and all this stuff, and you know, it hurt to lie to her.

> It scares the hell out of me that if they found out, you know, especially my dad, the man my mom's with right now. Like, he's been around since I was a real little kid and he's been married a couple of times. He has eleven children all together. I'm the only girl—the only one to go to University, so, I'm just like the apple of his eye, you know. He's so proud of me. And when I made honours, he was just, like, he couldn't have been prouder. He goes to the office and you know, he had a mechanic's shop, and he'd walk in and he'd tell all the boys, you know, "this is my daughter. She's an honours student." And he's you know, strict Irish Catholic, and he would just be crushed.

> It's really hard because, you know, you're lying to your parents. Well, I am, and I'm close to my family. And I was lying to my friends and to my boyfriend at the time.

Without a community of supportive others, these women have limited access to competing definitions of reality and are therefore more likely to feel some sort of guilt and shame for choosing to dance. Since it is through interacting with other subcultural members that people learn rationalizations for their behaviour, these women are likely to have limited access to the techniques of neutralization used by other dancers that are important for the maintenance of a positive sense of self.

LIMITATIONS OF THE DANCER'S SOCIALIZATION PROCESS

Although both career and goal-oriented dancers felt they were able to experience successful occupational socialization that enabled them to achieve competence as exotic dancers, most of the women interviewed talked about how the socialization process inadequately prepared them for some of the realities of the life of an exotic dancer. A Stripper's Handbook (1997), a booklet written by several dancers in the Toronto region, nicely illustrates the benefits and limitations of learning about exotic dancing through informal channels. Although the booklet contains helpful information and advice about the job (e.g., where to get a license, how much a license costs, how to save money on costumes, stage show rules, DJ fees, fines, freelancing vs. working on schedule, etc.), it also glosses over some of the negative effects the job can have on women's lives (e.g., relationship problems, inhibition of heterosexual desire, etc.). The tendency to overlook the negative is typical of the advice women reported being given by subcultural members, especially the women with limited ties to the subculture.

When discussing the limitations of their socialization experiences, the women we spoke with reported having little knowledge of, and therefore being unprepared for the impact of, dancing on their private lives. The area of impact most often mentioned was relationships. In terms of relationships, women spoke of the difficulties of having and sustaining heterosexual relationships with males outside of the industry. For some women, relationship difficulties were tied to the problems men they date tend to have with their occupation (see Prus, 1980):

> I'd suggest to any girl that ever dances, unless your boyfriend's a male dancer, don't date someone when you're stripping. Most guys say they can handle it. They can't and then they start coming into clubs and causing bull shit.

> My ex didn't like it. It wasn't because he didn't trust me, he just didn't like the whole idea. He didn't want me dancing not because he was jealous or anything, just because I think he knows it's stressful and it's just not good for you psychologically. None of the guys I dated ever were worried or anything, they knew that I don't really like it . . . They know I would never go out with anyone else that I met at work. My current boyfriend, he dances at a gay bar so he knows what it's all about.

Other women report that the difficulty of developing or sustaining heterosexual relationships was tied to the nature of their job (i.e., they usually work at night, in a bar, in a job that requires them be around and constantly interacting with customers, many of whom they don't like).

> Relationshipwise it's very hard. I think it's hard for someone to take a dancer seriously, it takes a certain type of guy that can, look beyond that and ah, if I'm involved I have a really hard time doing my job. If I'm single I'm better with my job. It's hard to meet people cuz I work nights all the time. When I was working full time I was there a good 5 nights a week. On my night off I don't want to go to a bar or anything, I'm in one every day, so you never get a chance to meet people. It's pretty much taboo to date someone you meet at work, cuz you don't know who they are outside of there and they've been giving you money to strip in front of them all night, and they are like, "Ooh yeah, I want to take you on a date." And you are thinking, "Yeah, sure you do. For what, why?" So that's hard. And it's hard if you have a boyfriend, it's hard for them to deal with it.

> Most of the time in the afternoon I like to spend my time alone. My boyfriend works and I like this. It's probably to do with my job. It's not because I don't

love my boyfriend, but sometimes, he gets on my nerves. Like, I work an eight hour shift and all night long everybody is bugging me. At least, I'm there for that. I bug people and people bug me. But, when you get out of there, it's just like you want to have silence everywhere.

Right now, my boyfriend doesn't live with me. We tried it and it didn't work. I'm not patient. I like to be by myself I really love to be by myself most of the time. And sometimes I think I would like to be with my boyfriend because I miss him, but as soon as he is here he gets on my nerves. Even if you have a boyfriend, sometimes you don't even want to talk with him . . . When I'm finished work, I really don't want to talk with anybody.

Despite the difficulties exotic dancers confront in terms of developing and sustaining relationships, some of the women interviewed expressed an interest/desire to have a stable intimate heterosexual relationship. Others, however, talked about being disinterested in men.

I'm kind of sick of, you know the men and, I just, I've always been a, you know, a big chested person. So, I always gotten the yee-haw's and stuff walking down the street and I just kinda had it after a while, you know?

I hate to be looked at. I don't like to be looked at by men. I don't like men very much.

One solution identified by dancers to the relationship difficulties and inhibited heterosexual desire dancers experience, is pursuing relationships with other women. According to the women we interviewed, it is not uncommon for female exotic dancers to develop lesbian relationships, either because of a disinterest in heterosexual relationships stemming from dancing, or because relationships with women are just easier to develop and sustain while they are working as exotic dancers (see Carey et al., 1974; McCaghy & Skipper, 1969; Prus, 1980).

I think a lot of girls end up bi . . . I think it's convenient because it's easier to go out with another dancer, another girl than go out with a guy. You know what I mean? They understand your likes and a lot of guys that date dancers are assholes. So why deal with the hassle of going out? Why not just date a girl? I would have [dated women] if I met a nice girl.

It's a lot easier to date a girl than to bother with going out. But I just happened to meet Paul who dances as well and fits into my lifestyle. But, if I wouldn't have met him I probably would date women. But I just never, I just never met any girl that I had enough in common with. A lot of the girls are [lesbian]. But a lot of people stereotype you. You know what I mean?

As noted by McCaghy and Skipper (1969), three conditions associated with the occupation are supportive of same sex relationships: "(1) isolation from affective social relationships; (2) unsatisfactory relationships with males; and (3) an opportunity structure allowing a wide range of sexual behavior" (p. 266).

Conclusion

As other studies of exotic dancers have found, there are various factors influencing occupational entry into exotic dancing. This study provides support for a combined career contingency/conversion model. According to this model, four factors influence entry into the exotic dancing: (1) knowledge and accessibility of an opportunity structure that makes exotic dancing an occupational alternative; (2) an awareness of the economic rewards associated with being an exotic dancer; (3) a recruitment process involving personal networks; and (4) financial need or a need for employment. For the women interviewed, these factors played a significant role in their anticipatory socialization process and their movement in the direction of exotic dancing.

Although similar to earlier studies of exotic dancers (this study found evidence of a combined career contingency/conversion model for entry into exotic dancing), there were also some differences between the findings of this study and that of previous research in the area. For example, contrary to earlier studies, we found little indication of dancers' having pre-job formal socialization experiences that involved professional training in entertainment-related fields, prior to entering dancing. This difference, however, may be tied to the evolution of stripping. Over the past 25 years or so, stripping has gone from a form of theatre or burlesque stage show, where complete nudity was rare and touching was prohibited, to the more raunchy table and lap dances performed today that often involve complete nudity, and sometimes physical and sexual contact between the dancer and the customer.

Despite some different findings in terms of the anticipatory socialization experiences of dancers, similar to other research in the area we found that once the decision to dance was made and they were employed as dancers, the women we interviewed continued to experience a socialization process through interacting with and observing other subcultural members. The on-the-job, informal occupational socialization the women reported experiencing enabled them to achieve job competence, even in a deviant occupation.

As social learning theories of deviance suggest, although most of us learn the norms and values of society, some of us also learn techniques for committing deviance and the specific motives, drives, rationalizations, and attitudes that allow us to neutralize our violation of normative codes. The socialization experiences of dancers fit with this framework. Learning occurs through observing and interacting with strip club employees, especially more experienced dancers. Through such observations and interactions, novice dancers learn techniques for rationalizing their involvement in the occupation, a process which enables them to stay in the job and succeed, while retaining a valued sense of self.

Although exotic dancers can experience socialization processes that result in job competence, their occupational socialization often inadequately prepares them for the potential impact of their job on their lives outside of the club. The most often mentioned area of concern was intimate relationships, due to the difficulties exotic dancers reported on developing and sustaining heterosexual relationships and desire.

NOTES

1. According to Ritzer and Walczak (1986, p. 374), "an occupation will be treated as deviant if it meets one or more of the following criteria: (1) it is illegal; (2) one or more of the

central activities of the occupation is a violation of nonlegalized norms and values; and (3) the culture, lifestyle, or setting associated with the occupation is popularly presumed to involve rule-breaking behaviour."

2. According to Matza (1992, p. 29), "drift is motion guided gently by underlying influences. The guidance is gentle and not constraining. The drift may be initiated or deflected by events so numerous as to defy codification. But underlying influences are operative nonetheless in that they make initiation to . . . [deviant behaviour] more probable, and they reduce the chances that an event will deflect the drifter from his [/her deviant] . . . path. Drift is a gradual process of movement, unperceived by the actor, in which the first stage may be accidental or unpredictable."

3. Appeal to higher loyalties involves rationalizing deviant behaviour by couching it within an altruistic framework.

4. Although money is part of the motivation for anyone seeking employment, for dancers, it was the amount of money that could be earned dancing, compared with the amount that could be earned in more legitimate jobs, that motivated them to try dancing.

REFERENCES

Boles, Jacqueline M. & Garbin, A.P. (1974a). The strip club and stripper-customer patterns of interaction. *Sociology and Social Research, 58,* 136–144.

Boles, Jacqueline M. & Garbin, A.P. (1974b). The choice of stripping for a living: An empirical and theoretical explanation. *Sociology of Work and Occupations, 1,* 110–123.

Boles, Jacqueline M. & Garbin, A.P. (1974c). "Stripping for a living: An occupational study of the night club stripper." In C.D. Bryant (Ed.), *Deviant Behavior: Occupational and Organizational Bases* (pp. 312–335). Chicago: Rand McNally.

Carey, S.H., Peterson, R.A., & Sharpe, L.K. (1974). A study of recruitment and socialization into two deviant female occupations. *Sociological Symposium, 8,* 11–24.

Dressel, P.L. & Petersen, D.M. (1982a). Gender roles, sexuality, and the male strip show: The structuring of sexual opportunity. *Sociological Focus, 15,* 151–162.

Dressel, P.L. & Petersen, D.M. (1982b). Becoming a male stripper: Recruitment, socialization and ideological development. *Work and Occupations, 9,* 387–406.

Enck, G.E. & Preston, J.D. (1988). Counterfeit intimacy: A dramaturgical analysis of an erotic performance. *Deviant Behavior, 9,* 369–381.

Forsyth, C.J. & Deshotels, T.H. (1997). The occupational milieu of the nude dancer. *Deviant Behavior, 18,* 125–142.

Hong, L.K. & Duff, R.W. (1977). Becoming a taxi-dancer: The significance of neutralization in a semi-deviant occupation. *Sociology of Work and Occupations, 4,* 327–342.

Lofland, J. & Stark, R. (1965). Becoming a world-saver: A theory of conversion to a deviant perspective. *American Sociological Association, 30,* 862–875.

McCaghy, C.H. & Skipper, J.K. (1969). Lesbian behavior as an adaptation to the occupation of stripping. *Social Problems, 17,* 262–270.

McCaghy, C.H. & Skipper, J.K. (1972). "Stripping: Anatomy of a deviant life style." In S. D. Feldman and G. W. Thielbar (Eds.), *Life Styles: Diversity in American Society* (pp. 362–373). Boston: Little, Brown.

Petersen, D. & Dressel, P.L. (1982). Equal time for women: Social notes on the male strip show. *Urban Life, 11,* 185–208.

Prus, R.C. & Sharper, C.R.D. (1977). *Road Hustler: The Career Contingencies of Professional Card and Dice Hustlers.* Toronto: Lexington Books.

Prus, R.C. & Styllianoss, I. (1980). *Hookers, Rounders, and Desk Clerks: The Social Organization of the Hotel Community.* Toronto: Gage Publishing Limited.

Reid, S.A., Epstein, J.S., & Benson, D.E. (1994). Role identity in a devalued occupation: The case of female exotic dancers. *Sociological Focus, 27,* 1–16.

Ronai, C.R. (1992). "The reflexive self through narrative: A night in the life of an erotic dancer/researcher." In C. Ellis and M.G. Flaherty (Eds.), *Investigating Subjectivity: Research on Lived Experience* (pp. 102–124). Newbury Park, CA: Sage Publications.

Ronai, C.R. & Ellis, C. (1989). Turn-ons for money: Interactional strategies of the table dancer. *Journal of Contemporary Ethnography, 118,* 271–298.

Sanders, C.R. (1974). Psyching out the crowd: Folk performers and their audiences. *Urban Life and Culture, 3,* 264–282.

Skipper, J.K. & McCaghy, C.H. (1971). "Stripteasing: A sex-oriented occupation." In James M. Henslin (Ed.), *Studies in the Sociology of Sex* (pp. 275–296). New York: Appleton-Century-Crofts.

Sykes, G.M. & Matza, D. (1957). Techniques of neutralization: A theory of delinquency. *American Sociological Review, 22,* 664–670.

Thompson, W.E. & Harred, J.L. (1992). Topless dancers: Managing stigma in a deviant occupation, *Deviant Behavior, 13,* 291–311.

QUESTIONS FOR DISCUSSION

1. Discuss how anticipatory socialization relates to dancing, according to the author.

2. Explain how a woman drifts into dancing. Give examples from the reading.

3. Discuss the author's explanation of the topology of dances.

QUESTIONS FOR PART 1

1. How do the readings in Chapter 1 address the meaning of deviance?

2. How do the readings in Chapter 2 specifically relate to deviance events and social control? Why is social control implemented in some cases but not in others?

3. How do the readings in Chapter 3 explain how a person becomes deviant?

EXPLAINING DEVIANT BEHAVIOR: MACRO PERSPECTIVES

D eviance is a property of groups. What is considered deviant in one group, culture, or society may not be in another. It follows that the meaning of deviance also changes from group to group. As a result, many sociologists believe that one must grasp the relationship between deviance and group structure in order to understand it fully. Macro perspectives of deviance explain deviance according to the "big picture" as applied to groups and society. The readings in Part 2 comprise some of the most important macro theories of deviance and criminal behavior. These theories explain a deviant act from group characteristics such as social class, geography, social structure, and race.

On February 29, 2000, a 6-year-old boy in Michigan was accused of fatally shooting his first-grade classmate. The boy found the weapon, a stolen .32-caliber handgun, on the floor in a bedroom in his house. The boy's home was in a poor neighborhood and his parents were struggling financially. The father was in and out of jail. The mother had been evicted from her home eight days before the shooting. She was working two part-time jobs 35 miles from home. Prosecutors believed the boy was too young to understand his actions and charges were not filed against him. The boy's uncle, a family friend, and the man who sold the gun were charged in the incident.

Can we understand this act by using theories that explain deviant acts from the broad perspective by looking at social structure, class, race, and ethnicity? In his classic article, "The Normal and the Pathological," Emile Durkheim identifies positive contributions of crime in a society, and claims that crime and deviance are fundamental conditions of social life and integral parts of a healthy, normal society. What is normal for Durkheim is simply the existence of deviance, provided it does not exceed a certain level for each social type. Durkheim enlightens us on why certain criminal acts, such as murder, are deemed to be inevitable activities in a free society. This is not to excuse murder, but to place it in the context of what we can realistically expect.

In Robert K. Merton's classic piece, "Social Structure and Anomie," deviance is the outcome of strain associated with anomie. Merton attributes deviance to social situations where individuals are unable to pursue socially acceptable goals through culturally approved means. Merton develops five possible ways people adapt to anomic situations.

One element of social structure in explaining deviance and crime is opportunity. While there are many everyday opportunities to commit crimes, most people do not take advantage of them. Perhaps they have learned that such behavior is morally wrong; perhaps they have no need to commit the crime; perhaps both of these are correct. Richard A. Cloward and Lloyd E. Ohlin's article is an integration of learning perspectives with opportunity perspectives, and it explores how breaking society's rules depends on one's location within the social structure. Differential opportunity structures enable individuals to identify differential access to legitimate and illegitimate means. If people do not have legitimate opportunities to succeed, they are likely to break the rules in order to succeed and meet their social needs. Some individuals will also have access to illegitimate opportunities and can compete in this illicit structure. Thus, individuals can look at others in legitimate and illegitimate systems.

Conflict theories of deviant behavior focus more on the explanation of the creation of deviance rather than explaining specific deviant acts. These theories address the central elements of power, rules and norms, and social control. Steven Spitzer's article focuses on three major problems: the definition of deviance, the etiology of deviance, and the etiology of control. Spitzer undertakes the development of a Marxian interpretation of deviance and social control. He explains deviance in relation to class productivity and the development of social control mechanisms.

Andrew L. Hochstetler and Neal Shover's article presents a Neo-Marxist perspective on conflict theory to predict that the use of punishment by capitalist states varies with economic conditions and the size of the labor surplus. Marxist theorists suggest that the poor are perceived as threats to a capitalist society's social order.

Richard Quinney's article addresses the creation of criminal laws, enforcement of the new laws, and the application of laws. Quinney recognizes that criminal laws are created, maintained, and controlled by the dominant class. The dominant class creates definitions of what are crimes, applies definitions to crimes, creates different behavioral patterns, and manufactures an ideology of crime, thus constructing a social reality of crime.

C H A P T E R 4

STRUCTURAL-FUNCTIONAL PERSPECTIVES

The Normal and the Pathological

Emile Durkheim

Crime is present not only in the majority of societies of one particular species but in all societies of all types. There is no society that is not confronted with the problem of criminality. Its form changes; the acts thus characterized are not the same everywhere; but, everywhere and always, there have been men who have behaved in such a way as to draw upon themselves penal repression. If, in proportion as societies pass from the lower to the higher types, the rate of criminality, i.e., the relation between the yearly number of crimes and the population, tended to decline, it might be believed that crime, while still normal, is tending to lose this character of normality. But we have no reason to believe that such a regression is substantiated. Many facts would seem rather to indicate a movement in the opposite direction. From the beginning of the [nineteenth] century, statistics enable us to follow the course of criminality. It has everywhere increased. In France the increase is nearly 300 percent. There is, then, no phenomenon that presents more indisputably all the symptoms of normality, since it appears closely connected with the conditions of all collective life. To make of crime a form of social morbidity would be to admit that morbidity is not something accidental, but, on the contrary, that in certain cases it grows out of the fundamental constitution of the living organism; it would result in wiping out all distinction between the physiological and the pathological. No doubt it is possible that crime itself will have abnormal forms, as, for example, when its rate is unusually high. This excess is, indeed, undoubtedly morbid in nature. What is normal, simply, is the existence of criminality, provided that it attains and does not exceed, for each social type, a certain level, which it is perhaps not impossible to fix in conformity with the preceding rules.[1]

Here we are, then, in the presence of a conclusion in appearance quite paradoxical. Let us make no mistake. To classify crime among the phenomena of normal sociology is not to say merely that it is an inevitable, although regrettable phenomenon, due to the incorrigible wickedness of men; it is to affirm that it is a factor in public health, an integral part of all healthy societies. This result is, at first glance, surprising enough to have puzzled even ourselves for a long time. Once this first surprise has been overcome, however, it is not difficult to find reasons explaining this normality and at the same time confirming it.

In the first place crime is normal because a society exempt from it is utterly impossible. Crime, we have shown elsewhere, consists of an act that offends certain very strong collective sentiments. In a society in which criminal acts are no longer committed, the sentiments they offend would have to be found without exception in all individual consciousnesses, and they must be found to exist with the same degree as sentiments contrary to them. Assuming that this condition could actually be realized, crime would not thereby disappear; it would only change its form, for the very cause which would thus dry up the sources of criminality would immediately open up new ones.

Indeed, for the collective sentiments which are protected by the penal law of a people at a specified moment of its history to take possession of the public conscience or for them to acquire a stronger hold where they have an insufficient grip, they must acquire an intensity greater than that which they had hitherto had. The community as a whole must experience them more vividly, for it can acquire from no other source the greater force necessary to control these individuals who formerly were the most refractory. For murderers to disappear, the horror of bloodshed must become greater in those social strata from which murderers are recruited; but, first it must become greater throughout the entire society. Moreover, the very absence of crime would directly contribute to produce this horror; because any sentiment seems much more respectable when it is always and uniformly respected.

One easily overlooks the consideration that these strong states of the common consciousness cannot be thus reinforced without reinforcing at the same time the more feeble states, whose violation previously gave birth to mere infraction of convention — since the weaker ones are only the prolongation, the attenuated form, of the stronger. Thus robbery and simple bad taste injure the same single altruistic sentiment, the respect for that which is another's. However, this same sentiment is less grievously offended by bad taste than by robbery; and since, in addition, the average consciousness has not sufficient intensity to react keenly to the bad taste, it is treated with greater tolerance. That is why the person guilty of bad taste is merely blamed, whereas the thief is punished. But, if this sentiment grows stronger, to the point of silencing in all consciousnesses the inclination which disposes man to steal, he will become more sensitive to the offenses which, until then, touched him but lightly. He will react against them, then, with more energy; they will be the object of greater opprobrium, which will transform certain of them from the simple moral faults that they were and give them the quality of crimes. For example, improper contracts, or contracts improperly executed, which only incur public blame or civil damages, will become offenses in law.

Imagine a society of saints, a perfect cloister of exemplary individuals. Crimes, properly so called, will there be unknown; but faults which appear venial to the layman will create there the same scandal that the ordinary offense does in ordinary consciousness. If, then, this society has the power to judge and punish, it will define these acts as criminal and will treat them as such. For the same reason, the perfect and upright man judges his smallest failings with a severity that the majority reserve for acts more truly in the nature of an offense. Formerly, acts of violence against persons were more frequent than they are today, because respect for individual dignity was less strong. As this has increased, these crimes have become more rare; and also, many acts violating this sentiment have been introduced into the penal law which were not included there in primitive times.[2]

In order to exhaust all the hypotheses logically possible, it will perhaps be asked why this unanimity does not extend to all collective sentiments without exception. Why should not even the most feeble sentiment gather enough energy to prevent all dissent? The moral consciousness of the society would be present in its entirety in all the individuals, with a vitality sufficient to prevent all acts offending it—the purely conventional faults as well as the crimes. But a uniformity so universal and absolute is utterly impossible; for the immediate physical milieu in which each one of us is placed, the hereditary antecedents, and the social influences vary from one individual to the next, and consequently diversify consciousnesses. It is impossible for all to be alike, if only because each one has his own organism and that these organisms occupy different areas in space. That is why, even among the lower peoples, where individual originality is very little developed, it nevertheless does exist.

Thus, since there cannot be a society in which the individuals do not differ more or less from the collective type, it is also inevitable that, among these divergences, there are some with a criminal character. What confers this character upon them is not the intrinsic quality of a given act but that definition which the collective conscience lends them. If the collective conscience is stronger, if it has enough authority practically to suppress these divergences, it will also be more sensitive, more exacting; and, reacting against the slightest deviations with the energy it otherwise displays only against more considerable infractions, it will attribute to them the same gravity as formerly to crimes. In other words, it will designate them as criminal.

Crime is, then, necessary; it is bound up with fundamental conditions of all social life, and by that very fact is useful, because these conditions of which it is part are themselves indispensable to the normal evolution of morality and law.

Indeed, it is no longer possible today to dispute the fact that law and morality vary from one social type to the next, nor that they change within the same type if the conditions of life are modified. But, in order that these transformations may be possible, the collective sentiments at the basis of morality must not be hostile to change, and consequently must have but moderate energy. If they were too strong, they would no longer be plastic. Every pattern is an obstacle to new patterns, to the extent that the first pattern is inflexible. The better a structure is articulated, the more it offers a healthy resistance to all modification; and this is equally true of functional, as of anatomical, organization. If there were no crimes, this condition could not have been fulfilled; for such a hypothesis presupposes that collective sentiments have arrived at a degree of intensity unexampled in history. Nothing is good indefinitely and to an unlimited extent. The authority which the moral conscience enjoys must not be excessive; otherwise no one would dare criticize it, and it would too easily congeal into an immutable form. To make progress, individual originality must be able to express itself. In order that the originality of the idealist whose dreams transcend his century may find expression, it is necessary that the originality of the criminal, who is below the level of his time, shall also be possible. One does not occur without the other.

Nor is this all. Aside from this indirect utility, it happens that crime itself plays a useful role in this evolution. Crime implies not only that the way remains open to necessary changes but that in certain cases it directly prepares these changes. Where crime exists, collective sentiments are sufficiently flexible to take on a new form, and crime sometimes helps to determine the form they will take. How many times, indeed, it is only an anticipation of future morality—a step toward what will be! According to Athenian law, Socrates was a criminal, and his condemnation was no more than just. However, his crime, namely, the independence of his thought, rendered a service not only to humanity but to his country. It served to prepare a new morality and faith which the Athenians needed, since the traditions by which they had lived until then were no longer in harmony with the current conditions of life. Nor is the case of Socrates unique; it is reproduced periodically in history. It would never have been possible to establish the freedom of thought we now enjoy if the regulations prohibiting it had not been violated before being solemnly abrogated. At that time, however, the violation was a crime, since it was an offense against sentiments still very keen in the average conscience. And yet this crime was useful as a prelude to reforms which daily became more necessary. Liberal philosophy had as its precursors the heretics of all kinds who were justly punished by secular authorities during the entire course of the Middle Ages and until the eve of modern times.

From this point of view the fundamental facts of criminality present themselves to us in an entirely new light. Contrary to current ideas, the criminal no longer seems a totally unsociable being, a sort of parasitic element, a strange and unassimilable body, introduced into the midst of society.[3] On the contrary, he plays a definite role in social life. Crime, for its part, must no longer be conceived as an evil that cannot be too much suppressed. There is no occasion for self-congratulation when the crime rate drops noticeably below the average level, for we may be certain that this apparent progress is associated with some social disorder. Thus, the number of assault cases never falls so low as in times of want.[4] With the drop in the crime rate, and as a reaction to it, comes a revision, or the need of a revision in the theory of punishment. If, indeed, crime is a disease, its punishment is its remedy and cannot be otherwise conceived; thus, all the discussions it arouses bear on the point of determining what the punishment must be in order to fulfill this role of remedy. If crime is not pathological at all, the object of punishment cannot be to cure it, and its true function must be sought elsewhere.

NOTES

1. From the fact that crime is a phenomenon of normal sociology, it does not follow that the criminal is an individual normally constituted from the biological and psychological points

of view. The two questions are independent of each other. This independence will be better understood when we have shown, later on, the difference between psychological and sociological facts.

2. Calumny, insults, slander, fraud, etc.

3. We have ourselves committed the error of speaking thus of the criminal, because of a failure to apply our rule (*Division du travail social*, pp. 395–96).

4. Although crime is a fact of normal sociology, it does not follow that we must not abhor it. Pain itself has nothing desirable about it; the individual dislikes it as a society does crime, and yet it is a function of normal physiology. Not only is it necessarily derived from the very constitution of every living organism, but it plays a useful role in life, for which reason it cannot be replaced. It would, then, be a singular distortion of our thought to present it as an apology for crime. We would not even think of protesting against such an interpretation, did we not know to what strange accusations and misunderstandings one exposes oneself when one undertakes to study moral facts objectively and to speak of them in a different language from that of the layman.

QUESTIONS FOR DISCUSSION

1. Explain why crime is normal and necessary, according to the author.

2. The author uses the examples of robbery and bad taste to show how one is blamed and the other is punished. Give a different example of an act that is blamed and another act that is punished. How does your response relate to the reading?

3. According to Athenian law, Socrates was a criminal due to the independence of his thought. Assuming crime plays a useful role in society, compare and contrast Socrates to the women's movement of independent thought in reference to breaking social rules.

Social Structure and Anomie

Robert K. Merton

PATTERNS OF CULTURAL GOALS AND INSTITUTIONAL NORMS

Among the several elements of social and cultural structures, two are of immediate importance. These are analytically separable although they merge in concrete situations. The first consists of culturally defined goals, purposes and interests, held out as legitimate objectives for all or for diversely located members of the society. The goals are more or less integrated—the degree is a question of empirical fact—and roughly ordered in some hierarchy of value. Involving various degrees of sentiment and significance, the prevailing goals comprise a frame of aspirational reference. They are the things "worth striving for." They are a basic, though not the exclusive, component of what Linton has called "designs for group living." And though some, not all, of these cultural goals are directly related to the biological drives of man, they are not determined by them.

A second element of the cultural structure defines, regulates, and controls the acceptable modes of reaching out for these goals. Every social group invariably couples its cultural objectives with regulations, rooted in the mores or institutions, of allowable procedures for moving toward these objectives. These regulatory norms are not necessarily identical with technical or efficiency norms. Many procedures which from the standpoint of particular individuals would be most efficient in securing desired values—the exercise of force, fraud, power—are ruled out of the institutional area of permitted conduct. At times, the disallowed procedures include some which would be efficient for the group itself—for example, historic taboos on vivisection, on medical experimentation, on the sociological analysis of "sacred" norms—since the criterion of acceptability is not technical efficiency but value-laden sentiments (supported by most members of the group or by those able to promote these sentiments through the composite use of power and propaganda). In all instances, the choice of expedients for striving toward cultural goals is limited by institutionalized norms.

We shall be primarily concerned with the first—a society in which there is an exceptionally strong emphasis upon specific goals without a corresponding emphasis upon institutional procedures. If it is not to be misunderstood, this statement must be elaborated. No society lacks norms governing conduct. But societies do differ in the degree to which the folkways, mores and institutional controls are effectively integrated with the goals which stand high in the hierarchy of cultural values. The culture may be such as to lead individuals to center their emotional convictions upon the complex of culturally acclaimed ends, with far less emotional support for prescribed methods of reaching out for these ends. With such differential emphases upon goals and institutional procedures, the latter may be so vitiated by the stress on goals as to have the behavior of many individuals limited only by considerations of technical expediency. In this context, the sole significant question becomes: Which of the available procedures is most efficient in netting the culturally approved value? The technically most effective procedure, whether culturally legitimate or not, becomes typically preferred to institutionally prescribed conduct. As this process of attenuation continues, the society becomes unstable and there develops what Durkheim called "anomie" (or normlessness).

The working of this process eventuating in anomie can be easily glimpsed in a series of familiar and instructive, though perhaps trivial, episodes. Thus, in competitive athletics, when the aim of victory is shorn of its institutional trappings and success becomes construed as "winning the game" rather than "winning under the rules of the game," a premium is implicitly set upon the use of illegitimate but technically efficient means. The star of the opposing football team is surreptitiously slugged; the wrestler incapacitates his opponent through ingenious but illicit techniques; university alumni covertly subsidize "students" whose talents are confined to the athletic field. The emphasis on the goal has so attenuated the satisfactions deriving from sheer participation in the competitive activity that only a successful outcome provides gratification. Through the same process, tension generated by the desire to win in a poker game is relieved by successfully dealing one's self four aces or, when the cult of success has truly flowered, by sagaciously shuffling the cards in a game of solitaire. The faint twinge of uneasiness in the last instance and the surreptitious nature of public delicts indicate clearly that the institutional rules of the game are *known* to those who evade them. But cultural (or idiosyncratic) exaggeration of the success-goal leads men to withdraw emotional support from the rules.

This process is of course not restricted to the realm of competitive sport, which has simply provided us with microcosmic images of the social macrocosm. The process whereby exaltation of the end generates a literal *demoralization,* that is, a de-institutionalization, of the means occurs in many groups where the two components of the social structure are not highly integrated.

Contemporary American culture appears to approximate the polar type in which great emphasis upon certain success-goals occurs without equivalent emphasis upon institutional means. It would of course be fanciful to assert that accumulated wealth stands alone as a symbol of success just as it would be fanciful to deny that Americans assign it a place high in their scale of values. In some large measure, money has been consecrated as a value in itself, over and above its expenditure for articles of consumption or its use for the enhancement of power. "Money" is peculiarly well adapted to become a symbol of prestige. As Simmel emphasized, money is highly abstract and impersonal. However acquired, fraudulently or institutionally, it can be used to purchase the same goods and services. The anonymity of an urban society, in conjunction with these peculiarities of money, permits wealth, the sources of which may be unknown to the community in which the plutocrat lives or, if known, to become purified in the course of time, to serve as a symbol of high status. Moreover, in the American Dream there is no final stopping point. The measure of "monetary success" is conveniently indefinite and relative. At each income level, as H. F. Clark found, Americans want just about 25 percent more (but of course this "just a bit more" continues to operate once it is obtained). In this flux of shifting standards, there is no stable resting point, or rather, it is the point which manages always to be "just ahead." An observer of a community in which annual salaries in six figures are not uncommon reports the anguished words of one victim of the American Dream: "In this town, I'm snubbed socially because I only get a thousand a week. That hurts."

To say that the goal of monetary success is entrenched in American culture is only to say that Americans are bombarded on every side by precepts which affirm the right or, often, the duty of retaining the goal even in the face of repeated frustration. Prestigeful representatives of the society reinforce the cultural emphasis. The family, the school and the workplace—the major agencies shaping the personality structure and goal formation of Americans—join to provide the intensive disciplining required if an individual is to retain intact a goal that remains elusively beyond reach, if he is to be motivated by the promise of a gratification which is not redeemed. As we shall presently see, parents serve as a transmission belt for the values and goals of the groups of which they are a part—above all, of their social class or of the class with which they identify themselves. And the schools are of course the official agency for the passing on of the prevailing values, with a large proportion of the textbooks used in city schools implying or stating explicitly "that education leads to intelligence and consequently to job and money success." Central to this process of disciplining people to maintain their unfulfilled aspirations are the cultural prototypes of success, the living documents testifying that the American Dream can be realized if one but has the requisite abilities.

Coupled with this positive emphasis upon the obligation to maintain lofty goals is a correlative emphasis upon the penalizing of those who draw in their ambitions. Americans are admonished "not to be a quitter" for in the dictionary of American culture, as in the lexicon of youth, "there is no such word as 'fail.'" The cultural manifesto is clear: one must not quit, must not cease striving, must not lessen his goals, for "not failure, but low aim, is crime."

Thus the culture enjoins the acceptance of three cultural axioms: First, all should strive for the same lofty goals since these are open to all; second, present seeming failure is but a way-station to ultimate success; and third, genuine failure consists only in the lessening or withdrawal of ambition.

In rough psychological paraphrase, these axioms represent, first a symbolic secondary reinforcement of incentive; second, curbing the threatened extinction of a response through an associated stimulus; third, increasing the motive-strength to evoke continued responses despite the continued absence of reward.

In sociological paraphrase, these axioms represent, first, the deflection of criticism of the social structure onto one's self among those so situated in the society that they do not have full and equal access to opportunity; second, the preservation of a structure of social power by having individuals in the lower social strata identify themselves, not with their compeers, but with those at the top (whom they will ultimately join); and third, providing pressures for conformity with the cultural dictates of unslackened ambition by the threat of less than full membership in the society for those who fail to conform.

It is in these terms and through these processes that contemporary American culture continues to be characterized by a heavy emphasis on wealth as a basic symbol of success, without a corresponding emphasis upon the legitimate avenues on which to march toward this goal. How do individuals living in this cultural context respond? And how do our observations bear upon the doctrine that deviant behavior typically derives from biological impulses breaking through the restraints imposed by culture? What, in short, are the consequences for the behavior of people variously situated in a social structure of a culture in which the emphasis on dominant success-goals has become increasingly separated from an equivalent emphasis on institutionalized procedures for seeking these goals?

TYPES OF INDIVIDUAL ADAPTATION

Turning from these culture patterns, we now examine types of adaptation by individuals within the culture-bearing society. Though our focus is still the cultural and social genesis of varying rates and types of deviant behavior, our perspective shifts from the plane of patterns of cultural values to the plane of types of adaptation to these values among those occupying different positions in the social structure.

We here consider five types of adaptation, as these are schematically set out in the following table, where (+) signifies "acceptance," (−) signifies "rejection," and (±) signifies "rejection of prevailing values and substitution of new values."

A Typology of Modes of Individual Adaptation   **TABLE 1**

Modes of Adaptation	Culture Goals	Institutionalized Means
I. Conformity	+	+
II. Innovation	+	−
III. Ritualism	−	+
IV. Retreatism	−	−
V. Rebellion	±	±

I. Conformity

To the extent that a society is stable, adaptation type I—conformity to both cultural goals and institutionalized means—is the most common and widely diffused. Were this not so, the stability and continuity of the society could not be maintained. . . .

II. Innovation

Great cultural emphasis upon the success-goal invites this mode of adaptation through the use of institutionally proscribed but often effective means of attaining at least the simulacrum of success—wealth and power. This response occurs when the individual has assimilated the cultural emphasis upon the goal without equally internalizing the institutional norms governing ways and means for its attainment. . . .

It appears from our analysis that the greatest pressures toward deviation are exerted upon the lower strata. Cases in point permit us to detect the sociological mechanisms involved in producing these pressures. Several researches have shown that specialized areas of vice and crime constitute a "normal" response to a situation where the cultural emphasis upon pecuniary success has been absorbed, but where there is little access to conventional and legitimate means for becoming successful. The occupational opportunities of people in these areas are largely confined to manual labor and the lesser white-collar crimes. Given the American stigmatization of manual labor *which has been found to hold rather uniformly in all social classes,* and the absence of realistic opportunities for advancement beyond this level, the result is a marked tendency toward deviant behavior. The status of unskilled labor and the consequent low income cannot readily compete *in terms of established standards of worth* with the promises of power and high income from organized vice, rackets and crime.

For our purposes, these situations exhibit two salient features. First, incentives for success are provided by the established values of the culture *and* second, the avenues available for moving toward this goal are largely limited by the class structure to those of deviant behavior. It is the *combination* of the cultural emphasis and the social structure which produces intense pressure for deviation. . . .

III. Ritualism

The ritualistic type of adaptation can be readily identified. It involves the abandoning or scaling down of the lofty cultural goals of great pecuniary success and rapid social mobility to the point where one's aspirations can be satisfied. But though one rejects the cultural obligation to attempt "to get ahead in the world," though one draws in one's horizons, one continues to abide almost compulsively by institutional norms. . . .

We should expect this type of adaptation to be fairly frequent in a society which makes one's social status largely dependent upon one's achievements. For, as has so often been observed, this ceaseless competitive struggle produces acute status anxiety. One device for allaying these anxieties is to lower one's level of aspiration—permanently. Fear produces inaction, or, more accurately, routinized action.

The syndrome of the social ritualist is both familiar and instructive. His implicit life-philosophy finds expression in a series of cultural clichés: "I'm not sticking *my* neck out," "I'm playing safe," "I'm satisfied with what I've got," "Don't aim high and you won't be disappointed." The theme threaded through these attitudes is that high ambitions invite frustration and danger whereas lower aspirations produce satisfaction and security. It is the perspective of the frightened employee, the zealously conformist bureaucrat in the teller's cage of the private banking enterprise or in the front office of the public works enterprise.

IV. Retreatism

Just as Adaptation I (conformity) remains the most frequent, Adaptation IV (the rejection of cultural goals and institutional means) is probably the least common. People who adapt (or maladapt) in this fashion are, strictly speaking, *in* the society but not *of* it. Sociologically these constitute the true aliens. Not sharing the common frame of values, they can be included as members of the *society* (in distinction from the *population*) only in a fictional sense.

In this category fall some of the adaptive activities of psychotics, autists, pariahs, outcasts, vagrants, vagabonds, tramps, chronic drunkards and drug addicts. They have relinquished culturally prescribed goals and their behavior does not accord with institutional norms. The competitive order is maintained but the frustrated and handicapped individual who cannot cope with this order drops out. Defeatism, quietism and resignation are manifested in escape mechanisms which ultimately lead him to "escape" from the requirements of the society. It is thus an expedient which arises from continued failure to near the goal by legitimate measures and from an inability to use the illegitimate route because of internalized prohibitions.

V. Rebellion

This adaptation leads men outside the environing social structure to envisage and seek to bring into being a new, that is to say, a greatly modified social structure. It presupposes alienation from reigning goals and standards. These come to be regarded as purely arbitrary. And the arbitrary is precisely that which can neither exact allegiance nor possess legitimacy, for it might as well be otherwise. In our society, organized movements for rebellion apparently aim to introduce a social structure in which the cultural standards of success would be sharply modified and provision would be made for a closer correspondence between merit, effort and reward.

THE STRAIN TOWARD ANOMIE

The social structure we have examined produces a strain toward anomie and deviant behavior. The pressure of such a social order is upon outdoing one's competitors. So long as the sentiments supporting this competitive system are distributed throughout the entire range of activities and are not confined to the final result of "success," the choice of means will remain largely within the ambit of institutional control. When, however, the cultural emphasis shifts from the satisfactions deriving from competition itself to almost exclusive concern with the outcome, the resultant stress makes for the breakdown of the regulatory structure.

QUESTIONS FOR DISCUSSION

1. How does the author address the belief that criminal behavior is frequently associated with the lower class or the poor?

2. Using Merton's typology, give examples of each adaptation.

3. Specifically, how do social structure and anomie theory explain deviant behavior in your geographical location?

Illegitimate Means and Delinquent Subcultures

Richard A. Cloward and Lloyd E. Ohlin

THE AVAILABILITY OF ILLEGITIMATE MEANS

Social norms are two-sided. A prescription implies the existence of a prohibition, and *vice versa*. To advocate honesty is to demarcate and condemn a set of actions which are dishonest. In other words, norms that define legitimate practices also implicitly define illegitimate practices. One purpose of norms, in fact, is to delineate the boundary between legitimate and illegitimate practices. In setting this boundary, in segregating and classifying various types of behavior, they make us aware not only of behavior that is regarded as right and proper but also of behavior that is said to be wrong and improper. Thus the criminal who engages in theft or fraud does not invent a new way of life; the possibility of employing alternative means is acknowledged, tacitly at least, by the norms of the culture.

This tendency for proscribed alternatives to be implicit in every prescription, and *vice versa*, although widely recognized, is nevertheless a reef upon which many a theory of delinquency has foundered. Much of the criminological literature assumes, for example, that one may explain a criminal act simply by accounting for the individual's readiness to employ illegal alternatives of which his culture, through its norms, has already made him generally aware. Such explanations are quite unsatisfactory, however, for they ignore a host of questions regarding the *relative availability* of illegal alternatives to various potential criminals. The aspiration to be a physician is hardly enough to explain the fact of becoming a physician; there is much that transpires between the aspiration and the achievement. This is no less true of the person who wants to be a successful criminal. Having decided that he "can't make it legitimately," he cannot simply choose among an array of illegitimate means, all equally available to him. . . . It is assumed in the theory of anomie that access to conventional means is differentially distributed, that some individuals, because of their social class, enjoy certain advantages that are denied to those elsewhere in the class structure. For example, there are variations in the degree to which members of various classes are fully exposed to and thus acquire the values, knowledge, and skills that facilitate upward mobility. It should not be startling, therefore, to suggest that there are socially structured variations in the availability of illegitimate means as well. In connection with delinquent subcultures, we shall be concerned principally with differentials in access to illegitimate means within the lower class.

Many sociologists have alluded to differentials in access to illegitimate means without explicitly incorporating this variable into a theory of deviant behavior. This is particularly true of scholars in the "Chicago tradition" of criminology. Two closely related theoretical perspectives emerged from this school. The theory of "cultural transmission," advanced by Clifford R. Shaw and Henry D. McKay, focuses on the development in some urban neighborhoods of a criminal tradition that persists from one generation to another despite constant changes in population.[1] In the theory of "differential association," Edwin H. Sutherland described the processes by which criminal values are taken over by the individual.[2] He asserted that criminal behavior is learned, and that it is learned in interaction with others who have already incorporated criminal values. Thus the first theory stresses the value systems of different areas; the second, the systems of social relationships that facilitate or impede the acquisition of these values.

Scholars in the Chicago tradition, who emphasized the processes involved in learning to be criminal, were actually pointing to differentials in the availability of illegal means—although they did not explicitly recognize this variable in their analysis. This can perhaps best be seen by examining Sutherland's classic work, *The Professional Thief*. "An inclination to steal," according to Sutherland, "is not a sufficient explanation of the genesis of the professional thief."[3] The "self-made" thief, lacking knowledge of the ways of securing immunity from prosecution and similar techniques of defense, "would quickly land in prison; . . . a person can be a professional thief only if he is recognized and received as such by other professional thieves." But recognition is not freely accorded: "Selection and tutelage are the two necessary elements in the process of acquiring recognition as a professional thief . . . A person cannot acquire recognition as a professional thief until he has had tutelage in professional theft, *and tutelage is given only to a few persons selected from the total population*." For one thing, "the person must be appreciated by the professional thieves. He must be appraised as having an adequate equipment of wits, front, talking-ability, honesty, reliability, nerve and determination." Furthermore, the aspirant is judged by high standards of performance, for only "a very small percentage of those who start on this process ever reach the stage of professional thief. . . ." Thus motivation and pressures toward deviance do not fully account for deviant behavior any more than motivation and pressures toward conformity account for conforming behavior. The individual must have access to a learning environment and, once having been trained, must be allowed to perform his role. Roles, whether conforming or deviant in content, are not necessarily freely available; access to them depends upon a variety of factors, such as one's socioeconomic position, age, sex, ethnic affiliation, personality characteristics, and the like. The potential thief, like the potential physician, finds that access to his goal is governed by many criteria other than merit and motivation.

What we are asserting is that access to illegitimate roles is not freely available to all, as is commonly assumed. Only those neighborhoods in which crime flourishes as a stable, indigenous institution are fertile criminal learning environments for the young. Because these environments afford integration of different age-levels of offender, selected young people are exposed to "differential association" through which tutelage is provided and criminal values and skills are acquired. To be prepared for the role may not, however, ensure that the individual will ever discharge it. One important limitation is that more youngsters are recruited into these patterns of differential associations than the adult criminal structure can possibly absorb. Since there is a surplus of contenders for these elite positions, criteria and mechanisms of selection must be evolved. Hence a certain proportion of those who aspire may not be permitted to engage in the behavior for which they have prepared themselves.

Thus we conclude that access to illegitimate roles, no less than access to legitimate roles, is limited by both social and psychological factors. We shall here be concerned primarily with socially structured differentials in illegitimate opportunities. Such differentials, we contend, have much to do with the type of delinquent subculture that develops.

LEARNING AND PERFORMANCE STRUCTURES

Our use of the term "opportunities," legitimate or illegitimate, implies access to both learning and performance structures. That is, the individual must have access to appropriate environments for the acquisition of the values and skills associated with the performance of a particular role, and he must be supported in the performance of the role once he has learned it.

Tannenbaum, several decades ago, vividly expressed the point that criminal role performance, no less than conventional role performance, presupposes a patterned set of relationships through which the requisite values and skills are transmitted by established practitioners to aspiring youth:

> It takes a long time to make a good criminal, many years of specialized training and much preparation. But training is something that is given to people. People learn in a community where the materials and the knowledge are to be had. A craft needs an atmosphere saturated with purpose and promise. The community provides the attitudes, the point of view, the philosophy of life, the example, the motive, the contacts, the friendships, the incentives. No child brings those into

the world. He finds them here and available for use and elaboration. The community gives the criminal his materials and habits, just as it gives the doctor, the lawyer, the teacher, and the candlestick-maker theirs.[4]

Sutherland systematized this general point of view, asserting that opportunity consists, at least in part, of learning structures. Thus "criminal behavior is learned" and, furthermore, it is learned "in interaction with other persons in a process of communication." However, he conceded that the differential-association theory does not constitute a full explanation of criminal behavior. In a paper circulated in 1944, he noted that "criminal behavior is partially a function of opportunities to commit specific classes of crime, such as embezzlement, bank burglary, or illicit heterosexual intercourse." Therefore, "while opportunity may be partially a function of association with criminal patterns and of the specialized techniques thus acquired, it is not determined entirely in that manner, and consequently differential association is not the sufficient cause of criminal behavior."[5]

To Sutherland, then, illegitimate opportunity included conditions favorable to the performance of a criminal role as well as conditions favorable to the learning of such a role (differential associations). These conditions, we suggest, depend upon certain features of the social structure of the community in which delinquency arises.

DIFFERENTIAL OPPORTUNITY: A HYPOTHESIS

We believe that each individual occupies a position in both legitimate and illegitimate opportunity structures. This is a new way of defining the situation. The theory of anomie views the individual primarily in terms of the legitimate opportunity structure. It poses questions regarding differentials in access to legitimate routes to success-goals; at the same time it assumes either that illegitimate avenues to success-goals are freely available or that differentials in their availability are of little significance. This tendency may be seen in the following statement by Merton:

> Several researches have shown that specialized areas of vice and crime constitute a "normal" response to a situation where the cultural emphasis upon pecuniary success has been absorbed, but where there is little access to conventional and legitimate means for becoming successful. The occupational opportunities of people in these areas are largely confined to manual labor and the lesser white-collar jobs. Given the American stigmatization of manual labor *which has been found to hold rather uniformly for all social classes,* and the absence of realistic opportunities for advancement beyond this level, the result is a marked tendency toward deviant behavior. The status of unskilled labor and the consequent low income cannot readily compete *in terms of established standards of worth* with the promises of power and high income from organized vice, rackets and crime. . . . [Such a situation] leads toward the gradual attenuation of legitimate, but by and large ineffectual, strivings and the increasing use of illegitimate, but more or less effective, expedients.[6]

The cultural-transmission and differential-association tradition, on the other hand, assumes that access to illegitimate means is variable, but it does not recognize the significance of comparable differentials in access to legitimate means. Sutherland's "ninth proposition" in the theory of differential association states:

> *Though criminal behavior is an expression of general needs and values, it is not explained by those general needs and values since non-criminal behavior is an expression of the same needs and values.* Thieves generally steal in order to secure money, but likewise honest laborers work in order to secure money. The attempts by many scholars to explain criminal behavior by general drives and values, such as the happiness principle, striving for social status, the money motive, or frustration, have been and must continue to be futile since they explain lawful behavior as completely as they explain criminal behavior.[7]

In this statement, Sutherland appears to assume that people have equal and free access to legitimate means regardless of their social position. At the very least, he does not treat access to legitimate means as variable. It is, of course, perfectly true that "striving for social status," "the money motive," and other socially approved drives do not fully account for either deviant or conforming behavior. But if goal-oriented behavior occurs under conditions in which there are socially structured obstacles to the satisfaction of these drives by legitimate means, the resulting pressures, we contend, might lead to deviance.

The concept of differential opportunity structures permits us to unite the theory of anomie, which recognizes the concept of differentials in access to legitimate means, and the "Chicago tradition," in which the concept of differentials in access to illegitimate means is implicit. We can now look at the individual, not simply in relation to one or the other system of means, but in relation to both legitimate and illegitimate systems. This approach permits us to ask, for example, how the relative availability of illegitimate opportunities affects the resolution of adjustment problems leading to deviant behavior. We believe that the way in which these problems are resolved may depend upon the kind of support for one or another type of illegitimate activity that is given at different points in the social structure. If, in a given social location, illegal or criminal means are not readily available, then we should not expect a criminal subculture to develop among adolescents. By the same logic, we should expect the manipulation of violence to become a primary avenue to higher status only in areas where the means of violence are not denied to the young. To give a third example, drug addiction and participation in subcultures organized around the consumption of drugs presuppose that persons can secure access to drugs and knowledge about how to use them. In some parts of the social structure, this would be very difficult; in others, very easy. In short, there are marked differences from one part of the social structure to another in the types of illegitimate adaptation that are available to persons in search of solutions to problems of adjustment arising from the restricted availability of legitimate means.[8] In this sense, then, we can think of individuals as being located in two opportunity structures—one legitimate, the other illegitimate. Given limited access to success-goals by legitimate means, the nature of the delinquent response that may result will vary according to the availability of various illegitimate means.[9]

NOTES

1. See esp. C. R. Shaw, *The Jack-Roller* (Chicago: University of Chicago Press, 1930); Shaw, *The Natural History of a Delinquent Career* (Chicago: University of Chicago Press, 1931); Shaw *et al., Delinquency Areas* (Chicago: University of Chicago Press, 1940); and Shaw and H. D. McKay, *Juvenile Delinquency and Urban Areas* (Chicago: University of Chicago Press, 1942).

2. E. H. Sutherland, ed. *The Professional Thief* (Chicago: University of Chicago Press, 1937); and Sutherland, *Principles of Criminology,* 4th Ed. (Philadelphia: Lippincott, 1947).

3. All quotations in this paragraph are from *The Professional Thief,* pp. 211–13. Emphasis added.

4. Frank Tannenbaum, "The Professional Criminal," *The Century,* Vol. 110 (May–Oct. 1925); p. 577.

5. See A. K. Cohen, Alfred Lindesmith, and Karl Schussler, eds., *The Sutherland Papers* (Bloomington, Ind.: Indiana University Press, 1956), pp. 31–35.

6. R. K. Merton, *Social Theory and Social Structure,* Rev. and Enl. Ed. (Glencoe, Ill.: Free Press, 1957), pp. 145–46.

7. *Principles of Criminology, op. cit.,* pp. 7–8.

8. For an example of restrictions on access to illegitimate roles, note the impact of racial definitions in the following case: "I was greeted by two prisoners who were to be my cell buddies. Ernest was a first offender, charged with being a 'hold-up' man. Bill, the other buddy, was an old offender, going through the machinery of becoming a habitual criminal, in and out of jail. . . . The first thing they asked me was, 'What are you in for?' I said, 'Jack-rolling.' The hardened one (Bill) looked at me with a superior air and said, 'A hoodlum, eh? An ordinary sneak thief. Not willing to leave jack-rolling to the niggers, eh? That's all they're good for. Kid, jack-rolling's not a white man's job.' I could see that he was disgusted with me, and I was too scared to say anything" (Shaw, *The Jack-Roller, op. cit.,* p. 101).

9. For a discussion of the way in which the availability of illegitimate means influences the adaptations of inmates to prison life, see R. A. Cloward, "Social Control in the Prison," *Theoretical Studies of the Social Organization of the Prison,* Bulletin No. 15 (New York: Social Science Research Council, March 1960), pp. 20–48.

QUESTIONS FOR DISCUSSION

1. Discuss the relationship between illegitimate means and delinquent subcultures, according to the authors. Give an example from your present geographical location or hometown that illustrates illegitimate means and delinquent subcultures.

2. How do the theories of anomie and differential association help explain illegitimate means and delinquent subcultures? Use examples to illustrate your response.

3. How does differential opportunity relate to social deviance in your geographical location?

C H A P T E R 5

CONFLICT PERSPECTIVES

Toward a Marxian Theory of Deviance[1]

Steven Spitzer

Within the last decade Americans sociologists have become increasingly reflective in their approach to deviance and social problems. They have come to recognize that interpretations of deviance are often ideological in their assumptions and implications, and that sociologists are frequently guilty of "providing the facts which make oppression more efficient and the theory which makes it legitimate to a larger constituency" (Becker and Horowitz, 1972: 48). To combat this tendency students of deviance have invested more and more energy in the search for a critical theory. This search has focused on three major problems: (1) the definition of evidence, (2) the etiology of deviance, and (3) the etiology of control.

TRADITIONAL THEORIES AND THEIR PROBLEMS

Traditional theories approached the explanation of deviance with little equivocation about the phenomenon to be explained. Prior to the 1960s the subject matter of deviance theory was taken for granted and few were disturbed by its preoccupation with "dramatic and predatory" forms of social behavior (Liazos, 1972). Only in recent years have sociologists started to question the consequences of singling out "nuts," "sluts," "perverts," "lames," "crooks," "junkies," and "juicers" for special attention. Instead of adopting conventional wisdom about *who* and *what* is deviant, investigators have gradually made the definitional problem central to the sociological enterprise. They have begun to appreciate the consequences of studying the powerless (rather than the powerful)—both in terms of the relationship between *knowledge of* and *control over* a group, and the support of the "hierarchy of credibility" (Becker, 1967) that such a focus provides. Sociologists have discovered the significance of the definitional process in their own, as well as society's response to deviance, and this discovery has raised doubts about the direction and purpose of the field.

Even when the definitional issue can be resolved critics are faced with a second and equally troublesome problem. Traditional theories of deviance are essentially *non-structural* and *ahistorical* in their mode of analysis. By restricting investigation to factors which are manipulable within existing structural arrangements these theories embrace a "correctional perspective" (Matza, 1969) and divert attention from the impact of the political economy as a whole. From this point of view deviance is *in* but not *of* our contemporary social order. Theories that locate the source of deviance in factors as diverse as personality structure, family systems, cultural transmission, social disorganization and differential opportunity share a common flaw—they attempt to understand deviance apart from historically specific forms of political and economic organization. Because traditional theories proceed without any sense of historical development, deviance is normally viewed as an episodic and transitory phenomenon rather than an outgrowth of long-term structural change. Sensitive sociologists have come to realize that critical theory must establish, rather than obscure, the relationship between deviance, social structure and social change.

A final problem in the search for a critical theory of deviance is the absence of a coherent theory of control. More than ever before critics have come to argue that deviance cannot be understood apart from the dynamics of control. Earlier theories devoted scant attention to the control process precisely because control was interpreted as a natural response to behavior generally assumed to be problematic. Since theories of deviance viewed control as a desideratum, no theory of control was required. But as sociologists began to question conventional images of deviance they revised their impressions of social control. Rather than assuming that societal reaction was necessarily defensive and benign, skeptics announced that controls could actually cause deviance. The problem was no longer simply to explain the independent sources of deviance and control, but to understand the reciprocal relationship between the two.

In elevating control to the position of an independent variable a more critical orientation has evolved. Yet this orientation has created a number of problems of its own. If deviance is simply a *status,* representing the outcome of a series of control procedures, should our theory of deviance be reduced to a theory of control? In what sense, if any, is deviance an achieved rather than an ascribed status? How do we account for the historical and structural sources of deviance apart from those shaping the development of formal controls?

Toward a Theory of Deviance Production

A critical theory must be able to account for both *deviance* and *deviants.* It must be sensitive to the process through which deviance is subjectively constructed and deviants are objectively handled, as well as the structural bases of the behavior and characteristics which come to official attention. It should neither beg the explanation of deviant behavior and characteristics by depicting the deviant as a helpless victim of oppression, nor fail to realize that his identification as deviant, the dimensions of his threat, and the priorities of the control system are a part of a broader social conflict. While acknowledging the fact that deviance is a *status* imputed to groups who share certain structural characteristics (e.g., powerlessness) we must not forget that these groups are defined by more than these characteristics alone.[2] We must not only ask why specific members of the underclass are selected for official processing, but also why they behave as they do. Deviant statuses, no matter how coercively applied, are in some sense achieved and we must understand this achievement in the context of political-economic conflict. We need to understand why capitalism produces both patterns of activity and types of people that are defined and managed as deviant.

In order to construct a general theory of deviance and control it is useful to conceive of a process of deviance production which can be understood in relationship to the development of class society. *Deviance production involves all aspects of the process through which populations are structurally generated, as well as shaped, channeled into, and manipulated within social categories defined as deviant.* This process includes the development of and changes in: (1) deviant definitions, (2) problem populations, and (3) control systems.

Most fundamentally, deviance production involves the development of and changes in deviant categories and images. A critical theory must examine where these images and definitions come from, what they reflect about the structure of and priorities in specific class societies, and how they are related to class conflict. If we are to explain, for example, how mental retardation becomes deviance and the feeble-minded deviant we need to examine the structural characteristics, economic and political dimensions of the society in which these definitions and images emerged. In the case of American society we must understand how certain correlates of capitalist development (proletarianization and nuclearization of the family) weakened traditional methods of assimilating these groups, how others (the emergence of scientific and meritocratic ideologies) sanctioned intellectual stratification and differential handling, and how still others (the attraction of unskilled labor and population concentrations) heightened concern over the "threat" that these groups were assumed to represent. In other words, the form and content of deviance definition must be assessed in terms of its relationship to both structural and ideological change.

A second aspect of deviance production is the development of and changes in problem behaviors and problem populations. If we assume that class societies are based on fundamental conflicts between groups, and that harmony is achieved through the dominance of a specific class, it makes sense to argue that deviants are culled from groups who create specific problems for those who rule. Although these groups may victimize or burden those outside of the dominant class, their problematic quality ultimately resides in their challenge to the basis and form of class rule. Because problem populations are not always "handled," they provide candidates for, but are in no sense equivalent to, official deviants. A sophisticated critical theory must investigate where these groups come from, why their behaviors and characteristics are problematic, and how they are transformed in a developing political economy. We must consider, for instance, why Chinese laborers in 19th century California and

Chicanos in the Southwest during the 1930s became the object of official concern, and why drug laws evolved to address the "problems" that these groups came to represent (Helmer and Vietorisz, 1973; Musto, 1973).

The changing character of problem populations is related to deviance production in much the same way that variations in material resources affect manufacturing. Changes in the quantity and quality of raw materials influence the scope and priorities of production, but the characteristics of the final product depend as much on the methods of production as the source material. These methods comprise the third element in deviance production—the development and operation of the control system. The theory must explain why a system of control emerges under specific conditions and account for its size, focus and working assumptions. The effectiveness of the system in confronting problem populations and its internal structure must be understood in order to interpret changes in the form and content of control. Thus, in studying the production of the "mentally ill" we must not only consider why deviance has been "therapeutized," but also how this development reflects the subtleties of class control. Under capitalism, for example, formal control of the mad and the birth of the asylum may be examined as a response to the growing demands for order, responsibility and restraint (cf. Foucault, 1965).

The Production of Deviance in Capitalist Society

The concept of deviance production offers a starting point for the analysis of both deviance and control. But for such a construct to serve as a critical tool it must be grounded in an historical and structural investigation of society. For Marx, the crucial unit of analysis is the mode of production that dominates a given historical period. If we are to have a Marxian theory of deviance, therefore, deviance production must be understood in relationship to specific forms of socio-economic organization. In our society, productive activity is organized capitalistically and it is ultimately defined by "the process that transforms on the one hand, the social means of subsistence and of production into capital, on the other hand the immediate producers into wage labourers" (Marx, 1967:714).

There are two features of the capitalist mode of production important for purposes of this discussion. First, as a mode of production it forms the foundation or infrastructure of our society. This means that the starting point of our analysis must be an understanding of the economic organization of capitalist societies and the impact of that organization on all aspects of social life. But the capitalist mode of production is an important starting point in another sense. It contains contradictions which reflect the internal tendencies of capitalism. These contradictions are important because they explain the changing character of the capitalist system and the nature of its impact on social, political and intellectual activity. The formulation of a Marxist perspective on deviance requires the interpretation of the process through which the contradictions of capitalism are expressed. In particular, the theory must illustrate the relationship between specific contradictions, the problems of capitalist development and the production of a deviant class.

The superstructure of society emerges from and reflects the ongoing development of economic forces (the infrastructure). In class societies this superstructure preserves the hegemony of the ruling class through a system of class controls. These controls, which are institutionalized in the family, church, private associations, media, schools and the state, provide a mechanism for coping with the contradictions and achieving the aims of capitalist development.

Among the most important functions served by the superstructure in capitalist societies is the regulation and management of problem populations. Because deviance processing is only one of the methods available for social control, these groups supply raw material for deviance production, but are by no means synonymous with deviant populations. Problem populations tend to share a number of social characteristics, but most important among these is the fact that their behavior, personal qualities, and/or position threaten the *social relations of production* in capitalist societies. In other words, populations become generally eligible for management as deviant when they disturb, hinder or call into question any of the following:

1. Capitalist modes of appropriating the product of human labor (e.g., when the poor "steal" from the rich)

2. The social conditions under which capitalist production take place (e.g., those who refuse or are unable to perform wage labor)

3. Patterns of distribution and consumption in capitalist society (e.g., those who use drugs for escape and transcendence rather than sociability and adjustment)

4. The process of socialization for productive and non-productive roles (e.g., youth who refuse to be schooled or those who deny the validity of "family life")[3]

5. The ideology which supports the functioning of capitalist society (e.g., proponents of alternative forms of social organization)

Although problem populations are defined in terms of the threat and costs that they present to the social relations of production in capitalist societies, these populations are far from isomorphic with a revolutionary class. It is certainly true that some members of the problem population may under specific circumstances possess revolutionary potential. But this potential can only be realized if the problematic group is located in a position of functional indispensability within the capitalist system. Historically, capitalist societies have been quite successful in transforming those who are problematic and indispensable (the protorevolutionary class) into groups who are either problematic and dispensable (candidates for deviance processing), or indispensable but not problematic (supporters of the capitalist order). On the other hand, simply because a group is manageable does not mean that it ceases to be a problem for the capitalist class. Even though dispensable problem populations cannot overturn the capitalist system, they can represent a significant impediment to its maintenance and growth. It is in this sense that they become eligible for management as deviants.

Problem populations are created in two ways—either directly through the expression of fundamental contradictions in the capitalist mode of production or indirectly through disturbances in the system of class rule. An example of the first process is found in Marx's analysis of the "relative surplus-population."

Writing on the "General Law of Capitalist Accumulation" Marx explains how increased social redundance is inherent in the development of the capitalist mode of production:

> With the extension of the scale of production, and the mass of the labourers set in motion, with the greater breadth and fullness of all sources of wealth, there is also an extension of the scale on which greater attraction of labourers by capital is accompanied by their greater repulsion. . . . The labouring population therefore produces, along with the accumulation of capital produced by it, the means by which itself is made relatively superfluous, . . . and it does this to an always increasing extent (Marx, 1967:631).

In its most limited sense the production of a relative surplus-population involves the creation of a class which is economically redundant. But insofar as the conditions of economic existence determine social existence, this process helps explain the emergence of groups who become both threatening and vulnerable at the same time. The marginal status of these populations reduces their stake in the maintenance of the system while their powerlessness and dispensability render them increasingly susceptible to the mechanisms of official control.

The paradox surrounding the production of the relative surplus-population is that this population is both useful and menacing to the accumulation of capital. Marx describes how the relative surplus-population "forms a disposable industrial army, that belongs to capital quite as absolutely as if the latter had bred it at its own cost," and how this army, "creates, for the changing needs of the self-expansion of capital, a mass of human material always ready for exploitation" (Marx, 1967:632).

On the other hand, it is apparent that an excessive increase in what Marx called the "lowest sediment" of the relative surplus-population, might seriously impair the growth of capital. The social expenses and threat to social harmony created by a large and economically stagnant surplus-population could jeopardize the preconditions for accumulation by undermining the ideology of equality so essential to the legitimation of production relations in bourgeois democracies, diverting revenues away from capital investment toward control and support operations, and providing a basis for political organization of the dispossessed.[4] To the extent that the relative surplus-population confronts the capitalist class as a threat to the social relations of production it reflects an important contradiction in modern capitalist societies: a surplus-population is a necessary product of and condition for the accumulation of wealth on a capitalist basis, but it also creates a form of social expense which must be neutralized or controlled if production relations and conditions for increased accumulation are to remain unimpaired.

Problem populations are also generated through contradictions which develop in the system of class rule. The institutions which make up the superstructure of capitalist society originate and are maintained to guarantee the interests of the capitalist class. Yet these institutions necessarily reproduce, rather than resolve, the contradictions of the capitalist order. In a dialectical fashion, arrangements which arise in order to buttress capitalism are transformed into their opposite—structures for the cultivation of internal threats. An instructive example of this process is found in the emergence and transformation of educational institutions in the United States.

The introduction of mass education in the United States can be traced to the developing needs of corporate capitalism (cf. Karier, 1973; Cohen and Lazerson, 1972;

Bowles and Gintis, 1972; Spring, 1972). Compulsory education provided a means of training, testing and sorting, and assimilating wage-laborers, as well as withholding certain populations from the labor market. The system was also intended to preserve the values of bourgeois society and operate as an "inexpensive form of police" (Spring, 1973:31). However, as Gintis (1973) and Bowles (1973) have suggested, the internal contradictions of schooling can lead to effects opposite of those intended. For the poor, early schooling can make explicit the oppressiveness and alienating character of capitalist institutions, while higher education can instill critical abilities which lead students to "bite the hand that feeds them." In both cases educational institutions create troublesome populations (i.e., drop outs and student radicals) and contribute to the very problems they were designed to solve.

After understanding how and why specific groups become generally bothersome in capitalist society, it is necessary to investigate the conditions under which these groups are transformed into proper objects for social control. In other words, we must ask what distinguishes the generally problematic from the specifically deviant. The rate at which problem populations are converted into deviants will reflect the relationship between these populations and the control system. This rate is likely to be influenced by the:

(1) *Extensiveness and Intensity of State Controls.* Deviance processing (as opposed to other control measures) is more likely to occur when problem management is monopolized by the state. As state controls are applied more generally the proportion of official deviants will increase.

(2) *Size and Level of Threat Presented by the Problem Population.* The larger and more threatening the problem population, the greater the likelihood that this population will have to be controlled through deviance processing rather than other methods. As the threat created by these populations exceeds the capacities of informal restraints, their management requires a broadening of the reaction system and an increasing centralization and coordination of control activities.

(3) *Level of Organization of the Problem Population.* When and if problem populations are able to organize and develop limited amounts of political power, deviance processing becomes increasingly less effective as a tool for social control. The attribution of deviant status is most likely to occur when a group is relatively impotent and atomized.

(4) *Effectiveness of Control Structures Organized through Civil Society.* The greater the effectiveness of the organs of civil society (i.e., the family, church, media, schools, sports) in solving the problems of class control, the less the likelihood that deviance processing (a more explicitly political process) will be employed.

(5) *Availability and Effectiveness of Alternative Types of Official Processing.* In some cases the state will be able effectively to incorporate certain segments of the problem population into specially created "pro-social" roles. In the modern era, for example, conscription and public works projects (Piven and Cloward, 1971) helped neutralize the problems posed by troublesome populations without creating new or expanding old deviant categories.

(6) *Availability and Effectiveness of Parallel Control Structures.* In many instances the state can transfer its costs of deviance production by supporting or at least tolerating the activities of independent control networks which operate in its interests. For example, when the state is denied or is reluctant to assert a monopoly over the use of

force it is frequently willing to encourage vigilante organizations and private police in the suppression of problem populations. Similarly, the state is often benefited by the policies and practices of organized crime, insofar as these activities help pacify, contain and enforce order among potentially disruptive groups (Schelling, 1967).

(7) *Utility of Problem Populations.* While problem populations are defined in terms of their threat and costs to capitalist relations of production, they are not threatening in every respect. They can be supportive economically (as part of a surplus labor pool or dual labor market), politically (as evidence of the need for state intervention) and ideologically (as scapegoats for rising discontent). In other words, under certain conditions capitalist societies derive benefits from maintaining a number of visible and uncontrolled "troublemakers" in their midst. Such populations are distinguished by the fact that while they remain generally bothersome, the costs that they inflict are most immediately absorbed by other members of the problem population. Policies evolve, not so much to eliminate or actively suppress these groups, but to deflect their threat away from targets which are sacred to the capitalist class. Victimization is permitted and even encouraged, as long as the victims are members of an expendable class.

Two more or less discrete groupings are established through the operations of official control. These groups are a product of different operating assumptions and administrative orientations toward the deviant population. On the one hand, there is *social junk* which, from the point of view of the dominant class, is a costly yet relatively harmless burden to society. The discreditability of social junk resides in the failure, inability or refusal of this group to participate in the roles supportive of capitalist society. Social junk is most likely to come to official attention when informal resources have been exhausted or when the magnitude of the problem becomes significant enough to create a basis for "public concern." Since the threat presented by social junk is passive, growing out of its inability to compete and its withdrawal from the prevailing social order, controls are usually designed to regulate and contain rather than eliminate and suppress the problem. Clear-cut examples of social junk in modern capitalist societies might include the officially administered aged, handicapped, mentally ill and mentally retarded.

In contrast to social junk, there is a category that can be roughly described as *social dynamite.* The essential quality of deviance managed as social dynamite is its potential actively to call into question established relationships, especially relations of production and domination. Generally, therefore, social dynamite tends to be more youthful, alienated and politically volatile than social junk. The control of social dynamite is usually premised on an assumption that the problem is acute in nature, requiring a rapid and focused expenditure of control resources. This is in contrast to the handling of social junk frequently based on a belief that the problem is chronic and best controlled through broad reactive, rather than intensive and selective measures. Correspondingly, social dynamite is normally processed through the legal system with its capacity for active intervention, while social junk is frequently (but not always)[5] administered by the agencies and agents of the therapeutic and welfare state.

Many varieties of deviant populations are alternatively or simultaneously dealt with as either social junk and/or social dynamite. The welfare poor, homosexuals, alcoholics and "problem children" are among the categories reflecting the equivocal nature of the control process and its dependence on the political, economic and ideological priorities of deviance production. The changing nature of these priorities and their implications for the future may be best understood by examining some of the tendencies of modern capitalist systems.

Monopoly Capital and Deviance Production

Marx viewed capitalism as a system constantly transforming itself. He explained these changes in terms of certain tendencies and contradictions immanent within the capitalist mode of production. One of the most important processes identified by Marx was the tendency for the organic composition of capital to rise. Simply stated, capitalism requires increased productivity to survive, and increased productivity is only made possible by raising the ratio of machines (dead labor) to men (living labor). This tendency is self-reinforcing since, "the further machine production advances, the higher becomes the organic composition of capital needed for an entrepreneur to secure the average profit" (Mandel, 1968:163). This phenomenon helps us explain the course of capitalist development over the last century and the rise of monopoly capital (Baran and Sweezy, 1966).

For the purposes of this analysis there are at least two important consequences of this process. First, the growth of constant capital (machines and raw material) in the production process leads to an expansion in the overall size of the relative surplus-population. The reasons for this are obvious. The increasingly technological character of production removes more and more laborers from productive activity for longer periods of time. Thus, modern capitalist societies have been required progressively to reduce the number of productive years in a worker's life, defining both young and old as economically superfluous. Especially affected are the unskilled who become more and more expendable as capital expands.

In addition to affecting the general size of the relative surplus-population, the rise of the organic composition of capital leads to an increase in the relative stagnancy of that population. In Marx's original analysis he distinguished between forms of superfluous population that were floating and stagnant. The floating population consists of workers who are "sometimes repelled, sometimes attracted again in greater masses, the number of those employed increasing on the whole, although in a constantly decreasing proportion to the scale of production" (1967:641). From the point of view of capitalist accumulation the floating population offers the greatest economic flexibility and the fewest problems of social control because they are most effectively tied to capital by the "natural laws of production." Unfortunately (for the capitalists at least), these groups come to comprise a smaller and smaller proportion of the relative surplus-population. The increasing specialization of productive activity raises the cost of reproducing labor and heightens the demand for highly skilled and "internally controlled" forms of wage labor (Gorz, 1970). The process through which unskilled workers are alternatively absorbed and expelled from the labor force is thereby impaired, and the relative surplus-population comes to be made up of increasing numbers of persons who are more or less permanently redundant. The boundaries between the "useful" and the "useless" are more clearly delineated, while standards for social disqualification are more liberally defined.

With the growth of monopoly capital, therefore, the relative surplus-population begins to take on the character of a population which is more and more absolute. At the same time, the market becomes a less reliable means of disciplining these populations and the "invisible hand" is more frequently replaced by the "visible fist." The implications for deviance production are twofold: (1) problem populations become gradually more problematic—both in terms of their size and their insensitivity to economic controls, and (2) the resources of the state need to be applied in greater proportion to protect capitalist relations of production and insure the accumulation of capital.

State Capitalism and New Forms of Control

The major problems faced by monopoly capitalism are surplus population and surplus production. Attempts to solve these problems have led to the creation of the welfare/warfare state (Baran and Sweezy, 1966; Marcuse, 1964; O'Connor, 1973; Gross, 1970). The warfare state attacks the problem of overconsumption by providing "wasteful" consumption and protection for the expansion of foreign markets. The welfare state helps absorb and deflect social expenses engendered by a redundant domestic population. Accordingly, the economic development of capitalist societies has come to depend increasingly on the support of the state.

The emergence of state capitalism and the growing interpenetration of the political and economic spheres have had a number of implications for the organization and administration of class rule. The most important effect of these trends is that control functions are increasingly transferred from the organs of civil society to the organs of political society (the state). As the maintenance of social harmony becomes more difficult and the contradictions of civil society intensify, the state is forced to take a more direct and extensive role in the management of problem populations. This is especially true to the extent that the primary socializing institutions in capitalist societies (e.g., the family and the church) can no longer be counted on to produce obedient and "productive" citizens.

Growing state intervention, especially intervention in the process of socialization, is likely to produce an emphasis on general-preventive (integrative), rather than selective-reactive (segregative) controls. Instead of waiting for troublemakers to surface and managing them through segregative techniques, the state is likely to focus more and more on generally applied incentives and assimilative controls. This shift is consistent with the growth of state capitalism because, on the one hand, it provides mechanisms and policies to nip disruptive influences "in the bud," and, on the other, it paves the way toward a more rational exploitation of human capital. Regarding the latter point, it is clear that effective social engineering depends more on social investment and anticipatory planning than coercive control, and societies may more profitably manage populations by viewing them as human capital, than as human waste. An investment orientation has long been popular in state socialist societies (Rimlinger, 1961, 1966), and its value, not surprisingly, has been increasingly acknowledged by many capitalist states.[6]

In addition to the advantages of integrative controls, segregative measures are likely to fall into disfavor for a more immediate reason—they are relatively costly to formulate and apply. Because of its fiscal problems the state must search for means of economizing control operations without jeopardizing capitalist expansion.

Segregative handling, especially institutionalization, has been useful in manipulating and providing a receptacle for social junk and social dynamite. Nonetheless, the per capita cost of this type of management is typically quite high. Because of its continuing reliance on segregative controls the state is faced with a growing crisis—the overproduction of deviance. The magnitude of the problem and the inherent weaknesses of available approaches tend to limit the alternatives, but among those which are likely to be favored in the future are:

(1) *Normalization.* Perhaps the most expedient response to the overproduction of deviance is the normalization of populations traditionally managed as deviant. Normalization occurs when deviance processing is reduced in scope without supplying specific alternatives, and certain segments of the problem population are "swept under the rug." To be successful this strategy requires the creation of invisible deviants who can be easily absorbed into society and disappear from view.

A current example of this approach is found in the decarceration movement which has reduced the number of inmates in prisons (BOP, 1972) and mental hospitals (NIMH, 1970) over the last fifteen years. By curtailing commitments and increasing turn-over rates the state is able to limit the scale and increase the efficiency of institutionalization. If, however, direct release is likely to focus too much attention on the shortcomings of the state a number of intermediate solutions can be adopted. These include subsidies for private control arrangements (e.g., foster homes, old age homes) and decentralized control facilities (e.g., community treatment centers, halfway houses). In both cases, the fiscal burden of the state is reduced while the dangers of complete normalization are avoided.

(2) *Conversion.* To a certain extent the expenses generated by problem and deviant populations can be offset by encouraging their direct participation in the process of control. Potential troublemakers can be recruited as policemen, social workers and attendants, while confirmed deviants can be "rehabilitated" by becoming counselors, psychiatric aides and parole officers. In other words, if a large number of the controlled can be converted into a first line of defense, threats to the system of class rule can be transformed into resources for its support.[7]

(3) *Containment.* One means of responding to threatening populations without individualized manipulation is through a policy of containment or compartmentalization. This policy involves the geographic segregation of large populations and the use of formal and informal sanctions to circumscribe the challenges that they present. Instead of classifying and handling problem populations in terms of the specific expenses that they create, these groups are loosely administered as a homogeneous class who can be ignored or managed passively as long as they remain in their place.

Strategies of containment have always flourished where social segregation exists, but they have become especially favored in modern capitalist societies. One reason for this is their compatibility with patterns of residential segregation, ghettoization, and internal colonialism (Blauner, 1969).

(4) *Support of Criminal Enterprise.* Another way the overproduction of deviance may be eased is by granting greater power and influence to organized crime. Although predatory criminal enterprise is assumed to stand in opposition to the goals of the state and the capitalist class, it performs valuable and unique functions in the service of class rule (McIntosh, 1973). By creating a parallel opportunity structure, organized crime

provides a means of support for groups who might otherwise become a burden on the state. The activities of organized crime are also important in the pacification of problem populations. Organized crime provides goods and services which ease the hardships and deflect the energies of the underclass. In this role the "crime industry" performs a cooling-out function and offers a control resource which might otherwise not exist. Moreover, insofar as criminal enterprise attempts to reduce uncertainty and risk in its operations, it aids the state in the maintenance of public order. This is particularly true to the extent that the rationalization of criminal activity reduces the collateral costs (i.e., violence) associated with predatory crime (Schelling, 1967).

CONCLUSION

A Marxian theory of deviance and control must overcome the weaknesses of both conventional interpretations and narrow critical models. It must offer a means of studying deviance which fully exploits the critical potential of Marxist scholarship. More than "demystifying" the analysis of deviance, such a theory must suggest directions and offer insights which can be utilized in the direct construction of critical theory. Although the discussion has been informed by concepts and evidence drawn from a range of Marxist studies, it has been more of a sensitizing essay than a substantive analysis. The further development of the theory must await the accumulation of evidence to refine our understanding of the relationships and tendencies explored. When this evidence is developed the contributions of Marxist thought can be more meaningfully applied to an understanding of deviance, class conflict and social control.

NOTES

1. Revised version of a paper presented at the American Sociological Association meetings, August, 1975. I would like to thank Cecile Sue Coren and Andrew T. Scull for their criticisms and suggestions.

2. For example, Turk (1969) defines deviance primarily in terms of the social position and relative power of various social groups.

3. To the extent that a group (e.g., homosexuals) blatantly and systematically challenges the validity of the bourgeois family it is likely to become part of the problem population. The family is essential to capitalist society as a unit for consumption, socialization and the reproduction of the socially necessary labor force (cf. Frankford and Snitow, 1972; Secombe, 1973; Zaretsky, 1973).

4. O'Connor (1973) discusses this problem in terms of the crisis faced by the capitalist state in maintaining conditions for profitable accumulation and social harmony.

5. It has been estimated, for instance, that ⅓ of all arrests in America are for the offense of public drunkenness. Most of these apparently involve "sick" and destitute "skid row alcoholics" (Morris and Hawkins, 1969).

6. Despite the general tendencies of state capitalism, its internal ideological contradictions may actually frustrate the adoption of an investment approach. For example, in discussing social welfare policy Rimlinger (1966:571) concludes that "in a country like the United States, which has a strong individualistic heritage, the idea is still alive that any kind of social

protection has adverse productivity effects. A country like the Soviet Union, with a centrally planned economy and a collectivist ideology, is likely to make an earlier and more deliberate use of health and welfare programs for purposes of influencing productivity and developing manpower."

7. In his analysis of the lumpenproletariat Marx (1964) clearly recognized how the underclass could be manipulated as a "bribed tool of reactionary intrigue."

REFERENCES

Baran, Paul, and Paul M. Sweezy. 1966: Monopoly Capital. New York; Monthly Review Press.

Becker, Howard S. 1967. "Whose side are we on?" Social Problems 14 (Winter): 239–247.

Becker, Howard S., and Irving Louis Horowitz. 1972. "Radical politics and sociological research: observations on methodology and ideology." American Journal of Sociology 78 (July): 48–66.

Blauner, Robert. 1969. "Internal colonialism and ghetto revolt." Social Problems 16 (Spring): 393–408.

Bowles, Samuel. 1973. "Contradictions in United States higher education." Pp. 165–199 in James H. Weaver (ed.), Modern Political Economy: Radical Versus Orthodox Approaches. Boston: Allyn and Bacon.

Bowles, Samuel, and Herbert Gintis. 1972. "I.Q. in the U.S. class structure." Social Policy 3 (November/December): 65–96.

Bureau of Prisons. 1972. National Prisoner Statistics. Prisoners in State and Federal Institutions for Adult Felons. Washington, D.C.: Bureau of Prisons.

Cohen, David K., and Marvin Lazerson. 1972. "Education and the corporate order." Socialist Revolution (March/April): 48–72.

Foucault, Michel. 1965. Madness and Civilization. New York: Random House.

Frankford, Evelyn, and Ann Snitow. 1972. "The trap of domesticity: notes on the family." Socialist Revolution (July/August): 83–94.

Gintis, Herbert. 1973. "Alienation and power." Pp. 431–465 in James H. Weaver (ed.), Modern Political Economy: Radical versus Orthodox Approaches. Boston: Allyn and Bacon.

Gorz, Andre. 1970. "Capitalist relations of production and the socially necessary labor force." Pp. 155–171 in Arthur Lothstein (ed.), All We Are Saying . . . New York: G. P. Putnam.

Gross, Bertram M. 1970. "Friendly fascism: a model for America." Social Policy (November/December): 44–52.

Helmer, John, and Thomas Vietorisz. 1973. "Drug use, the labor market and class conflict." Paper presented at Annual Meeting of the American Sociological Association.

Karier, Clarence J. 1973. "Business values and the educational state." Pp. 6–29 in Clarence J. Karier, Paul Violas, and Joel Spring (eds.), Roots of Crisis: American Education in the Twentieth Century. Chicago: Rand McNally.

Liazos, Alexander. 1972. "The poverty of the sociology of deviance: nuts, sluts and preverts." Social Problems 20 (Summer): 103–120.

Mandel, Ernest. 1968. Marxist Economic Theory (Volume I). New York: Monthly Review Press.

Marcuse, Herbert. 1964. One-Dimensional Man. Boston: Beacon Press.

Marx, Karl. 1964. Class Struggles in France 1848–1850. New York: International Publishers.

———. 1967. Capital (Volume I). New York: International Publishers.

Matza, David. 1969. Becoming Deviant. Englewood Cliffs: Prentice-Hall.

McIntosh, Mary. 1973. "The growth of racketeering." Economy and Society (February): 35–69.

Morris, Norval, and Gordon Hawkins. 1969. The Honest Politician's Guide to Crime Control. Chicago: University of Chicago Press.

Musto, David F. 1973. The American Disease: Origins of Narcotic Control. New Haven: Yale University Press.

National Institute of Mental Health. 1970. Trends in Resident Patients—State and County Mental Hospitals, 1950–1968. Biometry Branch, Office of Program Planning and Evaluation. Rockville, Maryland: National Institute of Mental Health.

O'Connor, James. 1973. The Fiscal Crisis of the State. New York: St. Martin's Press.

Piven, Frances, and Richard A. Cloward. 1971. Regulating the Poor: The Functions of Public Welfare. New York: Random House.

Rimlinger, Gaston V. 1961. "Social security, incentives, and controls in the U.S. and U.S.S.R." Comparison Studies in Society and History 4 (November): 104–124.

———. 1966. "Welfare policy and economic development: a comparative historical perspective." Journal of Economic History (December): 556–571.

Schelling, Thomas. 1967. "Economics and criminal enterprise." Public Interest (Spring): 61–78.

Secombe, Wally. 1973. "The housewife and her labour under capitalism." New Left Review (January–February): 3–24.

Spring, Joel. 1972. Education and the Rise of the Corporate State. Boston: Beacon Press.

———. 1973. "Education as a form of social control." Pp. 30–39 in Clarence J. Karier, Paul Violas, and Joel Spring (eds.), Roots of Crisis: American Education in the Twentieth Century. Chicago: Rand McNally.

Turk, Austin T. 1969. Criminality and Legal Order. Chicago: Rand McNally.

Zaretsky, Eli, 1973. "Capitalism, the family and personal life: parts 1 & 2." Socialist Revolution (January–April/May–June): 69–126, 19–70.

QUESTIONS FOR DISCUSSION

1. Explain the critical theory of deviance. Discuss how this applies to your college, university, or geographical location. How can the critical theory of deviance explain deviant behaviors in inner cities?

2. Discuss how populations become eligible for management as deviant when they disturb, hinder, or call into question "capitalist modes of appropriating the product of human labor" and "patterns of distribution and consumption in a capitalist society."

3. Discuss and give examples of how "social junk" and "social dynamite" relate to deviant behavior.

Street Crime, Labor Surplus, and Criminal Punishment, 1980–1990

Andrew L. Hochstetler and Neal Shover

There is enormous geographic and temporal variation in state use of punishment. In the United States, for example, there is well-documented regional and state-level variation in the use of imprisonment; in 1994, the incarceration rate (the number of imprisoned adults per 100,000 total population) was 462 for southern states but only 291 for the northeastern states (United States Bureau of Justice Statistics 1996). Geographic variation is apparent also in use of the death penalty; whereas some states do not permit capital punishment, others routinely and regularly execute offenders. As for evidence of temporal variation in punishment, we need look no farther than recent history. In the years after 1973, America's training school, jail and prison populations climbed to historically unprecedented levels. The adult imprisoned population alone grew by more than 300 percent between 1975 and 1994 (United States Bureau of Justice Statistics 1996). Explaining geographic and temporal variation in official use of imprisonment and other forms of punishment is a long-standing focal point of social problems theory and research. We continue this line of investigation by examining community-level determinants of change in the use of imprisonment by local courts in the United States during the 1980s.

Background

In conflict-theoretical explanations, crime control is portrayed as a process unusually sensitive to the interests and machinations of dominant classes and elites. Grounded in neo-Marxism, analysts sketch criminal punishment as a strategy and mechanism employed by the state to control a class whose interests potentially are threatening to capitalist structures and elites. Viewed in this way, the use of punishment may fluctuate with levels of street crime, but it also varies with prevailing economic conditions. When the economy is strong and the labor surplus shrinks, punishment is relaxed; in time of economic stagnation or crisis, when the labor surplus grows larger, official use of punishment rises. It is during these times that the structures of criminal justice draw off increasing numbers of those now rendered superfluous for production. This means that:

> increased use of imprisonment is not a direct response to any rise in crime, but is an ideologically motivated response to the perceived threat of crime posed by the swelling population of economically marginalized persons. This position does not deny the possibility of increasing crime accompanying unemployment, but

states instead that unemployment levels have an effect on the rate and severity of imprisonment *over and above* the changes in the volume and pattern of crime. (Box and Hale 1982:22)

With roots in pioneering work by Rusche and Kirchheimer (1939), there are several complementary theoretical explanations for the link between surplus labor and punishment. They variously emphasize economic, political and ideological forces, and they impute a variety of motives to elites and to criminal justice managers (Chiricos and Delone, 1992). Our theoretical point of departure is the general proposition that the unemployed are a threat or source of concern for dominant groups which is alleviated or otherwise managed by increased punishment. It is their presumed declining stake in conformity and their mounting desperation that make the unemployed the primary target of intensified punishment initiatives. Behind these crack-downs is elite anxiety, perhaps over potentially increasing political consciousness (Adamson 1984; Wallace 1980), class conflict (Melossi 1989), or rising levels of violent, expropriative street crime (Box and Hale 1982). A swelling mass of the unemployed is "social dynamite" (Spitzer 1975).

The structure of the American economy and the nature of American politics insure that both the shape and the dynamics of criminal justice reflect elite interests (Jacobs 1979). This requires neither the assumption that they conspire in the process or that they orchestrate the actions of criminal justice managers. The aggregate objective consequences of their anxiety are one thing; institutional dynamics and the motives of criminal justice practitioners are another. Remarkably little is known, however, about mechanisms and processes by which elite concerns may be communicated to and acted upon by control managers and apparatchiks. This is an area in which Marxist theories of social control lack specificity and precision.

Increasing anxiety and resentment in the ranks of criminal justice may also contribute to harsher punishment during economic downturns and times of rising unemployment. Squeezed fiscally between increases in the cost of living and their marginal, stagnant salaries, functionaries find new merit in the notion that severe penalties are needed to counteract the heightened temptations of illicit activity caused by hard times. The widespread belief that unemployment causes crime and that severe punishment deters underlies an increasing proportion of their decisions. Day-to-day they do what they can to increase the odds that crime does not become an alternative to economic hardship. The end result of their countless individual decisions is increased severity of punishment. Thus, part of the aggregate-level escalation of punishment may be an unintended consequence of employees in control bureaucracies applying conventional assumptions to crime control. Evidence suggests that for individual defendants, judicial decisions to incarcerate vary significantly by employment status (Chiricos and Bales 1991). Judges apparently view steady employment as an indicator of stability and unemployment as a sign of potential future trouble. Their actions may reassure elites even if this is not their intent. The relationship between labor surplus and penal sanctions requires assumptions about neither conspiracy nor specific direction.

The expanding crime-control apparatus that often accompanies the transformation and growth of punishment aids in the maintenance of stability and social order by providing jobs and a secure legitimate income for increasing numbers of the economically marginalized (Christie 1994). The criminal justice system, therefore, plays a dual role in managing the disadvantaged, desperate, and potentially lawless; in addition to incapacitation, employment opportunities provided by the expansion of criminal justice function as a relief valve for social discontent.

When the economy is strong and unemployment is low, institutional growth in crime control may level off, use of punishment is relaxed, and inclusionary crime-control approaches gain support from elected officials and state managers (Cohen 1985). This explanation for the changing use of punishment is consistent with the growth of rehabilitative ideologies and strategies in the United States during the years of post-World War II prosperity. It also helps explain why economic and structural transformations accompanying growth of the global economy, and the generalized anxiety they produce, has all but ended elected officials' public support for those "softer" crime-control approaches.

The preponderance of evidence from studies of the labor-surplus/punishment nexus supports conflict-theoretical explanations, even when fluctuation in street crime is controlled (Chiricos and Delone 1992). Supportive evidence is provided, first, by nation-level studies both in Europe and in the United States which operationalize punishment as the rate of imprisonment (Box and Hale 1982; 1985; Jankovic 1977; Laffargue and Godefroy 1989; Wallace 1980). Not all investigators report a significant relationship between unemployment and imprisonment (Jacobs and Helms 1996), but a substantial majority do. Results from state-level studies of the unemployment/imprisonment nexus are mixed; studies that employ longitudinal methods generally find the strongest support for the hypothesized relationship, while cross-sectional studies report more contradictory findings (Chiricos and Delone 1992).

Despite the generally confirmatory results of past research, there are reasons to question the labor-surplus/punishment relationship. To begin, methodological considerations suggest that nations and states may not be optimal units of analysis for investigating it. Since larger geographic units generally are more heterogeneous than smaller ones, national-level data are particularly likely to aggregate heterogeneity and mask substantial regional variation. This can confound and obscure empirical relationships of theoretical interest. In the United States, there is considerable intra-state variation in economic, demographic, crime and punishment variables. Analytically, state-level studies usually regress prison population variables on state demographic indicators despite the fact that inmates are not drawn randomly from its population, but largely from urbanized areas.

There is a theoretical reason as well for questioning the use of nations and states as units of analysis. Punishment policies generally are made at federal and state levels, but punishment is dispensed normally by *local* prosecutors and judges. Most serve local constituencies, and local political, structural and labor-market conditions likely constrain their decisions. Investigators are correct to use state-level data to examine variation in punishment policy (Link and Shover 1986; Barlow, Barlow, and Johnson 1996). Counties or SMSAs, however, may be a more appropriate unit of analysis for examining the relationship between labor surplus and punishment (Colvin 1990; Jankovic 1977; McCarthy 1990). Consistent with findings from national- and many state-level investigations, the small number of county-level studies published thus far report a positive relationship between unemployment and the use of imprisonment (Chiricos and Delone 1992).

Past county-level studies unfortunately are flawed by methodological shortcomings that limit confidence in theoretical understanding of the labor-surplus/punishment nexus. The theory linking historical change in punishment with change in the economy is a temporally dynamic one: as labor surplus increases criminal justice cracks down. Past county-level examinations of variation in use of imprisonment have employed cross-sectional analytic techniques that cannot assess these dynamic effects. Longitudinal techniques are required. Historically, problems of missing, inaccurate or inconsistently recorded information plagued county-level data. These shortcomings made investigators slow to use county-level data to examine justice issues. Complete and reliable county-level data became available in manageable format only recently. Despite an abundance of longitudinal national- and state-level studies of labor surplus and imprisonment, there are no dynamic spatial investigations at the county level.

The shortcomings of previous research diminish confidence in the underlying theoretical construction of the link between economic conditions and punishment. The present longitudinal study may help to rectify this. We test for a direct effect of *change* in the size of the labor surplus on *change* in the use of punishment while controlling for fluctuation in street crime and other variables. Thus, our methodology enables us to examine temporally dynamic causal relationships, and our use of counties as the units of analysis permits a test of the theoretical problem at the most appropriate aggregation.

Data and Methods

From correspondence with top-level criminal justice managers in the 50 states, we learned that 16 states could provide the requisite annual county-level prison commitment data. We began by selecting ten of these for inclusion in our state sample. We chose states from all regions of the United States, including only states with complete and apparently accurate data, and that would not require potentially time-consuming additional requirements to secure the needed data. The sample of states includes California, New Jersey, Ohio, Nebraska, Wisconsin, Illinois, Michigan, Mississippi, North Carolina, and South Carolina. For each state, we then selected from the listing of counties published in the *Uniform Crime Reports* all counties within designated Standard Metropolitan Statistical Areas and all counties with a 1980 total population of more than 25,000 (United States Department of Justice 1980). (Because we are interested in how closely our resulting sample of 269 counties approximates characteristics of United States counties, we compared them to all 1,409 counties of similar size in the United States in 1980 [United States Bureau of the Census 1994c].) As Table 1 shows, we found that, save for population size (sample counties were somewhat larger than the population), the sample compares closely with the population.

Nevertheless, the fact that the relationship between our sample of 269 counties and populations of theoretical or policy significance is unknown mandates caution in generalizing from the findings. Data were collected for the years 1980 and 1990, principally from official state and federal records.

As most investigators have done, we use the official rate of unemployment as our measure of labor surplus. In doing so, we are not unmindful of the belief that it is an

Comparison of Sample and All Counties with Populations 25,000 and Over   **TABLE 1**

Variable	Sample	Population
Mean population size	134,266	120,868
Proportion white	.86	.88
Proportion unemployed	.08	.08
Poverty rate	.13	.14
Income per capita 1989	$ 12,689	$ 12,367

unsatisfactory measure of the true level of unemployment in a community. It does not, for example, include unemployed men and women who have ceased searching actively for a job. The limitations of our data, however, do not permit us to construct an alternative measure of the size of the labor surplus.

Since variation in state use of punishment generally is attributed to variation in street crime, we included in the analysis measures of both violent and nonviolent crimes known to the police. We also included as control variables socio-demographic characteristics that may contribute to the rate of prison commitments, chief among them the proportion of young adult males in the population (Cohen and Land 1987; Inverarity and McCarthy 1988). Since crime and imprisonment are experiences disproportionately characteristic of young males, counties with a high percentage of young men in their population generally have higher crime rates and, consequently, more imprisonment (Blumstein 1983).

One of the most important changes in America's response to crime in the past 15 years is the dramatic increase of attention and resources devoted to drug-law enforcement. One indicator of this is a sharp increase in the proportion of the imprisoned population serving time for drug offenses (United States Bureau of Justice Statistics 1996). In light of this development, it would be useful to include as controls county-level arrests and prison commitments for drug crimes. The necessary data are not available. Myers and Inverarity (1992) show, however, that increasing arrests from drug crimes do not mediate unemployment's relationship to or explain changing rates in state-level imprisonment.

Inclusion of crime rates in our analysis controls for the effect of age on imprisonment that is mediated by the crime rate. The effect of age on imprisonment should be minimal if the conventional assumption that the age of the population affects imprisonment via crime is true. But age structure also may influence imprisonment directly, particularly if, as seems likely, the population perceived as most dangerous by political-economic elites is young males with restricted access to legitimate labor markets (Box and Hale 1982). Apart from any real threat from crime, large numbers of young males in a county may effect imprisonment by creating the perception of a threat from a population believed to be aggressive and difficult to control (Tittle and Curran 1988). The effect of a county's age structure on imprisonment after controlling for crime is interpreted as a reflection of this age-threat process. We used age data both to control for the proportion of the population composed of males ages 20–34 and to test for this direct effect.

The use of imprisonment varies directly with the size of the non-white population (Carroll and Doubet 1983; Joubert, Picou, and MacIntosh 1981). Like the young and the unemployed, non-whites may be perceived as particularly threatening, restive and potentially criminal. The presence of non-whites is viewed by some criminal justice officials as an indicator of a crime problem and, therefore, increases the use of crime control and imprisonment. Although there is some evidence that blacks receive longer sentences than whites for similar offenses (e.g., Spohn 1994), other studies suggest that the degree of discrimination against non-whites in sentencing and incarceration varies by social and economic context (Myers and Sabol 1987; Myers and Talarico 1986). For these reasons, we also controlled for the proportionate size of the non-white population.

Other measures of the economic health of a community may influence the use of imprisonment. Marxist theorists suggest that the poor are perceived to be a potential threat to social order. The effect of the impoverished working population on imprisonment is not reflected in unemployment rates. Increases in the proportionate size of the impoverished population may have an effect on change in imprisonment similar to unemployment. Consequently, poverty rates are included as a control variable, principally because they are a reasonable measure of an employed underclass. Poverty rates, however, represent only the percentage of a county's population who are officially poor and are not an indicator of the amount of wealth available to its families. Income is a better measure than poverty of the economic situation faced by them. A county's average personal income in 1980 dollars is included as a control variable to measure each county's economic health.

Given the methodological shortcomings of previous studies, we opted for statistical procedures that permit examination of changes in multiple cases at a few points in time. We used a panel design and analytic technique. Community values, traditions, cultures and institutional inertia all have an impact on both types and amounts of punishment employed against convicted offenders. Panel designs can account for both temporal and geographic variation. By observing the same counties at two points in time we insure that similar extraneous variables are in play in both time periods. We also can observe and analyze how change in some variables contributes to change in others and can even control for ongoing patterns of change common to all counties. A panel design permits us to investigate the effects of change in independent variables on change in imprisonment over the decade.

We used residual-change regression analysis to estimate changes in the level of variables in the panel from 1980 to 1990. This technique, which makes use of residual-change scores, has been employed to examine a variety of social problems (Bursik and Webb 1982; Chamlin 1992; Elliott and Voss 1974). To derive a residual-change score, the level of a variable in 1990 is regressed on its level in 1980. The equation then is used to predict the level of each variable in 1990. Subtracting the predicted value from the observed value in 1990 yields a measure of residual change. Residual-change scores have two properties useful for this research. First, they provide a measure of change that is statistically independent of a variable's initial levels, removing a variable's initial level's effect on the subsequent level of that same variable. The result represents change that is not expected on the basis of the variable's initial level alone (Bohrnstedt 1969). Residual-change regression permits an examination of how change in the levels of independent variables affect change in imprisonment.

Second, residual-change scores adjust for changes that other counties have undergone. They control the effects of trends common to all counties to determine change attributable to the variables of interest in a particular county. Since all 269 counties are used to estimate the regression equation which predicts the levels in 1990, the predicted values are automatically adjusted for change that other counties have undergone during the decade. Changes that occur across counties are controlled, leaving each county's unique change. We examine change by using two waves of data from 1980 and 1990 to determine change in socio-economic variable's contribution to change in imprisonment. This involves regressing the residual-change scores for imprisonment on the residual-change scores of the other variables. Changes in the independent variables, theoretically, should find expression in changes in imprisonment. This equation takes the form:

$$\text{Imprisonment}_{res:t,t-10} = f(\text{unemployment}_{res:t,t-10}, \text{violent crime}_{res:t,t-10}, \text{property crime}_{res:t,t-10}, \text{percent non-white}_{res:t,t-10}, \text{age}_{res:t,t-10}, \text{income}_{res:t,t-10}, \text{poverty}_{res:t,t-10})$$

The subscripts for independent variables indicate that they are residual-change transformations. $\text{Imprisonment}_{res:t,t-10}$ is the residual-change score of prison commitment during the 10-year period.

Failure to find a positive, significant contribution of unemployment to county prison commitments would cast doubt on the notion of imprisonment as a response to the unemployed's threats to elites. The presence of significant effects for crime in the absence of significance for unemployment will not support the hypothesis of an independent effect of unemployment on imprisonment.

Results

Recall that we expect imprisonment to covary positively with all the independent variables. To test this, we begin by examining the relationships between change in the rate of prison commitments, in unemployment, in crime, and in the other control variables between 1980 and 1990. As predicted, the results reported in Table 2 show that change in the rate of unemployment (b = .169), violent crime rates (b = .147), and the proportion of males age 20–34 in counties' population (b = .267) are related positively to change in imprisonment rates. The finding which has the greatest bearing on our research question, of course, is that change in unemployment is an independently significant predictor of change in imprisonment. Change in unemployment is related to change in use of imprisonment over and above the effects of crime and the other control variables. Change in property crime, percent non-white, poverty rates, and average income do not produce change in imprisonment. Because some evidence suggests very high non-white population rates may decrease crime control aimed at minorities (Liska and Chamlin 1984), we also tested for a curvilinear effect for non-white. None was found.

To determine if our results were biased by collinearity, a common problem in research using economic predictors, we examined zero-order correlations and performed standard regression diagnostics (Belsley, Kuh, and Welsh 1980). Some high zero-order correlations between predictor variables were identified, particularly between violent and property crime (r = .45) and between poverty and income (r = .49). These correlations do not necessarily indicate the presence of collinearity, but they did warrant further investigation. Subsequent collinearity diagnostics yielded

TABLE 2   Summary Table for Regression of Changes in Imprisonment on Changes in Independent Variables, 1980–1990

Variable	B	Beta
Unemployment	.010	.169**
Violent crime	.054	.147*
Property crime	−.003	−.035
Males 20–34	.013	.267***
Non-white	.002	.051
Poverty	−.004	−.102
Income	0.000	.034
$R^2 = .139$	Constant = .084*	

*p ≤ .05   **p ≤ .01   ***p ≤ .005

low VIF values and low condition indices, which indicates multicollinearity is not a concern in this equation.

The explained variance of our model is low ($R^2 = .14$). One possible explanation of this is that there is little change left to explain because imprisonment levels in 1980 explain most of the variation in 1990. We checked to see if this was the source of our small R^2 by regressing 1990 imprisonment on the full set of predictors and on 1980 imprisonment. The results were clear; the single best predictor by far was 1980 imprisonment ($R^2 = .74$). By comparison, other effects were negligible. This suggests either that there is very little change from 1980 to 1990 or that the pattern of change is similar across counties. Since 1990 imprisonment levels are accounted for by 1980 levels, only a small amount of change remains unexplained, and the low explained variance in our results can still be viewed as substantively important.

Conclusions and Implications

Our findings can be summarized briefly. Change in violent street crime, in the proportionate size of the young male population, and in labor surplus contribute to change in the use of imprisonment while changing levels of property crime do not. These relationships persist even when street-crime rates and other presumed correlates of imprisonment are controlled. Our analysis, therefore, confirms findings from earlier investigations of the relationship between labor surplus and punishment. The criminal justice system grows increasingly punitive as labor surplus increases. The fact that our findings were achieved using both a unit of analysis more appropriate theoretically than measures employed by most investigators and a longitudinal design only strengthen confidence in them.

The observed relationship between violent street crime and punishment is consistent with results obtained by other investigators (e.g., Inverarity and McCarthy 1988). That our findings differed for violent crime and property crime reinforces the importance of disaggregating crime rates in macro-level research. The relationship between the proportionate size of the young male population and punishment is not surprising. The fact that young males commit the majority of street crime means that in the aggregate they probably symbolize the threat of crime and disorder.

Although our principal objective has been a conflict interpretation of the relationship among street crime, labor surplus and punishment, the significance of our investigation is more than theoretical. At a time when public schools in many regions of the United States are under severe budgetary constraints, when major components of the nation's infrastructure have eroded, and millions of citizens cannot secure quality health care, expenditure of tax revenues for crime control has skyrocketed. It is only through a better understanding of the sources of these changes that we can predict their likely development or have any hope of controlling them. The findings of this study and others like it suggest an explanation for why past predictions about fluctuations in punishment that failed to include projected rates of unemployment or other economic measures have proven inaccurate.

REFERENCES

Adamson, Christopher. 1984. "Toward a Marxian penology: Captive criminal populations as economic threats and resources." *Social Problems* 31: 435–458.

Barlow, David E., Melissa Hickman Barlow, and W. Wesley Johnson. 1996. "The political economy of criminal justice policy: A time-series analysis of economic conditions, crime and federal criminal justice legislation, 1948–1987." *Justice Quarterly* 13: 223–242.

Belsley, David A., Edwin Kuh, and Roy E. Welsh. 1980. *Regression Diagnostics: Identifying Influential Data and Sources of Collinearity.* New York: John Wiley & Sons.

Blumstein, Alfred. 1983. "Prisons: Population, capacity, and alternatives." In *Crime and Public Policy,* J. Q. Wilson (ed.), 229–250. San Francisco: ICS.

Bohrnstedt, George W. 1969. "Observations on the measurement of change." In *Sociological Methodology 1969,* E. F. Borgata and G. W. Bohrnstedt (eds.), 113–136. San Francisco: Jossey-Bass.

Box, Steven, and Chris Hale. 1982. "Economic crisis and the rising prisoner population in England and Wales." *Crime and Social Justice* 17: 20–35.

———. 1985. "Unemployment, imprisonment and prison overcrowding." *Contemporary Crises* 9: 209–228.

Bursik, Robert J., and Jim Webb. 1982. "Community change and patterns of delinquency." *American Journal of Sociology* 88: 24–42.

Carroll, Leo, and Mary Beth Doubet. 1983. "U.S. social structure and imprisonment." *Criminology* 21: 449–456.

Chamlin, Mitchell B. 1992. "Intergroup threat and social control: Welfare expansion among states during the 1960s and 1970s." In *Social Threat and Social Control,* A. E. Liska (Ed.), 151–164. Albany: State University of New York Press.

Chiricos, Theodore G., and William D. Bales. 1991. "Unemployment and punishment: An empirical assessment." *Criminology* 29: 701–724.

Chiricos, Theodore G., and Miriam A. Delone. 1992. "Labor surplus and punishment: A review and assessment of theory and evidence." *Social Problems* 39: 421–446.

Christie, Nils. 1994. *Crime Control as Industry: Toward GULAGS Western Style.* New York: Routledge.

Cohen, Stanley. 1985. *Visions of Social Control.* Cambridge, U.K.: Polity.

Cohen, Lawrence E., and Kenneth L. Land. 1987. "Age structure and crime: Symmetry versus asymmetry and the projection of crime rates through the 1990s." *American Sociological Review* 52: 170–183.

Colvin, Mark. 1990. "Labor markets, industrial monopolization, welfare and imprisonment: Evidence from a cross section of U.S. counties." *Sociological Quarterly* 31: 440–456.

Elliot, Delbert S., and Harwin L. Voss. 1974. *Delinquency and Dropout.* Lexington, Mass.: Heath.

Inter-university Consortium for Political and Social Research. 1991. *Uniform Crime Report: County Level Arrest and Offenses Data.* Ann Arbor: University of Michigan.

Inverarity, James, and Daniel McCarthy. 1988. "Punishment and social structure revisited: Unemployment and imprisonment in the U.S., 1948–1984." *Sociological Quarterly* 29: 263–279.

Jacobs, David. 1979. "Inequality and police force strength: Conflict theory and coercive control in metropolitan areas." *American Sociological Review* 44: 913–925.

Jacobs, David, and Ronald E. Helms. 1996. "Toward a political model of incarceration: A time-series examination of multiple explanations for prison admission rates." *American Journal of Sociology* 102: 323–357.

Jankovic, Ivan. 1977. "Labor market and imprisonment." *Crime and Social Justice* 8: 17–31.

Joubert, Paul E., J. Steven Picou, and Alex McIntosh. 1981. "U.S. social structure, crime, and imprisonment." *Criminology* 19: 344–359.

Laffargue, Bernard, and Thiery Godefroy. 1989. "Economic cycles and punishment." *Contemporary Crises* 13: 371–404.

Link, Christopher T., and Neal Shover. 1986. "The origins of criminal sentencing reforms." *Justice Quarterly* 3: 329–341.

Liska, Allen E., and Mitchell B. Chamlin. 1984. "Social structure and crime control among macrosocial units." *American Journal of Sociology* 90: 383–395.

McCarthy, Belinda. 1990. "A micro-level analysis of social structure and social control: Intrastate use of jail and prison confinement." *Justice Quarterly* 7: 325–340.

Melossi, Dario. 1989. "An introduction: Fifty years later, punishment and social structure in comparative analysis." *Contemporary Crises* 13: 311–326.

Myers, Greg, and James Inverarity. 1992. "Strategies of disaggregation in imprisonment rate research." Presented at the annual meeting of the American Society of Criminology.

Myers, Martha A., and Susette M. Talarico. 1986. "The social context of racial discrimination in sentencing." *Social Problems* 33: 236–251.

Myers, Samuel L., Jr., and William J. Sabol. 1987. "Business cycles and racial disparities in punishment." *Contemporary Policy Issues* 5: 46–58.

Rusche, Georg, and Otto Kirchheimer. 1939. *Punishment and Social Structure.* New York: Columbia University Press.

Spitzer, Steven. 1975. "Toward a Marxian theory of deviance." *Social Problems* 22: 638–651.

Spohn, Cassia. 1994. "Crime and the social control of blacks: Offender/victim race and the sentencing of violent offenders." In *Inequality, Crime, and Social Control,* G. S. Bridges and M. Myers (eds.), 249–268. Boulder, Colo.: Westview.

Tittle, Charles R., and Debra A. Curran. 1988. "Contingencies for dispositional disparities in juvenile justice." *Social Forces* 67: 23–58.

U.S. Bureau of the Census. 1994a. *Revised Estimates of County Population Characteristics 1980–1989.* Washington, D.C.: Estimates Division, U.S. Bureau of the Census.

———. 1994b. *Modified Age, Race and Sex.* Washington, D.C.: Estimates Division, U.S. Bureau of the Census.

———. 1994c. County and City Data Book: U.S.A. Counties. (CD-ROM). Washington, D.C.: U.S. Government Printing Office.

U.S. Bureau of Economic Analysis. 1994. Regional Economic Information System 1969–1993. (CD-ROM). Washington, D.C.: U.S. Government Printing Office.

U.S. Bureau of Justice Statistics. 1996. *Correctional Populations in the United States.* 1994. Washington, D.C.: U.S. Government Printing Office.

U.S. Bureau of Labor Statistics. 1992. *The Consumer Price Index: Questions and Answers.* Washington, D.C.: U.S. Government Printing Office.

U.S. Department of Justice. 1980. *Uniform Crime Reports for the United States.* Washington, D.C.: U.S. Government Printing Office.

Wallace, Don. 1980. "The political economy of incarceration trends in late U.S. capitalism." *Insurgent Sociologist* 9: 59–65.

QUESTIONS FOR DISCUSSION

1. From a neo-Marxist view, the punishment of offenders is relaxed when the economy is strong and the labor surplus shrinks and punishment of offenders increases when the economy is stagnant and the labor surplus increases. Does this apply today?

2. How do the author's findings reflect a neo-Marxist perspective?

3. What is the relationship among street crimes, labor surplus, and criminal punishment? What evidence can you provide to support your response?

The Social Reality of Crime

Richard Quinney

A theory that helps us begin to examine the legal order critically is the one I call the *social reality of crime*. Applying this theory, we think of crime as it is affected by the dynamics that mold the society's social, economic, and political structure. First, we recognize how criminal law fits into capitalist society. The legal order gives reality to the crime problem in the United States. Everything that makes up crime's social reality, including the application of criminal law, the behavior patterns of those who are defined as criminal, and the construction of an ideology of crime, is related to the established legal order. The social reality of crime is constructed on conflict in our society.

The theory of the social reality of crime is formulated as follows.

    I.   THE OFFICIAL DEFINITION OF CRIME: *Crime as a legal definition of human conduct is created by agents of the dominant class in a politically organized society.*

The essential starting point is a definition of crime that itself is based on the legal definition. Crime, as *officially* determined, is a *definition* of behavior that is conferred on some people by those in power. Agents of the law (such as legislators, police, prosecutors, and judges) are responsible for formulating and administering criminal law. Upon *formulation* and *application* of these definitions of crime, persons and behaviors become criminal.

Crime, according to this first proposition, is not inherent in behavior, but is a judgment made by some about the actions and characteristics of others. This proposition allows us to focus on the formulation and administration of the criminal law as it applies to the behaviors that become defined as criminal. Crime is seen as a result of the class-dynamic process that culminate in defining persons and behaviors as criminal. It follows, then, that the greater the number of definitions of crime that are formulated and applied, the greater the amount of crime.

II.   FORMULATING DEFINITIONS OF CRIME: *Definitions of crime are composed of behaviors that conflict with the interests of the dominant class.*

Definitions of crime are formulated according to the interests of those who have the power to translate their interests into public policy. Those definitions are ultimately incorporated into the criminal law. Furthermore, definitions of crime in a society change as the interests of the dominant class change. In other words, those who are able to have their interests represented in public policy regulate the formulation of definitions of crime.

The powerful interests are reflected not only in the definitions of crime and the kinds of penal sanctions attached to them, but also in the *legal policies* on handling those defined as criminals. Procedural rules are created for enforcing and administering the criminal law. Policies are also established on programs for treating and punishing the criminally defined and programs for controlling and preventing crime. From the initial definitions of crime to the subsequent procedures, correctional and penal programs, and policies for controlling and preventing crime, those who have the power regulate the behavior of those without power.

III.   APPLYING DEFINITIONS OF CRIME: *Definitions of crime are applied by the class that has the power to shape the enforcement and administration of criminal law.*

The dominant interests intervene in all the stages at which definitions of crime are created. Because class interests cannot be effectively protected merely by formulating criminal law, the law must be enforced and administered. The interests of the powerful, therefore, also operate where the definitions of crime reach the *application* stage. As Vold has argued, crime is "political behavior and the criminal becomes in fact a member of a 'minority group' without sufficient public support to dominate the control of the police power of the state." Those whose interests conflict with the ones represented in the law must either change their behavior or possibly find it defined as criminal.

The probability that definitions of crime will be applied varies according to how much the behaviors of the powerless conflict with the interests of those in power. Law enforcement efforts and judicial activity are likely to increase when the interests of the dominant class are threatened. Fluctuations and variations in applying definitions of crime reflect shifts in class relations.

Obviously, the criminal law is not applied directly by those in power; its enforcement and administration are delegated to authorized *legal* agents. Because the groups responsible for creating the definitions of crime are physically separated from the groups that have the authority to enforce and administer law, local conditions determine how the definitions will be applied. In particular, communities vary in their expectations of law enforcement and the administration of justice. The application of definitions is also influenced by the visibility of offenses in a community and by the public's norms about reporting possible violations. And especially important in enforcing and administering the criminal law are the legal agents' occupational organization and ideology.

The probability that these definitions will be applied depends on the actions of the legal agents who have the authority to enforce and administer the law. A definition of crime is applied depending on their evaluation. Turk has argued that during "criminalization," a criminal label may be affixed to people because of real or fancied attributes: "Indeed, a person is evaluated, either favorably or unfavorably, not because he *does* something, or even because he *is* something, but because others react to their perceptions of him as offensive or inoffensive." Evaluation by the definers is affected by the way in which the suspect handles the situation, but ultimately the legal agents' evaluations and subsequent decisions are the crucial factors in determining the criminality of human acts. As legal agents evaluate more behaviors and persons as worthy of being defined as crimes, the probability that definitions of crime will be applied grows.

> IV. How Behavior Patterns Develop in Relation to Definitions of Crime: *Behavior patterns are structured in relation to definitions of crime, and within this context people engage in actions that have relative probabilities of being defined as criminal.*

Although behavior varies, all behaviors are similar in that they represent patterns within the society. All persons—whether they create definitions of crime or are the objects of these definitions—act in reference to *normative systems* learned in relative social and cultural settings. Because it is not the quality of the behavior but the action taken against the behavior that gives it the character of criminality, that which is defined as criminal is relative to the behavior patterns of the class that formulates and applies definitions. Consequently, people whose behavior patterns are not represented when the definitions of crime are formulated and applied are more likely to act in ways that will be defined as criminal than those who formulate and apply the definitions.

Once behavior patterns become established with some regularity within the segments of society, individuals have a framework for creating *personal action patterns*. These continually develop for each person as he moves from one experience to another. Specific action patterns give behavior an individual substance in relation to the definitions of crime.

People construct their own patterns of action in participating with others. It follows, then, that the probability that persons will develop action patterns with a high potential for being defined as criminal depends on (1) structured opportunities, (2) learning experiences, (3) interpersonal associations and identifications, and (4) self-conceptions. Throughout the experiences, each person creates a conception of self as a human social being. Thus prepared, he behaves according to the anticipated consequences of his actions.

In the experiences shared by the definers of crime and the criminally defined, personal-action patterns develop among the latter because they are so defined. After they have had continued experience in being defined as criminal, they learn to manipulate the application of criminal definitions.

Furthermore, those who have been defined as criminal begin to conceive of themselves as criminal. As they adjust to the definitions imposed upon them, they learn to play the criminal role. As a result of others' reactions, therefore, people may develop personal-action patterns that increase the likelihood of their being defined as criminal in the future. That is, increased experience with definitions of crime increases the probability of their developing actions that may be subsequently defined as criminal.

Thus, both the definers of crime and the criminally defined are involved in reciprocal action patterns. The personal-action patterns of both the definers and the defined are shaped by their common, continued, and related experiences. The fate of each is bound to that of the other.

V.   CONSTRUCTING AN IDEOLOGY OF CRIME: *An ideology of crime is constructed and diffused by the dominant class to secure its hegemony.*

This ideology is created in the kinds of ideas people are exposed to, the manner in which they select information to fit the world they are shaping, and their way of interpreting his information. People behave in reference to the *social meanings* they attach to their experiences.

Among the conceptions that develop in a society are those relating to what people regard as crime. The concept of crime must of course be accompanied by ideas about the nature of crime. Images develop about the relevance of crime, the offender's characteristics, the appropriate reaction to crime, and the relation of crime to the social order. These conceptions are constructed by communication, and, in fact, an ideology of crime depends on the portrayal of crime in all personal and mass communication. This ideology is thus diffused throughout the society.

One of the most concrete ways by which an ideology of crime is formed and transmitted is the official investigation of crime. The President's Commission on Law Enforcement and Administration of Justice is the best contemporary example of the state's role in shaping an ideology of crime. Not only are we as citizens more aware of crime today because of the President's Commission, but official policy on crime has been established in a crime bill, the Omnibus Crime Control and Safe Streets Act of 1968. The crime bill, itself a reaction to the growing fears of class conflict in American society, creates an image of a severe crime problem and, in so doing, threatens to negate some of our basic constitutional guarantees in the name of controlling crime.

Consequently, the conceptions that are most critical in actually formulating and applying the definitions of crime are those held by the dominant class. These conceptions are certain to be incorporated into the social reality of crime. The more the government acts in reference to crime, the more probable it is that definitions of crime will be created and that behavior patterns will develop in opposition to those definitions. The formulation of definitions of crime, their application, and the development of behavior patterns in relation to the definitions, are thus joined in full circle by the construction of an ideological hegemony toward crime.

VI.   CONSTRUCTING THE SOCIAL REALITY OF CRIME: *The social reality of crime is constructed by the formulation and application of definitions of crime, the development of behavior patterns in relation to these definitions, and the construction of an ideology of crime.*

The first five propositions are collected here into a final composition proposition. The theory of the social reality of crime, accordingly, postulates creating a series of phenomena that increase the probability of crime. The result, holistically, is the social reality of crime.

Because the first proposition of the theory is a definition and the sixth is a composite, the body of the theory consists of the four middle propositions. These form a model of crime's social reality. The model relates the proposition units into a theoretical system. Each unit is related to the others. The theory is thus a system of interacting developmental propositions. The phenomena denoted in the propositions and their relationships culminate in what is regarded as the amount and character of crime at any time—that is, in the social reality of crime.

The theory of the social reality of crime as I have formulated it is inspired by a change that is occurring in our view of the world. This change, pervading all levels of society, pertains to the world that we all construct and from which, at the same time, we pretend to separate ourselves in our human experiences. For the study of crime, a revision in thought has directed attention to the criminal process: All relevant phenomena contribute to creating definitions of crime, development of behaviors by those involved in criminal-defining situations, and constructing an ideology of crime. The result is the social reality of crime that is constantly being constructed in society.

QUESTIONS FOR DISCUSSION

1. According to the author, how do the official definition of crime and the formulating definitions of crime develop?
2. The theory of the social reality of crime is formulated into six propositions. Discuss each proposition, using examples, in relation to how crime has developed in America.
3. What are the general processes that Quinney identifies in the creation of law?

QUESTIONS FOR PART 2

1. Social deviance can be found at every college and university. Discuss how Chapter 4 could explain deviant behavior at your college, university, or geographical location.
2. Some people believe that social deviance is a lower class phenomenon. Discuss how Chapter 5 may support this assertion.

PART 3

EXPLAINING DEVIANT BEHAVIOR: MICRO PERSPECTIVES

While the macro approach looks at deviance from a broad perspective, the micro approach looks at face-to-face encounters, human interaction, and small groups to explain deviant behavior. In Part 3, a number of selections are chosen to address the labeling, learning, control, and feminist perspectives.

Some theories incorporate a structural element (emphasizing a relationship to certain structural conditions within society) whereas others use a processual element (the process in which individuals commit deviant acts). The selections chosen for Part III illustrate both structural and processual elements. Labeling theory focuses on processual elements (causes of deviant acts), control theory focuses more on structural elements (the distribution of deviance in time and space), and both learning and feminists' perspectives incorporate structural and processual elements.

Whether we use structural or processual elements, there are still debates in sociology of deviance on why an individual becomes deviant. Some observers concentrate on what kinds of people become deviant whereas others concentrate on the process by which a person becomes deviant. Edwin M. Lemert develops the idea that an individual becomes deviant due to being labeled. In other words, if labels are successfully applied and an individual's legitimate roles are reorganized toward deviant activities, a deviant role is adopted. Lemert refers to the latter roles as secondary deviation, which is developed as a means of adjustment to societal reaction to the original (primary) deviation.

Many deviant acts can be viewed as learned behavior. Edwin H. Sutherland and Donald R. Cressey contribute a concise statement of one of the most popular learning theories of deviance. This classic article purports to explain criminal and noncriminal behavior of individuals in nine propositions. Sutherland and Cressey explore how a sociable, gregarious, active, and athletic young male can become delinquent while an isolated, introverted, and psychopathic young male does not engage in delinquent behavior.

Some deviance can be explained by looking at properties of subcultures. The subculture of violence thesis is one of the least tested and most cited perspectives in criminological and sociological research. Marvin Wolfgang and Franco Ferracuti developed a subculture of violence thesis to explain African Americans' violent behavior. In Liqun Cao, Anthony Adams, and Vickie J. Jensen's article, the authors test Wolfgang and Ferracuti's thesis of some widespread, violent values among African Americans. Contrary to Wolfgang and Ferracuti's thesis, Cao, Adams, and Jensen found white males were more likely to express violent tendencies. While this hardly resolves the debate, the absence of unambiguous empirical data supporting the subculture of violence thesis is noteworthy.

Control theorists add a different piece to the deviance puzzle. Control theorists assume that delinquent behavior is a result of weak or broken bonds with society. The most celebrated statement of modern Durkheimian themes on the importance of social integration and control is Travis Hirschi's control theory. Hirschi does not look at why individuals are delinquent, but rather why they are not delinquent. His explanation is that individuals are not delinquent due to their attachment, commitment, involvement, and belief in and to society.

In a later article, Michael Gottfredson and Hirschi focus on an individual cause of criminal behavior: low self-control. Gottfredson and Hirschi integrate social learning and psychological predisposition in the explanation of deviance. They expand the general argument that delinquent and criminal behavior result from inadequate social control by emphasizing a lack of personal self-control. According to Gottfredson and Hirschi, delinquent and criminal behavior is an immediate and often gratifying behavior with few long-term benefits.

Kimberly L. Kempf's article summarizes much of the literature on control theory. She addresses various issues, including the generalizability of the elements of social bonds, whether the theory explains both general deviance and specific forms of delinquency, and whether social control is equally applicable across socioeconomic groups, race, and age categories.

The feminist perspective of deviance and the study of females have traditionally been neglected. In Meda Chesney-Lind's article, the importance of public settings, the role played by the juvenile justice system, and historical aspects of female roles as they apply to delinquency are discussed. She also addresses androcentric biases in delinquency theories, compares female and male offending, and develops a feminist model of female delinquency.

The crime of rape is frequently associated with the development of a feminist model of female delinquency. Lois Copeland and Leslie R. Wolfe's article outlines and discusses statistical data on rape in a patriarchal society. The authors also discuss sexism and violence against women, flaws in the criminal justice system, and new definitions of rape. In contrast to Copeland and Wolfe's article, Neil Gilbert's article points out the flaws in the statistics and definitions of rape. He suggests that not only has the problem of sexual assaults been magnified but the data has also been misinterpreted. Gilbert explores the methodological flaws in the frequently cited Mary Koss study, how discrepancies in rape research have been ignored, and an issue with the "take back the night" movement.

CHAPTER 6

INTERACTIONIST AND LEARNING PERSPECTIVES

Primary and Secondary Deviation

Edwin M. Lemert

SOCIOPATHIC INDIVIDUATION

The deviant person is a product of differentiating and isolating processes. Some persons are individually differentiated from others from the time of birth onward, as in the case of a child born with a congenital physical defect or repulsive appearance, and as in the case of a child born into a minority racial or cultural group. Other persons grow to maturity in a family or in a social class where pauperism, begging, or crime are more or less institutionalized ways of life for the entire group. In these latter instances the person's sociopsychological growth may be normal in every way, his status as a deviant being entirely caused by his maturation within the framework of social organization and culture designated as "pathological" by the larger society. This is true of many delinquent children in our society.[1]

> It is a matter of great significance that the delinquent child, growing up in the delinquency areas of the city, has very little access to the cultural heritages of the larger conventional society. His infrequent contacts with this larger society are for the most part formal and external. Quite naturally his conception of moral values is shaped and molded by the moral code prevailing in his play groups and

the local community in which he lives . . . the young delinquent has very little
appreciation of the meaning of the traditions and formal laws of society. . . .
Hence the conflict between the delinquent and the agencies of society is, in its
broader aspects, a conflict of divergent cultures.

The same sort of gradual, unconscious process which operates in the socialization of the deviant child may also be recognized in the acquisition of socially unacceptable behavior by persons after having reached adulthood. However, with more
verbal and sophisticated adults, step-by-step violations of societal norms tend to be
progressively rationalized in the light of what is socially acceptable. Changes of this
nature can take place at the level of either overt or covert behavior, but with a greater
likelihood that adults will preface overt behavior changes with projective symbolic
departures from society's norms. When the latter occur, the subsequent overt changes
may appear to be "sudden" personality modifications. However, whether these
changes are completely radical ones is to some extent a moot point. One writer holds
strongly to the opinion that sudden and dramatic shifts in behavior from normal to
abnormal are seldom the case, that a sequence of small preparatory transformations
must be the prelude to such apparently sudden behavior changes. This writer is impressed by the day-to-day growth of "reserve potentialities" within personalities of
all individuals, and he contends that many normal persons carry potentialities for abnormal behavior, which, given proper conditions, can easily be called into play.[2]

Personality Changes Not Always Gradual

This argument is admittedly sound for most cases, but it must be taken into consideration that traumatic experiences often speed up changes in personality.[3] Nor can
the "trauma" in these experiences universally be attributed to the unique way in
which the person conceives of the experience subjectively. Cases exist to show that
personality modifications can be telescoped or that there can be an acceleration of
such changes caused largely by the intensity and variety of the social stimulation.
Most soldiers undoubtedly have entirely different conceptions of their roles after
intensive combat experience. Many admit to having "lived a lifetime" in a relatively short period of time after they have been under heavy fire in battle for the first
time. Many generals have remarked that their men have to be a little "shooted" or
"blooded" in order to become good soldiers. In the process of group formation,
crises and interactional amplification are vital requisites to forging true, role-oriented
group behavior out of individuated behavior.[4]

The importance of the person's conscious symbolic reactions to his or her own
behavior cannot be overstressed in explaining the shift from normal to abnormal behavior or from one type of pathological behavior to another, particularly where
behavior variations become systematized or structured into pathological roles. This
is not to say that conscious choice is a determining factor in the differentiating process. Nor does it mean that the awareness of the self is a purely conscious perception.
Much of the process of self-perception is doubtless marginal from the point of view
of consciousness.[5] But however it may be perceived, the individual's self-definition is
closely connected with such things as self-acceptance, the subordination of minor to
major roles, and with the motivation involved in learning the skills, techniques, and
values of a new role. *Self-definitions or self-realizations are likely to be the result of*

sudden perceptions and they are especially significant when they are followed imme-diately by overt demonstrations of the new role they symbolize. The self-defining junctures are critical points of personality genesis and in the special case of the atypi-cal person they mark a division between two different types of deviation.

Primary and Secondary Deviation

There has been an embarrassingly large number of theories, often without any rela-tionship to a general theory, advanced to account for various specific pathologies in human behavior. For certain types of pathology, such as alcoholism, crime, or stut-tering, there are almost as many theories as there are writers on these subjects. This has been occasioned in no small way by the preoccupation with the origins of patho-logical behavior and by the fallacy of confusing *original* causes with *effective* causes. All such theories have elements of truth, and the divergent viewpoints they contain can be reconciled with the general theory here if it is granted that original causes or antecedents of deviant behaviors are many and diversified. This holds especially for the psychological processes leading to similar pathological behavior, but it also holds for the situational concomitants of the initial aberrant conduct. A person may come to use excessive alcohol not only for a wide variety of subjective reasons but also be-cause of diversified situational influences, such as the death of a loved one, business failure, or participating in some sort of organized group activity calling for heavy drinking of liquor. Whatever the original reasons for violating the norms of the com-munity, they are important only for certain research purposes, such as assessing the extent of the "social problem" at a given time or determining the requirements for a rational program of social control. From a narrower sociological viewpoint the devi-ations are not significant until they are organized subjectively and transformed into ac-tive roles and become the social criteria for assigning status. The deviant individuals must react symbolically to their own behavior aberrations and fix them in their socio-psychological patterns. The deviations remain primary deviations or symptomatic and situational as long as they are rationalized or otherwise dealt with as functions of a socially acceptable role. Under such conditions normal and pathological behav-iors remain strange and somewhat tensional bedfellows in the same person. Undeni-ably a vast amount of such segmental and partially integrated pathological behavior exists in our society and has impressed many writers in the field of social pathology.

Just how far and for how long a person may go in dissociating his sociopathic tendencies so that they are merely troublesome adjuncts of normally conceived roles is not known. Perhaps it depends upon the number of alternative definitions of the same overt behavior that he can develop; perhaps certain physiological factors (lim-its) are also involved. However, if the deviant acts are repetitive and have a high visibility, and if there is a severe societal reaction, which, through a process of identification is incorporated as part of the "me" of the individual, the probability is greatly increased that the integration of existing roles will be disrupted and that reorganization based upon a new role or roles will occur. (The "me" in this context is simply the subjective aspect of the societal reaction.) Reorganization may be the adoption of another normal role in which the tendencies previously defined as "pathological" are given a more acceptable social expression. The other general pos-sibility is the assumption of a deviant role, if such exists; or, more rarely, the person may organize an aberrant sect or group in which he creates a special role of his own. *When a person begins to employ his deviant behavior or a role based upon it as a means of defense, attack, or adjustment to the overt and covert problems created by the consequent societal reaction to him, his deviation is secondary.* Objective evidences of this change will be found in the symbolic appurtenances of the new role, in clothes, speech, posture, and mannerisms, which in some cases heighten social visibility, and which in some cases serve as symbolic cues to professionalization.

Role Conceptions of the Individual Must Be Reinforced by Reactions of Others

It is seldom that one deviant act will provoke a sufficiently strong societal reaction to bring about secondary deviation, unless in the process of introjection the individual imputes or projects meanings into the social situation which are not present. In this case anticipatory fears are involved. For example, in a culture where a child is taught sharp distinctions between "good" women and "bad" women, a single act of questionable morality might conceivably have a profound meaning for the girl so indulging. However, in the absence of reactions by the person's family, neighbors, or the larger community, reinforcing the tentative "bad-girl" self-definition, it is questionable whether a transition to secondary deviation would take place. It is also doubtful whether a temporary exposure to a severe punitive reaction by the community will lead a person to identify himself with a pathological role, unless, as we have said, the experience is highly traumatic. Most frequently there is a progressive reciprocal relationship between the deviation of the individual and the societal reaction, with a compounding of the society reaction out of the minute accretions in the deviant behavior, until a point is reached where in-grouping and outgrouping between society and the deviant is manifest.[6] At this point a stigmatizing of the deviant occurs in the form of name calling, labeling, or stereotyping.

The sequence of interaction leading to secondary deviation is roughly as follows: (1) primary deviation; (2) social penalties; (3) further primary deviation; (4) stronger penalties and rejections; (5) further deviation, perhaps with hostilities and resentment beginning to focus upon those doing the penalizing; (6) crisis reached in the tolerance quotient, expressed in formal action by the community stigmatizing of the deviant; (7) strengthening of the deviant conduct as a reaction to the stigmatizing and penalties; (8) ultimate acceptance of deviant social status and efforts at adjustment on the basis of the associated role.

As an illustration of this sequence the behavior of an errant schoolboy can be cited. For one reason or another, let us say excessive energy, the schoolboy engages in a classroom prank. He is penalized for it by the teacher. Later, due to clumsiness, he creates another disturbance and again he is reprimanded. Then, as something happens, the boy is blamed for something he did not do. When the teacher uses the tag "bad boy" or "mischief maker" or other invidious terms, hostility and resentment are excited in the boy, and he may feel that he is blocked in playing the role expected of him. Thereafter, there may be a strong temptation to assume his role in the class as defined by the teacher, particularly when he discovers that there are rewards as well as penalties deriving from such a role. There is, of course, no implication here that such boys go on to become delinquents or criminals, for the mischief-maker role may later become integrated with or retrospectively rationalized as part of a role more acceptable to school authorities.[7] If such a boy continues this unacceptable role and becomes delinquent, the process must be accounted for in the light of the general theory of this volume. There must be a spreading corroboration of a sociopathic self-conception and societal reinforcement at each step in the process.

The most significant personality changes are manifest when societal definitions and their subjective counterpart become generalized. When this happens, the range of major role choices becomes narrowed to one general class.[8] This was very obvious in the case of a young girl who was the daughter of a paroled convict and who was attending a small Middle Western college. She continually argued with herself and with the author, in whom she had confided, that in reality she belonged on the "other side of the railroad tracks" and that her life could be enormously simplified by acquiescing in this verdict and living accordingly. While in her case there was a tendency to dramatize her conflicts, nevertheless there was enough societal reinforcement of her self-conception by the treatment she received in her relationship with her father and on dates with college boys to lend it a painful reality. Once these boys took her home to the shoddy dwelling in a slum area where she lived with her father, who was often in a drunken condition, they abruptly stopped seeing her again or else became sexually presumptive.

NOTES

1. Shaw, C., *The Natural History of a Delinquent Career,* Chicago, 1941, pp. 75–76. Quoted by permission of the University of Chicago Press, Chicago.

2. Brown, L. Guy, *Social Pathology,* 1942, pp. 44–45.

3. Allport, G., *Personality, A Psychological Interpretation,* 1947, p. 57.

4. Slavson, S. R., *An Introduction to Group Psychotherapy,* 1943, pp. 10, 229*ff.*

5. Murphy, G., *Personality,* 1947, p. 482.

6. Mead, G., "The Psychology of Punitive Justice," *American Journal of Sociology,* 23 March, 1918, pp. 577–602.

7. Evidence for fixed or inevitable sequences from predelinquency to crime is absent. Sutherland, E. H., *Principles of Criminology,* 1939, 4th ed., p. 202.

8. Sutherland seems to say something of this sort in connection with the development of criminal behavior. *Ibid.,* p. 86.

QUESTIONS FOR DISCUSSION

1. Discuss the author's conception of sociopathic individuation. Give an example of sociopathic individuation to show your understanding of the concept.

2. How does the author illustrate secondary deviation? Give an example of secondary deviation to show your understanding of the concept.

3. Apply the author's eight-step sequence of interaction leading to secondary deviance to a real-life situation.

Differential Association Theory

Edwin H. Sutherland and Donald R. Cressey

The following statements refer to the process by which a particular person comes to engage in criminal behavior.

1. *Criminal behavior is learned.* Negatively, this means that criminal behavior is not inherited, as such; also, the person who is not already trained in crime does not invent criminal behavior, just as a person does not make mechanical inventions unless he has had training in mechanics.

2. *Criminal behavior is learned in interaction with other persons in a process of communication.* This communication is verbal in many respects but includes also "the communication of gestures."

3. *The principal part of the learning of criminal behavior occurs within intimate personal groups.* Negatively, this means that the impersonal agencies of communication, such as movies and newspapers, play a relatively unimportant part in the genesis of criminal behavior.

4. *When criminal behavior is learned, the learned includes (a) techniques of committing the crime, which are sometimes very complicated, sometimes very simple; (b) the specific direction of motives, drives, rationalizations, and attitudes.*

5. *The specific direction of motives and drives is learned from definitions of the legal codes as favorable or unfavorable.* In some societies an individual is surrounded by persons who invariably define the legal codes as rules to be observed, while in others he is surrounded by persons whose definitions are favorable to the violation of the legal codes. In our American society these definitions are almost always mixed, with the consequence that we have culture conflict in relation to the legal codes.

6. *A person becomes delinquent because of an excess of definitions favorable to violation of law over definitions unfavorable to violation of law.* This is the principle of differential association. It refers to both criminal and anticriminal associations and has to do with counteracting forces. When persons become criminal, they do so because of contacts with criminal patterns and also because of isolation from anticriminal patterns. Any person inevitably assimilates the surrounding culture unless other patterns are in conflict; a southerner does not pronounce *r* because other southerners do not pronounce *r*. Negatively, this proposition of differential association means that associations which are neutral so far as crime is concerned have little or no effect on the genesis of criminal behavior. Much of the experience of a person is neutral in this sense, for example, learning to brush one's teeth. This behavior has no negative or positive effect on criminal behavior except as it may be related to associations which are concerned with the legal codes. This neutral behavior is important especially as an occupier of the time of a child so that he is

"Differential Association Theory" from *Criminology*, 9th edition, by Edwin H. Sutherland and Donald R. Cressey, pp. 75–77. Philadelphia: Lippincott, 1974. Reprinted with the permission of the estate of Donald R. Cressey.

not in contact with criminal behavior during the time he is so engaged in the neutral behavior.

7. *Differential associations may vary in frequency, duration, priority, and intensity.* This means that associations with criminal behavior and also associations with anticriminal behavior vary in those respects. "Frequency" and "duration" as modalities of associations are obvious and need no explanation. "Priority" is assumed to be important in the sense that lawful behavior developed in early childhood may persist throughout life, and also that delinquent behavior developed in early childhood may persist throughout life. This tendency, however, has not been adequately demonstrated, and priority seems to be important principally through its selective influence. "Intensity" is not precisely defined, but it has to do with such things as the prestige of the source of a criminal or anticriminal pattern and with emotional reactions related to the associations. In a precise description of the criminal behavior of a person, these modalities would be rated in quantitative form and a mathematical ratio reached. A formula in this sense has not been developed, and the development of such a formula would be extremely difficult.

8. *The process of learning criminal behavior by association with criminal and anticriminal patterns involves all of the mechanisms that are involved in any other learning.* Negatively, this means that the learning of criminal behavior is not restricted to the process of imitation. A person who is seduced, for instance, learns criminal behavior by association, but this process would not ordinarily be described as imitation.

9. *While criminal behavior is an expression of general needs and values, it is not explained by those general needs and values, since noncriminal behavior is an expression of the same needs and values.* Thieves generally steal in order to secure money, but likewise honest laborers work in order to secure money. The attempts by many scholars to explain criminal behavior by general drives and values, such as the happiness principle, striving for social status, the money motive, or frustration, have been, and must continue to be, futile, since they explain lawful behavior as completely as they explain criminal behavior. They are similar to respiration, which is necessary for any behavior, but which does not differentiate criminal from noncriminal behavior.

It is not necessary, at this level of explanation, to explain why a person has the associations he has; this certainly involves a complex of many things. In an area where the delinquency rate is high, a boy who is sociable, gregarious, active, and athletic is very likely to come in contact with the other boys in the neighborhood, learn delinquent behavior patterns from them, and become a criminal; in the same neighborhood the psychopathic boy who is isolated, introverted, and inert may remain at home, not become acquainted with the other boys in the neighborhood, and not become delinquent. In another situation, the sociable, athletic, aggressive boy may become a member of a scout troop and not become involved in delinquent behavior. The person's associations are determined in a general context of social organization. A child is ordinarily reared in a family; the place of residence of the family is determined largely by family income; and the delinquency rate is in many respects related to the rental value of the houses. Many other aspects of social organization affect the kind of associations a person has.

The preceding explanation of criminal behavior purports to explain the criminal and noncriminal behavior of individual persons. It is possible to state sociological theories of criminal behavior which explain the criminality of a community, nation, or other group. The problem, when thus stated, is to account for variations in crime rates and involves a comparison of the crime rates of various groups or the crime rates of a particular group at different times. The explanation of a crime rate must be consistent with the explanation of the criminal behavior of the person, since the crime rate is a summary statement of the number of persons in the group who commit crimes and the frequency with which they commit crimes. One of the best explanations of crime rates from this point of view is that a high crime rate is due to social disorganization. The term *social disorganization* is not entirely satisfactory, and it seems preferable to substitute for it the term *differential social organization*. The postulate on which this theory is based, regardless of the name, is that crime is rooted in the social organization and is an expression of that social organization. A group may be organized for criminal behavior or organized against criminal behavior. Most communities are organized for both criminal and anticriminal behavior, and, in that sense the crime rate is an expression of the differential group organization. Differential group organization as an explanation of variations in crime rates is consistent with the differential association theory of the processes by which persons become criminals.

QUESTIONS FOR DISCUSSION

1. How does the author explain the concepts of criminal and noncriminal?
2. Can differential association explain deviant and criminal activities at your college, university, or geographic location? Explain your answer.
3. Take a deviant or criminal act and explain this behavior by applying all nine propositions of differential association theory.

A Test of the Black Subculture of Violence Thesis: A Research Note

Liqun Cao, Anthony Adams, and Vickie J. Jensen

The subculture of violence thesis is perhaps one of the most cited, but one of the least tested, propositions in the sociological and criminological literature (Adler et al., 1994; Barlow, 1996; Coser et al., 1990; Siegel, 1992). And when subcultures of violence have been studied, the focus has usually been on regional variations (see Hawley and Messner, 1989) rather than on race.

The subculture of violence thesis has been most fully developed by Wolfgang and Ferracuti (1967). Based on research conducted in inner-city Philadelphia in the mid-

"A Test of the Black Subculture of Violence Thesis: A Research Note" by Liqun Cao, Anthony Adams, and Vickie J. Jensen from *Criminology*, vol. 35, no. 2 (1997), pp. 367–379. Reprinted with permission of American Society of Criminology.

1950s (Wolfgang, 1958), Wolfgang and Ferracuti (1967: 161) attempted to bring together "psychological and sociological constructs to aid in the explanation of the concentration of violence in specific socio-economic groups and ecological areas." They argued that certain segments of society have adopted distinctively violent subcultural values. This value system provides its members with normative support for their violent behavior, thereby increasing the likelihood that hostile impulses will lead to violent action. Relying on official data on violent crime, Wolfgang and Ferracuti (1967) speculate that there is a subculture of violence among blacks. They wrote, "Our subculture-of-violence thesis would, therefore, expect to find a large spread to the learning of, resort to, and criminal display of the violence value among minority groups such as Negroes" (1967: 264).

This speculation was later incorporated in the literature as the black subculture of violence thesis. In addition, Wolfgang and Ferracuti (1967) argued that the subculture of violence varies according to gender, age, social class, employment status, region, and urban environment. The research reported in this article tests their black subculture of violence thesis.

According to Wolfgang and Ferracuti's theory (1967), a subculture is a value system that can transcend geographic areas in a society. Wolfgang and Ferracuti are not alone in asserting that the subculture of violence is a value system (Erlanger, 1976). Reed (1972), for example, contends more explicitly that the existence of a subculture of violence can be established with attitudinal data measuring beliefs in violence.

Thus far, most empirical studies have focused on testing the subculture of violence thesis's predictions about southern violence. Race has entered these analyses mainly as a control variable. A number of studies claiming to have investigated a black subculture of violence have used the macrolevel variable of percentage blacks in an areas (Messner, 1983; Williams and Flewelling, 1988) as a proxy. While these studies have found a relationship between percentage black and violence, it is difficult to tell if subcultural beliefs are responsible. Thus, the association between the subculture of violence among blacks and violent behavior remains largely inconclusive.

The subculture of violence thesis has also been tested using attitudinal measures of approval of violence. These efforts are more appropriate than using aggregate group membership because Wolfgang and Ferracuti (1967) focused on values as the defining feature of subcultures. In this regard, a number of studies that have tested the subculture of violence thesis as a value and belief system have helped to inform our study. Using cross-tabular techniques, Ball-Rokeach (1973) and Poland (1978) tested the hypothesis that violent behavior results from a commitment to subcultural values condoning violence. Both found little evidence in support of the subculture of violence thesis.

Erlanger (1974) provided the only direct test of Wolfgang and Ferracuti's contention that there is a subculture of violence among blacks. Using tabular analysis, he found that there is "an absence of major difference by race" in approval of interpersonal violence (p. 283). Further, using data on peer esteem and social psychological correlates of fighting among males aged 21 to 64 who resided in Milwaukee, Wisconsin, he reported that "poor whites are more likely to fight than poor blacks" (p. 285).

Doerner (1978) developed a multiple regression model and Hartnagel (1980) used a path model to assess the southern subculture of violence. Both controlled race in their models, but neither found that blacks are more likely than whites to approve of assaultive behavior. In fact, Doerner (1978) reported that whites are significantly more likely to approve of assaultive behavior than blacks. Austin's research (1980) provided some support of the subcultural explanation of violence as measured by beliefs. He, however, failed to control race in his models.

The most sophisticated study of the subculture of violence is that by Dixon and Lizotte (1987), who examined gun ownership and its relationship to the southern subculture of violence using two direct index measures of subcultural values and beliefs. Their analyses revealed that white males are significantly more supportive of the use of violence in both hypothetical defensive and offensive situations. While Dixon and Lizotte directly measured individuals' beliefs in violence, they did not control all of the independent variables proposed by Wolfgang and Ferracuti, such as employment and violent history. A more recent study by Ellison (1991) tested the acceptance of violence in defensive situations using one measure developed by Dixon and Lizotte. The results confirmed Dixon and Lizotte's finding that whites are more likely than nonwhites to condone interpersonal violence in retaliatory situations.

Thus, existing studies on the subculture of violence indicate that race has been seriously neglected in direct test and that beliefs in violence among southerners have received excessive attention. When race has been addressed, whites are often contrasted with nonwhites, which is imprecise. In addition, there have been weaknesses in samples, techniques of statistical analyses, and use of control variables.

This study examines beliefs in violence among African-Americans and tests one implication of Wolfgang and Ferracuti's hypothesis—that violent values are widespread among African-Americans. It improves on previous studies in several ways. First, we examine blacks, instead of non-whites. Second, we use direct measures of subcultural values. Third, we use a nationwide representative sample to test the thesis. Finally, our model includes the key variables identified by Wolfgang and Ferracuti (1967) and thus potentially offers the most complete examination to date of the factors affecting beliefs in violence.

METHODS

Sample

The data for this study were drawn from the General Social Survey. We include the years from 1983 to 1991, although 1985 data were not included because of the yearly item rotation. People not born in this country and who identified themselves neither as whites nor as blacks were eliminated from our analyses.[1] Further, only males and people under the age of 65 were selected for this analysis.

The total number of people in our final analysis was 3,218, and 395 (12.3%) were African-Americans. Some of the sample characteristics by race are presented in Table 1. On average, blacks and whites in our sample were similar in ages although whites had received more education, had higher family incomes, and were more likely to be employed full-time than blacks (75% of whites vs. 61% of blacks claimed that they were employed full-time at the time of interview). Blacks were more likely than whites to be found in the census South[2] (47% vs. 32%) and in cities with more than 50,000 residents (58% vs. 22%).

Selected Sample Characteristics   TABLE 1

Characteristics	Mean	S.D.	Cases
Age			
Blacks	38.17	13.59	395
Whites	38.69	12.61	2,823
Education			
Blacks	11.94	3.05	395
Whites	13.24	3.00	2,823
Income			
Blacks	3.55	1.39	395
Whites	4.12	1.18	2,823
Cities greater than 50,000			
Blacks	0.58	0.50	395
Whites	0.22	0.42	2,823
Region			
Blacks	0.47	0.50	395
Whites	0.32	0.47	2,823
Employment			
Blacks	0.61	0.49	395
Whites	0.75	0.43	2,823

Dependent Variables

The dependent variable in this study—the subculture of violence—is operationalized using three measures. One is an index of violent defensive values and the other
two are violent offensive values. We use the index of violent defensive values developed by Dixon and Lizotte (1987) from the General Social Survey. This index taps
approval for violence undertaken to protect children, women, and property from an
assailant. It is composed of three items. Respondents were asked, "Would you approve of a man punching a stranger who (1) had hit the man's child after the child
accidentally damaged the stranger's car? (2) was beating up a woman and the man
saw it? (3) had broken into the man's house?" We coded the answers to each item according to Dixon and Lizotte's (1987) scale: 0 = no, 1 = don't know and not sure,
and 2 = yes. The final index variable varies from 0 = do not approve of punching at
any of the above situations to 6 = approve of punching at all of the above situations.
The reliability of our measure is .553.

As for the violent offensive values, we selected two items from the General Social
Survey. These two items asked, "Would you approve of a man punching a stranger if
the stranger (1) . . . was in a protest march showing opposition to the other man's
view? (2) . . . was drunk and bumped into the man and his wife on the street?" These
questions attempt to capture support for violent responses to nonphysical threat or
unintentional conduct. Because these two items cannot be formed into an index variable, we coded them into two binary variables with 1 = yes and 0 = no. The logistic regression technique was used to analyze each item separately.

We did not use Dixon and Lizotte's four-item scale of offensive subculture of violence because that scale confounds two types of conceptually different items within
it. In two of the items (as shown in the previous paragraph), the subject is a generalized *other* striking another man, while in the other two items the subject is a more

specific other—*a policeman*—striking another male civilian.[3] Many empirical studies have shown that blacks in general have a significantly lower evaluation of the police than whites do (Brandl et al., 1994; Peek et al., 1981; but also see Cao et al., 1996). Thus, the word *policeman* might cause a reaction that is different from the reaction to the word *man,* which would tend to confuse the racial difference toward acceptance of violence.

These variables capture the most important component of Wolfgang and Ferracuti's concept of a subculture of violence. They contend that in such a subculture people do not define personal assaults as wrong or antisocial. We believe this is the crucial element, although Wolfgang (1958: 329) also implicates in his concept of subculture an idea of shared norms "in which quick resort to physical aggression is a socially approved and expected concomitant of certain stimuli."

Independent Variables

Wolfgang and Ferracuti (1967) have specifically inferred that the demographic groups with the "most intense" subculture of violence would be among black, young males; those from the lower and working classes; and those from the South. We include all of these variables in our analysis.

The major independent variable of race is coded so that black = 1 and white = 0. The black subculture of violence thesis hypothesizes that race is positively related to values in favor of violent reaction to both defensive and offensive situations.

The other demographic variables are coded as follows. Age is coded as the respondent's age in the year interviewed. Education is coded as years of formal schooling respondents reported, ranging between 0 (no formal education) and 20 years of schooling. Family income is coded as an ordinal variable: 1 = $3,999 and below, 2 = $4,000 to $6,999, 3 = $7,000 to $14,999, 4 = $15,000 to $24,999, and 5 = $25,000 and above. All three variables—age, education, and family income—are expected to be negatively associated with our dependent variables.

Other independent variables identified by Wolfgang and Ferracuti (1967) as affecting the subculture of violence are cities with more than 50,000 residents, region, and employment status. All of these variables are coded as dummy variables: those who reside in central cities larger than 50,000 people are coded as 1 and those living outside as 0; the census South is coded as 1 and the other regions as 0; those who are employed full-time are coded as 1 and other employment statuses as 0. It is expected that living in central cities and in the South are positively related to holding violent subcultural values while employment status is negatively related to holding such values.

We control for drinking habits because Erlanger (1974: 289) proposed that "the use of liquor may be part of a broader social configuration which generates situations conducive to violence." We coded the degree of alcohol consumption as 0 for those who do not drink, 1 for those who say they drink sometimes, and 2 for those who say they sometimes drink more than they think they should. We also controlled for family income when the respondent was 16 years old, coding it as 1 = far below average to 5 = far above average. Finally, violent history is controlled using a combined variable composed of three items: a respondent's experiences of being hit as a child, as an adult, and both; and being threatened with a gun as a child, as an adult, and both.[4]

Race and Defensive Subculture   TABLE 2

Variables	*b*	β
Race (Black = 1)	−.45*	−.09
Age	−.01*	−.09
Education	.02	.03
Income	.11*	.08
Cities	−.31*	−.08
Region (South = 1)	.11	.03
Employment (yes = 1)	.15	.04
Drinking	.12*	.06
Income at 16	.12*	.06
Violent history	.05*	.08
Constant	4.00*	
R^2		.07
F		22.79
Significance		.00

*$p \leq 0.01$.

The index varies from 0 = no such experiences to 8 = being hit in both childhood and adulthood and being threatened with a gun in both childhood and adulthood. These socialization variables are important because Wolfgang (1958) argues that subcultures of violence are characterized by social values that are transmitted during socialization and govern behavior in a variety of structurally induced situations.

RESULTS

Table 2 presents the results of the ordinary least squares regression analysis of race predicting beliefs in defensive use of violence. Whites are significantly more likely than blacks to express their support for the use of violence in defensive situations. This result contradicts the expectation from the subculture of black violence thesis set forth by Wolfgang and Ferracuti. It is consistent, however, with previous findings using a similar dependent variable (Dixon and Lizotte, 1987; Ellison, 1991).

Consistent with our prediction, expressed support for interpersonal violence in defensive situations declines with age. In addition, drinking and violent history significantly increase support for the use of violence in defensive situations. Contrary to our expectations, support for interpersonal violence in defensive situations increases with current family income and with family income at age 16, but it is less for those residing in large cities than those residing elsewhere. Other variables in our model, notably residence in the South, are not statistically significant.

The results of the logistic regressions concerning race and violent reaction to nonthreatening situations are presented in Table 3. They show no significant race effect on expressed support of interpersonal violence in either of the offensive situations. Again, the findings are inconsistent with the subculture of violence predictions concerning race, but they are consistent with previous findings on the subject (Doerner, 1978; Erlanger, 1974; Felson et al., 1994; Hartnagel, 1980).

TABLE 3   Race and Violent Reaction to Opposing Protestors and Unintentional Bumps

	Reaction to	
Variable	Unintentional Bumps Logit Coefficients (S.E.)	Protestors Logit Coefficients (S.E.)
Race (Black = 1)	−.11 (.19)	.04 (.25)
Age	.00 (.01)	.01 (.01)
Education	−.05* (.02)	−.11* (.03)
Income	−.10 (.05)	−.21* (.07)
Cities	−.34 (.15)	−.13 (.21)
Region (South = 1)	.70* (.12)	.46* (.18)
Employment	−.20 (.14)	.09 (.20)
Drinking	−.16 (.08)	−.15 (.11)
Income at 16	−.05 (.07)	−.24 (.10)
Violent history	.00 (.02)	.09* (.04)
Constant	−.98 (.42)	−.75 (.60)
−2 log likelihood	2058.23	1133.91

*$p \leq 0.01$.

Other results from Table 3 are consistent with our predictions: Education and family income predict less support of interpersonal violence in offensive situations, and residence in the South and having a history of violent victimization predict greater support. The remaining independent variables in our model did not significantly affect the dependent measures.

DISCUSSION

Fischer (1975:1334) calls ethnic subculture "the most difficult test case for subculture theory." We took up this challenge and tested Wolfgang and Ferracuti's thesis of the black subculture of violence. With a number of control variables in our models, we explored whether there is a difference between whites and blacks in the distribution of individual attitudes (or values) concerning violence in defensive and offensive situations. Contrary to the expectations of the black subculture of violence thesis, our results indicate that white males express significantly more violent beliefs in defensive or retaliatory situations than blacks and that there is no significant difference between white and black males in beliefs in violence in offensive situations. These results, however, are consistent with a number of previous findings on the issue (Dixon and Lizotte, 1987; Doerner, 1978; Ellison, 1991; Erlanger, 1974; Felson et al., 1994; Hartnagel, 1980).

Our data are limited to the general public, which may be inadequate to test subcultural theories (Matsueda et al., 1992). Specifically including criminals in our sample, however, would trigger a larger debate on the topic of whether criminals, as a group, regardless of race, have a different value system than law-abiding citizens (Agnew, 1994; Sykes and Matza, 1957). Another limitation of our data is that we were unable to locate the neighborhoods of our respondents. It is possible that a subculture of violence may involve belief systems that characterize a particular urban community (Cloward and Ohlin, 1960; Fischer, 1995). Inclusion of this ecological elements, thus, would shift focus from subcultural beliefs in violence, which could transcend place, to a more complicated interaction between community and value system. Such an interactive relationship between place and values could potentially explain our failure to find subcultural beliefs in violence in the general population sample of African-Americans. It is clear, however, that the assumption that African-Americans generally endorse violence more than whites is questionable.

We do not claim that we have offered a complete test of Wolfgang and Ferracuti's theory (1967). Our dependent variables measure only individual's verbal approval of defensive and offensive assaults. We are not able to measure the connections of these beliefs to violent behavior, nor is it clear whether this support would be carried over to more lethal forms of violence. Wolfgang and Ferracuti's original development of the subculture of violence thesis was stimulated by an interest in both assault and lethal violence. Further, their theory is broad, incorporating general elements of structural explanations of crime in addition to cultural explanations of crime. For example, they (1967: 263–264) explain black involvement with crime by arguing that "restricted and isolated from the institutionalized means to achieve the goals of the dominant culture, many more Negroes than whites are caught in what Merton, Cloward and Ohlin, and others refer to as the differential opportunity structure, and are more likely to commit crime." However, Wolfgang and Ferracuti do insist that the main component of a subculture of violence is a value system that can be separated from its structural carrier and that being African-American should predict violent values.

Our data and analyses indicate that blacks in the general U.S. population are no more likely than whites to embrace values favorable to violence. This, however, cannot be interpreted as a rejection of the subculture of violence as a useful concept to explain violent behavior. It simply means that being black does not imply a greater probability of embracing a subculture of violence as measured by individual's beliefs and attitudes.

A more fruitful search for the root causes of black violence may lie, as Sampson (1987), Parker (1989), and Shihadeh and Steffensmeier (1994) have suggested, in the structurally disadvantaged position of blacks in the U.S. society. At the same time that inner-city African-Americans have few legitimate opportunities for social advancement (Wilson, 1986; 1996), they have ample opportunities to act violently. Thus, differential opportunities in conjunction with racial inequality may be the key to understanding black violence (Cloward and Ohlin, 1960; Cullen, 1983). Violent behavior may constitute "the normal reaction of normal people to abnormal conditions" (Plant, 1937: 248).

NOTES

1. Nearly 6% of our original sample who chose the "other" category in race and ethnicity were deleted. Most of them were Hispanics. We argue that those Hispanics who were born in this country and have a strong Hispanic identity would have chosen this category (thus, being excluded from our analysis) and those whose Hispanic identity is not strong (because they were born and raised here) may choose either white or black as their dominant "new" identity.

2. The census South is slightly different from the more traditional confederal South. It includes Delaware, Maryland, West Virginia, Virginia, North Carolina, South Carolina, Georgia, Florida, District of Columbia, Kentucky, Tennessee, Alabama, Mississippi, Arkansas, Oklahoma, Louisiana, and Texas. See Ellison (1991) for a discussion.

3. One asked, "Would you approve of a *policeman* striking an adult male citizen if the male citizen had said vulgar and obscene things to the policeman?" The other asked, "Would you approve of a *policeman* striking an adult male citizen if the male citizen was being questioned as a suspect in a murder case?"

4. We also had variables of anomie, native South (a combination of growing up in the South and currently living in the South), and TV watching, but we dropped them in our final model because none of them was statistically significant in predicting the subculture of violence, because their theoretical implication is not strong (TV watching and residence) or because we do not have information on every year (anomie and TV watching). Our results do not change with or without these variables.

REFERENCES

Adler, Freda, Gerhard O. W. Mueller, and William S. Laufer. 1994. *Criminal Justice*. New York: McGraw-Hill.

Agnew, Robert. 1994. The techniques of neutralization and violence. *Criminology* 32: 555–580.

Austin, Roy L. 1980. Adolescent subcultures of violence. *The Sociological Quarterly* 21: 545–561.

Ball-Rokeach, Sandra J. 1973. Values and violence: A test of the subculture of violence thesis. *American Sociological Review*. 38: 736–749.

Barlow, Hugh D. 1996. *Criminology*. New York: HarperCollins College Publishers.

Brandl, Steven, James Frank, Robert E. Worden, and Timothy S. Bynum. 1994. Global and specific attitudes toward the police: Disentangling the relationship. *Justice Quarterly* 11: 119–134.

Cao, Liqun, James Frank, and Francis T. Cullen. 1996. Race, community context, and confidence in the police. *American Journal of Police*. 15: 3–22.

Cloward, Richard A. and Lloyd E. Ohlin. 1960. *Delinquency and Opportunity: A Theory of Delinquent Gangs*. New York: Free Press.

Coser, Lewis A., Steven L. Nock, Patricia A. Steffan, and Daphne Spain. 1990. *Introduction to Sociology*. San Diego: Harcourt Brace Jovanovich.

Cullen, Francis T. 1983. Rethinking Crime and Deviance Theory: The Emergence of a Structuring Tradition. Totowa, N.J.: Rowman & Allanheld.

Dixon, Jo and Alan J. Lizotte. 1987. Gun ownership and the southern subculture of violence. *American Journal of Sociology* 93: 383–405.

Doerner, William G. 1978. The index of southernness revisited: The influence of wherefrom upon whodunnit. *Criminology* 16: 47–65.

Ellison, Christopher G. 1991. An eye for an eye? A note on the southern subculture of violence thesis. *Social Forces* 69: 1223–1239.

Erlanger, Howard S. 1974. The empirical status of the sub-cultures of violence thesis. *Social Problems* 22: 280–292.

———. 1976. Is there a subculture of violence in the South? *Journal of Criminal Law and Criminology* 66: 483–490.

Felson, Richard B., Allen E. Liska, Scott J. South, and Thomas L. McNulty. 1994. The subculture of violence and delinquency: Individual vs. school context effects. *Social Forces* 73: 155–173.

Fischer, Claude S. 1975. Toward a subcultural theory of urbanism. *American Journal of Sociology* 80: 1319–1341.

———. 1995. The subcultural theory of urbanism: A twentieth-year assessment. *American Journal of Sociology* 101: 543–577.

Hartnagel, Timothy F. 1980. Subculture of violence: Further evidence. *Pacific Sociological Review* 23: 217–242.

Hawley, F. Frederick and Steven F. Messner. 1989. The southern violence construct: A review of arguments, evidence, and the normative context. *Justice Quarterly* 6: 481–511.

Matsueda, Ross L., Rosemary Gartner, Irving Piliavin, and Michael Polakowski. 1992. The prestige of criminal and conventional occupations: A subcultural model of criminal activity. *American Sociological Review.* 57: 752–770.

Messner, Steven F. 1983. Regional and racial effects on the urban homicide rate: The subculture of violence revisited. *American Journal of Sociology.* 88: 997–1007.

Parker, Robert Nash. 1989. Poverty, subculture of violence, and type of homicide. *Social Forces.* 67: 983–1007.

Peek, Charles W., George D. Lowe, and Jon P. Alston. 1981. Race and attitudes toward local police. *Journal of Black Studies.* 11: 361–374.

Plant, James S. 1937. *Personality and the Cultural Pattern.* London: Oxford University Press.

Poland, James M. 1978. Subculture of violence: Youth offender value systems. *Criminal Justice and Behavior* 5: 159–164.

Reed, John Shelton. 1972. *The Enduring South: Subcultural Persistence in Mass Society.* Lexington, Mass.: D.C. Heath.

Sampson, Robert. 1987. Urban black violence: The effect of male joblessness and family disruption. *American Journal of Sociology* 93: 348–382.

Shihadeh, Edward S. and Darrell J. Steffensmeier. 1994. Economic inequality, family disruption, and urban black violence: Cities as units of stratification and social control. *Social Forces* 73: 729–751.

Siegel, Larry J. 1992. *Criminology.* New York: West Publishing.

Sykes, Gresham M. and David Matza. 1957. Techniques of neutralization: A theory of delinquency. *American Sociological Review* 22: 664–670.

Williams, Kirk and Robert Flewelling. 1988. The social production of criminal homicide: A comparative study of disaggregated rates in American cities. *American Sociological Review* 53: 421–431.

Wilson, William J. 1986. The urban underclass in advanced industrial society. In P. E. Peterson (ed.), *The New Urban Realty.* Washington, D.C. Brookings Institution.

———. 1996. *When Work Disappears: The World of the New Urban Poor.* New York: Random House.

Wolfgang, Marvin E. 1958. *Patterns of Criminal Homicide.* Philadelphia: University of Pennsylvania Press.

Wolfgang, Marvin E. and Franco Ferracuti. 1967. *The Subculture of Violence, Towards an Integrated Theory in Criminology.* New York: Tavistock.

QUESTIONS FOR DISCUSSION

1. What is the black subculture of violence thesis? According to the reading, is there support for this thesis? Explain your response.

2. If we developed a white subculture of violence thesis, what would be the tenets? Use the reading to develop this thesis. According to the reading, would there be support for this new thesis? Explain your response.

3. A newspaper reporter has asked you to write an article on the subculture of violence. Based on the reading, write an article explaining the subculture of violence.

CHAPTER 7

CONTROL PERSPECTIVES

Control Theory

Travis Hirschi

Control theories assume that delinquent acts result when an individual's bond to society is weak or broken . . . [Elements of the bond are as follows].

ATTACHMENT

It can be argued that all of the characteristics attributed to the psychopath follow from, are effects of, his lack of attachment to others. To say that to lack attachment to others is to be free from moral restraints is to use lack of attachment to explain the guiltlessness of the psychopath, the fact that he apparently has no conscience or superego. In this view, lack of attachment to others is not merely a symptom of psychopathy, it *is* psychopathy; lack of conscience is just another way of saying the same thing; and the violation of norms is (or may be) a consequence.

For that matter, given that man is an animal, "impulsivity" and "aggressiveness" can also be seen as natural consequences of freedom from moral restraints. However, since the view of man as endowed with natural propensities and capacities like other animals is peculiarly unpalatable to sociologists, we need not fall back on such a view to explain the amoral man's aggressiveness. The process of becoming alienated from others often involves or is based on active interpersonal conflict. Such conflict could easily supply a reservoir of *socially derived* hostility sufficient to account for the aggressiveness of those whose attachments to others have been weakened.

"Control Theory" from *Causes of Delinquency* by Travis Hirschi (Berkeley: University of California Press, 1969), pp. 16–26. Reprinted by permission of the author.

Durkheim said it many years ago: "We are moral beings to the extent that we are social beings." This may be interpreted to mean that we are moral beings to the extent that we have "internalized the norms" of society. But what does it mean to say that a person has internalized the norms of society? The norms of society are by definition shared by the members of society. To violate a norm is, therefore, to act contrary to the wishes and expectations of other people. If a person does not care about the wishes and expectations of other people—that is, if he is insensitive to the opinion of others—then he is to that extent not bound by the norms. He is free to deviate.

The essence of internalization of norms, conscience, or superego thus lies in the attachment of the individual to others. This view has several advantages over the concept of internalization. For one, explanations of deviant behavior based on attachment do not beg the question, since the extent to which a person is attached to others can be measured independently of his deviant behavior. Furthermore, change or variation in behavior is explainable in a way that it is not when notions of internalization or superego are used. For example, the divorced man is more likely after divorce to commit a number of deviant acts, such as suicide or forgery. If we explain these acts by reference to the superego (or internal control), we are forced to say that the man "lost his conscience" when he got a divorce; and, of course, if he remarries, we have to conclude that he gets his conscience back. . . .

Commitment

"Of all passions, that which inclineth men least to break the laws, is fear. Nay, excepting some generous natures, it is the only thing, when there is the appearance of profit or pleasure by breaking the laws, that makes men keep them." Few would deny that men on occasion obey the rules simply from fear of the consequences. This rational component in conformity we label commitment. What does it mean to say that a person is committed to conformity? . . . [It means] that the person invests time, energy, himself, in a certain line of activity—say, getting an education, building up a business, acquiring a reputation for virtue. When or whenever he considers deviant behavior, he must consider the costs of this deviant behavior, the risk he runs of losing the investment he has made in conventional behavior.

If attachment to others is the sociological counterpart of the superego or conscience, commitment is the counterpart of the ego or common sense. To the person committed to conventional lines of action, risking one to ten years in prison for a ten-dollar holdup is stupidity, because to the committed person the costs and risks obviously exceed ten dollars in value. (To the psychoanalyst, such an act exhibits failure to be governed by the "reality-principle.") In the sociological control theory, it can be and is generally assumed that the decision to commit a criminal act may well be rationally determined—that the actor's decision was not irrational given the risks and costs he faces. . . .

Involvement

Many persons undoubtedly owe a life of virtue to a lack of opportunity to do otherwise. Time and energy are inherently limited: "Not that I would not, if I could, be both handsome and fat and well dressed, and a great athlete, and make a million a year, be a wit, a bon vivant, and a lady killer, as well as a philosopher, a philanthropist, a statesman, warrior, and African explorer, as well as a 'tone-poet' and saint. But the thing is simply impossible." The things that William James here says he would like to be or do are all, I suppose, within the realm of conventionality, but if he were to include illicit actions he would still have to eliminate some of them as simply impossible.

Involvement or engrossment in conventional activities is thus often part of a control theory. The assumption, widely shared, is that a person may be simply too busy doing conventional things to find time to engage in deviant behavior. The person involved in conventional activities is tied to appointments, deadlines, working hours, plans, and the like, so the opportunity to commit deviant acts rarely arises. To the extent that he is engrossed in conventional activities, he cannot even think about deviant acts, let alone act out his inclinations. . . .

Belief

The control theory assumes the existence of a common value system within the society or group whose norms are being violated. If the deviant is committed to a value system different from that of conventional society, there is, within the context of the theory, nothing to explain. The question is, "Why does a man violate the rules in which he believes?" It is not, "Why do men differ in their beliefs about what constitutes good and desirable conduct?" The person is assumed to have been socialized (perhaps imperfectly) into the group whose rules he is violating; deviance is not a question of one group imposing its rules on the members of another group. In other words, we not only assume the deviant *has* believed the rules, we assume he believes the rules even as he violates them.

How can a person believe it is wrong to steal at the same time he is stealing? In the strain theory, this is not a difficult problem. (In fact, the strain theory was devised specifically to deal with this question.) The motivation to deviance adduced by the strain theorist is so strong that we can well understand the deviant act even assuming the deviator believes strongly that it is wrong. However, given the control theory's assumptions about motivation, if both the deviant and the nondeviant believe the deviant act is wrong, how do we account for the fact that one commits it and the other does not?

Control theories have taken two approaches to this problem. In one approach, beliefs are treated as mere words that mean little or nothing. . . . The second approach argues that the deviant rationalizes his behavior so that he can at once violate the rule and maintain his belief in it. . . . We assume, however, that there is *variation* in the extent to which people believe they should obey the rules of society, and furthermore, that the less a person believes he should obey the rules, the more likely he is to violate them.

QUESTIONS FOR DISCUSSION

1. Take a recent deviant or criminal situation and explain this behavior using control theory.

2. How do the four elements of social bond explain delinquency or deviance?

3. Discuss major issues that would arise in applying social bond to a culturally diverse society.

A General Theory of Crime

Michael Gottfredson and Travis Hirschi

Theories of crime lead naturally to interest in the propensities of individuals committing criminal acts. These propensities are often labeled "criminality." In pure classical theory, people committing criminal acts had no special propensities. They merely followed the universal tendency to enhance their own pleasure. If they differed from noncriminals, it was with respect to their location in or comprehension of relevant sanction systems. For example, the individual cut off from the community will suffer less than others from the ostracism that follows crime; the individual unaware of the natural or legal consequences of criminal behavior cannot be controlled by these consequences to the degree that people aware of them are controlled; the atheist will not be as concerned as the believer about penalties to be exacted in a life beyond death. Classical theories on the whole, then, are today called *control* theories, theories emphasizing the prevention of crime through consequences painful to the individual.

Although, for policy purposes, classical theorists emphasize legal consequences, the importance to them of moral sanctions is so obvious that their theories might well be called underdeveloped *social control* theories. In fact, Bentham's list of the major restraining motives—motives acting to prevent mischievous acts—begins with goodwill, love of reputation, and the desire for amity (1970: 134–36). He goes on to say that fear of detection prevents crime in large part because of detection consequences for "reputation, and the desire for amity" (p. 138). Put another way, in Bentham's view, the restraining power of legal sanctions in large part stems from their connection to social sanctions.

If crime is evidence of the weakness of social motives, it follows that criminals are less social than noncriminals and that the extent of their asociality may be determined by the nature and number of their crimes. Calculations of the extent of an individual's mischievousness is a complex affair, but in general the more mischievous or depraved the offenses, and the greater their number, the more mischievous or depraved the offender (Bentham 1970: 134–42). (Classical theorists thus had reason to be interested in the seriousness of the offense. The relevance of seriousness to current theories of crime is not so clear.)

Because classical or control theories infer that offenders are not restrained by social motives, it is common to think of them as emphasizing an asocial human nature. Actually, such theories make people only as asocial as their acts require. Pure or consistent control theories do not add criminality (i.e., personality concepts or attributes such as "aggressiveness" or "extraversion") to individuals beyond that found in their criminal acts. As a result, control theories are suspicious of images of an antisocial, psychopathic, or career offender, or of an offender whose motives to crime are somehow larger than those given in the crimes themselves. Indeed, control theories are compatible with the view that the balance of the total control structure favors conformity, even among offenders:

> For in every man, be his disposition ever so depraved, the social motives are those which . . . regulate and determine the general tenor of his life. . . . The general and standing bias of every man's nature is, therefore, towards that side to which the force of the social motives would determine him to adhere. This being the case, the force of the social motives tends continually to put an end to that of the dissocial ones; as, in natural bodies, the force of friction tends to put an end to that which is generated by impulse. Time, then, which wears away the force of the dissocial motives, adds to that of the social. [Bentham 1970: 141]

Positivism brought with it the idea that criminals differ from non-criminals in ways more radical than this, the idea that criminals carry within themselves properties peculiarly and positively conducive to crime. [Previously] we examined the efforts of the major disciplines to identify these properties. Being friendly to both the classical and positivist traditions, we expected to end up with a list of individual properties reliably identified by competent research as useful in the description of "criminality"—such properties as aggressiveness, body build, activity level, and intelligence. We further expected that we would be able to connect these individual-level correlates of criminality directly to the classical idea of crime. As our review progressed, however, we were forced to conclude that we had overestimated the success of positivism in establishing important differences between "criminals" and "non-criminals" beyond their tendency to commit criminal acts. Stable individual differences in the tendency to commit criminal acts were clearly evident, but many or even most of the other differences between offenders and nonoffenders were not as clear or pronounced as our reading of the literature had led us to expect.[1]

If individual differences in the tendency to commit criminal acts (within an overall tendency for crime to decline with age) are at least potentially explicable within classical theory by reference to the social location of individuals and their comprehension of how the world works, the fact remains that classical theory cannot shed much light on the positivistic finding (denied by most positivistic theories . . .) that these differences *remain reasonably stable with change in the social location of individuals and change in their knowledge of the operation of sanction systems.* This is the problem of self-control, the differential tendency of people to avoid criminal acts whatever the circumstances in which they find themselves. Since this difference among people has attracted a variety of names, we begin by arguing the merits of the concept of self-control.

Self-Control and Alternative Concepts

Our decision to ascribe stable individual differences in criminal behavior to self-control was made only after considering several alternatives, one of which (criminality) we had used before (Hirschi and Gottfredson 1986). A major consideration was consistency between the classical conception of crime and our conception of the criminal. It seemed unwise to try to integrate a choice theory of crime with a deterministic image of the offender, especially when such integration was unnecessary. In fact, the compatibility of the classical view of crime and the idea that people differ in self-control is, in our view, remarkable. As we have seen, classical theory is a theory of social or external control, a theory based on the idea that the costs of crime depend on the individual's current location in or bond to society. What classical theory lacks is an explicit idea of self-control, the idea that people also differ in the extent to which they are vulnerable to the temptations of the moment. Combining the two ideas thus merely recognizes the simultaneous existence of social and individual restraints on behavior.

An obvious alternative is the concept of criminality. The disadvantages of that concept, however, are numerous. First, it connotes causation or determinism, a positive tendency to crime that is contrary to the classical model and, in our view, contrary to the facts. Whereas self-control suggests that people differ in the extent to which they are restrained from criminal acts, criminality suggests that people differ in the extent to which they are compelled to crime. The concept of self-control is thus consistent with the observation that criminals do not require or need crime, and the concept of criminality is inconsistent with this observation. By the same token, the idea of low self-control is compatible with the observation that criminal acts require no special capabilities, needs, or motivation; they are, in this sense, available to everyone. In contrast, the idea of criminality as a special tendency suggests that criminal acts require special people for their performance and enjoyment. Finally, lack of restraint or low self-control allows almost any deviant, criminal, exciting, or dangerous act; in contrast, the idea of criminality covers only a narrow portion of the apparently diverse acts engaged in by people at one end of the dimension we are now discussing.

The concept of conscience comes closer than criminality to self-control, and is harder to distinguish from it. Unfortunately, that concept has connotations of compulsion (to conformity) not, strictly speaking, consistent with a choice model (or with the operation of conscience). It does not seem to cover the behaviors analogous to crime that appear to be controlled by natural sanctions rather than social or moral sanctions, and in the end it typically refers to how people feel about their acts rather than to the likelihood that they will or will not commit them. Thus accidents and employment instability are not usually seen as produced by failures of conscience, and writers in the conscience tradition do not typically make the connection between moral and prudent behavior. Finally, conscience is used primarily to summarize the results of learning via negative reinforcement, and even those favorably disposed to its use have little more to say about it (see, e.g., Eysenck 1977; Wilson and Herrnstein 1985).

We are now in position to describe the nature of self-control, the individual characteristic relevant to the commission of criminal acts. We assume that the nature of this characteristic can be derived directly from the nature of criminal acts. We thus infer from the nature of crime what people who refrain from criminal acts are like before they reach the age at which crime becomes a logical possibility. We then work back further to the factors producing their restraint, back to the causes of self-control. In our view, lack of self-control does not require crime and can be counteracted by situational conditions or other properties of the individual. At the same time, we suggest that high self-control effectively reduces the possibility of crime—that is, those possessing it will be substantially less likely at all periods of life to engage in criminal acts.

The Elements of Self-Control

Criminal acts provide *immediate* gratification of desires. A major characteristic of people with low self-control is therefore a tendency to respond to tangible stimuli in the immediate environment, to have a concrete "here and now" orientation. People with high self-control, in contrast, tend to defer gratification.

Criminal acts provide *easy or simple* gratification of desires. They provide money without work, sex without courtship, revenge without court delays. People lacking self-control also tend to lack diligence, tenacity, or persistence in a course of action.

Criminal acts are *exciting, risky, or thrilling*. They involve stealth, danger, speed, agility, deception, or power. People lacking self-control therefore tend to be adventuresome, active, and physical. Those with high levels of self-control tend to be cautious, cognitive, and verbal.

Crimes provide *few or meager long-term benefits*. They are not equivalent to a job or a career. On the contrary, crimes interfere with long-term commitments to jobs, marriages, family, or friends. People with low self-control thus tend to have unstable marriages, friendships, and job profiles. They tend to be little interested in and unprepared for long-term occupational pursuits.

Crimes require *little skill or planning*. The cognitive requirements for most crimes are minimal. It follows that people lacking self-control need not possess or value cognitive or academic skills. The manual skills required for most crimes are minimal. It follows that people lacking self-control need not possess manual skills that require training or apprenticeship.

Crimes often result in *pain or discomfort for the victim*. Property is lost, bodies are injured, privacy is violated, trust is broken. It follows that people with low self-control tend to be self-centered, indifferent, or insensitive to the suffering and needs of others. It does not follow, however, that people with low self-control are routinely unkind or antisocial. On the contrary, they may discover the immediate and easy rewards of charm and generosity.

Recall that crime involves the pursuit of immediate pleasure. It follows that people lacking self-control will also tend to pursue immediate pleasures that are *not* criminal: they will tend to smoke, drink, use drugs, gamble, have children out of wedlock, and engage in illicit sex.

Crimes require the interaction of an offender with people or their property. It does not follow that people lacking self-control will tend to be gregarious or social. However, it does follow that, other things being equal, gregarious or social people are more likely to be involved in criminal acts.

The major benefit of many crimes is not pleasure but relief from momentary irritation. The irritation caused by a crying child is often the stimulus for physical abuse. That caused by a taunting stranger in a bar is often the stimulus for aggravated assault. It follows that people with low self-control tend to have minimal tolerance for frustration and little ability to respond to conflict through verbal rather than physical means.

Crimes involve the risk of violence and physical injury, of pain and suffering on the part of the offender. It does not follow that people with low self-control will tend to be tolerant of physical pain or to be indifferent to physical discomfort. It does follow that people tolerant of physical pain or indifferent to physical discomfort will be more likely to engage in criminal acts whatever their level of self-control.

The risk of criminal penalty for any given criminal act is small, but this depends in part on the circumstances of the offense. Thus, for example, not all joyrides by teenagers are equally likely to result in arrest. A car stolen from a neighbor and returned unharmed before he notices its absence is less likely to result in official notice than is a car stolen from a shopping center parking lot and abandoned at the convenience of the offender. Drinking alcohol stolen from parents and consumed in the family garage is less likely to receive official notice than drinking in the parking lot outside a concert hall. It follows that offenses differ in their validity as measures of self-control: those offenses with large risk of public awareness are better measures than those with little risk.

In sum, people who lack self-control will tend to be impulsive, insensitive, physical (as opposed to mental), risk-taking, short-sighted, and nonverbal, and they will tend therefore to engage in criminal and analogous acts. Since these traits can be identified prior to the age of responsibility for crime, since there is considerable tendency for these traits to come together in the same people, and since the traits tend to persist through life, it seems reasonable to consider them as comprising a stable construct useful in the explanation of crime.

The Many Manifestations of Low Self-Control

Our image of the "offender" suggests that crime is not an automatic or necessary consequence of low self-control. It suggests that many non-criminal acts analogous to crime (such as accidents, smoking, and alcohol use) are also manifestations of low self-control. Our image therefore implies that no specific act, type of crime, or form of deviance is uniquely required by the absence of self-control.

Because both crime and analogous behaviors stem from low self-control (that is, both are manifestations of low self-control), they will all be engaged in at a relatively high rate by people with low self control. Within the domain of the crime, then, there will be much versatility among offenders in the criminal acts in which they engage.

Research on the versatility of deviant acts supports these predictions in the strongest possible way. The variety of manifestations of low self-control is immense. In spite of years of tireless research motivated by a belief in specialization, no credible evidence of specialization has been reported. In fact, the evidence of offender versatility is overwhelming (Hirschi 1969; Hindelang 1971: Wolfgang, Figlio, and Sellin 1972; Petersilia 1980; Hindelang, Hirschi, and Weis 1981; Rojek and Erikson 1982; Klein 1984).

By versatility we mean that offenders commit a wide variety of criminal acts, with no strong inclination to pursue a specific criminal act or a pattern of criminal acts to the exclusion of others. Most theories suggest that offenders tend to specialize, whereby such terms as robber, burglar, drug dealer, rapist, and murderer have predictive or descriptive import. In fact, some theories create offender specialization as part of their explanation of crime. For example, Cloward and Ohlin (1960) create distinctive subculture of delinquency around particular forms of criminal behavior, identifying subcultures specializing in theft, violence, or drugs. In a related way, books are written about white-collar crime as though it were a clearly distinct specialty requiring a unique explanation. Research projects are undertaken for the study of drug use, or vandalism, or teen pregnancy (as though every study of delinquency were not a study of drug use and vandalism and teenage sexual behavior). Entire schools of criminology emerge to pursue patterning, sequencing, progression, escalation, onset, persistence, and desistance in the career of offenses or offenders. These efforts survive largely because their proponents fail to consider or acknowledge the clear evidence to the contrary. Other reasons for survival of such ideas may be found in the interest of politicians and members of the law enforcement community who see policy potential in criminal careers or "career criminals" (see, e.g., Blumstein et al. 1986).

Occasional reports of specialization seem to contradict this point, as do everyday observations of repetitive misbehavior by particular offenders. Some offenders rob the same store repeatedly over a period of years, or an offender commits several rapes over a (brief) period of time. Such offenders may be called "robbers" or "rapists." However, it should be noted that such labels are retrospective rather than predictive and that they typically ignore a large amount of delinquent or criminal behavior by the same offenders that is inconsistent with their alleged specialty. Thus, for example, the "rapist" will tend also to use drugs, to commit robberies and burglaries (often in concert with the rape), and to have a record for violent offenses other than rape. There is a perhaps natural tendency on the part of observers (and in official accounts) to focus on the most serious crimes in a series of events, but this tendency should not be confused with a tendency on the part of the offender to specialize in one kind of crime.

Recall that one of the defining features of crime is that it is simple and easy. Some apparent specialization will therefore occur because obvious opportunities for an easy score will tend to repeat themselves. An offender who lives next to a shopping area that is approached by pedestrians will have repeat opportunities for purse snatching, and this may show in his arrest record. But even here the specific "criminal career" will tend to quickly run its course and to be followed by offenses whose content and character is likewise determined by convenience and opportunity (which is the reason why some form of theft is always the best bet about what a person is likely to do next).

The evidence that offenders are likely to engage in noncriminal acts psychologically or theoretically equivalent to crime is, because of the relatively high rates of these "noncriminal" acts, even easier to document. Thieves are likely to smoke, drink, and skip school at considerably higher rates than nonthieves. Offenders are considerably more likely than nonoffenders to be involved in most types of accidents, including household fires, auto crashes, and unwanted pregnancies. They are also considerably more likely to die at an early age (see, e.g., Robins 1966; Eysenck 1977; Gottfredson 1984).

Good research on drug use and abuse routinely reveals that the correlates of delinquency and drug use are the same. As Akers (1984) has noted, "compared to the abstaining teenager, the drinking, smoking, and drug-taking teen is much more likely to be getting into fights, stealing, hurting other people, and committing other delinquencies." Akers goes on to say, "but the variation in the order in which they take up these things leaves little basis for proposing the causation of one by the other." In our view, the relation between drug use and delinquency is not a causal question. The correlates are the same because drug use and delinquency are both manifestations of an underlying tendency to pursue short-term, immediate pleasure. This underlying tendency (i.e., lack of self-control) has many manifestations, as listed by Harrison Gough (1948):

unconcern over the rights and privileges of others when recognizing them would interfere with personal satisfaction in any way; impulsive behavior, or apparent incongruity between the strength of the stimulus and the magnitude of the behavioral response; inability to form deep or persistent attachments to other persons or to identify in interpersonal relationships; poor judgment and planning in attaining defined goals; apparent lack of anxiety and distress over social maladjustment and unwillingness or inability to consider maladjustment qua maladjustment; a tendency to project blame onto others and to take no responsibility for failures; meaningless prevarication, often about trivial matters in situations where detection is inevitable; almost complete lack of dependability . . . and willingness to assume responsibility; and, finally, emotional poverty. [p. 362]

This combination of characteristics has been revealed in the life histories of the subjects in the famous studies by Lee Robins. Robins is one of the few researchers to focus on the varieties of deviance and the way they tend to go together in the lives of those she designates as having "antisocial personalities." In her words: "We refer to someone who fails to maintain close personal relationships with anyone else, [who] performs poorly on the job, who is involved in illegal behaviors (whether or not apprehended), who fails to support himself and his dependents without outside aid, and who is given to sudden changes of plan and loss of temper in response to what appear to others as minor frustrations" (1978: 255).

For 30 years Robins traced 524 children referred to a guidance clinic in St. Louis, Missouri, and she compared them to a control group matched on IQ, age, sex, and area of the city. She discovered that, in comparison to the control group, those people referred at an early age were more likely to be arrested as adults (for a wide variety of offenses), were less likely to get married, were more likely to be divorced, were more likely to marry a spouse with a behavior problem, were less likely to have children (but if they had children were likely to have more children), were more likely to have children with behavior problems, were more likely to be unemployed, had considerably more frequent job changes, were more likely to be on welfare, had fewer contacts with relatives, had fewer friends, were substantially less likely to attend church, were less likely to serve in the armed forces and more likely to be dishonorably discharged if they did serve, were more likely to exhibit physical evidence of excessive alcohol use, and were more likely to be hospitalized for psychiatric problems (1966: 42–73).

Note that these outcomes are consistent with four general elements of our notion of low self-control: basic stability of individual differences over a long period of time; great variability in the kinds of criminal acts engaged in; conceptual or causal equivalence of criminal and non-criminal acts; and inability to predict the specific forms of deviance engaged in, whether criminal or noncriminal. In our view, the idea of an antisocial personality defined by certain behavioral consequences is too positivistic or deterministic, suggesting that the offender must do certain things given his antisocial personality. Thus we would say only that the subjects in question are *more likely* to commit criminal acts (as the data indicate they are). We do not make commission of criminal acts part of the definition of the individual with low self-control.

Be this as it may, Robins's retrospective research shows that predictions derived from a concept of antisocial personality are highly consistent with the results of prospective longitudinal and cross-sectional research: offenders do not specialize; they tend to be involved in accidents, illness, and death at higher rates than the general population; they tend to have difficulty persisting in a job regardless of the particular characteristics of the job (no job will turn out to be a good job); they have difficulty acquiring and retaining friends; and they have difficulty meeting the demands of long-term financial commitments (such as mortgages or car payments) and the demands of parenting.

Seen in this light, the "costs" of low self-control for the individual may far exceed the costs of his criminal acts. In fact, it appears that crime is often among the least serious consequences of a lack of self-control in terms of the quality of life of those lacking it.

The Causes of Self-Control

We know better what deficiencies in self-control lead to than where they come from. One thing is, however, clear; low self-control is not produced by training, tutelage, or socialization. As a matter of fact, all of the characteristics associated with low self-control tend to show themselves in the absence of nurturance, discipline, or training. Given the classical appreciation of the causes of human behavior, the implications of this fact are straightforward: the causes of low self-control are negative rather than positive; self-control is unlikely in the absence of effort, intended or unintended, to create it. (This assumption separates the present theory from most modern theories of crime, where the offender is automatically seen as a product of positive forces, a creature of learning, particular pressures, or specific defect. We will return to this comparison once our theory has been fully explicated.)

At this point it would be easy to construct a theory of crime causation, according to which characteristics of potential offenders lead them ineluctably to the commission of criminal acts. Our task at this point would simply be to identify the likely sources of impulsiveness, intelligence, risk-taking, and the like. But to do so would be to follow the path that has proven so unproductive in the past, the path according to which criminals commit crimes irrespective of the characteristics of the setting or situation.

We can avoid this pitfall by recalling the elements inherent in the decision to commit a criminal act. The object of the offense is clearly pleasurable, and universally so. Engaging in the act, however, entails some risk of social, legal, and/or natural sanctions. Whereas the pleasure attained by the act is direct, obvious and immediate, the pains risked by it are not obvious, or direct, and are in any event at greater remove from it. It follows that, though there will be little variability among people in their ability to see the pleasures of crime, there will be considerable variability in their ability to calculate potential pains. But the problem goes further than this: whereas the pleasures of crime are reasonably equally distributed over the population, this is not true for the pains. Everyone appreciates money; not everyone dreads parental anger or disappointment upon learning that the money was stolen.

So, the dimensions of self-control are, in our view, factors affecting calculation of the consequences of one's acts. The impulsive or short-sighted person fails to consider the negative or painful consequences of his acts; the insensitive person has fewer negative consequences to consider; the less intelligent person also has fewer negative consequences to consider (has less to lose).

No known social group, whether criminal or noncriminal, actively or purposefully attempts to reduce the self-control of its members. Social life is not enhanced by low self-control and its consequences. On the contrary, the exhibition of these tendencies undermines harmonious group relations and the ability to achieve collective ends. These facts explicitly deny that a tendency to crime is a product of socialization, culture, or positive learning of any sort.

The traits composing low self-control are also not conducive to the achievement of long-term individual goals. On the contrary, they impede educational and occupational achievement, destroy interpersonal relations, and undermine physical health and economic well-being. Such facts explicitly deny the notion that criminality is an alternative route to the goals otherwise obtainable through legitimate avenues. It follows that people who care about the interpersonal skill, educational and occupational achievement, and physical and economic well-being of those in their care will seek to rid them of these traits.

Two general sources of variation are immediately apparent in this scheme. The first is the variation among children in the degree to which they manifest such traits to begin with. The second is the variation among caretakers in the degree to which they recognize low self-control and its consequences and the degree to which they are willing and able to correct it. Obviously, therefore, even at this threshold level the sources of low self-control are complex.

There is good evidence that some of the traits predicting subsequent involvement in crime appear as early as they can be reliably measured, including low intelligence, high activity level, physical strength, and adventuresomeness (Glueck and Glueck 1950; West and Farrington 1973). The evidence suggests that the connection between these traits and commission of criminal acts ranges from weak to moderate. Obviously, we do not suggest that people are born criminals, inherit a gene for criminality, or anything of the sort. In fact, we explicitly deny such notions. . . . What we do suggest is that individual differences may have an impact on the prospects for effective socialization (or adequate control). Effective socialization is, however, always possible whatever the configuration of individual traits.

Other traits affecting crime appear later and seem to be largely products of ineffective or incomplete socialization. For example, differences in impulsivity and insensitivity become noticeable later in childhood when they are no longer common to all children. The ability and willingness to delay immediate gratification for some larger purpose may therefore be assumed to be a consequence of training. Much parental action is in fact geared toward suppression of impulsive behavior, toward making the child consider the long-range consequences of acts. Consistent sensitivity to the needs and feelings of others may also be assumed to be a consequence of training. Indeed, much parental behavior is directed toward teaching the child about the rights and feelings of others, and of how these rights and feelings ought to constrain the child's behavior. All of these points focus our attention on childrearing.

Child-Rearing and Self-Control: The Family

The major "cause" of low self-control thus appears to be ineffective child-rearing. Put in positive terms, several conditions appear necessary to produce a socialized child. Perhaps the place to begin looking for these conditions is the research literature on the relation between family conditions and delinquency. This research (e.g., Glueck and Glueck 1950; McCord and McCord 1959) has examined the connection between many family factors and delinquency. It reports that discipline, supervision, and affection tend to be missing in the homes of delinquents, that the behavior of the parents is often "poor" (e.g., excessive drinking and poor supervision [Glueck and Glueck 1950: 110–11]); and that the parents of delinquents are unusually likely to have criminal records themselves. Indeed, according to Michael Rutter and Henri Giller, "of the parental characteristics associated with delinquency, criminality is the most striking and most consistent" (1984: 182).

Such information undermines the many explanations of crime that ignore the family, but in this form it does not represent much of an advance over the belief of the general public (and those who deal with offenders in the criminal justice system) that "defective upbringing" or "neglect" in the home is the primary cause of crime.

To put these standard research findings in perspective, we think it necessary to define the conditions necessary for adequate child-rearing to occur. The minimum conditions seem to be these: in order to teach the child self-control, someone must (1) monitor the child's behavior; (2) recognize deviant behavior when it occurs; and (3) punish such behavior. This seems simple and obvious enough. All that is required to activate the system is affection for *or* investment in the child. The person who cares for the child will watch his behavior, see him doing things he should not do, and correct him. The result may be a child more capable of delaying gratification, more sensitive to the interests and desires of others, more independent, more willing to accept restraints on his activity, and more unlikely to use force or violence to attain his ends.

When we seek the causes of low self-control, we ask where this system can go wrong. Obviously, parents do not prefer their children to be unsocialized in the terms described. We can therefore rule out in advance the possibility of positive socialization to unsocialized behavior (as cultural or subcultural deviance theories suggest). Still, the system can go wrong at any one of four places. First, the parents may not care for the child (in which case none of the other conditions could be met); second, the parents, even if they care, may not have the time or energy to monitor the child's behavior; third, the parents, even if they care *and* monitor, may not see anything wrong with the child's behavior; finally, even if everything else is in place, the parents may not have the inclination or the means to punish the child. So, what may appear at first glance to be nonproblematic turns out to be problematic indeed. Many things can go wrong. According to much research in crime and delinquency, in the homes of problem children many things have gone wrong: "Parents of stealers do not track ([they] do not interpret stealing . . . as deviant); they do not punish; and they do not care" (Patterson 1980: 88–89; see also Glueck and Glueck 1950; McCord and McCord 1959; West and Farrington 1977).

NOTE

1. We do not mean to imply that stable individual differences between offenders and non-offenders are nonexistent. The fact of the matter is, however, that substantial evidence documenting individual differences is not as clear to us as it appears to be to others. The evidence on intelligence is an exception. Here differences favoring nonoffenders have been abundantly documented (cf. Wilson and Herrnstein 1985).

REFERENCES

Akers, Ronald L. 1984. "Delinquent Behavior, Drugs, and Alcohol: What Is the Relationship?" *Today's Delinquent,* 3: 19–47.

Bentham, Jeremy. 1970 [1789]. *An Introduction to the Principles of Morals and Legislation.* London: The Athlone Press.

Blumstein, Alfred, Jacqueline Cohen, Jeffery Roth, and Christy Visher. 1986. *Criminal Careers and "Career Criminals."* Washington, D.C.: National Academy Press.

Cloward, Richard, and Lloyd Ohlin. 1960. *Delinquency and Opportunity.* New York: The Free Press.

Eysenck, Hans. 1977. *Crime and Personality.* Rev. ed. London: Paladin.

Glueck, Sheldon, and Eleanor Glueck. 1950. *Unraveling Juvenile Delinquency.* Cambridge, Mass.: Harvard University Press.

Gottfredson, Michael. 1984. *Victims of Crime: The Dimensions of Risk.* London: HMSO.

Gough, Harrison G. 1948. "A Sociological Theory of Psychopathy." *American Journal of Sociology,* 53: 359–66.

Hindelang, Michael J. 1971. "Age, Sex, and the Versatility of Delinquent Involvements." *Social Problems,* 18: 522–35.

Hirschi, Travis. 1969. *Causes of Delinquency.* Berkeley: University of California Press.

———. 1986. "The Distinction Between Crime and Criminality." In *Critique and Explanation: Essays in Honor of Gwynne Nettler,* edited by T. F. Hartnagel and R. Silverman (pp. 55–69). New Brunswick, N.J.: Transaction.

Klein, Malcolm. 1984. "Offense Specialization and Versatility Among Juveniles." *British Journal of Criminology,* 24: 185–94.

McCord, William, and Joan McCord. 1959. *Origins of Crime: A New Evolution of the Cambridge-Somerville Study.* New York: Columbia University Press.

Patterson, Gerald R. 1980. "Children Who Steal." In *Understanding Crime,* edited by T. Hirschi and M. Gottfredson (pp. 73–90). Beverly Hills, Calif.: Sage.

Petersilia, Joan. 1980. "Criminal Career Research: A Review of Recent Evidence." In *Crime and Justice: An Annual Review of Research,* vol. 2, edited by M. Tonry and N. Morris (pp. 321–79). Chicago: University of Chicago Press.

Robins, Lee. 1966. *Deviant Children Grown Up.* Baltimore: Williams and Wilkins.

Rojek, Dean, and Maynard Erickson. 1982. "Delinquent Careers." *Criminology,* 20: 5–28.

West, Donald, and David Farrington. 1973. *Who Becomes Delinquent?* London: Heinemann.

———. 1977. *The Delinquent Way of Life.* London: Heinemann.

Wilson, James Q., and Richard Herrnstein. 1985. *Crime and Human Nature.* New York: Simon and Schuster.

Wolfgang, Marvin, Robert Figlio, and Thorsten Sellin. 1972. *Delinquency in a Birth Cohort.* Chicago: University of Chicago Press.

QUESTIONS FOR DISCUSSION

1. According to the authors, why do individuals commit criminal acts? Do you agree or disagree with the authors? Explain your response.

2. Discuss the elements of self-control. Give elements from your geographical location to support each element.

3. Discuss the manifestations and causes of low self-control. Create a program that will address these issues and help families produce nondelinquent children.

The Empirical Status of Hirschi's Control Theory

Kimberly L. Kempf

Travis Hirschi introduced his theory of social control in 1969 in *Causes of Delinquency* and it has been gaining popularity for nearly twenty-five years. It was argued recently that, "After a number of years of being dominated by strain theories, criminological theory and research has come increasingly under the influence of control theories, particularly as formulated by Hirschi" (Vold and Bernard 1986, 247). There are several indications that social control theory is perhaps the most popular criminological theory, and yet to date there has been no systematic critique of control theory research.

The absence of an evaluation summarizing the research achievements of control theory has not, however, precluded the "wheel of science" from amassing results. As evidence, the theory is a popular subject for dissertations.[1] Scholarly papers either cite the relevance of social control or purport to be formal tests of the theory with perhaps unrivaled frequency. This favored status of control theory has been explained by its testable nature, because the theory lends itself to self-report survey techniques, and because it has achieved support from empirical research (Vold and Bernard 1986, 247; citing Nettler 1984; Arnold and Brungardt 1983; Kornhauser 1978). It also has been argued that "at the beginning of the 1970s, [Hirschi's theory] was the only theoretical formulation that tried to synthesize in a coherent and complex theoretical plan a great deal of information on the causes of delinquent behavior" (LeBlanc 1983, 40; LeBlanc, Ouimet, and Tremblay 1988, 164). The likely acceptance of this theory for future policy development has been noted as well (Williams and McShane 1988, 131).

There are concerns, however, about the scope of social control theory. Just how generalizable are the elements of the social bond? Does the theory explain specific forms of delinquency (e.g., theft, damage, violence), as well as general deviance (LeBlanc et al. 1988, 175)? Is the theory equally applicable across age, race and socioeconomic groups? Thio (1978, 49–53), for example, contended that control theory is oversimplified and applicable to unsophisticated delinquent behavior only. Vold and Bernard (1987, 248) argue that neither adult criminality nor truly offensive delinquency actually have been examined (1987, 248). In a similar vein, Albert Cohen commented that Hirschi's theory of social control may be "fertile but not yet fecund" (1985).

"If social bonding theory is to be an 'enduring contribution to criminology' (Gibbons 1979, 121), then the basis for such longevity should be the research results which support, qualify and lead to modification in the perspective" (Krohn and Massey 1980, 539). It is likely that Hirschi would agree to this standard because he has questioned the continued dominance of differential association theory based on its scientific adequacy (Hirschi and Gottfredson 1980, 9). Hirschi also previously acknowledged the need for empirical tests to build on one another to achieve theoretical development. In doing so he stated, "it is easier to construct theories 'twenty years ahead of their time' than theories grounded on and consistent with data currently available" (Hirschi 1969, preface). This paper attempts to determine the extent to which these objectives have been accomplished by the empirical tests of the theory of social control. Has causation been established and, if so, to what degree are the findings generalizable? Hirschi's theory will be outlined briefly and its connections to earlier theories highlighted. An overview of empirical tests of Hirschi's theory will be presented. The current state of the theory research will be identified for scholars interested in testing control theory today. Not only does the volume of published social control research merit an evaluation of this type, but it is important to determine the knowledge gained in twenty years.

The Theory of Social Control

"A person is free to commit delinquent acts because his ties to the conventional order have somehow been broken" (Hirschi 1969, 3). This perspective of human behavior offered by Hirschi developed from much earlier ideas, including Hobbes' comments on the contradictions between laws and human nature in *Leviathan*. The influence of Durkheim's anomie, or the weakened collective conscience, and the resulting breakdown of social restraints on behavior also is apparent. Durkheim (1951, 1965ab) discussed the need for rules and the unpredictable behavior of children. Similarly, Piaget (1936) held that behavior is the response to pressures and attempts to influence them.

Control theory seems to accept, at least partially, the Freudian idea that deviant impulses can come forth naturally. For example, Aichhorn (1963) viewed delinquency as the result of unrepressed id freed from restraints by inadequate superego development. Reiss (1951) accepted weak ego and superego development as an explanation for poor personal controls among delinquents. Eysenck's (1964) typology of autonomous nervous system response also included conditions under which punishment fails to inhibit natural tendencies.

Toby (1957) and Briar and Piliavin (1965) introduced the concept "stakes in conformity" to control theory by suggesting that all youths are tempted to violate the law, but some have more to lose in doing so. This contribution clarified that self-interest is an important dimension of the element of commitment in control theory.

Targeting family as the primary source of control, Nye (1958) accepted the Freudian psychology that society must exercise control over our "animal instincts." Internal control occurs when parents socialize children to accept society's values, and thereby develop their conscience. The affection and respect children feel for parents serves as an indirect control. Direct control comes from the presence of parents, friends, police, among others, or from the threat of punishment from these conforming persons. Society also provides legitimate outlets for satisfying the inherent needs of persons in Nye's social control theory.

The theory of delinquency tested by Reckless (1961) included outer containment forces (including poverty, conflict, external restraint, minority group affiliation, and limited access to legitimate opportunities) and inner containment (such as drives, motives, frustrations, rebellion, and hostility).

In *Delinquency and Drift*, Matza (1964) provided a popular explanation for delinquency by integrating earlier ideas of subterranean values, drift, and techniques of neutralization (Matza and Sykes 1961; Sykes and Matza 1957). Drifting to and from delinquency involves freedom from restraint in the social structure. Techniques of neutralization provide delinquents with the opportunity to rationalize their behavior a priori, and thereby neutralize their guilt. Hirschi adapted these techniques of neutralization directly within his element of belief, but he rejected the notion that neutralization motivates delinquents. Instead, control theory contends that the probability of delinquency increases as persons' beliefs in the validity of social norms weakens (Hirschi 1969, 26, 199). The assumption of variation in belief is critical to control theory, and Hirschi incorporates the techniques of neutralization as components of his measures of belief.

The theory of social control offered by Hirschi evolved from many previous contributions. The ability to deviate from normative behavior is considered universal by control theory. Most people do not indulge in deviant behavior because of their bond to society. The social bond was conceptualized by Hirschi through the following four elements:

1. attachment of the individual to others (caring about others, their opinions and expectations),
2. commitment to conventional lines of action (the rational component including risk, energy and self investment in conventional behaviors),
3. involvement (time engrossed in conventional activities), and
4. belief in legitimate order (attribution of moral validity to social norms).

These components of the bond to society were viewed by Hirschi as independent and as having a generally negative association with the likelihood of engaging in delinquent behavior. Hirschi postulated that as the elements of the social bond become weakened, the probability of delinquency will increase. Readers not familiar with Hirschi's theory are encouraged to review the original publication.

Hirschi tested his theory through a survey administered in 1965 to a stratified random sample of 3,605 adolescent males drawn as part of the Richmond Youth Project in California. Based on a series of tabular analyses, Hirschi concluded support for his theoretical model. Hirschi conceded, however, that the theory underestimated the importance of peers and the Richmond data failed to tap all domains of this concept. He also accepted that too much importance had been bestowed on the concept of involvement. He deferred revision of the theory until more is learned about the processes that effect the elements of the bond and whether delinquency itself motivates commitment (Hirschi 1969, 229–31). Hirschi has not sought these answers through additional empirical investigation.

The intent of this paper is to examine empirical investigations of the theory of social control since Hirschi's 1969 contribution. Among subsequent studies, attention will be given to the intent of the inquiry, the nature of the subjects tested, conceptualization, technique of investigation, and relevance of findings. This quasi-meta analysis will provide new information about the status of control theory and the contribution it has made to our understanding of crime and delinquency.

Descriptions of the Tests of Hirschi's Theory

Empirical tests of control theory published between 1970 and 1991 were identified through a two-stage process. Studies initially were located through a search of the *Social Science Citation Indices, Criminal Justice Abstracts* and *Psychological Literature Indices* using "control," "social control," "social bond," and "Hirschi" as key words. The bibliographies of studies identified in this manner were examined to locate additional investigations with which to buttress the initial sample. The latter effort was useful for locating books and articles published in journals perhaps not reviewed in the three indices.

Three criteria were adopted for screening the research. First, there must be an acknowledged test of control theory. Second, Hirschi (1969) must be cited, thereby increasing the likelihood that the test was of Hirschi's version of the theory. And, third, the study must be published, utilizing the assumption that the publication process lends credibility to the scholarly contribution of the article. Although the selection process was rigorous, it was not presumed to be exhaustive and undoubtedly some tests of Hirschi's theory are regrettably omitted. There is, however, no reason to believe the sample studies are not representative of control theory research so they should enable us to ascertain the general level of growth achieved since 1969.

Ultimately, seventy-one empirical tests of control theory were identified (shown in Table 1). These investigations included a variety of specific research objectives. There were twenty-eight replication efforts, of which many attempted to extend the theory to demographic groups not examined by Hirschi, such as company executives (Lasley 1988), rural youth (Gardner and Shoemaker 1989; Hindelang 1973; Krohn and Massey 1980; Lyerly and Skipper 1981), females (Hindelang 1973) and minority youths (Gardner and Shoemaker 1989; Robbins 1984) and other domains of deviance, including sixteen studies of only drug or alcohol use, four studies of adult crime, two studies of sexual behavior, and one study of mental health disorders. There also were sixteen studies in which social control was compared to another theory. Sixteen additional studies attempted to test theoretical models integrating social control with another theory.[2] The integration efforts with social control most often included social learning or differential association, followed second by deterrence. In a few cases, elements of social control also were combined with strain, social disorganization, culture conflict and conflict of interest theories. Eleven attempts to initiate contributions to theory development were reported. Potential improvements in these eleven studies included the application of new statistical procedures, analyses of unique data, and innovative sample groups.

With the exception of two studies that relied on official adult records (Linquist et al. 1985 and Minor 1977) and one study of aggregated data (Caldas and Pounder 1990), control theory has been tested with survey data collected primarily through self-administered questionnaires. Although names were recorded in the Richmond data, most surveys assured anonymity. Studies have relied almost exclusively on one-time cross sectional designs. Exceptions include twelve studies based on longitudinal data, however, the majority of these analyzed the data as though they were cross sectional. Data were shared by several studies. For example, six studies used Hirschi's own Richmond data and five studies were based on the Youth in Transition data (Bachman et al. 1975). There also were multiple studies of original data by the same researchers (e.g., Hagan et al. 1985, 1987; Caplan and LeBlanc 1985; LeBlanc et al. 1988). Tests of control theory have included as few as seventy-two subjects and as many as over 4,000. White adolescent students, a majority of whom are male, have been the usual target of the control theory research.[3] In view of the overrepresentation of students, it is not surprising to discover that random selection of any type[4] was not included in most studies.

The apparent lack of variation in design elements precludes this investigation from pursuing an empirically based meta analysis, such as those used by Bridges and Weis (1988) to examine the accumulation of research on criminal violence or the structural covariates of homicide rates by Land et al. (1985). The similarity in research designs may serve as an advantage in this investigation of the empirical status of the theory, however, because differences among the empirical findings may be more readily attributed to operational definitions of the social bond.

Conceptual and Operational Definitions

Because these studies were self-acknowledged tests of control theory (rather than subjectively designated as such for our purpose), elements of the social bond were classified according to the original researchers' intent (e.g., if they used the label of attachment, so did I). In the rare event that the study did not mention the social bond specifically, an attempt was made to be inclusive and to categorize variables according to other studies with comparable indicators.

This classification revealed that all four elements of the social bond were included in only seventeen studies. Attachment was the most frequently included element, but there was *no* single element that appeared in every study. The volume and variety of additional measures in these studies also is noteworthy. Many of the variables classified by some researchers as concepts from other theories or spurious factors were defined in other studies as elements of the social bond.[5]

Not only were they omitted from study, but it is equally surprising that justification for the missing elements was rarely provided. One exception, the element of involvement in conventional activities, was subsumed in the measure of commitment (Krohn and Massey 1980; Krohn et al. 1984) and attachment and commitment (Minor 1977) because involvement was viewed as "too conceptually ambiguous to be of theoretical utility" (Minor 1977, 122). Although a significant effect for involvement was shown in Agnew (1985), this investigator chose to ignore the concept in a subsequent test (Agnew 1991). This a priori decision to omit involvement is even more questionable because it is based on findings from Elliott et al. (1985), a study that utilized the same data as Agnew (1991), and which he also criticized. Involvement is the least often measured element in the collection of studies. It is plausible that many investigators dismissed it based on Hirschi's interpretation of its lesser importance or they may have adhered to Minor's argument that involvement is an ambiguous term.

Studies on the Relationship between Social Control & Crime

TABLE 1

Author(s)	Data Collection Year & Location	Usable Cases	% Total	Age/grade	% Male	$ White	Population	Test Type
Hirschi (1969)	1965, Richmond, CA	3,605	65	7–12 g.	100[1]	59	student	original
Agnew (1985)	1966, 68, USA	1,886	83	10, 11th g.	100	88	student	replication
Agnew (1991)	1976, 77, USA	1,655	96	ages 11–17	—	—	youth	replication
Aultman (1979)	1976, Tallahassee	1,500	—	6–12th g.	53	75	student	comparison
Bishop (1984)	——, Virginia	2,147	71	6–12th g.	100	—	student	expansion
Buffalo & Rodgers (1971)	1966, Topeka, KS	164	96	ages 13–18	50	—	delinquent	replication
Burkett & Jensen (1975)	——, Seattle	1,056	—	12th g.	—	—	student	replication
Caldas & Pounder (1990)	1980, Louisiana	64[2]	—	—	100	—	—	replication
Caplan & LeBlanc (1985)	1974, Montreal	1,441	98	—	100	—	student	integration
Cernkovich (1978)	1964, Seattle & midwest	412	—	ages 14–18	100	100	student	integration
Conger (1976)	1965, Richmond, CA	374	—	7th g.	100	28	student	integration
Dembo et al. (1985)	1976, New York City	1,493	94	7–12th g.	100[1]	100	student	integration
Elliott et al. (1988)	1976–83, USA	1,501	23	7–9th g.	53	0	youth	replication
Eve (1978)	1971, southeast	300	97	ages 11–12	45	94	student	integration
Finnell & Jones (1975)	1972, south	265	79	11–12th g.	—	79	student	replication
Friedman & Rosenbaum (1988)	1985, midwestern city	2,926	95	college seniors	49	—	student	expansion
Gardner & Shoemaker (1989)	1984, Virginia	733	91	high school	47	79	student	replication
Ginsberg & Greenley (1978)	1971, 74, Madison, WI	274	86	ages 13–20	—	56	student	comparison
Gottfredson et al. (1991)	1982, 84 4 cities	3,729	93	college	—	—	student	integration
Hagan & Simpson (1978)	1971, Toronto	302	72	junior high	50	27	student	replication
Hagan et al. (1985)	1979, Toronto	458	—	college	50	—	student	integration
Hagan et al. (1987)	1979, Toronto	463	69	9–12th g.	100	—	student	integration
Hepburn (1977)	1972, midwest city	139	—	9–12th g.	50	100	stu. & del.	comparison
Hindelang (1973)	rural NY	941	96	ages 14–17	—	98	student	replication
Jensen & Eve (1976)	1965, Richmond, CA	>4,000	—	6–12th g.	48	—	student	comparison
Johnson, B. (1973)	1970, 21 NY colleges	3,500	—	7–12th g.	44	89	student	comparison
Johnson, K. (1984)	——, northwest town	345	—	college	51	83	student	replication
Johnson R. (1987)	——, Seattle	734	43	10–12th g.	—	71	student	replication
Kaplan et al. (1984)	1971–73, Houston	3,052	92	10th g.	100	—	student	integration
Kelly & Pink (1973)	1967, northwest	284	67	7–9th g.	50	—	student	replication
Krohn & Massey (1980)	——, 3 midwest states	2,054	85	10th g.	—	—	student	replication
Krohn et al. (1984)	——, 4 midwest states	695	—	7–12th g.	100	99	student	comparison
LaGrange & White (1985)	——, New Jersey	341	67	ages 12, 15, 18	77	—	general	integration
Lasley (1988)	——, multinational corp.	435	51	ages 33–78	100[1]	77	executives	replication
LeBlanc (1983)	1974, 76, Montreal	458	27	ages 12–18	56	—	student	replication
LeBlanc et al. (1988)	1974, 76, Montreal &	825	72	ages 14–18	—	—	student	comparison
	1985, Montreal	797	—	ages 14–15	—	—	student	expansion
Levine & Kozak (1979)	1977, Chicago	796	32	5–12th g.	—	—	student	replication

Author(s)	Data Collection Year & Location	Usable Cases	% Total	Age/grade	% Male	$ White	Population	Test Type
Linden, E. & Hackler (1973)	1966, Seattle	200	—	ages 13–15	100	—	l.c. housing	comparison
Linden, R. (1978)	1965, Richmond, CA	990	62	7–12th g.	100	100	student	replication
Lindquist et al. (1985)	——, 3 Florida co.	328	63	adults	71	—	offender	replication
Liska & Reed (1985)	1966, 68, USA	1,886	83	10, 12th g.	100	88	student	expansion
Lyerly & Skipper (1981)	——, Virginia	100	—	teenage	100	100	delinquent	replication
Mak (1990)	——, Canberra, Australia	793	85	8–12th g.	51	—	student	expansion
Mannle & Lewis (1979)	——, southeast state	267	85	age 15	47	53	delinquent	replication
Marcos et al. (1986)	——, southeast	2,676	97	ages 14–19	47	83	student	integration
Mathur & Dodder (1985)	——, southwest	722	—	youth	44	94	stu. & del.	expansion
Matsueda (1982)	1965, Richmond, CA	1,140	72	7–12th g.	100	100	student	comparison
Matsueda & Heimer (1987)	1965, Contra Costa co, CA	2,589	75	high school	100	61	student	comparison
Matsueda (1989)	1966, 68, USA	1,912	—	high school	100	100	students	comparison
Meier & Johnson (1977)	1971, Cook co, IL	632	—	ages 15–18	—	—	student	integration
Menard & Morse (1984)	——, San Diego	257	39	9th g.	40	—	student	replication
Mercer et al. (1978)	1974, Ontario	286	—	—	43	—	student	replication
Minor (1977)	——, Tallahassee	260	95	ages 18–50	50	100	general	integration
Paternoster & Triplett (1988)	1982, southeast	1,544	—	10–11th g.	—	—	student	comparison
Paternoster et al. (1983)	1975, 76, ——	300	—	college	51	90	student	comparison
Poole & Regoli (1979)	1972, midwest	72	69	ages 14–17	100	100	student	integration
Rankin (1977)	1974, Wayne co, MI	385	79	7–11th g.	50	100	student	expansion
Robbins (1984)	1978–80, 3 reservations	129	70	ages 10–17	—	0	youth	replication
Rosenbaum (1987)	1977–79, Seattle	1,612	—	ages 15–18	—	69	stu. & del.	replication
Segrave & Hastad (1985)	——, east coast city	1,776	—	ages 14–18	51	—	student	integration
Shover et al. (1979)	1977, southeast	1,002	39	8–12th g.	39	—	student	comparison
Singer & Levine (1988)	1987, suburban	560	38	high school	—	—	student	integration
Smith (1979)	——, NJ, IA, OR	1,950	98	ages ≥15	42	—	general	comparison
Thompson et al. (1984)	——, southwest	724	—	youth	44	94	stu. & del.	replication
Thornton & Voight (1984)	——, Louisiana city	3,500	—	ages 11–17	50	75	student	expansion
Torstensson (1990)	1953–1983, Stockholm	791	11	female cohort	—	0	delinquent	replication
Udry (1988)	1982, southern city	201	75	ages 13–16	51	100	student	integration
Wells & Rankin (1983)	1966–70, USA	1,799	79	10–12th g.	100	—	student	expansion
Wiatrowski et al. (1981)	1966, 68, USA	1,886	83	10, 12th g.	100	88	student	expansion
Williams & Hawkins (1989)	1985–87, USA	483	40	ages ≥18	100	—	married	replication
Williams (1985)	——, Texas	1,342	70	ages ≥16	—	—	licensed driver	comparison
Winfree et al. (1981)	——, rural rocky mtn	605	67	6–12th g.	49	83	student	integration

1. Complete data were obtained but not analyzed for 1,940 girls (Hirschi 1969, 35–36). Females were omitted from the actual analysis because they were expected to be "less at risk of using various drugs" (Dembo et al. 1985, 273) and to "increase clarity of the interpretation" (La Grange and White 1985, 20).

2. This study used aggregate level data for sixty-four parishes and school districts.

The absence of arguments defending the omission of the other three elements of the social bond remains perplexing.

A more parsimonious theoretical model of social control than that presented by Hirschi might have been achieved in the control theory research, and thereby explain the missing elements. In view of this possibility, this inquiry next examines the operational definitions of each element of the social bond adhered to in the sample of studies.

Attachment to Conventional Others (Parents, Peers, School)

In Hirschi's own test, the number of measures of attachment exceeded those for the other elements of the social bond. He identified parents, the school, and peers as separate foci for attachment, and multiple dimensions were noted for each target. Based on some of the results, however, Hirschi also commented "that distinctions between the dimensions of attachment to the parent may be artificial" (1969, 93). The law also has been specified as a source of attachment (Jensen and Eve 1976), although Hirschi (1969) had specified this as an element of belief.

Attachment to Parents

For attachment to parents, Hirschi was concerned about the direct and psychological holds the parent(s) had on the child, the intimacy of parent-child communication and affection or respect. Virtual supervision was operationalized through a summated index of these questions: *does your mother (father) know where you are when you are away from home? does your mother (father) know whom you are with when you are away from home?* Two intimacy of communication indices were constructed. For the child to parent index, responses to the following items were summed: *do you share your thoughts and feelings with your mother (father)? how often have you talked over your future plans with your mother (father)?* The parent to child index combined these items: *when you don't know why your mother (father) makes a rule, will she explain the reason? when you come across things you don't understand, does your mother (father) help you with them? does your mother (father) ever explain why she feels the way she does?* Affection was noted as the crucial element by Hirschi, although problems were mentioned with available measures (Hirschi 1969, 90–93). Affectional identification was measured by the item: *would you like to be the kind of person your father (mother) is?* Equivalent importance was given to mothers and fathers for all dimensions.

Attachment to parents was the most frequent focus of attachment, as well as the most often examined element in the social bond among the studies reviewed. Fifty studies (70 percent of all studies) included attachment to parents, sometimes specified narrowly as only mother, or broadly as family. Hirschi's items indicating affection for parents, and parental supervision have been replicated the most often. New variables have been most frequently substituted for the parent to child communication index; this occurred in seven studies intended to replicate or improve the theory. Only the virtual supervision measures have been substituted for other concepts.

There were a few unique indicators of attachment to family as well. For example, residing with family was considered attachment by Hagan and Simpson (1978). In their study based on probation files (Linquist et al. 1985), marital status and number of dependents measured attachment to family.

Attachment to Peers

Hirschi measured attachment to peers according to the following items: *would you like to be the kind of person your best friends are? a friend's reaction would be the worst thing about getting caught for stealing; do you respect your best friend's opinions about the important things in life? the number of close friends who have been picked up by the police.*

There were thirty-nine studies (55 percent of all reviewed) with indicators of attachment to peers (or friends, associates). Delinquent peer items were by far the most common. Delinquent peer items often differed from Hirschi's use of number of peers and usually specified the type of delinquency involvement or type of feelings for these friends. Interest in being like friends and respect for friends' opinions were equally represented among the studies. Hirschi's item concerning friends' reaction to a theft was rarely studied. Fourteen studies, including nine replication efforts, tested unique peer attachment variables, such as dating, importance of time with family (Agnew 1985), would prefer other friends (Conger 1976), school peers reject me (Johnson 1987), and my friends are among the most popular (Linden 1978).

Attachment to School

Affection for the school, measured by general like or dislike for school, was Hirschi's variable. There were other school related variables,[6] but they were mentioned in connection with "the causal chain." According to Hirschi, "The causal chain runs from academic incompetence to poor school performance to disliking of school to rejection of the school's authority to the commission of delinquent acts" (Hirschi 1969, 132).

Twenty-nine studies (41 percent of the total) incorporated attachment to school as an element of the social bond, but none of them used Hirschi's measure to do so. Instead, eleven similar variables were substituted. Some of the other school variables Hirschi used, particularly those related to teachers, have since been considered within attachment. Attachment to school was measured by GPA (Jensen and Eve 1976; Wiatrowski et al. 1981) and educational and occupational goals (Aultman 1979). Finally, twelve studies rejected the attachment construct in their tests of the variable like or dislike school; eight of these studies attempted to expand the scope of the theory.

Involvement in Conventional Activities

Variables which identify activity type or length of time spent on certain activities account for most of the measures of involvement in conventional activities. Hirschi's items included the following: *currently working for pay; number of hour spent on homework per day; ever feel that 'there's nothing to do'; number of hours spent talking with friends; number of hours spent riding around in a car.* Data also were obtained but results were not reported by Hirschi for involvement in other school activities, clubs, and community organizations.

Involvement was adapted in only twenty-five (35 percent) subsequent investigations of control theory. Among these studies, hours spent on homework, followed by time working most often served to measure the concept. These two indicators also represented other concepts, however, and with nearly the same frequency. Time spent in school organizations, clubs and community activities also were reported as involvement. Church attendance and religiosity were identified as involvement in three studies. Scales or multiple variables were used frequently, although exceptions merit concern, such as a dichotomous measure of church attendance within last year used as the sole criterion of involvement (Hagan and Simpson 1978). Other anomalies among the involvement measures include items more often found as variables for other elements of the bond (e.g., value of good grades, parents know location with whom, like and respect friends, perception of opportunities). Many of the social control measures available in the National Youth Survey data have been equated as indicators of involvement (Jensen 1986, cited in Agnew 1991).

Commitment to Conventional Lines of Actions

Hirschi viewed achievement orientation, passage to "adult status," aspirations and expectations as dimensions of commitment. His index of achievement orientation included the following items: *how important is getting good grades to you personally? whatever I do, I try hard; I try hard in school.* A summated index with activities considered adult-like included *smoking, drinking, and dating.* Students were asked which of these they did and at what age they began. The importance of having a car and attitude about school intervention in smoking outside the classroom also were identified. Aspirations and expectations first were measured by Hirschi for education and employment. Individual items included the following: *how much schooling would you like to get eventually? how much schooling do you expect to get? do your parents want you to go to college? the only reason to have a job is for money; you should not expect too much out of life; an easy life is a happy life; what type of job would you like to get eventually? what type of job do you expect to get? do you think that either your competence or racial discrimination will keep you from getting the kind of job you want to have eventually?*

In the items specified as commitment in forty-one studies (58 percent of all reviewed), each of Hirschi's four concepts is represented, although there was less consensus among recent studies regarding the measurement of commitment than was shown for either the elements of attachment or involvement. Thirteen studies included an item tapping the importance of grades, with five of them specifying the target of grade achievement. Aspiration for school level also was used, even more frequently than any of Hirschi's expectation items. Dating, more often than smoking or drinking, typified "adult status" items. Three of Hirschi's variables were never utilized. Moreover, many of Hirschi's variables specified other concepts nearly as often as they did commitment. There also were twenty commitment variables not used by Hirschi and the diversity and potential for overlapping with other elements can be seen through them. For example, commitment to education was identified through amount of time spent on homework, GPA and liking or disliking of school. Enjoyment of risk-taking behaviors and value orientations ("conventional," and those directed at grades, education, the law, and avoiding delinquency) were other criteria for commitment.

Belief

Hirschi identified values relative to law and the legal system, techniques of neutralization, and fatalistic statements as the dimensions of belief. He used items concerning *respect for the local police and concern for teacher's opinions* as indicative of respect for impersonal authority. Acceptance of the normative system was measured by the item: *it is alright to get around the law if you can get away with it.* The most techniques of neutralization were operationalized as follows: *most criminals shouldn't really be blamed for the things they have done; I can't seem to stay out of trouble no matter how hard I try; most things that people call 'delinquency' don't really hurt anyone; the man who leaves the keys in his car is as much to blame for its theft as the man who steals it; police try to give all kids an even break; suckers deserve to be taken advantage of.* Three fatalistic statements also were tested: *what is going to happen to me will happen, no matter what I do; there is no sense looking ahead since no one knows what the future will be like; a person should live for today and let tomorrow take care of itself.*

In comparison to commitment, the more recent measurement of belief in the legitimate order in thirty-seven studies (52 percent) offers a high level of consistency. The majority of the measures include value orientations toward the law and legal system, although twenty of them differed considerably from Hirschi's variables and six shared the focus of the perceived legitimacy, obedience or likelihood of sanctions. Each of Hirschi's variables for the techniques of neutralization were mentioned, the most frequent item "most things called 'delinquency' don't really hurt anyone." Hirschi's concept of fatalistic statements was never identified.

Explained Behavior

Finally, the outcome of the control model should be considered in reviewing the conceptual and operational definitions used in tests of the theory. Hirschi argued, however, that the findings of control theory research should not be dependent on the operational definition of delinquency (Hirschi 1969, 55).

There were several items on the Richmond questionnaire that dealt with delinquency, but Hirschi chose to test only six. Two of the items came from the Nye-Short (1957) scale of delinquency: *have you ever taken little things (worth less than $2) that did not belong to you? have you ever banged up something that did not belong to you on purpose?* The remaining four items were taken from the scale of theft developed by Dentler and Monroe (1961): *have you ever taken things of some value (between $2 and $50) that did not belong to you? have you ever taken things of large value (worth over $50) that did not belong to you? have you ever taken a car for a ride without the owner's permission? not counting fights you may have had with a brother or sister, have you ever beaten up on anyone or hurt anyone on purpose?* Hirschi initially created three summated indices from the six items. The standard index measured the total number of acts ever committed. The persistence index took number and recency into account. The recency index tapped only the number of acts committed during the previous year. Despite its skewness and lower predictive capability, only the recency scale was analyzed because it "was considered most appropriate as an operationalization of delinquency in terms of the theory" (Hirschi 1969, 62–3). This argument that better conceptual clarity is obtained from the recent past was the only rationale given for ignoring the other scales.

The theory of social control has been viewed by its researchers as capable of explaining delinquency, crime, adolescent sexual behavior, mental health disorders and conformity. The domain of items cover a full range of deviant behavior, from skipping a school rally or theft of under $2 to armed robbery. Several studies adapted modified versions of the original Nye-Short scale of delinquency;[7] others made a point of including dimensions of crime viewed as serious. A majority of the studies relied on summated scales of the number of infractions reported for each offense category. The recall period was generally limited to activities within the previous one, and sometimes three years. There also were crime-specific criteria, the focus of which was most often alcohol or drug use.

Discussion of the Findings

The measurement of concepts in the theory of social control has been examined and inconsistencies have been observed. Several studies introduced measurement improvements, yet the research is weakened because little or no attention has been given to the construct validity of the four elements of the social bond. Many studies relied upon factor analyses to identify composite indices, some of which were replicated in several studies. The potential for apparent criterion validity in these efforts is not a substitute for research concern about the integrity of theoretical concepts. This investigation will now turn to the empirical findings and conclusions of the reviewed studies to determine the current state of knowledge on control theory.

Although it achieved most significance in Hirschi's test, attachment was consistently identified as the weakest predictor of delinquency in the research by Krohn and Massey (1980, 538). Johnson (1987) reported no gender differences in attachment. But, perhaps the most consistent finding was the identified need for refinements in the sources of attachment (Conger 1976; Johnson 1987; Linden and Hackler 1973; Mathur and Dodder 1985; Wiatrowski et al. 1981).

Among the attachment foci, the most important indicators were affection for parents (Hindelang 1973, 180), mothers (Krohn and Massey 1980), regulation or pressure from parents and peers (Aultman 1979, 163), and school (Lyerly and Skipper 1981; Mathur and Dodder 1985; Wiatrowski et al. 1981). Johnson (1987, 313) argued that attachment to parents is too broad, misdirected, or nonlinear and "detachment" is more appropriate. Menard and Morse (1984, 1362) argued that school alienation causes family alienation. Finally, although it remains controversial,[8] attachment to delinquent friends contributed to the explanation of crime and delinquency in several studies (Buffalo and Rodgers 1971; Conger 1976; Hepburn 1977; Hindelang 1973; Kaplan et al. 1984; Krohn et al. 1984; Linden and Hackler 1973[9]; Mannle and Lewis 1979; Meier and Johnson 1977; Winfree et al. 1981).

Involvement in conventional activities was shown to be fairly unimportant unless it was concerned with commitment (Hindelang 1973, 481–84). Obviously, this sentiment was echoed by those researchers who omitted this element from their models and by Conger, who argued that "involvement can be dispensed with as a crucial element" (1976, 20). On the other hand, involvement in homework and extracurricular school activities was identified as a good predictor in the control models tested by Agnew (1985) and Wiatrowski et al. (1981).

Commitment, also considered "stakes in conformity,"[10] was identified as the element with the strongest explanatory value in only the control model examined by Krohn and Massey (1980). Their index and GPA were the most powerful commitment variables; gender differences in commitment also were shown (Krohn and Massey 1980, 538). For Meier and Johnson (1977), age, the proxy they actually used for commitment, was the most influential extralegal variable. Commitment achieved a low explanatory value in Wiatrowski et al. (1981). Lyerly and Skipper (1981) found a significant commitment effect, but not in the direction hypothesized by control theory.

Belief in the law had the greatest explanatory value in only one study (Aultman 1979), but the measure was consistently important in each of the five crime specific models she examined. A moderate effect for belief was reported by Paternoster and Triplett (1988). Weak or no effect for belief also was reported (Agnew 1985; Paternoster et al. 1983).

Demographic differences were observed in the findings of several of the control models. Age effects were shown (Agnew 1985; LaGrange and White 1985; Meier and Johnson 1977); peers were more important to older adolescents (LaGrange and White 1985) and the theory fared better for younger subjects (Agnew 1985). Gender differences were identified (Eve 1978; Segrave and Hastad 1985). Race also was important in the study by Eve (1978). Lyerly and Skipper (1981) found no significant variation according to rural or urban residences among their delinquent subjects. During one year, Linden (1978) reported a nonsignificant socioeconomic status effect, while Eve (1978) found socioeconomic status was important.

The noted importance of specifying interactions (Hagan and Simpson 1978; Krohn et al. 1984; Menard and Morse 1984; Poole and Regoli 1979; Thornton and Voight 1984) and establishing causal order among the elements of the social bond (Cernkovich 1978; Hepburn 1977; Johnson 1987; Kelly and Pink 1973; Menard and Morse 1984; Minor 1977; Paternoster et al. 1983; Thompson et al. 1984) is also evident in the multivariate, and especially path analytic models examined. Arguments in favor of longitudinal designs contend that cross sectional data "may have exaggerated the effect [of social control]" (Agnew 1985, 58) and led to erroneous conclusions (Paternoster et al. 1983, 476).

In sum, the empirical models of social control discussed above were tested for a variety of deviant behavior, and different, and sometimes contrary, results were found. When crime specific models were examined, higher levels of explanation were found for the general crime model (Minor 1977), minor delinquency (Aultman 1979; Krohn and Massey 1980) and school infractions (Eve 1978).

NOTES

1. Dissertations on control theory identified from *Dissertation Abstracts* include Antonio 1972; Bishop 1982; Caplan 1979; Chapman 1986; Creechan 1982; Gardner 1984; Gibbons 1981; Lauritsen 1989; Massey 1983; Matsueda 1984; Meier 1974; Minor 1975; Murcos 1985; Okada, 1987; Rankin, 1978; Riley, 1986; Rosenbaum 1983; Sellers 1987; Slaght 1985; Smith, C. 1983; Smith M. 1980; Smith, W. 1984; Stuck 1985; Taub 1986; Thornton 1977; Tieman 1976; Wiatrowski 1978.

2. Hirschi, however, is critical of integrated theoretical models, particularly the model developed by Elliott et al. (1979) (Hirschi 1979).

3. As with Hirschi (1969), there were a few studies for which data on female and/or minority subjects were available but analyzed and notations about their inclusion in future investigations were made, but apparently not yet published.

4. This includes two-stage sampling designs in which schools were typically chosen nonrandomly, then individual classrooms within the schools were randomly selected.

5. For example, Elliott et al. (1988) used GPA, education, and employment expectations as measures of strain theory; whereas, Hirschi defined them as commitment. Parental and peer reinforcement, religiosity, gender role expectations, perceived academic and social success, television viewing, dating, school performance were included in other studies as variables of theories other than social control.

6. Academic competence was measured by grades in English and mathematics (overall GPA was unavailable), Differential Aptitude test score, and self-ratings of school ability. Concern for teacher's opinion of them, another school variable, was identified within the element of belief.

7. Of the twenty-three original items in the Short and Nye questionnaire, eleven were identified as unidimensional (and therefore reliable). Only seven items were included in the final investigation. Use of Guttman scaling to identify unidimensionality among the items was severely criticized by Hindelang et al. (1981, 49–50).

8. Even nine years after his own test of the theory, it was Hirschi's contention that "the ties among delinquents are not equal in quality to those among other peer groups" (1978, 337).

9. They discussed this attachment in terms of "affective ties."

10. This usage follows Toby (1957) and Briar and Piliavin (1965).

11. "Avoid the fallacy fallacy. When a theorist or a methodologist tells you you cannot do something, do it anyway. Breaking rules can be fun." (Hirschi 1973, 171–72)

12. Lagged endogenous variables are included in regression models of change. Lagged variables measure prior behavior as a predictor of current behavior of the same type.

13. Analysis of the effects of change in predictor variables requires the inclusion of a lagged predictor variable for each of the variables of interest. Agnew's interest in the effects of change in many social control variables made his model impossible to analyze statistically.

REFERENCES

Aichhorn, August. 1963. *Wayward Youth,* New York: Viking.

Agnew, Robert. 1985. "Social control theory and delinquency: a longitudinal test," *Criminology* 23 (1): 47–61.

———. 1991. "A longitudinal test of social control theory and delinquency." *Journal of Research in Crime and Delinquency* 28 (2): 126–56.

Aultman, Madeline G. 1979. "Delinquency causation: a typological comparison of path models." *Journal of Criminal Law and Criminology* 70: 152–63.

Bishop, Donna M. 1982. "Deterrence and social control: a longitudinal study of the effects of sanctioning and social bonding on the prevention of delinquency" (unpublished doctoral dissertation, State University of New York at Albany). Ann Arbor, Mich.: University Microfilms #ADG82-18700.

Briar, Scott and Irving Piliavin. 1965. "Delinquency, situational inducements and commitment to conformity." *Social Problems:* 35–45.

Bridges, George and Joseph G. Weis. 1989. "Measuring violent behavior: effects of study design on correlates of interpersonal violence." In Neil Weiner and Marvin Wolfgang (eds.) *Violent Crime, Violent Criminals,* Newbury Park, Cal.: Sage.

Buffalo, M.D. and Joseph W. Rogers. 1971. "Behavioral norms, moral norms and attachment: problems of deviance and conformity." *Social Problems* 19 (1): 101–13.

Burkett, Steven R. and Eric L. Jensen. 1975. "Conventional ties, peer influence, and the fear of apprehension: a study of adolescent marijuana use." *The Sociological Quarterly* 16 (Autumn): 522–33.

Caldas, Stephen J. and Diana G. Pounder. 1990. "Teenage fertility and its social integration correlates: a control theory explanation." *Sociological Spectrum* 10: 541–60.

Caplan, Aaron and Marc LeBlanc. 1985. "A cross-cultural verification of a social control theory." *International Journal of Comparative and Applied Criminal Justice* 9 (2): 123–33.

Cernkovich, Stephen A. 1978. "Evaluating 2 models of delinquency causation." *Criminology* 16 (3): 335–53.

Chapman, William R. 1986. "Delinquency theory and attachment to peers" (unpublished doctoral dissertation, State University of New York at Albany). Ann Arbor, Mich.: University Microfilms #ADG86-14180.

Cohen, Albert. 1985. "Crime and criminal justice in Japan." Presentation at the Center for Studies in Criminology and Criminal Law, University of Pennsylvania.

Conger, Rand. 1976. "Social control and social learning models of delinquency: a synthesis." *Criminology* 14 (1): 17–40.

Dembo, Richard, Gary Grandon, Robert W. Taylor, Lawrence La Voie, William Burgos, and Sames Schmeidler. 1985. "The influence of family relationships on marijuana use among a sample of inner city youth." *Deviant Behavior* 6 (3): 267–68.

Dentler, Robert A. and Lawrence J. Monroe. 1961. "Social correlates of early adolescent theft." *American Sociological Review* 26: 733–43.

Durkheim, Emile. 1951. *Suicide,* New York: Free Press.

———. 1965a. *The Division of Labor in Society,* New York: Free Press.

———. 1965b. *The Rules of the Sociological Method,* New York: Free Press.

Elliott, Delbert S., David Huzinga, and Scott Menard. 1988. *Multiple Problem Youth: Delinquency, Substance Use, and Mental Health Problems.* N.Y.: Springer-Verlag.

Eve, Raymond A. 1978. "A study of the efficacy and interactions of several theories for explaining rebelliousness among high school students." *Journal of Criminal Law and Criminology* 69: 115–25.

Eysenck, Hans J. 1964. *Crime and Personality.* Boston: Houston Mifflin.

Finnell, William S. and John D. Jones. 1975. "Marijuana, alcohol and academic performance." *Journal of Drug Education* 5 (1): 13–21.

Friedman, Jennifer and Dennis R. Rosenbaum. 1988. "Social control theory: the salience of components by age, gender, and type of crime." *Journal of Quantitative Criminology* 4 (4): 363–81.

Gardner, LeGrande and Donald J. Shoemaker. 1989. "Social bonding and delinquency: a comparative analysis." *The Sociological Quarterly* 30 (3): 481–500.

Ginsberg, Irving J. and James R. Greenley. 1979. "Competing theories of marijuana use: a longitudinal study." *Journal of Health and Social Behavior* 19 (March): 22–34.

Gottfredson, Denise C., Richard J. McNeil, III, and Gary D. Gottfredson. 1991. "Social area influences on delinquency: a multilevel analysis." *Journal of Research in Crime and Delinquency* 28 (2): 197–226.

Hagan, John and John H. Simpson. 1978. "Ties that bind: conformity and the social control of student discontent." *Sociology and Social Research* 61 (4): 520–38.

Hagan, John, A. R. Gillis, and John Simpson. 1985. "The class structure of gender and delinquency: toward a power-control theory of common delinquent behavior." *American Journal of Sociology* 90 (6): 1151–78.

Hagan, John, John Simpson, and A. R. Gillis. 1987. "Class in the household: a power-control theory of gender and delinquency." *American Journal of Sociology* 92 (4): 788–816.

Hepburn, John R. 1977. "Criminology: testing alternative models of delinquency causation." *The Journal of Criminal Law and Criminology* 67 (4): 450–60.

Hindelang, Michael J. 1973. "Causes of delinquency: a partial replication and extension." *Social Problems* 21: 471–87.

Hirschi, Travis. 1969. *Cause of Delinquency*. Berkeley: University of California Press.

Hirschi, Travis and Michael Gottfredson (eds.). 1980. *Understanding Crime: Current Theory and Research*. Newbury, Cal.: Sage.

Jensen, Gary F. and Raymond Eve. 1976. "Sex differences in delinquency." *Criminology* 13: 427–48.

Johnson, Bruce D. 1973. *Marihuana Users and Drug Subcultures*. New York: John Wiley & Sons.

Johnson, Kirk Alan. 1984. "The applicability of social control theory in understanding adolescent alcohol use." *Sociological Spectrum* 4: 275–94.

Johnson, Richard. 1987. "Mother's versus father's role in causing delinquency." *Adolescence* 22 (86): 305–15.

Kaplan, Howard B., Steven S. Martin, and Cynthia Robbins. 1984. "Pathways to adolescent drug use: self-derogation, peer influences, weakening of social controls and early substance use." *Journal of Health and Social Behavior:* 25 (September): 270–89.

Kelly Delos H. and William T. Pink. 1973. "School commitment, youth rebellion and delinquency," *Criminology,* 10 (4): 473–85.

Krohn, Marvin D. and James L. Massey. 1980. "Social control and delinquent behavior: an examination of the elements of the social bond." *The Sociological Quarterly* 21 (4): 529–543.

Krohn, Marvin D., Lonn Lanza-Kaduce and Ronald L. Akers. 1984. "Community context and theories of deviant behavior: an examination of social learning and social bonding theories." *The Sociological Quarterly* 25 (Summer): 353–71.

LaGrange, Randy I. and Helene Raskin White. 1985. "Age differences in delinquency: a test of theory." *Criminology* 23 (1): 19–45.

Land, Kenneth C., Patricia L. McCall and Lawrence E. Cohen. 1985. "Structural covariates of homicide rates: are there any invariances across time and social space?" *American Journal of Sociology* 95 (4): 922–63.

LeBlanc, Marc. 1983. "Delinquency as an eiphenomenon of adolescence." in R. Corrado, M. LeBlanc and J. Tre'panier (eds.). *Current Issues in Juvenile Justice*. Butterworths: Toronto.

LeBlanc, Marc, Marc Ouimet and Richard E. Tremblay. 1988. "An integrated control theory of delinquent behavior: a validation 1976–1985." *Psychiatry* 51 (May): 164–76.

Levine, Edward M. and Conrad Kozak. 1979. "Drug and alcohol use, delinquency, and vandalism among upper middle class pre- and post-adolescents. *Journal of Youth and Adolescence* 8 (1): 91–101.

Linden, Eric and James C. Hackler. 1973. "Affective ties and delinquency." *Pacific Sociological Review* 16 (1): 27–47.

Linden, Rick. 1978. "Myths of middle class delinquency: a test of the generalizability of social control theory." *Youth and Society* 9: 407–32.

Lindquist, Charles A., Terry D. Smusz and William Doerner. 1985. "Causes of conformity: an application of control theory to adult misdemeanant probationers." *Journal of Offender Therapy and Comparative Criminology* 29 (1): 1–13.

Liska, Allen E. and Mark D. Reed. 1985. "Ties to conventional institutions and delinquency: estimating reciprocal effects." *American Sociological Review* 50 (4): 547–60.

Lyerly, Robert R. and James K. Skipper. 1981. "Rates of rural-urban delinquency." *Criminology* 19 (3): 385–99.

Mak, Anita S. 1990. "Testing a psychological control theory of delinquency." *Criminal Justice and Behavior* 17 (2): 215–30.

Mannle, Henry W. and Peter W. Lewis. 1979. "Control theory reexamined: race and the use of neutralizations among institutionalized delinquents." *Criminology* 17: 58–74.

Marcos, Anastasios C., Stephen J. Bahr, and Richard E. Johnson. 1986. "Test of a bonding/association theory of adolescent drug use." *Social Forces* 65 (1): 135–44.

Mathur, Minu and Richard A. Dodder. 1985. "Delinquency and the attachment bond in Hirschi's control theory." *Free Inquiry Into Creative Sociology* 13 (1): 99–103.

Matsueda, Ross L. 1982. "Testing control theory and differential association. *American Sociological Review* 47 (4): 489–504.

———. 1989. "The dynamics of moral beliefs and minor deviance." *Social Forces* 68 (2): 428–57.

Matsueda, Ross L. and Karen Heimer. 1987. "Race, family structure, and delinquency: a test of differential association and social control theories." *American Sociological Review* 52 (December): 826–40.

Matza, David. 1964. *Delinquency and Drift*. New York: Wiley.

Matza, David and Gresham M. Sykes. 1961. "Juvenile delinquency and subterranean values." *American Sociological Review* 26 (October): 712–719.

Meier, Robert F. and Weldon T. Johnson. 1977. "Deterrence as social control: the legal and extra-legal production of conformity." *American Sociological Review* 42 (4): 292–304.

Menard, Scott and Barbara J. Morse. 1984. "A structural critique of the IQ-delinquency hypothesis: theory and evidence." *American Journal of Sociology* 89 (6): 1347–78.

Mercer, G. William, John D. Hundleby and Richard A. Carpenter. 1978. "Adolescent drug use and attitudes toward the family." *Canadian Journal of Behavior Science* 10 (1): 79–90.

Minor, W. William. 1977. "A deterrence control theory of crime." In Robert F. Meier (ed.), *Theory in Criminology: Contemporary Views*. Sage Publications: Beverly Hills, Cal.: 117–138.

Nye, F. Ivan. 1958. *Family Relationships and Delinquent Behavior*. New York: Wiley.

Nye, F. Ivan and James F. Short, Jr. 1957. "Scaling delinquent behavior." *American Sociological Review* 22, 3 (June), 326–31.

Paternoster, Raymond and Ruth Triplett. 1988. "Disaggregating self-reported delinquency and its implications for theory." *Criminology* 26 (4): 591–625.

Paternoster, Raymond, Linda E. Saltzman, Gordon P, Waldo and Theodore G. Chiricos. 1983. "Perceived risk and social control: do sanctions really deter? *Law & Society* 17 (3): 425–56.

Poole, and Obert M. Regoli. 1979. "Parental support, delinquent friends and delinquency: a test of interaction effects." *Journal of Criminal Law and Criminology* 70 (2): 188–93.

Rankin, Joseph H. 1978. "Investigating the interrelations among social control variables and conformity; changing attitudes toward capital punishment; schools and delinquency" (unpublished doctoral dissertation, University of Arizona), Ann Arbor, Mich." University Microfilms #ADG78-11515.

Reckless, Walter C. 1961. *The Crime Problem* (third edition). New York: Appleton-Century Crofts.

Reiss, Albert J. Jr. 1951. "Delinquency as the failure of personal and social controls." *American Sociological Review* 16: 196–207.

Robbins, Susan. P. 1984. "Anglo concepts and Indian reality: a study of juvenile delinquency." *Social Casework: The Journal of Contemporary Social Work* 65 (4): 235–41.

Rosenbaum, Jill L. 1987. "Social control, gender, and delinquency: an analysis of drug, property and violent offenders." *Justice Quarterly* 4 (I): 117–32.

Segrave, Jeffery O. and Douglas N. Hastad. 1985. "Evaluating three models of delinquency causation for males and females: strain theory, subculture theory and control theory." *Sociological Focus* 18 (1): 1–17.

Shover, Neal, Stephen Norland, Jennifer James, and William E. Thornton. 1979. "Gender roles and delinquency." *Social Forces* 58: 126–75.

Singer, Simon and Murray Levine. 1988. "Power-control theory, gender, and delinquency: a partial replication with additional evidence on the effects of peers." *Criminology* 26 (4) 627–48.

Smith, Douglas A. 1979. "Sex and deviance: an assessment of major sociological variables." *The Sociological Quarterly* 20 (2): 183–86.

Thio, Alex. 1978. *Deviant Behavior*. Boston: Houghton Mifflin.

Thompson, William E., Jim Mitchell and Richard A. Dodder. 1984. "An empirical test of Hirschi's control theory of delinquency." *Deviant Behavior* 5 (1): 11–22.

Thornton, William and Lydia Voight. 1984. "Television and delinquency: a neglected dimension of social control." *Youth and Society* 15 (4): 445–68.

Toby, Jackson. 1957. "Social disorganization and stake in conformity: complementary factors in the predatory behavior of hoodlums." *Journal of Criminal Law, Criminology & Police Science* 48: 12–17.

Torstensson, Marie. 1990. "Female delinquents in a birth cohort: tests of some aspects of control theory." *Journal of Quantitative Criminology* 6 (1): 101–15.

Udry, J. Richard. 1988. "Biological predispositions and social control in adolescent sexual behavior." *American Sociological Review* 53 (October): 709–22.

Vold, George B. and Thomas J. Bernard. 1986. *Theoretical Criminology*. Oxford: Oxford University Press, third edition.

Wells, L. Edward and Joseph H. Rankin. 1983. "Self-concept as a mediating factor in delinquency." *Social Psychology Quarterly* 46 (1): 11–22.

Wiatrowski, Michael D., David B. Griswold and Mary K. Robbins. 1981. "Social control theory and delinquency." *American Sociological Review* 46 (October): 525–41.

Williams, Frank P., III. 1985. "Deterrence and social control: rethinking the relationship." *Journal of Criminal Justice*. 13: 141–51.

Williams, Frank E., III and Marilyn D. McShane. 1988. *Criminology Theory*. Englewood Cliffs, N.J.: Prentice Hall.

Williams, Kirk R. and Richard Hawkins. 1989. "Controlling male aggression in intimate relationships. *Law & Society Review* 23 (4): 591–612.

Winfree, L, Thomas, Harold E. Theis and Kurt T. Griffith. 1981. "Cross cultural examination of complementary social deviance." *Youth and Society* 12: 465–89.

QUESTIONS FOR DISCUSSION

1. Outline the historical and theoretical contributions to control theory.

2. Discuss the author's finding on the empirical status of Hirschi's control theory.

3. Outline the four elements of control theory. Which element best predicts crime and delinquency? Which element is worst at predicting crime and delinquency?

C H A P T E R 8

FEMINIST PERSPECTIVES

Girls' Crime and Woman's Place: Toward a Feminist Model of Female Delinquency

Meda Chesney-Lind

I ran away so many times. I tried anything man, and they wouldn't believe me. . . . As far as they are concerned they think I'm the problem. You know, runaway, bad label. (Statement of a 16-year-old girl who, after having been physically and sexually assaulted, started running away from home and was arrested as a "runaway" in Hawaii.)

You know, one of these days I'm going to have to kill myself before you guys are gonna listen to me. I can't stay at home. (Statement of a 16-year-old Tucson runaway with a long history of physical abuse [Davidson, 1982, p. 26].)

Who is the typical female delinquent? What causes her to get into trouble? What happens to her if she is caught? These are questions that few members of the general public could answer quickly. By contrast, almost every citizen can talk about "delinquency," by which they generally mean male delinquency, and can even generate some fairly specific complaints about, for example, the failure of the juvenile justice system to deal with such problems as "the alarming increase in the rate of serious juvenile crime" and the fact that the juvenile courts are too lenient on juveniles found guilty of these offenses (Opinion Research Corporation, 1982).

This situation should come as no surprise since even the academic study of delinquent behavior has, for all intents and purposes, been the study of male delinquency.

"The delinquent is a rogue male" declared Albert Cohen (1955, p. 140) in his influential book on gang delinquency. More than a decade later, Travis Hirschi, in his equally important book entitled *The Causes of Delinquency,* relegated women to a footnote that suggested, somewhat apologetically, that "in the analysis that follows, the 'non-Negro' becomes 'white,' and the girls disappear."

This pattern of neglect is not all that unusual. All areas of social inquiry have been notoriously gender blind. What is perhaps less well understood is that theories developed to describe the misbehavior of working- or lower-class male youth fail to capture the full nature of delinquency in America; and, more to the point, are woefully inadequate when it comes to explaining female misbehavior and official reactions to girls' deviance.

To be specific, delinquent behavior involves a range of activities far broader than those committed by the stereotypical street gang. Moreover, many more young people than the small visible group of "troublemakers" that exist on every intermediate and high school campus commit some sort of juvenile offense and many of these youth have brushes with the law. One study revealed, for example, that 33% of all the boys and 14% of the girls born in 1958 had at least one contact with the police before reaching their eighteenth birthday (Tracy, Wolfgang, and Figlio, 1985, p. 5). Indeed, some forms of serious delinquent behavior, such as drug and alcohol abuse, are far more frequent than the stereotypical delinquent behavior of gang fighting and vandalism and appear to cut across class and gender lines.

Studies that solicit from youth themselves the volume of their delinquent behavior consistently confirm that large numbers of adolescents engage in at least some form of misbehavior that could result in their arrest. As a consequence, it is largely trivial misconduct, rather than the commission of serious crime, that shapes the actual nature of juvenile delinquency. One national study of youth aged 15–21, for example, noted that only 5% reported involvement in a serious assault, and only 6% reported having participated in a gang fight. In contrast, 81% admitted to having used alcohol, 44% admitted to having used marijuana, 37% admitted to having been publicly drunk, 42% admitted to having skipped classes (truancy), 44% admitted having had sexual intercourse, and 15% admitted to having stolen from the family (McGarrell and Flanagan, 1985, p. 363). Clearly, not all of these activities are as serious as the others. It is important to remember that young people can be arrested for all of these behaviors.

Indeed, one of the most important points to understand about the nature of delinquency, and particularly female delinquency, is that youth can be taken into custody for both criminal acts and a wide variety of what are often called "status offenses." These offenses, in contrast to criminal violations, permit the arrest of youth for a wide range of behaviors that are violations of parental authority: "running away from home," "being a person in need of supervision," "minor in need of supervision," being "incorrigible," "beyond control," truant, in need of "care and protection," and so on. Juvenile delinquents, then, are youths arrested for either criminal or noncriminal status offenses; and, as this discussion will establish, the role played by uniquely juvenile offenses is by no means insignificant, particularly when considering the character of female delinquency.

Examining the types of offenses for which youth are actually arrested, it is clear that again most are arrested for the less serious criminal acts and status offenses. Of

Rank Order of Adolescent Male and Female   **TABLE 1**
Arrests for Specific Offenses, 1977 and 1986

	Male			Female			
1977	% of Total Arrests	1986	% of Total Arrests	1977	% of Total Arrests	1986	% of Total Arrests
(1) Larceny-theft	18.4	(1) Larceny-theft	20.4	(1) Larceny-theft	27.0	(1) Larceny-theft	25.7
(2) Other offenses	14.5	(2) Other offenses	16.5	(2) Runaway	22.9	(2) Runaway	20.5
(3) Burglary	13.0	(3) Burglary	9.1	(3)Other offenses	14.2	(3) Other offenses	14.8
(4) Drug abuse violations	6.5	(4) Vandalism	7.0	(4) Liquor laws	5.5	(4) Liquor laws	8.4
(5) Vandalism	6.4	(5) Vandalism	6.3	(5) Curfew & loitering violations	4.0	(5) Curfew & loitering violations	4.7

	1977	1986	% N change		1977	1986	% N change
Arrests for serious violent offenses[a]	4.2%	4.7%	2.3	Arrests for serious violent offenses	1.8%	2.0%	+1.7
Arrests of all violent offenses[b]	7.6%	9.6%	+10.3	Arrests of all violent offenses	5.1%	7.1%	+26.0
Arrests for status offenses[c]	8.8%	8.3%	−17.8	Arrests for status offenses	26.9%	25.2%	−14.7

Source: Compiled from Federal Bureau of Investigation (1987, p. 169).

[a] Arrests for murder and nonnegligent manslaughter, robbery, forcible rape, and aggravated assault.

[b] Also includes arrests for other assaults.

[c] Arrests for curfew and loitering law violation and runaway.

the one and a half million youth arrested in 1983, for example, only 4.5% of these arrests were for such serious violent offenses as murder, rape, robbery, or aggravated assault (McGarrell and Flanagan, 1985, p. 479). In contrast, 21% were arrested for a single offense (larceny, theft) much of which, particularly for girls, is shoplifting (Sheldon and Horvath, 1986).

Table 1 presents the five most frequent offenses for which male and female youth are arrested and from this it can be seen that while trivial offenses dominate both male and female delinquency, trivial offenses, particularly status offenses, are more significant in the case of girls' arrests; for example the five offenses listed in Table 1 account for nearly three-quarters of female offenses and only slightly more than half of male offenses.

More to the point, it is clear that, though routinely neglected in most delinquency research, status offenses play a significant role in girls' official delinquency. Status offenses accounted for about 25.2% of all girls' arrests in 1986 (as compared to 26.9% in 1977) and only about 8.3% of boys' arrests (compared to 8.8% in 1977). These figures are somewhat surprising since dramatic declines in arrests of youth for these offenses might have been expected as a result of the passage of the Juvenile Justice and Delinquency Prevention Act in 1974, which, among other things, encouraged jurisdictions to divert and deinstitutionalize youth charged with noncriminal offenses. While the figures in Table 1 do show a decline in these arrests, virtually all of this decline occurred in the 1970s. Between 1982 and 1986 girls' curfew arrests increased by 5.1% and runaway arrests increased by a striking 24.5%. And the upward trend continues; arrests of girls for running away increased by 3% between 1985 and 1986 and arrests of girls for curfew violations increased by 12.4% (Federal Bureau of Investigation, 1987, p. 171).

Looking at girls who find their way into juvenile court populations, it is apparent that status offenses continue to play an important role in the character of girls' official delinquency. In total, 34% of the girls, but only 12% of the boys, were referred to court in 1983 for these offenses (Snyder and Finnegan, 1987, pp. 6–20). Stating these figures differently, they mean that while males constituted about 81% of all delinquency referrals, females constituted 46% of all status offenders in courts (Snyder and Finnegan, 1987, p. 20). Similar figures were reported for 1977 by Black and Smith (1981). Fifteen years earlier, about half of the girls and about 20% of the boys were referred to court for these offenses (Children's Bureau, 1965). These data do seem to signal a drop in female status offense referrals, though not as dramatic a decline as might have been expected.

For many years statistics showing large numbers of girls arrested and referred for status offenses were taken to be representative of the different types of male and female delinquency. However, self-report studies of male and female delinquency do not reflect the dramatic differences in misbehavior found in official statistics. Specifically, it appears that girls charged with these noncriminal status offenses have been and continue to be significantly over-represented in court populations.

Teilmann and Landry (1981) compared girls' contribution to arrests for runaway and incorrigibility with girls' self-reports of these two activities, and found a 10.4% overrepresentation of females among those arrested for runaway and a 30.9% over-representation in arrests for incorrigibility. From these data they concluded that girls are "arrested for status offenses at a higher rate than boys, when contrasted to their self-reported delinquency rates" (Teilmann and Landry, 1981, pp. 74–75). These findings were confirmed in another recent self-report study. Figueira-McDonough (1985, p. 277) analyzed the delinquent conduct of 2,000 youths and found "no evidence of greater involvement of females in status offenses." Similarly, Canter (1982) found in the National Youth Survey that there was no evidence of greater female involvement, compared to males, in any category of delinquent behavior. Indeed, in this sample, males were significantly more likely than females to report status offenses.

Utilizing Canter's national data on the extensiveness of girls' self-reported delinquency and comparing these figures to official arrests of girls (see Table 2) reveals that girls are underrepresented in every arrest category with the exception of status offenses and larceny theft. These figures strongly suggest that official practices tend to exaggerate the role played by status offenses in girls' delinquency.

Delinquency theory, because it has virtually ignored female delinquency, failed to pursue anomalies such as these found in the few early studies examining gender differences in delinquent behavior. Indeed, most delinquency theories have ignored status offenses. As a consequence, there is considerable question as to whether existing theories that were admittedly developed to explain male delinquency can adequately explain female delinquency. Clearly, these theories were much influenced by the notion that class and protest masculinity were at the core of delinquency. Will the

Comparison of Sex Differences in Self-Reported   **TABLE 2**
and Official Delinquency for Selected Offenses

	Self-Report[a] M/F Ratios (1976)	Official Statistics[b] M/F Arrest Ratio	
		1976	1986
Theft	3.5:1 (Felony theft) 3.4:1 (Minor theft)	2.5:1	2.7:1
Drug violation	1:1 (Hard drug use)	5.1:1	6.0:1 (Drug abuse violation)
Vandalism	5.1:1	12.3:1	10.0:1
Disorderly conduct	2.8:1	4.5:1	4.4:1
Serious assault	3.5:1 (Felony assault)	5.6:1	5.5:1 (Aggravated assault)
Minor assault	3.4:1	3.8:1	3.4:1
Status offense	1.6:1	1.3:1	1.1:1 (Runaway, curfew)

[a] Extracted from Rachelle Canter (1982, p. 383).
[b] Compiled from Federal Bureau of Investigation (1986, p. 173).

"add women and stir approach" be sufficient? Are these really theories of delinquent behavior as some (Simons, Miller, and Aigner, 1980) have argued?

This article will suggest that they are not. The extensive focus on male delinquency and the inattention the role played by patriarchal arrangements in the generation of adolescent delinquency and conformity has rendered the major delinquency theories fundamentally inadequate to the task of explaining female behavior. There is, in short, an urgent need to rethink current models in light of girls' situation in patriarchal society.

To understand why such work must occur, it is first necessary to explore briefly the dimensions of the androcentric bias found in the dominant and influential delinquency theories. Then the need for a feminist model of female delinquency will be explored by reviewing the available evidence on girls' offending. This discussion will also establish that the proposed overhaul of delinquency theory is not, as some might think, solely an academic exercise. Specifically, it is incorrect to assume that because girls are charged with less serious offenses, they actually have few problems and are treated gently when they are drawn into the juvenile justice system. Indeed, the extensive focus on disadvantaged males in public settings has meant that girls' victimization and the relationship between that experience and girls' crime has been systematically ignored. Also missed has been the central role played by the juvenile justice system in the sexualization of girls' delinquency and the criminalization of girls' survival strategies. Finally, it will be suggested that the official actions of the juvenile justice system should be understood as major forces in girls' oppression as they have historically served to reinforce the obedience of all young women to demands of patriarchal authority no matter how abusive and arbitrary.

THE ROMANCE OF THE GANG
OR THE WEST SIDE STORY SYNDROME

From the start, the field of delinquency research focused on visible lower-class male delinquency, often justifying the neglect of girls in the most cavalier of terms. Take, for example, the extremely important and influential work of Clifford R. Shaw and Henry D. McKay who beginning in 1929, utilized an ecological approach to the study of juvenile delinquency. Their impressive work, particularly *Juvenile Delinquency in Urban Areas* (1942) and intensive biographical case studies such as Shaw's *Brothers in Crime* (1938) and *The Jack-Roller* (1930), set the stage for much of the subcultural research on gang delinquency. In their ecological work, however, Shaw and McKay analyzed only the official arrest data on male delinquents in Chicago and repeatedly referred to these rates as "delinquency rates" (though they occasionally made parenthetical reference to data on female delinquency) (see Shaw and McKay, 1942, p. 356). Similarly, their biographical work traced only male experiences with the law; in *Brothers in Crime,* for example, the delinquent and criminal careers of five brothers were followed for fifteen years. In none of these works was any justification given for the equation of male delinquency with delinquency.

Early fieldwork on delinquent gangs in Chicago set the stage for another style of delinquency research. Yet here too researchers were interested only in talking to and following the boys. Thrasher studied over a thousand juvenile gangs in Chicago during roughly the same period as Shaw and McKay's more quantitative work was being done. He spent approximately one page out of 600 on the five of six female gangs he encountered in his field observation of juvenile gangs. Thrasher (1927, p. 228) did mention, in passing, two factors he felt accounted for the lower number of girl gangs: "First, the social patterns for the behavior of girls, powerfully backed by the great weight of tradition and custom, are contrary to the gang and its activities; and secondly, girls, even in urban disorganized areas, are much more closely supervised and guarded than boys and usually well incorporated into the family groups or some other social structure."

Another major theoretical approach to delinquency focuses on the subculture of lower-class communities as a generating milieu for delinquent behavior. Here again, noted delinquency researchers concentrated either exclusively or nearly exclusively on male lower-class culture. For example, Cohen's work on the subculture of delinquent gangs, which was written nearly twenty years after Thrasher's, deliberately considers only boys' delinquency. His justification for the exclusion of the girls is quite illuminating:

> My skin has nothing of the quality of down or silk, there is nothing limpid or flute-like about my voice, I am a total loss with needle and thread, my posture and carriage are wholly lacking in grace. These imperfections cause me no distress—if anything, they are gratifying—because I conceive myself to be a man and want people to recognize me as a full-fledged, unequivocal representative of my sex. My wife, on the other hand, is not greatly embarrassed by her inability to tinker with or talk about the internal organs of a car, by her modest attainments in arithmetic or by her inability to lift heavy objects. Indeed, I am reliably informed that many women—I do not suggest that my wife is among them— often affect ignorance, frailty and emotional instability because to do otherwise

would be out of keeping with a reputation for indubitable femininity. In short, people do not simply want to excel; they want to excel as a man or as a woman. [Cohen, 1955, p. 138]

From this Cohen (1955, p. 140) concludes that the delinquent response "however it may be condemned by others on moral grounds has at least one virtue; it contestably confirms, in the eyes of all concerned, his essential masculinity." Much the same line of argument appears in Miller's influential paper on the "focal concerns" of lower-class life with its emphasis on importance of trouble, toughness, excitement, and so on. These, the author concludes, predispose poor youth (particularly male youth) to criminal misconduct. However, Cohen's comments are notable in their candor and probably capture both the allure that male delinquency has had for at least some male theorists as well as the fact that sexism has rendered the female delinquent as irrelevant to their work.

Emphasis on blocked opportunities (sometimes the "strain" theories) emerged out of the work of Robert K. Merton (1938) who stressed the need to consider how some social structures exert a definite pressure upon certain persons in the society to engage in nonconformist rather than conformist conduct. His work influenced research largely through the efforts of Cloward and Ohlin who discussed access to "legitimate" and "illegitimate" opportunities for male youth. No mention of female delinquency can be found in their *Delinquency and Opportunity* except that women are blamed for male delinquency. Here, the familiar notion is that boys, "engulfed by a feminine world and uncertain of their own identification . . . tend to 'protest' against femininity" (Cloward and Ohlin, 1960, p. 49). Early efforts by Ruth Morris to test this hypothesis utilizing different definitions of success based on the gender of respondents met with mixed success. Attempting to assess boys' perceptions about access to economic power status while for girls the variable concerned itself with the ability or inability of girls to maintain effective relationships, Morris was unable to find a clear relationship between "female" goals and delinquency (Morris, 1964).

The work of Edwin Sutherland emphasized the fact that criminal behavior was learned in intimate personal groups. His work, particularly the notion of differential association, which also influenced Cloward and Ohlin's work, was similarly male oriented as much of his work was affected by case studies he conducted of male criminals. Indeed, in describing his notion of how differential association works, he utilized male examples (e.g., "In an area where the delinquency rate is high a boy who is sociable, gregarious, active, and athletic is very likely to come in contact with the other boys, in the neighborhood, learn delinquent behavior from them, and become a gangster" [Sutherland, 1978, p. 131]). Finally, the work of Travis Hirschi on the social bonds that control delinquency ("social control theory") was, as was stated earlier, derived out of research on male delinquents (though he, at least, studied delinquent behavior as reported by youth themselves rather than studying only those who were arrested).

Such a persistent focus on social class and such an absence of interest in gender in delinquency is ironic for two reasons. As even the work of Hirschi demonstrated, and as later studies would validate, a clear relationship between social class position and delinquency is problematic, while it is clear that gender has a dramatic and consistent effect on delinquency causation (Hagan, Gillis, and Simpson, 1985). The second irony, and one that consistently eludes even contemporary delinquency theorists, is the fact that while the academics had little interest in female delinquents, the same could not be said for the juvenile justice system. Indeed, work on the early history of the separate system for youth, reveals that concerns about girls' immoral conduct were really at the center of what some have called the "childsaving movement" (Platt, 1969) that set up the juvenile justice system.

"The Best Place to Conquer Girls"

The movement to establish separate institutions for youthful offenders was part of the larger Progressive movement, which among other things was keenly concerned about prostitution and other "social evils" (white slavery and the like) (Schlossman and Wallach, 1978; Rafter, 1985, p. 54). Childsaving was also a celebration of women's domesticity, though ironically women were influential in the movement (Platt, 1969; Rafter, 1985). In a sense, privileged women found, in the moral purity crusades and the establishment of family courts, a safe outlet for their energies. As the legitimate guardians of the moral sphere, women were seen as uniquely suited to patrol the normative boundaries of the social order. Embracing rather than challenging these stereotypes, women carved out for themselves a role in the policing of women and girls (Feinman, 1980; Freedman, 1981; Messerschmidt, 1987). Ultimately, many of the early childsavers' activities revolved around the monitoring of young girls', particularly immigrant girls', behavior to prevent their straying from the path.

This state of affairs was the direct consequence of a disturbing coalition between some feminists and the more conservative social purity movement. Concerned about female victimization and distrustful of male (and to some degree female) sexuality, notable women leaders, including Susan B. Anthony, found common cause with the social purists around such issues as opposing the regulation of prostitution and raising the age of consent (see Messerschmidt, 1987). The consequences of such a partnership are an important lesson for contemporary feminist movements that are, to some extent, faced with the same possible coalitions.

Girls were the clear losers in this reform effort. Studies of early family court activity reveal that virtually all the girls who appeared in these courts were charged for immorality or waywardness (Chesney-Lind, 1971; Schlossman and Wallach, 1978; Shelden, 1981). More to the point, the sanctions for such misbehavior were extremely severe. For example, in Chicago (where the first family court was founded), one-half of the girl delinquents, but only one-fifth of the boy delinquents, were sent to reformatories between 1899–1909. In Milwaukee, twice as many girls as boys were committed to training schools (Schlossman and Wallach, 1978, p. 72); and in Memphis females were twice as likely as males to be committed to training schools (Shelden, 1981, p. 70).

In Honolulu, during the period 1929–1930, over half of the girls referred to court were charged with "immorality," which meant evidence of sexual intercourse. In addition, another 30% were charged with "waywardness." Evidence of immorality was vigorously pursued by both arresting officers and social workers through lengthy questioning of the girl and, if possible, males with whom she was suspected of having sex. Other evidence of "exposure" was provided by gynecological examinations that were routinely ordered in virtually all girls' cases. Doctors, who understood the purpose of such examinations, would routinely note the condition of the hymen: "admits intercourse hymen rupture," "no laceration," "hymen ruptured" are typical of the notations on the forms. Girls during this period were also twice as likely as males to be detained where they spent five times as long on the average as their male counterparts. They were also nearly three times more likely to be sentenced to the training school (Chesney-Lind, 1971). Indeed, girls were half of those committed to training schools in Honolulu well into the 1950s (Chesney-Lind, 1973).

Not surprisingly, large numbers of girls' reformatories and training schools were established during this period as well as places of "rescue and reform." For example, Schlossman and Wallach note that 23 facilities for girls were opened during the 1910–1920 decade (in contrast to the 1850–1910 period where the average was 5 reformatories per decade [Schlossman and Wallach, 1985, p. 70]), and these institutions did much to set the tone of official response to female delinquency. Obsessed with precocious female sexuality, the institutions set about to isolate the females from all contact with males while housing them in bucolic settings. The intention was to hold the girls until marriageable age and to occupy them in domestic pursuits during their sometimes lengthy incarceration.

The links between these attitudes and those of juvenile courts some decades later are, of course, arguable; but an examination of the record of the court does not inspire confidence. A few examples of the persistence of what might be called a double standard of juvenile justice will suffice here.

A study conducted in the early 1970s in a Connecticut training school revealed large numbers of girls incarcerated "for their own protection." Explaining this pattern, one judge explained, "Why most of the girls I commit are for status offenses, I figure if a girl is about to get pregnant, we'll keep her until she's sixteen and then ADC (Aid to Dependent Children) will pick her up" (Rogers, 1972). For more evidence of official concern with adolescent sexual misconduct, consider Linda Hancock's (1981) content analysis of police referrals in Australia. She noted that 40% of the referrals of girls to court made specific mention of sexual and moral conduct compared to only 5% of the referrals of boys. These sorts of results suggest that all youthful female misbehavior has traditionally been subject to surveillance for evidence of sexual misconduct.

Gelsthorpe's (1986) field research on an English police station also revealed how everyday police decision making resulted in disregard of complaints about male problem behavior in contrast to active concern about the "problem behavior" of girls. Notable, here, was the concern about the girls' sexual behavior. In one case, she describes police persistence in pursuing a "moral danger" order for a 14-year-old picked up in a truancy run. Over the objections of both the girl's parents and the Social Services Department and in the face of a written confirmation from a surgeon that the girl was still premenstrual, the officers pursued the application because, in one officer's words, "I know her sort . . . free and easy. I'm still suspicious that she might be pregnant. Anyway, if the doctor can't provide evidence we'll do her for being beyond the care and control of her parents, no one can dispute that. Running away is proof" (Gelsthorpe, 1986, p. 136). This sexualization of female deviance is highly significant and explains why criminal activities by girls (particularly in past years) were overlooked so long as they did not appear to signal defiance of parental control (see Smith, 1978).

In their historic obsession about precocious female sexuality, juvenile justice workers rarely reflected on the broader nature of female misbehavior or on the sources of this misbehavior. It was enough for them that girls' parents reported them out of control. Indeed, court personnel tended to "sexualize" virtually all female defiance that lent itself to that construction and ignore other misbehavior (Chesney-Lind, 1973, 1977; Smith, 1978). For their part, academic students of delinquency were so entranced with the notion of the delinquent as a romantic rogue male challenging a rigid and unequal class structure, that they spent little time on middle-class delinquency, trivial offenders, or status offenders. Yet it is clear that the vast bulk of delinquent behavior is of this type.

Some have argued that such an imbalance in theoretical work is appropriate as minor misconduct, while troublesome, is not a threat to the safety and well-being of the community. This argument might be persuasive if two additional points could be established. One, that some small number of youth "specialize" in serious criminal behavior while the rest commit only minor acts, and, two, that the juvenile court rapidly releases those youth that come into its purview for these minor offenses, thus reserving resources for the most serious youthful offenders.

The evidence is mixed on both of these points. Determined efforts to locate the "serious juvenile offender" have failed to locate a group of offenders who specialize only in serious violent offenses. For example, in a recent analysis of a national self-report data set, Elliott and his associates noted "there is little evidence for specialization in serious violent offending; to the contrary, serious violent offending appears to be embedded in a more general involvement in a wide range of serious and non-serious offenses" (Elliott, Huizinga, and Morse, 1987). Indeed, they went so far as to speculate that arrest histories that tend to highlight particular types of offenders reflect variations in police policy, practices, and processes of uncovering crime as well as underlying offending patterns.

More to the point, police and court personnel are, it turns out, far more interested in youth they charge with trivial or status offenses than anyone imagined. Efforts to deinstitutionalize "status offenders," for example, ran afoul of juvenile justice personnel who had little interest in releasing youth guilty of non-criminal offenses (Chesney-Lind, 1988). As has been established, much of this is a product of the system's history that encouraged court officers to involve themselves in the noncriminal behavior of youth in order to "save" them from a variety of social ills.

Indeed, parallels can be found between the earlier Progressive period and current national efforts to challenge the deinstitutionalization components of the Juvenile Justice and Delinquency Prevention Act of 1974. These come complete with their celebration of family values and concerns about youthful independence. One of the arguments against the act has been that it allegedly gave children the "freedom to run away" (Office of Juvenile Justice and Delinquency Prevention, 1985) and that it has hampered "reunions" of "missing" children with their parents (Office of Juvenile Justice, 1986). Suspicions about teen sexuality are reflected in excessive concern about the control of teen prostitution and child pornography.

Opponents have also attempted to justify continued intervention into the lives of status offenders by suggesting that without such intervention, the youth would "escalate" to criminal behavior. Yet there is little evidence that status offenders escalate to criminal offenses, and the evidence is particularly weak when considering female delinquents (particularly white female delinquents) (Datesman and Aickin, 1984). Finally, if escalation is occurring, it is likely the product of the justice system's insistence on enforcing status offense laws, thereby forcing youth in crisis to live lives of escaped criminals.

The most influential delinquency theories, however, have largely ducked the issue of status and trivial offenses and, as a consequence, neglected the role played by the agencies of official control (police, probation officers, juvenile court judges, detention home workers, and training school personnel) in the shaping of the "delinquency problem." When confronting the less than distinct picture that emerges from the actual distribution of delinquent behavior, however, the conclusion that agents of social control have considerable discretion in labeling or choosing not to label particular behavior as "delinquent" is inescapable. This symbiotic relationship between delinquent behavior and the official response to that behavior is particularly critical when the question of female delinquency is considered.

Toward a Feminist Theory of Delinquency

To sketch out completely a feminist theory of delinquency is a task beyond the scope of this article. It may be sufficient, at this point, simply to identify a few of the most obvious problems with attempts to adapt male-oriented theory to explain female conformity and deviance. Most significant of these is the fact that all existing theories were developed with no concern about gender stratification.

Note that this is not simply an observation about the power of gender roles (though this power is undeniable). It is increasingly clear that gender stratification in patriarchal society is as powerful a system as is class. A feminist approach to delinquency means construction of explanations of female behavior that are sensitive to its patriarchal context. Feminist analysis of delinquency would also examine ways in which agencies of social control—the police, the courts, and the prisons—act in ways to reinforce woman's place in male society (Harris, 1977; Chesney-Lind, 1986). Efforts to construct a feminist model of delinquency must first and foremost be sensitive to the situations of girls. Failure to consider the existing empirical evidence on girls' lives and behavior can quickly lead to stereotypical thinking and theoretical dead ends.

An example of this sort of flawed theory building was the early fascination with the notion that the women's movement was causing an increase in women's crime; a notion that is now more or less discredited (Steffensmeier, 1980; Gora, 1982). A more recent example of the same sort of thinking can be found in recent work on the "power-control" model of delinquency (Hagan, Simpson, and Gillis, 1987). Here, the authors speculate that girls commit less delinquency in part because their behavior is more closely controlled by the patriarchal family. The authors' promising beginning quickly gets bogged down in a very limited definition of patriarchal control (focusing on parental supervision and variations in power within the family). Ultimately, the authors' narrow formulation of patriarchal control results in their arguing that mother's work force participation (particularly in high status occupations) leads to increases in daughters' delinquency since these girls find themselves in more "egalitarian families."

This is essentially a not-too-subtle variation on the earlier "liberation" hypothesis. Now, mother's liberation causes daughter's crime. Aside from the methodological problems with the study (e.g., the authors argue that female-headed households are equivalent to upper-status "egalitarian" families where both parents work, and they measure delinquency using a six-term scale that contains no status offense items), there is a more fundamental problem with the hypothesis. There is no evidence to suggest that as women's labor force participation accelerated and the number of female-headed households soared, aggregate female delinquency measured both by self-report and official statistics either declined or remained stable (Ageton, 1983; Chilton and Datesman, 1987; Federal Bureau of Investigation, 1987).

By contrast, a feminist model of delinquency would focus more extensively on the few pieces of information about girls' actual lives and the role played by girls' problems, including those caused by racism and poverty, in their delinquency behavior. Fortunately, a considerable literature is now developing on girls' lives and much of it bears directly on girls' crime.

Criminalizing Girls' Survival

It has long been understood that a major reason for girls' presence in juvenile courts was the fact that their parents insisted on their arrest. In the early years, conflicts with parents were by far the most significant referral source; in Honolulu 44% of the girls who appeared in court in 1929 through 1930 were referred by parents.

Recent national data, while slightly less explicit, also show that girls are more likely to be referred to court by "sources other than law enforcement agencies" (which would include parents). In 1983, nearly a quarter (23%) of all girls but only 16% of boys charged with delinquent offenses were referred to court by non-law enforcement agencies. The pattern among youth referred for status offenses (for which girls are overrepresented) was even more pronounced. Well over half (56%) of the girls charged with these offenses and 45% of the boys were referred by sources other than law enforcement (Snyder and Finnegan, 1987, p. 21; see also Pope and Feyerherm, 1982).

The fact that parents are often committed to two standards of adolescent behavior is one explanation for such a disparity—and one that should not be discounted as a major source of tension even in modern families. Despite expectations to the contrary, gender-specific socialization patterns have not changed very much and this is especially true for parents' relationships with their daughters (Katz, 1979). It appears that even parents who oppose sexism in general feel "uncomfortable tampering with existing traditions" and "do not want to risk their children becoming misfits" (Katz, 1979, p. 24). Clearly, parental attempts to adhere to and enforce these traditional notions will continue to be a source of conflict between girls and their elders. Another important explanation for girls' problems with their parents, which has received attention only in more recent years, is the problem of physical and sexual abuse. Looking specifically at the problem of childhood sexual abuse, it is increasingly clear that this form of abuse is a particular problem for girls.

Girls are, for example, much more likely to be the victims of child sexual abuse than are boys. Finkelhor and Baron estimate from a review of community studies that roughly 70% of the victims of sexual abuse are female (Finkelhor and Baron, 1986,

p. 45). Girls' sexual abuse also tends to start earlier than boys (Finkelhor and Baron, 1986, p. 48); they are more likely than boys to be assaulted by a family member (often a stepfather) (DeJong, Hervada, and Emmett, 1983; Russell, 1986), and as a consequence, their abuse tends to last longer than male sexual abuse (DeJong, Hervada, and Emmett, 1983). All of these factors are associated with more severe trauma—causing dramatic short- and long-term effects in victims (Adams-Tucker, 1982). The effects noted by researchers in this area move from the more well known "fear, anxiety, depression, anger and hostility, and inappropriate sexual behavior" (Browne and Finkelhor, 1986, p. 69) to behaviors of greater familiarity to criminologists, including running away from home, difficulties in school, truancy, and early marriage (Browne and Finkelhor, 1986).

Herman's study of incest survivors in therapy found that they were more likely to have run away from home than a matched sample of women whose fathers were "seductive" (33% compared to 5%). Another study of women patients found that 50% of the victims of child sexual abuse, but only 20% of the nonvictim group, had left home before the age of 19 (Meiselman, 1978).

Not surprisingly, then, studies of girls on the streets or in court populations are showing high rates of both physical and sexual abuse. Silbert and Pines (1981, p. 409) found, for example, that 60% of the street prostitutes they interviewed had been sexually abused as juveniles. Girls at an Arkansas diagnostic unit and school who had been adjudicated for either status or delinquency offenses reported similarly high levels of sexual abuse as well as high levels of physical abuse; 53% indicated they had been sexually abused, 25% recalled scars, 38% recalled bleeding from abuse, and 51% recalled bruises (Mouzakitas, 1981).

A sample survey of girls in the juvenile justice system in Wisconsin (Phelps et al., 1982) revealed that 79% had been subjected to physical abuse that resulted in some form of injury, and 32% had been sexually abused by parents or other persons who were closely connected to their families. Moreover, 50% had been sexually assaulted ("raped" or forced to participate in sexual acts) (Phelps et al., 1982, p. 66). Even higher figures were reported by McCormack and her associates (McCormack, Janus, and Burgess, 1986) in their study of youth in a runaway shelter in Toronto. They found that 73% of the females and 38% of the males had been sexually abused. Finally, a study of youth charged with running away, truancy, or listed as missing persons in Arizona found that 55% were incest victims (Reich and Gutierres, 1979).

Many young women, then, are running away from profound sexual victimization at home, and once on the streets they are forced further into crime in order to survive. Interviews with girls who have run away from home show, very clearly, that they do not have a lot of attachment to their delinquent activities. In fact, they are angry about being labeled as delinquent, yet all engaged in illegal acts (Koroki and Chesney-Lind, 1985). The Wisconsin study found that 54% of the girls who ran away found it necessary to steal money, food, and clothing in order to survive. A few exchanged sexual contact for money, food, and/or shelter (Phelps et al., 1982, p. 67). In their study of runaway youth, McCormack, Janus, and Burgess (1986, pp. 392–393) found that sexually abused female runaways were significantly more likely than their nonabused counterparts to engage in delinquent or criminal activities such as substance abuse, petty theft, and prostitution. No such pattern was found among male runaways.

Research (Chesney-Lind and Rodriguez, 1983) on the backgrounds of adult women in prison underscores the important links between women's childhood victimizations and their later criminal careers. The interviews revealed that virtually all of this sample were the victims of physical and/or sexual abuse as youngsters; over 60% had been sexually abused and about half had been raped as young women. This situation prompted these women to run away from home (three-quarters had been arrested for status offenses) where once on the streets they began engaging in prostitution and other forms of petty property crime. They also begin what becomes a lifetime problem with drugs. As adults, the women continue in these activities since they possess truncated educational backgrounds and virtually no marketable occupational skills (see also Miller, 1986).

Confirmation of the consequences of childhood sexual and physical abuse on adult female criminal behavior has also recently come from a large quantitative study of 908 individuals with substantiated and validated histories of these victimizations. Widom (1988) found that abused or neglected females were twice as likely as a matched group of controls to have an adult record (16% compared to 7.5). The difference was also found among men, but it was not as dramatic (42% compared to 33%). Men with abuse backgrounds were also more likely to contribute to the "cycle of violence" with more arrests for violent offenses as adult offenders than the control group. In contrast, when women with abuse backgrounds did become involved with the criminal justice system, their arrests tended to involve property and order offenses (such as disorderly conduct, curfew, and loitering violations) (Widon, 1988, p. 17).

Given this information, a brief example of how a feminist perspective on the causes of female delinquency might look seems appropriate. First, like young men, girls are frequently the recipients of violence and sexual abuse. But unlike boys, girls' victimization and their response to that victimization is specifically shaped by their status as young women. Perhaps because of the gender and sexual scripts found in patriarchal families, girls are much more likely than boys to be victims of family-related sexual abuse. Men, particularly men with traditional attitudes toward women, are likely to define their daughters or stepdaughters as their sexual property (Finkelhor, 1982). In a society that idealizes inequality in male/female relationships and venerates youth in women, girls are easily defined as sexually attractive by older men (Bell, 1984). In addition, girls' vulnerability to both physical and sexual abuse is heightened by norms that require that they stay at home where their victimizers have access to them.

Moreover, their victimizers (usually males) have the ability to invoke official agencies of social control in their efforts to keep young women at home and vulnerable. That is to say, abusers have traditionally been able to utilize the uncritical commitment of the juvenile justice system toward parental authority to force girls to obey them. Girls' complaints about abuse were, until recently, routinely ignored. For this reason, statutes that were originally placed in law to "protect" young people have, in the case of girls' delinquency, criminalized their survival strategies. As they run away from abusive homes, parents have been able to employ agencies to enforce their return. If they persisted in their refusal to stay in that home, however intolerable, they were incarcerated.

Young women, a large number of whom are on the run from homes character-ized by sexual abuse and parental neglect, are forced by the very statutes designed to protect them into the lives of escaped convicts. Unable to enroll in school or take a job to support themselves because they fear detection, young female runaways are forced into the streets. Here they engage in panhandling, petty theft, and occasional prostitution in order to survive. Young women in conflict with their parents (often for very legitimate reasons) may actually be forced by present laws into petty crimi-nal activity, prostitution, and drug use.

In addition, the fact that young girls (but not necessarily young boys) are defined as sexually desirable and, in fact, more desirable than their older sisters due to the double standard of aging, means that their lives on the streets (and their survival strategies) take on unique shape—one again shaped by patriarchal values. It is no ac-cident that girls on the run from abusive homes, or on the streets because of profound poverty, get involved in criminal activities that exploit their sexual object status. American society has defined as desirable youthful, physically perfect women. This means that girls on the streets, who have little else of value to trade, are encouraged to utilize this "resource" (Campagna and Poffenberger, 1988). It also means that the criminal subculture views them from this perspective (Miller, 1986).

Female Delinquency, Patriarchal Authority, and Family Courts

The early insights into male delinquency were largely gleaned by intensive field ob-servation of delinquent boys. Very little of this sort of work has been done in the case of girls' delinquency, though it is vital to an understanding of girls' definitions of their own situations, choices, and behavior (for exceptions to this see Campbell, 1984; Peacock, 1981; Miller, 1986; Rosenberg and Zimmerman, 1977). Time must be spent listening to girls. Fuller research on the settings, such as families and schools, that girls find themselves in and the impact of variations in those settings should also be undertaken (see Figueira-McDonough, 1986). A more complete understanding of how poverty and racism shape girls' lives is also vital (see Messerschmidt, 1986; Campbell, 1984). Finally, current qualitative research on the reaction of official agen-cies to girls' delinquency must be conducted. This latter task, admittedly more dif-ficult, is particularly critical to the development of delinquency theory that is as sensitive to gender as it is to race and class.

It is clear that throughout most of the court's history, virtually all female delin-quency has been placed within the larger context of girls' sexual behavior. One expla-nation for this pattern is that familial control over girls' sexual capital has historically been central to the maintenance of patriarchy (Lerner, 1986). The fact that young women have relatively more of this capital has been one reason for the excessive con-cern that both families and official agencies of social control have expressed about youthful female defiance (otherwise much of the behavior of criminal justice person-nel makes virtually no sense). Only if one considers the role of women's control over their sexuality at the point in their lives that their value to patriarchal society is so pronounced, does the historic pattern of jailing of huge numbers of girls guilty of mi-nor misconduct make sense.

This framework also explains the enormous resistance that the movement to curb the juvenile justice system's authority over status offenders encountered. Supporters of the change were not really prepared for the political significance of giving youth the freedom to run. Horror stories told by the opponents of deinstitutionalization about victimized youth, youthful prostitution, and youthful involvement in pornography (Office of Juvenile Justice and Delinquency Prevention, 1985) all neglect the unpleasant reality that most of these behaviors were often in direct response to earlier victimization, frequently by parents, that officials had, for years, routinely ignored. What may be at stake in efforts to roll back deinstitutionalization efforts is not so much "protection" of youth as it is curbing the right of young women to defy patriarchy.

In sum, research in both the dynamics of girls' delinquency and official reactions to that behavior is essential to the development of theories of delinquency that are sensitive to its patriarchal as well as class and racial context.

REFERENCES

Adams-Tucker, Christine. 1982. "Proximate Effects of Sexual Abuse in Childhood." *American Journal of Psychiatry* 193: 1252–1256.

Ageton, Suzanne S. 1983. "The Dynamics of Female Delinquency, 1976–1980.," *Criminology* 21: 555–584.

Bell, Inge Powell. 1984. "The Double Standard: Age." in *Women: A Feminist Perspective,* edited by Jo Freeman. Palo Alto, CA: Mayfield.

Black, T. Edwin and Charles P. Smith. 1981. *A Preliminary National Assessment of the Number and Characteristics of Juveniles Processed in the Juvenile Justice System.* Washington, DC: Government Printing Office.

Browne, Angela and David Finkelhor. 1986. "Impact of Child Sexual Abuse: A Review of Research," *Psychological Bulletin* 99: 66–77.

Campagna, Daniel S. and Donald I. Poffenberger. 1988. *The Sexual Trafficking in Children,* Dover, DE; Auburn House.

Campbell, Ann. 1984. *The Girls in the Gang.* Oxford: Basil Blackwell.

Canter, Rachelle J. 1982. "Sex Differences in Self-Report Delinquency," *Criminology* 20: 373–393.

Chesney-Lind, Meda. 1971. *Female Juvenile Delinquency in Hawaii,* Master's thesis, University of Hawaii.

———. 1973. "Judicial Enforcement of the Female Sex Role," *Issues in Criminology* 3: 51–71.

———. 1978. "Young Women in the Arms of the Law," In *Women, Crime and the Criminal Justice System,* edited by Lee H. Bowker, Boston: Lexington.

———. 1986. "Women and Crime: the Female Offender," *Signs* 12: 78–96.

———. 1988. "Girls and Deinstitutionalization: Is Juvenile Justice Still Sexist?" *Journal of Criminal Justice Abstracts* 20: 144–165.

——— and Noelie Rodriguez 1983. "Women Under Lock and Key," *Prison Journal* 63: 47–65.

Children's Bureau, Department of Health, Education and Welfare. 1965. *1964 Statistics on Public Institutions for Delinquent Children.* Washington, DC; Government Printing Office.

Chilton, Roland and Susan K. Datesman. 1987. "Gender, Race and Crime: An Analysis of Urban Arrest Trends, 1960–1980," *Gender and Society* 1:152–171.

Cloward, Richard A. and Lloyd E. Ohlin. 1960. *Delinquency and Opportunity,* New York: Free Press.

Cohen, Albert K. 1955. *Delinquent Boys: The Culture of the Gang,* New York: Free Press.

Datesman, Susan and Mikel Aickin. 1984. "Offense Specialization and Escalation Among Status Offenders," *Journal of Criminal Law and Criminology,* 75: 1246–1275.

Davidson, Sue, Ed. 1982. *Justice for Young Women.* Tucson, AZ; New Directions for Young Women.

DeJong, Allan R., Arturo R. Hervada, and Gary A. Emmett. 1983. "Epidemiologic Variations in Childhood Sexual Abuse," *Child Abuse and Neglect* 7: 155–162.

Elliott, Delbert, David Huizinga, and Barbara Morse. 1987. "A Career Analysis of Serious Violent Offenders," In *Violent Juvenile Crime: What Can We Do About It?* edited by Ira Schwartz, Minneapolis, MN: Hubert Humphrey Institute.

Federal Bureau of Investigation. 1987. *Crime in the United States 1986,* Washington DC; Government Printing Office.

Feinman, Clarice. 1980. *Women in the Criminal Justice System,* New York: Praeger.

Figueira-McDonough, Josefina. 1985. "Are Girls Different? Gender Discrepancies Between Delinquent Behavior and Control," *Child Welfare* 64: 273–289.

———— 1986. "School Context, Gender, and Delinquency," *Journal of Youth and Adolescence* 15: 79–98.

Finkelhor, David. 1982. "Sexual Abuse: A Sociological Perspective," *Child Abuse and Neglect* 6: 95–102.

———— and Larry Baron. 1986. "Risk Factors for Child Sexual Abuse," *Journal of Interpersonal Violence* 1: 43–71.

Freedman, Estelle. 1981. *Their Sisters' Keepers,* Ann Arbor; University of Michigan Press.

Geltshorpe, Loraine. 1986. "Towards a Skeptical Look at Sexism," *International Journal of the Sociology of Law* 14: 125–152.

Gora, Joann. 1982. *The new Female Criminal: Empirical Reality or Social Myth,* New York: Praeger.

Hagan, John, A. R. Gillis, and John Simpson. 1985. "The Class Structure of Gender and Delinquency: Toward a Power-Control Theory of Common Delinquent Behavior," *American Journal of Sociology* 90: 1151–1178.

Hagan, John, John Simpson, and A. R. Gillis. 1987. "Class in the Household: A Power-Control Theory of Gender and Delinquency," *American Journal of Sociology* 92: 788–816.

Hancock, Linda. 1981. "The Myth that Females are Treated More Leniently than Males in the Juvenile Justice System." *Australian and New Zealand Journal of Criminology* 16: 4–14.

Harris, Anthony. 1977. "Sex and Theories of Deviance," *American Sociological Review* 42: 3–16.

Herman, Julia L. 1981. *Father-Daughter Incest.* Cambridge, MA; Harvard University Press.

Katz, Phyllis A. 1979. "The Development of Female Identity," In *Becoming Female: Perspectives on Development,* edited by Claire B. Kopp, New York: Plenum.

Koroki, Jan and Meda Chesney-Lind. 1985. *Everything Just Going Down the Drain.* Hawaii; Youth Development and Research Center.

Lerner, Gerda. 1986. *The Creation of Patriarchy.* New York: Oxford.

McCormack, Arlene, Mark-David Janus, and Ann Wolbert Burgess. 1986. "Runaway Youths and Sexual Victimization: Gender Differences in an Adolescent Runaway Population," *Child Abuse and Neglect* 10: 387–395.

McGarrell, Edmund F. and Timothy J. Flanagan. 1985. *Sourcebook of Criminal Justice Statistics—1984.* Washington, DC: Government Printing Office.

Meiselman, Karen. 1978. *Incest.* San Francisco: Jossey-Bass.

Merton, Robert K. 1938. "Social Structure and Anomie." *American Sociological Review* 3 (October): 672–782.

Messerschmidt, James. 1986. *Capitalism, Patriarchy, and Crime: Toward a Socialist Feminist Criminology,* Totowa, NJ: Rowman 7 Littlefield.

—— 1987. "Feminism, Criminology, and the Rise of the Female Sex Delinquent, 1880–1930," *Contemporary Crises* 11: 243–263.

Miller, Eleanor. 1986. *Street Woman,* Philadelphia: Temple University Press.

Miller, Walter B. 1958. "Lower Class Culture as the Generating Milieu of Gang Delinquency," *Journal of Social Issues* 14: 5–19.

Morris, Ruth. 1964. "Female Delinquency and Relational Problems," *Social Forces* 43: 82–89.

Mouzakitas, C M. 1981. "An Inquiry into the Problem of Child Abuse and Juvenile Delinquency," In *Exploring the Relationship Between Child Abuse and Delinquency,* edited by R. J. Hunner and Y. E. Walkers, Montclair, NJ: Allanheld, Osmun.

National Female Advocacy Project. 1981. *Young Women and the Justice System: Basic Facts and Issues.* Tucson, AZ; New Directions for Young Women.

Office of Juvenile Justice and Delinquency Prevention. 1985. *Runaway Children and the Juvenile Justice and Delinquency Prevention Act: What is the Impact?* Washington, DC; Government Printing Office.

Opinion Research Corporation. 1982. "Public Attitudes Toward Youth Crime: National Public Opinion Poll." Mimeographed. Minnesota; Hubert Humphrey Institute of Public Affairs, University of Minnesota.

Peacock, Carol. 1981. *Hand Me Down Dreams.* New York: Shocken.

Phelps, R. J. et al. 1982. *Wisconsin Female Juvenile Offender Study Project Summary Report,* Wisconsin: Youth Policy and Law Center, Wisconsin Council of Juvenile Justice.

Platt, Anthony M. 1969. *The Childsavers,* Chicago: University of Chicago Press.

Pope, Carl and William H. Feyerherm. 1982. "Gender Bias in Juvenile Court Dispositions," *Social Service Review* 6: 1–17.

Rafter, Nicole Hahn. 1985. *Partial Justice.* Boston: Northeastern University Press.

Reich, J. W. and S. E. Gutierres. 1979. "Escape/Aggression Incidence in Sexually Abused Juvenile Delinquents," *Criminal Justice and Behavior* 6: 239–246.

Rogers, Kristine. 1972. "For Her Own Protection. . . . Conditions of Incarceration for Female Juvenile Offenders in the State of Connecticut," *Law and Society Review* (Winter): 223–246.

Rosenberg, Debby and Carole Zimmerman. 1977. *Are My Dreams Too Much to Ask For?* Tucson, AZ.: New Directions for Young Women.

Russell, Diane E. 1986. *The Secret Trauma: Incest in the Lives of Girls and Women,* New York: Basic Books.

Schlossman, Steven and Stephanie Wallach. 1978. "The Crime of Precocious Sexuality: Female Juvenile Delinquency in the Progressive Era," *Harvard Educational Review* 48: 65–94.

Shaw, Clifford R. 1930. *The Jack-Roller,* Chicago: University of Chicago Press.

—— 1938. *Brothers in Crime,* Chicago: University of Chicago Press.

—— and Henry D. McKay, 1942. *Juvenile Delinquency in Urban Areas,* Chicago: University of Chicago Press.

Shelden, Randall. 1981. "Sex Discrimination in the Juvenile Justice System: Memphis, Tennessee, 1900–1917." In *Comparing Female and Male Offenders,* edited by Marguerite Q. Warren. Beverly Hills, CA: Sage.

—— and John Horvath, 1986. "Processing Offenders in a Juvenile Court: A Comparison of Males and Females." Paper presented at the annual meeting of the Western Society of Criminology, Newport Beach, CA, February 27–March 2.

Silbert, Mimi and Ayala M. Pines. 1981. "Sexual Child Abuse as an Antecedent to Prostitution," *Child Abuse and Neglect* 5: 407–411.

Simons, Ronald L., Martin G. Miller, and Stephen M. Aigner. 1980. "Contemporary Theories of Deviance and Female Delinquency: An Empirical Test," *Journal of Research in Crime and Delinquency* 17: 42–57.

Smith, Lesley Shacklady. 1978. "Sexist Assumptions and Female Delinquency," In *Women, Sexuality and Social Control,* edited by Carol Smart and Barry Smart, London: Routledge & Kegan Paul.

Snyder, Howard N. and Terrence A. Finnegan. 1987. *Delinquency in the United States.* Washington, DC: Department of Justice.

Steffensmeier, Darrell J. 1980. "Sex Differences in Patterns of Adult Crime, 1965–1977," *Social Forces* 58: 1080–1109.

Sutherland, Edwin. 1978. "Differential Association." In *Children of Ishmael: Critical Perspectives on Juvenile Justice,* edited by Barry Krisberg and James Austin. Palo Alto, CA: Mayfield.

Teilmann, Katherine S. and Pierre H. Landry, Jr. 1981. "Gender Bias in Juvenile Justice." *Journal of Research in Crime and Delinquency* 18: 47–80.

Thrasher, Frederic M. 1927. *The Gang.* Chicago: University of Chicago Press.

Tracy, Paul E., Marvin E. Wolfgang, and Robert M. Figlio. 1985. *Delinquency in Two Birth Cohorts: Executive Summary.* Washington, DC: Department of Justice.

Widom, Cathy Spatz. 1988. "Child Abuse, Neglect, and Violent Criminal Behavior." Unpublished manuscript.

Questions for Discussion

1. Discuss the author's feminist model of female delinquency. Give examples of how the feminist model can be supported in your geographical location.

2. The author suggests that existing delinquency theories are inadequate to explain female deviant acts. Discuss how two theories, according to the author, are inadequate in explaining female deviance.

3. Are the author's points about female delinquency, patriarchal authority and the juvenile justice system valid? Explain your response. How could we revamp the juvenile justice system to address the author's concerns?

A Radical Feminist View of Rape

Lois Copeland and Leslie R. Wolfe

The question of rape is synergistically related to the more fundamental issue of society's being patriarchal, with misogyny its supporting cornerstone. Violence (especially rape) and the threat of violence against women are useful instruments for maintaining male dominance. It follows that unless laws, male behavioral patterns, and societal perceptions are radically altered to create a genuine gender equity, being female will always remain a risk factor.

- "You're a bunch of fucking feminists!" the gunman shouted. He then ordered the women to line up against one wall and opened fire with a 22-caliber automatic rifle. Fourteen women were killed (Russell and Caputi, 1990).
- He hammered his girlfriend's left temple once with a claw hammer, he then swung the hammer in an arc twice more; this time "her head split open like a watermelon." Thirty-five days later he was out on bail, taking college courses (Harrison, 1982).
- "I remember countless episodes of how my husband blackened my eyes, bloodied my lips, how he dislocated my shoulder . . . how I miscarried our child after he beat me. I left three times, each time he threatened to injure or kill members of my family to get me to return home. I stayed with him out of fear" (Baer, 1990).
- She has a few drinks and flirts with some patrons at a bar. Suddenly she finds herself held down while six young men rape her as onlookers cheer them on. Their defense was that she acted seductively and "led them on" (Karmen, 1990).
- Rachel attends a party in her dorm at a big university. A football player, an acquaintance, asks her to come to his room down the hall. He assaults her, not listening to her protests to stop. The next day he comes to her room and asks her out. He felt the behavior the preceding evening was normal (Warshaw, 1988).
- The female student is invited to a frat party. Alcohol is plentiful. She is encouraged to drink and is taken to a room where a "train"[1] of men have sex with her (Sanday, 1990).

SEXISM AND VIOLENCE AGAINST WOMEN: CRIMES OF MISOGYNY

These cases are snapshots of the kind of violence against women that occurs daily throughout the United States and the world. To understand the nature and extent of this violence, we first must look at the culture of domination and patriarchy against which women's movements worldwide have struggled for many years; indeed, "the U.S. social change movement on behalf of abused women is entering its third decade" (Chapman, 1990). Throughout these past two decades, feminist theorists have written extensively about violence against women,[2] which is seen as the quintessential example of sex discrimination and sexual oppression—as the most powerful tool of male domination and patriarchal control.[3] As Charlotte Bunch has recently reiterated:

> Sex discrimination kills women daily. When combined with race, class, and other forms of oppression, it constitutes a deadly denial of women's right to life and liberty on a large scale throughout the world. The most pervasive violation of females is violence against women in all its manifestations, from wife battery, incest, and rape, to dowry deaths, genital mutilation, and female sexual slavery (Bunch, 1990).

Violence—and the internalized and constant threat of violence—is seen as an instrument of control, keeping women "in their place." As with other hate crimes, violence against each woman terrorizes and intimidates the entire class—all women.

The threat of violence permeates every aspect of women's lives. It alters where women live, work, and study, as they try to be safe by staying within certain prescribed bounds.

Women learn these rules when they are girls; they learn to "protect" themselves by restricting their lives, to "be careful" and to accept the blame when their precautions fail. In short, women learn their "place" and their fear very early.

Women—whether they are white or women of color, heterosexual or lesbian, young or old—know that they cannot go to places men can go without the fear of being attacked and violated. Campuses, parking lots, libraries, shopping centers, parks, jogging trails—all are possible danger zones. And, even when women stay within society's prescribed bounds their safety is not assured; studies have shown that women often are most at risk with their intimate partners or friends (Browne, 1987).

Women also learn (as do men) the cultural myths about violence against women which continue to victimize women and which, in large part, still shape our attitudes. These myths suggest that woman battering and rape are "crimes of passion," that wife abuse is a "private, family affair," and that women who are battered, raped, or killed "had it coming" because of some fault or error of their own. Even when violence against women is defined as a societal rather than personal problem, it does not receive the level of serious attention that other violations of individual freedom or of civil and human rights receive.[4]

Such attitudes are woven throughout the fabric of our society, and violence against women still is portrayed as acceptable and inevitable in many subtle and overt ways. These attitudes are so deeply ingrained that a Rhode Island Rape Crisis Center survey of 1,700 sixth- to ninth-grade students in 1988 found that a substantial percentage of these pre-adolescents and adolescents believed that a man has the right to kiss or have sexual intercourse with a woman against her will, particularly if he has "spent money on her." Half of the students said that a woman who walks alone at night and dresses "seductively" is "asking to be raped" (Mann, 1988).

As men absorb and accept their patriarchal rights of ownership of women and children, they also may assume their right to control and demand obedience from their partners (and often their unmarried female relatives as well) and even to use force to ensure it. Indeed, the English Common Law's "rule of thumb"[5] is evidence of longstanding legal and political support for such violence (Holtzman, 1986). Cultural support reaffirms it, as "men who assault their wives are actually living up to cultural prescriptions that are cherished in Western society—aggressiveness, male dominance, and female subordination—and they are using physical force as a means to enforcing that dominance" (Dobash and Dobash, 1979).

Battered women who seek to break the cycle and free themselves from abusive relationships still confront sexist assumptions that further victimize them. In court, battered women often are blamed for the abuse and its seriousness is minimized—suggesting that such violence is a normal expression of male dominance. Examples are legion, as the women who operate shelters for battered women and the attorneys who represent them can attest; in one case, for instance, the judge hearing her divorce case told a battered woman—who had suffered physical and mental abuse during 23 years of marriage—"that he thought she was lying and that he could not believe that her husband, an upstanding citizen, would beat her unless she 'had it coming'" (Supreme Court [MI], 1989).

Women who are raped are further victimized by cultural myths infused into the legal system (Smart, 1989) and by rape laws that were based on patriarchal assumptions about female sexuality and men's rights. Women who are raped must defend themselves from the suggestion that they consented to violent sexual intercourse by "contributory behavior"—by saying "no" but meaning "yes," by wearing "seductive" clothes, by having had a prior sexual history, by going to "dangerous" places (a bar, a campus fraternity party, Central Park). In short, women must confront the assumption that most men do not rape and that most women bring it on themselves.

Acquaintance rape, like wife abuse, is still defined as a woman's personal problem. The myth that rape is a crime of sexually aroused and violent strangers—not "normal" men, not friends or dates or partners—further punishes women. It is assumed that the existence of any prior relationship (even if it is a recent acquaintanceship) suggests consent and that a man is entitled to sexual control of a woman. Acquaintance and date rapes thus are trivialized and hidden. Indeed, young women who report such assaults on campuses often find themselves doubly punished for raising the issue.[6]

While stranger rape and acquaintance rape are both considered rape, the differences in attitude and prosecution are monumental. Men who know their victims are least likely to be arrested, prosecuted, and convicted. Where there is so-called "contributory behavior" by the woman, juries are less likely to convict. An aggravated rape—defined as one in which there are multiple assailants, the rape is accompanied by additional violence, or the rapist uses a weapon—will more likely result in a conviction. Indeed, while theoretically all rapes are investigated and prosecuted to the fullest extent of the law, in reality if a woman is raped by a stranger—especially if a weapon has been used and/or the alleged rapist is a man of color while the victim is white, thus reflecting the extent of race and sex bias combined—she is more likely to be believed and to see him prosecuted and convicted (Estrich, 1987).

Finally, while male murder victims are likely to be murdered as a result of felonious activities or during alcohol- or drug-influenced brawls (Federal Bureau of Investigation, 1990), women are most likely to be murdered simply because they are women. Many murders of women are classic examples of gender-biased hate crimes;

the term "femicide," used most recently by Jane Caputi and Diana Russell (1990), refers to the political murders of women who are killed solely because they are women and reflects a theoretical analysis of these crimes as gender-biased hate crimes. Perhaps the most obvious case of such a political crime is the Montreal murders committed by Marc Lepine; before he murdered fourteen women engineering students, he expressed his hatred of them as "feminists" who had gained entry into the male-dominated field that he could not. Their visibility in this field increased his anger and hatred of all women; not only did he resent their success, he also blamed their presence for his failure to be admitted to the engineering program. It is a classic case.

But it is not the only one. Although murder, the ultimate crime of violence, affects both women and men, many murders of women can be seen as the final expression of patriarchal values of sexual domination (Caputi, 1987). For example, all recorded serial murderers have been men and the large majority kill women (Caputi, 1987); in addition, they frequently bind, rape, and torture their victims before they murder them.

Feminist analysts and activists against violence all insist that violence against women must no longer be defined solely as a crime against an individual who happens to be female and is unfortunate enough to become a victim. Rather, this violence must be seen for what it is—a crime of misogyny, of hatred of women. While not every man beats his female partner or rapes women, feminist theorists would suggest that society's acceptance of patriarchal assumptions and structures also accepts and condones these violations of women's autonomy. The evidence is in the fact that women worldwide "are routinely subject to torture, starvation, terrorism, humiliation, mutilation, and even murder simply because they are female" (Bunch, 1990).

"The message," Charlotte Bunch reminds us, "is domination; stay in your place or be afraid. Contrary to the argument that such violence is only personal or cultural, it is profoundly political. It results from the structural relationships of power, domination, and privilege between men and women in society. Violence against women is central to maintaining those political relations at home, at work, and in all public spheres" (Bunch, 1990).

This viewpoint takes the next step in this analysis—placing violence against women in the context of widely accepted definitions of bias-motivated hate crimes. We seek to show that acts of violence based on gender—like acts of violence based on race, ethnicity, national origin, religion, and sexual identity—are not random, isolated acts. Rather, these are crimes against individuals that are meant to intimidate and terrorize the larger group or class of people—women.

DATA COLLECTION AND THE MEANING OF STATISTICS

The very pervasiveness of violence against women—reflected in available statistics, inadequate as they may be—documents the extent to which "the risk factor is being female" (Heise, 1989). Department of Justice figures show overwhelmingly that reported crimes against women are increasing while crimes against men are decreasing. According to a recent study published by the Committee on the Judiciary of the US Senate, the rate of sexual assaults is now increasing four times faster than the overall crime rate, and the number of reported rapes reached 100,000 in the United States in 1990 (Majority Staff, Committee on the Judiciary, 1991). Since 1974, assaults against

Women Are Not Safe on the Streets
- In 1984, 2.3 million violent crimes (rape, assault, and robbery) were committed against women over the age of 12 (Select Committee on Children, Youth, and Families, 1987).
- In 1988 the FBI received reports of 92,486 forcible rapes of women over the age of 12; only 20 percent to 40 percent were stranger rapes (FBI, 1987).
- In 1982, 4,118 serial murders were reported; the majority of the victims were women (Caputi, 1987).
- Two hundred three rape cases, many involving prostitutes or women who used drugs, were dropped by the Oakland (CA) Police Department without even minimal investigation (Gross, 1990).
- Since 1974, the rate of reported assaults against women ages 20 to 24 has risen by 48 percent, but assault rates against young men in the same age group declined by 12 percent (Committee on the Judiciary, 1990).

Women Are Not Safe in Their Homes
- Every 15 seconds a woman is beaten by her husband or boyfriend (Bureau of Justice Statistics, 1986).
- Thirty percent of women who are homicide victims are killed by their husbands or boyfriends (FBI, 1986).

- Each year, 4,000 women are killed in the context of domestic violence situations—by husbands or partners who have abused them (Stark, 1981).
- One in four—25 percent—of women who attempted suicide had been victims of family violence (Browne, 1987).
- One in seven women in a San Francisco sample reportedly were raped by their husbands (Russell, 1990).
- Nearly nine percent of college women who are raped are raped by family members (Select Committee on Children, Youth, and Families, 1990).

Women Are Not Always Safe with Their Friends
- Rape crisis centers report that 70 to 80 percent of all rapes are committed by acquaintances of the women (Warshaw, 1985).
- A study of 2,291 adult working women found that 39 percent of rapes were committed by husbands, partners, or relatives; only 17 percent were committed by total strangers (Koss, 1990).
- According to a national study, 84 percent of women who had been raped knew their attackers (Ageton, 1983).

young women have risen by an astounding 50 percent, while assaults against young men have dropped by 12 percent (Congressional Caucus on Women's Issues, 1990). Many women live in fear of violence not only on the streets but in their own homes.

Statistics do more than demonstrate how widespread violence against women truly is in our society. They tell us that women are not safe anywhere, from anyone. In the larger sense, as Susan Schechter has stated, "violence against women robs women of possibilities, self-confidence, and self-esteem. In this sense, violence is more than a physical assault; it is an attack on women's dignity and freedom" (Schechter, 1982). And each act of violence against a single woman intimidates and terrifies all women.

These are not isolated random instances of violence in our homes and on the streets. Yet as compelling as the statistics on reported crimes are, they represent a substantial undercount of actual violence against women, for a variety of reasons.

Women Are Not Safe on College Campuses

- In a survey of 3,187 college women, 478 reported having been raped; of these, 10.6 percent were raped by strangers, 24.9 percent were raped by non-romantic acquaintances, 21 percent were raped by casual dates, and 30 percent were raped by steady dates (Koss, Gidycz, and Wisniewski, 1987).
- One out of every four college women is attacked by a rapist before she graduates; one in seven will be raped (Koss, 1990).
- More than half of college rape victims are attacked by dates (Koss, 1990), and studies of high school and college students conducted during the 1980s reported rates of dating violence ranging from 12 to 65 percent (Levy, 1990).
- The number of college women raped in 1986 was 14 times higher than officially reported in the National Crime Survey (Koss, 1990).

Women Are Not Safe from Sexual Harassment at School, at Work, and on the Streets

- A government survey revealed that 42 percent of female respondents working in federal government agencies reported that they had been harassed during a two-year period from 1985 to 1987 (Chapman, 1988).
- A survey of women psychology graduates found that 17 percent had sexual contact with a professor while they were working towards their degree (Project on the Status and Education of Women, 1986).
- Virtually every woman has been subject to some form of street harassment, in which individual men or groups of men whistle, make sexual comments or slurs, issue sexual invitations, or yell obscenities at women passing by (Hughes and Sandler, 1988).

Many women's anti-violence groups believe that violence against women is minimized because women's lives are not valued and the violence is so commonplace (Pharr, 1990). Perhaps for similar reasons, women often do not report incidents of rape and battering.

National studies indicate that as many as four million women are battered each year, but only about 48 percent of cases are reported to police (Langan and Innes, 1986). Battered women do not report abuse for many reasons. Women face the legitimate fear that their partners will carry out their threats of retaliatory violence or loss of access to their children if they report the crime or leave the relationship. Indeed, much evidence suggests that a woman is in the greatest danger when she tries to leave her abuser (Browne, 1987). In addition, as substantial research has reported during the past two decades, many battered women are economically dependent on their abusive partners and many still suffer from the constellation of beliefs and feelings now defined as "battered woman syndrome" (King, 1984). And finally, a woman may believe (with justification) that the criminal justice system will trivialize her reports of abuse and be unable or unwilling to protect her (National Center for State Courts, 1990).

Furthermore, only 7 percent of all rapes are reported to police and fewer than 5 percent of college women report incidents of rape to the police; more than half of raped college women tell *no one* of their assaults. According to Koss, Woodruff, and Koss (1990), FBI studies corroborate these findings, reporting that 10 percent of rapes and sexual assaults are reported to police; by comparison, the reporting rate for robbery is 53 percent, for assault it is 46 percent, and for burglary it is 52 percent.

FEDERAL DATA COLLECTION: FLAWS IN THE SYSTEM

But even if all women who had been raped or battered were to report the crimes to police, accurate data still would be unavailable, because of the manner in which the federal government compiles crime statistics. The US Department of Justice administers two statistical programs to measure crime in the United States: the Uniform Crime Reporting (UCR) Program and the National Crime Survey (NCS). Because of differences in methodology and crime coverage, the results from these two data collection programs are not strictly comparable or consistent (Klaus and Rand, 1984).

The Federal Bureau of Investigation (FBI) administers the UCR Program, which collects information on crimes reported to law enforcement authorities in the following categories: homicide, forcible rape, robbery, aggravated assault, burglary, larceny-theft, motor vehicle theft, and arson.

These reports, however, do not include a category for reporting cases of domestic violence, thus suggesting that it is not a "crime" in the eyes of the FBI. In addition, offense data on crimes that may be family violence, such as non-aggravated assault, are not reported to the FBI. Also, data about relationships between the victim and the offender, including familial relationships, are collected only for the crime of homicide.[7] Thus, it is impossible to use national statistics from police departments to determine the extent of family violence, including rape.

The Bureau of Justice Statistics' National Crime Survey (NCS) collects detailed information on the frequency and nature of crimes, whether or not those crimes are reported to law enforcement agencies. The NCS survey is based on an extensive, scientifically selected sample of approximately 100,000 American households (Bureau of Justice Statistics, 1989). Theoretically, the survey could provide statistical information on various aspects of violence against women. But, because the survey is not specifically designed to address the sensitive nature of sexual assault and woman abuse, this is virtually impossible; for example, the NCS does not code "spouse" and "ex-spouse" separately (Klaus and Rand, 1984), thus making it impossible to determine how many divorced or separated women continue to be abused once they leave the marriage.

> *Women who are raped are further victimized . . . by rape laws that were based on patriarchal assumptions about female sexuality and men's rights.*

In addition, problems with the methodology of the NCS data collection system affect the accuracy of all data, but particularly data on rape (Koss, 1990). Koss suggests that those problems originate with the interview methodology: interviews are not conducted in private; the interviewers are not matched by gender or race/ethnicity to interviewees; the rape screening question and the follow-up questions are inadequate; the UCR definition of rape that serves as the foundation of the NCS definition is too narrow and is inconsistent with state statutes and federal rape law; and multiple incidents of rape involving the same persons are excluded from the calculation of victimization rates in the NCS. This distorts the picture of rape because acquaintance rape is more likely than stranger rape to involve multiple incidents; thus, the NCS survey exaggerates the incidence of stranger rape and interracial rape (Koss, 1990), as compared to acquaintance and familial rape.

According to Steven R. Schlesinger, Director of the Bureau of Justice Statistics, "neither of these two methods is particularly well-suited to estimating the incidence of family violence, so the figures presented here (about 450,000) cannot and, in fact, should not be used to estimate directly the extent of family violence in the United States." On the subject of rape, he suggested that "it is indeed unfortunate that even with the benefits of the elegant and expensive survey technology employed in the NCS an accurate picture of rape fails to emerge from the NCS. Rather than being revealed, the true incidence of rape is covered up by these data" (Koss, 1990).

Although the NCS does not plan any changes in methodology or definitions,[8] the Justice Department has begun to address some data collection problems in the FBI's Uniform Crime Reporting (UCR) Program, by implementing the National Incident-Based Reporting System (NIBRS). This system should increase the degree of detail in the reporting of criminal offenses. According to the Justice Department, more will be known about when and where crime takes place, what form it takes, and the characteristics of its victims and perpetrators (Federal Bureau of Investigation, 1988). Data are expected to be more accurate and detailed. For example, domestic violence will be added to the list of reportable offenses; further, the age, sex, race, ethnicity and relationship of the victim will be correlated with those of the offender.

Along with improving data collection methods, the UCR is revising several definitions listed in the Crime Index. Rape, for example, was defined narrowly as "the carnal knowledge of a female forcibly and against her will." The definition has been broadened in NIBRS to include "the carnal knowledge of a person, forcibly and/or against that person's will; or, not forcibly or against the person's will where the victim is incapable of giving consent because of his/her temporary or permanent mental or physical incapacity." Although not ideal, the expanded definition allows a larger number of sexual assaults, including same sex rapes, to be recorded as such.

Each act of violence against a single woman intimidates and terrifies all women.

In addition, new definitions have been devised for rape offenses which had not been previously reported, such as "forcible sodomy," "sexual assault with an object," and "forcible fondling." The revised definitions of rape and the new definitions incorporated in NIBRS may improve data collection in sexual assault and domestic violence cases. However, this improvement will come slowly, when each state is ready to comply; while a few states already are prepared and are waiting for the program to be put in place, others may not be ready for eight to ten years. Unfortunately, until all states have changed over to NIBRS the UCR (FBI) will continue to publish its annual report, *Crimes in the United States,* with incomplete data; a second report of NIBRS data also will be published.

In short, despite some optimism about these proposed changes, we still must rely upon official statistics on crimes against women published by the FBI and the Bureau of Justice Statistics that underestimate, do not accurately report, or do not report at all on the extent of violence against women. While revisions have been made to update the reporting of crimes in general and crimes against women in particular, much more needs to be done to portray the scope of anti-woman violence.

NOTES

1. "Pulling train" or "gang banging" refers to a group of men lining up like train cars to take turns having sex with the same woman. "Bernice Sandler recently reported that she had found more than seventy-five documented cases of gang rape (pulling train or gang banging) on college campuses in recent years" (Sanday, 1990, p. 1).

2. See Eva Figes (1970), who suggests that the "motivation for male domination is the idea of paternity . . . [as man] saw himself as the physical father of the child [woman] bears . . . himself as creator, woman [as] mere vessel. Since no man can control all men, it is primarily the woman he must control, mentally/physically" (p. 34). Susan Brownmiller (1975) links rape and patriarchy; also see Andrea Dworkin (1974), Catharine MacKinnon (1989), and Kate Millett (1970), for example. bell hooks (1990) writes that "both groups [white men and Black men] have been socialized to condone patriarchal affirmation of rape as an acceptable way to maintain male domination. It is the merging of sexuality with male domination within patriarchy that informs the construction of masculinity for men of all races and classes" (p. 59).

3. "Patriarchy is the ideology and sexism the system that holds it in place" (Pharr, 1988, p. 8).

4. See Charlotte Bunch's excellent analysis: "Despite a clear record of deaths and demonstrable abuse, women's rights are not commonly classified as human rights. This is problematic both theoretically and practically, because it has grave consequences for the way society views and treats the fundamental issues of women's lives" (Bunch, 1990, p. 486).

5. "In addition to sexual domination and control, men were granted ownership rights to physically abuse their wives. The expression 'rule of thumb' comes from the tradition embodied in common law that made it legal for a man to beat his wife as long as the stick used was not bigger around than his thumb" (Holtzman, 1986, p. 1435).

6. Testifying before the Senate Judiciary Committee, a young woman described being raped on her college campus and being further victimized by the institution's response. While a first-year student at a small college, she was raped by a male friend who had offered to walk her home after a fraternity party. For a week following the assault she talked to no one, skipping classes and meals; later, with the help of a friend, she sought help from a school counselor, who told her to "keep her grades up" so she could transfer colleges and not to press charges. As a result, the rapist's name was never revealed and he went on to rape several other "friends" while in college (testimony before Senate Judiciary Committee, August 29, 1990).

7. According to a spokesperson from the National Victims Resource Center, "a Bureau of Justice Statistics statistician has been asking for years to compile accurate statistics on domestic violence victims."

8. The wording of some of the screening questions pertaining to rape was changed several years ago; NCS is waiting for the results of these changes to show up on 1990 data before making further changes.

REFERENCES

Ageton, S. S. (1983). *Sexual assault among adolescents: A national study.* Final report submitted to the National Institute of Mental Health.

Baer, S. (1990, October 2). Freedom from fear. *Baltimore Sun,* C-1.

Browne, A. (1987). *When battered women kill.* New York: The Free Press.

Brownmiller, S. (1975). *Against our will: Men, women, and rape.* New York: Simon and Schuster.

Bunch, C. (1990). Women's rights as human rights: Toward a re-vision of human rights. *Human Rights Quarterly, 12,* 486–98.

Bureau of Justice Statistics. (1989). *Criminal victimization in the United States.* Washington, DC: US Department of Justice.

Caputi, J. (1987). *The age of sex crime.* Bowling Green, OH: Bowling Green State University Popular Press.

Caputi, J. and Russell, D. E. H. (1990, September/October). "Femicide": Speaking the unspeakable. *Ms.,* 35.

Chapman, G. (1988). Sexual harassment of women in federal employment. *Response, 11,* (2), 26.

Chapman, J. R. (1990, Summer). Violence against women as a violation of human rights. *Social Justice, 17,* (2), 54–70.

Committee on the Judiciary, US Senate. (1990, October 2). *Violence Against Women Act of 1990,* Report 101–545. Washington, DC: Committee on the Judiciary.

Congressional Caucus on Women's Issues. (1990, July). *Violence against women.* Washington, DC: CCWI.

Dobash, R. E. and Dobash, R. (1979). *Violence against wives.* New York: The Free Press.

Dworkin, A. (1974). *Woman hating.* New York: Dutton.

Estrich, S. (1987). *Real rape.* Cambridge, MA: Harvard University Press.

Federal Bureau of Investigation. (1990, 1987). *Uniform crime reports for the United States.* Washington, DC: US Department of Justice.

Federal Bureau of Investigation. (1988, July). Uniform crime reporting: National incident-based reporting system, Vol 1, Data Collection Guidelines. Washington, DC: US Department of Justice.

Federal Bureau of Investigation. (1986). *Crime in the United States.* Washington, DC: US Department of Justice.

Figes, E. (1970). *Patriarchal attitudes.* Greenwich, CT: Fawcett.

Gross, J. (1990, September 20). 203 rape cases reopened in Oakland as the police chief admits mistakes. *The New York Times,* 14.

Harrison, B. G. (1982, September). The Yale murder: Poor Richard and "the girl he couldn't share." *Ms.,* 85.

Heise, L. (1989). International dimensions of violence against women. *Response, 12,* (1), 3–11.

Holtzman, E. (1986, October). Women and the law. *Villanova Law Review, 31,* 1429–1438.

hooks, b. (1990). *Yearning.* Boston: South End Press.

Hughes, J. O. and Sandler. B. R. (1988). *Peer harassment: Hassles for women on campus.* Washington, DC: Association of American Colleges.

Karmen, A. (1990). *Crime victims.* Pacific Grove, CA: Brooks/ Cole.

King, N. R. (1984, Fall). Book review: The battered woman syndrome. *Response, 7,* (4), 28.

Klaus, P. A. and Rand, M. R. (1984, April). *Family violence.* Washington, DC: Bureau of Justice Statistics.

Koss, M. P. (1990, August 29). Rape incidence: A review and assessment of the data. Testimony presented to the Committee on the Judiciary, US Senate.

Koss, M. P., Gidycz, C. A., and Wisniewski, N. (1987). The scope of rape: Incidence and prevalence of sexual aggression and victimization in a national sample of higher education students. *Journal of Consulting and Clinical Psychology, 55,* 162–170.

Koss, M. P., Woodruff, W. V., and Koss, P. (1990, August). *A criminological study: Statistics on sexual violence against women.* Unpublished manuscript.

Langan, P. and Innes, C. (1986, August). *Preventing domestic violence against women, special report*. Washington, DC: Bureau of Justice Statistics, US Department of Justice.

Levy, B. (1990). Abusive teen dating relationships. *Response, 13*, (1), 5.

MacKinnon, C. (1989). *Toward a feminist theory of the state*. Cambridge, MA: Harvard University Press.

Majority Staff, Committee on the Judiciary, United States Senate. (1991, March 21). *Violence against women: The increase of rape in America 1990*. Washington, DC: Committee on the Judiciary.

Mann, J. (1988, May 6). Twisted attitudes taint youth. *Washington Post*, D3.

Millett, K. (1970). *Sexual politics*. Garden City, NY: Doubleday.

National Center for State Courts. (1990). *Proceedings of the national conference on gender bias in the courts*. Williamsburg, VA: NCSC.

Pharr, S. (1988). *Homophobia: A weapon of sexism*. Inverness, CA: Chardon Press.

Pharr, S. (1990, January). Hate violence against women. *Transformation, 5*, (1), 1.

Project on the Status and Education of Women, Association of American Colleges. (1986, Summer). Sex with professor not a good idea in retrospect. *On Campus With Women, 16*, (1), 3.

Russell, D. E. H. (1990). *Rape in marriage*. Bloomington: Indiana University Press.

Russell, D. E. H. and Caputi, J. (1990, March–April). Canadian massacre. *New Directions for Women, 19*, (2), 17.

Sanday, P. (1990). *Fraternity gang rape*. New York: New York University Press.

Schechter, S. (1982). *Women and male violence*. Boston: South End Press.

Select Committee on Children, Youth, and Families. (1987, September 16). *Hearing on women, violence, and the law*. Washington, DC: US House of Representatives.

Smart, C. (1989). *Feminism and the power of law*. London and New York: Routledge.

Stark, E. and others. (1981). Wife abuse in the medical setting: An introduction for health personnel. *National Clearinghouse on Domestic Violence, Monograph Series No. 7*. Washington, DC: US Government Printing Office.

Supreme Court of Michigan. (1989, December). *Final report of the Michigan Supreme Court task force on gender issues in the courts*. Lansing, MI: Supreme Court.

Warshaw, R. (1988). *I never called it rape*. New York: Harper and Row.

QUESTIONS FOR DISCUSSION

1. Specifically, what is the radical feminist view of rape? Is there any validity to this point of view? Explain your response. Explain what the authors mean by "violence against each woman terrorizes and intimidates the entire class—all women."

2. According the authors, how are sexism and violence against women crimes of misogyny? What crimes of misogyny have you noticed at your college, university, or geographical location?

3. Discuss statistics from the reading that specifically address assaults (physical and sexual) on college campuses. From your own perspective, are the statistics valid in explaining assaults? Explain your response.

Was It Rape? An Examination of Sexual Abuse Statistics

Neil Gilbert

- At least one-quarter of young women will be sexually assaulted before they leave high school.
- Twenty-seven percent of female college students have been victims of rape or attempted rape.
- Almost half of all women will be victims of rape or attempted rape sometime in their lives.

These alarming statistics have informed many media reports about the prevalence of sexual assault against women in our society. By whatever measure, the problem is a serious one. But should we trust the statistics? A careful review of some of the research that has been done suggests that the problem of sexual assault has been magnified and the data misinterpreted, in part because the issue of sexual violence against women has become enmeshed in politics.

One should not underestimate the difficulty of research in this area. Not only is it often hard to get people to talk about their deeply traumatic experiences, but deciding what to include in a definition of the problem is particularly tricky. In determining the prevalence of child sexual abuse, for example, should a one-time encounter with an exhibitionist be lumped together with repeated forcible rape? A review of the surveys reveals more about the ambiguities of definition than the magnitude of the problem. Legal definitions of these crimes vary from state to state and thus provide no clear reference points for researchers. Beyond this, the subject of sexual violence against women is an emotional one and this makes impartial research very difficult. Finally, researchers must deal with wide discrepancies between reported crimes and the projections their surveys yield.

The estimated prevalence rates cited above, for example, are considerably higher than what one would infer from the official figures on child sexual abuse and rape. Of course, since many if not most incidents of child sexual abuse are never reported to authorities and rape is also underreported, official figures yield conservative estimates of the problem. Still, the size of the discrepancies should make us skeptical of the research findings cited above. In 1992, for example, an estimated 499,120 cases of child sexual abuse were reported to child protective services, of which approximately 40 percent were substantiated. An annual incidence rate of about 200,000 substantiated cases (or 3 in 1,000 children) is serious, but it is only a fraction of the prevalence rate of 25 percent plus cited on the previous page. Similarly, Bureau of

Justice Statistics findings reveal that in 1990 approximately 130,000 women or about 1.2 women in 1,000 over 12 years of age were victims of rape or attempted rape. No trivial number, of course, but that annual figure translates into a 4 to 7 percent lifetime prevalence rate—not the nearly 50 percent rate cited previously. While the officially reported cases underestimate the full extent of sexual violence against women, the highly publicized prevalence rates overestimate it. What follows is an analysis of some of the research cited most often.

PROBLEMS OF DEFINITION
Child Sexual Abuse

The different views of the full range of offenses that constitute child sexual abuse are apparent in a review of 15 surveys conducted since 1976 that attempt to estimate the prevalence of this problem. According to these surveys, the proportion of females sexually molested as children ranges from 6 percent to 62 percent of the population (for males the figures range from 3 percent to 31 percent); half of the studies showed a female prevalence rate of 6 percent to 15 percent. Discrepancies among these studies are due in large measure to differences in the researchers' operational definitions of sexual abuse. Sexual abuse in the studies that yielded the highest results included everything from sexual propositions, exposure to an exhibitionist, and unwanted touches and kisses and fondling to sexual intercourse and other physical contact.

Two of the largest and most widely cited surveys of the prevalence of child sexual abuse illustrate how expansively the problem is defined by researchers. Diana Russell (professor emeritus of sociology at Mills College) surveyed 930 women in San Francisco in the late 1970s and reported that 54 percent of her respondents were victims of incestuous or extrafamilial sexual abuse at least once before the age of 18. There were, however, sampling problems with this survey. Although efforts were made to achieve a random sample of participants, the fact that researchers were able to complete interviews with less than 50 percent of the original sample (930 out of 2,000) makes generalizations to the population of the United States or even San Francisco extremely speculative. Another problem with the survey was the way sexual abuse was defined. The 54 percent prevalence rate reflects a definition of child sexual abuse under which children who receive unwanted hugs and kisses are classified as victims, as are others who are not touched at all (children who encounter exhibitionists). Using a slightly narrower definition, Russell also calculated a lower rate of 38 percent. This narrower measure eliminated cases that did not involve physical contact, but it did include unwanted sexual experiences ranging from a single incident of attempted petting to repeated rape.

Rape

The most highly publicized figures on rape do not hold up under close examination any better than the numbers just cited on child sexual abuse. For example, in her 1984 book, in addition to finding a child sexual abuse prevalence rate of 38 percent to 54 percent, Diana Russell also found that 44 percent of the women in her survey were victims of rape or attempted rape an average of twice in their lives. Beyond sampling bias there is also the question of how rates of rape and attempted rape were measured. Russell's estimates were derived from responses to 38 questions, of which only one of these questions asked respondents whether they had been a victim of rape or attempted rape any time in their lives. And to this question, only 22 percent of the sample, or one-half of women defined as victims by Russell, answered in the affirmative.

A considerable proportion of the cases counted as rape and attempted rape in this study are based on the researchers' interpretation of experiences described by respondents. Thus, in assessing these interpretations, one might bear in mind Russell's perceptions about what constitutes rape. She says, "If one were to see sexual behavior as a continuum with rape at one end and sex liberated from sex-role stereotyping at the other, much of what passes for normal heterosexual intercourse would be seen as close to rape."

Rape on Campus

Quoted in newspapers and journals, on television, and during the 1990 and 1993 Senate hearings on the Violence Against Women Act, the *Ms.* Magazine Campus Project on Sexual Assault directed by Mary Koss is the most widely cited study of rape on college campuses.

Based on a survey of 6,159 students at 32 colleges, the *Ms.* study reported that 15 percent of female college students have been victims of rape and 12 percent of attempted rape, an average of two times between the ages of 14 and 21. The vast majority of offenders were acquaintances, often dates. Using questions that Koss claimed represented a strict legal description of the crime, she also calculated that during a 12-month period 6.6 percent of college women were victims of rape and 10 percent attempted rape with more than one-half of all these victims assaulted twice. If the victimization reported in the *Ms.* study continued at this annual rate over four years, one would expect well over half of all college women to suffer an incident of rape or attempted rape during that period, and more than one-quarter of them to be victimized twice.

The *Ms.* project was funded by the National Institute of Mental Health and its findings published in several professional journals, which is some indication of the clout the project has. Nevertheless, there are compelling reasons to question the magnitude of rape and attempted rape conveyed by this study. To begin with, there is a notable discrepancy between Koss's definition of rape and the way most of the women she labeled as victims interpreted their experiences: almost three-quarters of the students whom Koss categorized as victims of rape did not think that they had been raped. Moreover, in support of the students' own account of their experiences, 42 percent of those identified as victims had sex again with the man whom Koss (but not the students) claimed had raped them; of those whom Koss categorized as victims of *attempted* rape, 35 percent later had sex with their purported offender.

As Koss reported in Judiciary Committee hearings in 1990, "Among college women who had an experience that met legal requirements for rape, only a quarter labeled their experience as rape. Another quarter thought their experience was some kind of crime, but not rape. The remaining half did not think their experience qualified as any type of crime."

A somewhat different account of how these same students labeled their experience was reported in 1988 by Koss and others:

a. 11 percent of the students said they "don't feel victimized,"
b. 49 percent labeled the experience "miscommunication,"
c. 14 percent labeled it, "crime, but not rape," and
d. 27 percent said it was "rape."

The fact that 42 percent of students classified as victims by Koss had sex again with the men who supposedly raped them was not mentioned in her testimony before the Judiciary Committee. In 1988 she originally reported: "Surprisingly, 42 percent of the women indicated that they had sex again with the offender on a later occasion, but it is not known if this was forced or voluntary; most relationships (87 percent) did eventually break up subsequent to the victimization."

But in response to questions raised elsewhere, Koss again offered different accounts of her data. In 1991, in a letter to the *Wall Street Journal,* Koss was no longer surprised by this finding and claimed to know that when the students had sex again with the offenders they were raped a second time and that the relationship broke up not "eventually" (as do most college relationships), but immediately after the second rape. In this revised version of her findings Koss explained, "Many victims reacted to the first rape with self-blame and thought that if they tried harder to be clear they could influence the man's behavior. Only after the second rape did they realize the problem was the man, not themselves. Afterwards, 87 percent of the women ended the relationship with the man who raped them." Koss went on to suggest that these students did not know they were raped because many of them were sexually inexperienced and "lacked familiarity with what consensual intercourse should be like."

One part of the problem in the definition of rape used in the *Ms.* study is the vagueness of a couple of the questions in the survey. Although the exact legal definition of rape varies by state, most definitions involve sexual penetration accomplished against a person's will by means of physical force, threat of bodily harm, or when the victim is incapable of giving consent; the latter condition usually includes cases in which the victim is mentally ill, developmentally disabled, or intentionally incapacitated through the administration of intoxicating or anesthetic substances. Rape and attempted rape were operationally defined in the *Ms.* study by five questions, three of which referred to the threat or use of "some degree of physical force." But the other two questions asked were: "Have you had a man attempt sexual intercourse (get on top of you, attempt to insert his penis) when you didn't want to by giving you alcohol or drugs?" and "Have you had sexual intercourse when you didn't want to because a man gave you alcohol or drugs?" But what does it mean to have sex when you don't want to "because" a man gives you alcohol or drugs? How does one attempt intercourse "by" giving alcohol or drugs? Positive responses to these survey questions do not indicate whether duress, intoxication, force, or the threat of force was present; whether the woman's judgment or control was substantially impaired; or whether the man purposefully got the woman drunk in order to prevent her from resisting his sexual advances. They could conceivably mean that a woman was trading sex for drugs or that a few drinks lowered the respondent's inhibitions and she consented to an act she later regretted. While the items could have been clearly worded to denote the legal standard of "intentional incapacitation of the victim," as the questions stand there is no way to detect whether an affirmative response corresponds to the legal definition of rape. In 1993, Koss acknowledged that the drug and alcohol questions were ambiguous. When these questions are removed, according to Koss's revised estimates, the prevalence rate of rape and attempted rape declines by one-third.

Ignoring Discrepancies

One of the reasons the *Ms.* study has gained so much credibility is that Koss claims her findings are corroborated by many other studies. This creates the appearance that the *Ms.* findings are independently verified by a cumulative block of scientific evidence. However, an examination of these other studies reveals the highly variable methodology and questionable definitions employed in them, along with the tremendous discrepancies among the findings. Koss and Sarah Cook (a doctoral candidate in community psychology at the University of Virginia) refer to one such study this way: "In a just released telephone survey of more than 4,000 women in a nationally representative sample, the rate of rape was reported to be 14 percent, although this rate excluded rapes of women unable to consent." This, of course, is close to the 15 percent rate of rape claimed by the *Ms.* study. What Koss fails to tell the reader, however, is that approximately 40 percent of the rapes reported in this study affected women between the ages of 14 and 24, resulting in about a 5 percent rape prevalence estimate for that age group. According to the *Ms.* study, 15 percent of college women are victims of rape between the ages of 14 to 21. Thus, for almost the same age group, the *Ms.* study found a rate of rape *three times higher* than that detected by the National Victim Center survey. Only by sweeping the relevant details under the carpet of "supporting literature" can the results of studies such as these be construed as independently confirming the *Ms.* findings.

In addition, advocates claim that data in this field all tend in the same direction. All the data do not tend in the same direction. If the work of Koss, Russell, and others suggests that prevalence rates of rape and attempted rape rise considerably above 10 percent to as much as one out of every two women, there are other studies that point to rates well below 10 percent. These studies, of course, rarely make headlines. Linda George (professor of psychiatry and sociology at Duke University), Idee Winfield (assistant professor of sociology at Louisiana State University), and Dan Blazer (Gibbons Professor of Psychiatry at Duke University), for example, found a 5.9 percent lifetime prevalence rate of sexual assault in North Carolina. This finding was based on responses to the broad question, "Has someone ever pressured you against your will into forced contact with the sexual parts of your body or their body?" While regional differences may account for some of the disparity between the *Ms.* findings and those of the Duke University team, in 1990 the rate of rapes reported to the police in North Carolina was 83 percent of the national average. An even lower prevalence rate was detected by Stephanie Riger and Margaret Gordon (policy researchers at Northwestern University), who reported that among 1,620 respondents randomly selected in Chicago, San Francisco, and Philadelphia, only 2 percent had been raped or sexually assaulted in their lifetime.

The most startling disparity is between the *Ms.* study's finding on the annual incidence of rape and attempted rape and the number of these offenses actually reported to the authorities on college campuses. Using her survey questions, Koss found that 166 women in 1,000 were victims of rape and attempted rape in just one year on campuses across the country (each victimized an average of 1.5 times). In sharp contrast, the 1993 FBI figures show that in 1992 at about 500 major colleges and universities with an overall population of 5 million students, only 408 cases of rape and attempted rape were reported to the police, less than one incident of rape and attempted rape per campus. This number yields an annual rate of .16 in 1,000 for female students, which is 1,000 times smaller than Koss's finding. Although it is generally agreed that many rape victims do not report their ordeal, no one to my knowledge publicly claims that the problem is 1,000 times greater than the cases reported to the police—a rate at which almost every woman in the country would be raped at least once every year.

Demonizing Men / Infantilizing Women / Trivializing Violence

By promoting their figures on rape and violence, researchers who are also advocates seek to raise public consciousness about this issue. Raising consciousness can sometimes be positive. But they also seek to alter public consciousness by molding perceptions of the basic nature of the problem and of what constitutes the common experience of heterosexual relations. Thus, because they distort the issue, the effects can be harmful.

If it is true that one-third of female children are sexually abused and almost half of all women will suffer an average of two incidents of rape or attempted rape at some time in their lives, one is ineluctably driven to conclude that most men are pedophiles or rapists. This view of men is repeatedly expressed by some prominent activities in the child sexual abuse prevention movement and the rape crisis movement. As Diana Russell puts it, "Efforts to explain rape as a psychopathological phenomenon are inappropriate. How could it be that all these rapes are being perpetrated by a tiny segment of the male population?" Her explanation of "the truth that must be faced is that this culture's notion of masculinity—particularly as it is applied to male sexuality—predisposes men to violence, to rape, to sexually harass, and to sexually abuse children." In a similar vein, Koss notes that her findings support the view that sexual violence against woman "rests squarely in the middle of what our culture defines as 'normal' interaction between men and women." University of Michigan law professor Catherine MacKinnon offers a vivid rendition of the theme that rape is a social disease afflicting most men. Writing in the *New York Times,* she advises that when men charged with the crime of rape come to trial the court should ask, "Did this member of a group sexually trained to woman-hating aggression commit this particular act of woman-hating sexual aggression?"

Given such perspectives, the sexual politics of advocacy research on violence against women demonizes men and defines the common experience in heterosexual relations as inherently violent and menacing. This is the message that is being delivered on college campuses, and a frightening atmosphere is the result. As social critic Louis Menand observes: "The assumption that sexual relations among students at a progressive liberal-arts college should be thought of as per se fraught with the potential for violence is now taken for granted by just about everyone." A similar message is being delivered to students at a very early age by sexual abuse prevention programs in the schools. Many of these programs make a point of teaching kids as young as three that male family members, particularly fathers, uncles, and grandfathers may sexually abuse children.

Taking Back the Night

A puzzling question remains: If advocacy research trivializing violence against woman and its estimates of rape on campus are so greatly exaggerated compared to the cases reported to police and counseling centers, why are thousands of female students across the country marching by candlelight to take back the night? Something is going on here. Those in the rape crisis movement may take this anguished behavior as confirmation of widespread violence against college women. But there are several alternative explanations that suggest a climate in which statistics are used and abused and heavily influenced by the social dynamics of fear, power, conflict, and sex.

Inflated Fear   In light of the statistics on sexual violence being bandied about, the view that take-back-the-night protests confirm an epidemic of rape can, of course, be turned on its head. Rather than responding to real violence against women on campus, these demonstrations may reflect a level of fear inflated by the exaggerated reports of rape, which create a premonition of danger that is vastly out of proportion to the actual risks of campus life. According to Katie Roiphe, a significant number of students are walking around with the alarming belief that 50 percent of women are raped. "This hyperbole," she notes, "contains within it a state of perpetual fear."

Power of Victimization   One might argue as well that demonstrations to take back the night are incited less by irrational fear than by the sense of righteousness and power that often accompanies the invocation of victim status. "Even the privileged," as political scientist Charles Sykes tells us, "have found that being oppressed has its advantage."

Increased Gender Conflict   Over the last few decades more women have entered college and joined the paid labor force than ever before. In 1960, most women were married before their twenty-first birthday. Today at that age they are competing with men for graduate school slots in almost every field and for jobs throughout the economy. An unprecedented number of women have also become heads of single-parent families, to which absent fathers often contribute minimal support. These developments enlarge occasions for friction and resentment in economic and social relations between the sexes, particularly among the college age group. The anxieties and animosities expressed in rallies to take back the night, in part, may reflect a real increase in tensions between men and women. This explanation suggests that "sexual assault" has come to represent a broad and vaguely formulated category of offenses, which ranges from rape to the incivilities of heightened gender conflict.

Sexual Turmoil   Explanations that refer to fear, power, and conflict downplay the extent to which sexual activities influence the highly emotional responses to the threat of rape on campus. Sexual relations between young men and women, never exactly serene, have become more troublesome and confused in recent times. This is because there is more of it to manage, sex has become more dangerous, and it is harder to say no. Since 1960, as the median age for premarital intercourse declined and the median age for first marriage rose, the period of premarital sexual activity has grown from an average of 1 to 7.5 years for women. The sexual revolution in the late 1960s discounted the moral strictures against premarital sex. Without the shield of morality, it became more socially awkward to avoid having sex in situations where one's feelings were ambiguous, or even in situations where sex was totally unwanted. (It is easier to assert morality and say no because premarital sex is wrong, than to say no for all the other reasons: I don't love [or even like] you enough; I am not ready yet.) Shortly after the time span for sexual activity began to increase and the shield of morality fell, moreover, sexual relations became more dangerous as the risks of contracting AIDS and other sexually transmitted diseases rose to alarming levels.

These explanations are not mutually exclusive. They describe a cultural climate of strained gender relations under which advocacy research has been relatively successful in furthering radical sexual politics. The statistics fuel a powerful crusade. Theirs is a politics based on the convictions that violence is the norm in heterosexual relations, that all women are oppressed and deserve special treatment by government, and that from an early age women need to be empowered by publicly supported training and prevention programs. The radical feminist agenda coincides with the interests of some sexual abuse prevention professionals. This agenda does not, however, serve the interests of poor women, minorities, teenage boys, and other groups most in need of social protection from violence.

Misdirected Policy

In the deliberations on federal policy regarding violence against women, advocacy research on rape was accepted almost at face value. In the opening statement during Senate hearings on the Violence Against Women Act in 1990, Chairman Joseph Biden explained:

> One out of every four college women will have been attacked by a rapist before they graduate, and one in seven will have been raped. Less than 5 percent of these women will report these rapes to the police. Rape remains the least reported of all major crimes. . . . Dr. Koss will tell us today that the actual number of college women raped is more than 14 times the number reported by official governmental statistics. Indeed, while studies suggest that about 1,275 women were raped at America's three largest universities last year, only three rapes—only three—were reported to the police.

The panel was informed that unreported rapes of college women are more than 400 times the number of incidents reported to the police. In fact, as noted earlier, according to Koss's figures the unreported rate is about 1,000 times higher than the rate of rape and attempted rape reported on college campuses. And Koss's conclusions were never seriously challenged.

Due to the credibility attributed to advocacy research, under Title IV (Safe Campuses for Women), the Violence Against Women Act of 1993 proposes to appropriate $20 million for rape education and prevention programs to make college campuses safe for women. Extrapolating from the 408 reported cases of rape and attempted rape on 500 campuses in 1992 yields approximately 1,000 reported incidents among all colleges in the United States. The appropriation proposed under Title IV thus amounts to $20,000 per reported case.

Compared to the $20 million designated for college campuses, the Violence Against Women Act proposes to appropriate $65 million (approximately $650 per reported case) for prevention and education programs to serve the broader community. Under this arrangement, a disproportionate sum will be distributed for education and prevention programs on college campuses, where most victims, according to the *Ms.* study, do not know they have been raped. And comparatively meager resources will be invested in similar programs for low-income and minority communities. Whatever the value of the college programs, the cost is remarkably high compared to the funds allocated per reported case of victims outside of college campuses, among whom poor and minority women are vastly overrepresented, according to the Bureau of Justice Statistics. According to the 1993 FBI data, overall there were 109,062 reported cases of rape and attempted rape in 1992. Based on these rates, to make the rest of society as safe as college campuses would require an expenditure of $2 billion on educational and prevention programs.

It is difficult to criticize advocacy research without giving the impression that one cares less about the problems of victims than do those who magnify the size of problems such as rape and child abuse. Advocacy researchers who uncover a problem, measure it with reasonable accuracy, and bring it to public attention perform a valuable service by raising public consciousness. But the current trend in research on sexual violence against women is to inflate the problem and redefine it in line with the advocates' ideological preferences. The few impose their definition of social ills on the many. By creating sensational headlines, this type of advocacy research invites the formulation of social policies that are likely to be neither effective nor fair.

QUESTIONS FOR DISCUSSION

1. Explain what the author means when he says that "the problem of sexual assault has been magnified and the data misinterpreted, in part because the issue of sexual violence against women has become enmeshed in politics." Use examples from the reading to support your response.

2. What are some methodological flaws of Mary Koss's studies on rape? Do you agree or disagree that these are flaws? Explain your response.

3. The author suggests that a puzzling question remains for sexual victims: "If advocacy research trivializing violence against woman and its estimates of rape on campus are so greatly exaggerated compared to the cases reported to police and counseling centers, why are thousands of female students across the country marching by candlelight to take back the night?" What is your answer to this question? Explain how the social dynamics

of fear, power, conflict, and sex, according to the author, apply to your college, university, or geographical location.

QUESTIONS FOR PART 3

1. How do the readings in Chapter 6 address processual elements of deviance?

2. How do the readings in Chapter 7 address structural elements of deviance?

3. How do the readings in Chapter 8 address the issues with females and deviance? In addition, according to the readings, how clear is the issue with rape and rape research?

P A R T 4

FORMS OF DEVIANT BEHAVIOR

Deviance takes many forms. Homicide, aggravated assault, rape, theft, hacking, marijuana use, cocaine use, alcoholism, prostitution, cyberporn, and suicide are all considered deviant. Some deviant acts, such as homicide, have always been defined as deviant. The nature of others, such as drug use, has changed over time to reflect changing circumstances, values, and conceptions of what is acceptable in certain situations. Still others, such as computer hacking, are new and recently defined deviant acts. The range of deviant acts is very wide and each act elicits different degrees of disapproval.

In Part 4, there are articles that reflect a wide range of different forms of deviant behavior. The intent of the articles is not to cover every viewpoint and example of deviance, but to illustrate the variety of deviant acts from both classic and contemporary perspectives. Since no one is deviant all the time, deviant roles, events, and acts committed over time become important when expounding on deviant behavior. Like all social roles, deviant roles are learned, and some individuals perform the role of deviant more than others.

Deviant events take many forms. All deviant events breach some social norm. To understand deviant events, one must begin with the history and the immediate situation in which the deviant behavior occurred. Moreover, to understand deviant acts over time, one would need to analyze the temporal relations with deviant acts. Some deviant individuals may have progressed into a life of deviant activities, while others do not progress. Some deviants may participate in established deviant activities, such as drug use, while still other individuals contribute to newly formed deviant acts, such as the various new forms of deviance using computers.

When observers explore violent behavior, some ask why the incident transpired. When studying the sociology of deviance, some look at deviant roles, events, and acts of violent behavior. These observers would ask about certain roles for violent behaviors. For example, have violent crimes changed over time? Is there a certain situation where violent behavior is more prevalent? Has the acceptance of a violent crime like rape changed over time? What is rape? Are there justifications for rape? The selection by Terance D. Miethe and Richard McCleary summarizes much that is known about murder and assault, and Diana Scully and Joseph Marolla's classic paper shows how some offenders can explain the crime of rape to themselves and others. The meaning of rape has broadened in the last two decades and there is continued confusion regarding this serious crime. The selection by Shelly Schaefer Hinck and Richard W. Thomas shows the continued confusion about rape and the persistence of rape myths.

When engaging in the sociology of deviance, one also looks at the roles, events, and acts of white-collar crime. For example, what causes white-collar crime? How do we account for white-collar crime? What role do employees play in a white-collar crime? Have white-collar crimes changed over time? Who are computer hackers? What is the sociology of hacking? These questions are raised in the chapter on white-collar crime. Michael L. Benson's article examines the ability of white-collar offenders to excuse their conduct. Tim Jordan and Paul Taylor explore a new form of white-collar crime: computer hacking.

Drug use continues to be a controversial topic in the sociology of deviance. Some individuals ask why cigarette and alcohol use are legal when they are the leading cause of drug deaths while the use of substances like marijuana, heroin, cocaine, and crack cocaine is illegal? What are the deviant roles of a drinker? What are different cultural patterns for drinking behaviors? Have drinking behaviors changed over time? If drinking is legal and worse than marijuana, how does one learn to become a drinker and marijuana user? Has the problem with drugs changed over time? The selection by Howard S. Becker illustrates the importance of social, not pharmacological, factors in the continued use of drugs. Although directed specifically toward marijuana use, his insights apply to other drugs as well.

Negative media images of drug dealers who deliberately exploit drug users for economic reasons may be correct, but such images often miss the connection between deviant and nondeviant worlds. The selection by Sheigla Murphy, Dan Waldorf, and Craig Reinarman shows that there are many paths to drug dealing and that dealers often maintain at least some contact with conventional society. Nevertheless, dealers do experience a change of self concept from one who *has* a good connection for drugs to one who *is* a good connection for drugs. Occasionally, the use of new substances is thought to be part of the drug problem and a legitimate candidate for public policy. James A. Inciardi, Hilary L. Surratt, Dale D. Chitwood, and Clyde B. McCoy show the origins of crack cocaine in the United States and the reasons this substance came to occupy a high priority on the public policy agenda.

Alcohol is a drug that raises as many questions about regulation as currently illegal drugs. The negative effects of alcohol are easily documented. Norman K. Denzin's account of the alcoholic's divided self illustrates how the alcoholic can experience a downward spiral in attempting to keep the two selves (drunk and sober) apart. It is a losing battle for many alcoholics. Craig MacAndrew and Robert B. Edgerton explore the common belief that one's behavior while intoxicated reflects a weakening of inhibitions brought about through drinking. They find, however, that drunken behavior is culturally determined. Even being drunk is a social process that involves learning group expectations about how to behave when intoxicated. The paper by Henry Wechsler, Beth E. Molnar, Andrea E. Davenport, and John S. Baer addresses alcohol use among a specific population: college students. This recent study has been widely discussed, supported, and criticized.

Questions similar to those asked about drug use can be asked about deviant sexual behavior. For example, is there a deviant role for sexual deviants? How does one become a prostitute? What deviant roles do prostitutes play in society? Is there a demand for prostitution? Are computers contributing to pornography? Will cyberporn cause an increase in sexual assaults? With sexual deviation in society, how should parents talk to young children about sex? In what ways, if any, have societal values about sex and sexuality changed over time? The paper by Barbara Sherman Heyl provides an excellent starting point for a sociological understanding of becoming and being a sex worker. The learning process is a key component here because most women are socialized to avoid sex work. The last two selections in this chapter deal with how computers and the Internet have changed the nature of pornography and sex. The computer provides a new system for the consumption of pornography. Anyone with access to the Internet can find sexually explicit content, as Philip Elmer-Dewitt's selection shows. The computer can also be used to deliver sexual advice and values. The selection by Susie Bright shares her message to her daughter on sexuality.

And last, suicide can be viewed not only as an act of desperation but also as a rational act. Although rates of suicide are higher among older age groups, it is particularly difficult to understand suicide with young people who have most of their lives in front of them. In the paper by Jennifer Langhinrichsen-Rohling, Peter Lewinsohn, Paul Rohde, John Seeley, Candice M. Monson, Kathryn A. Meyer, and Richard Langford, the focus is on gender differences not only in actual suicides, but also in suicide-related behavior. The selection from Emile Durkheim presents a classic statement on this form of deviance. Durkheim deliberately chose what appears to be a highly individualistic act to illustrate the importance of a sociological viewpoint.

In Part 4, we address classical and contemporary roles, events, and acts of deviance. We are aware that the following articles are not mutually exclusive for all classical and contemporary articles, but the articles start a discussion on deviant roles, events, and acts of deviant behavior.

C H A P T E R 9

VIOLENT BEHAVIOR

Homicide and Aggravated Assault

Terance D. Miethe and Richard McCleary

Violence has been a common theme in the United States and other countries. Acts of civil unrest such as worker strikes, political protests, and civil rights demonstrations often erupt into violence, causing death and physical injury. Longstanding terrorist campaigns and isolated instances of terrorism against state and civilian targets are a daily occurrence on the international scene. The Oklahoma City bombing that killed 168 people and the letter bomb attacks by the Unabomber have made it clear that the U.S. is not immune to these violent acts. Although collective acts of violence and terrorist attacks are more visible and receive greater media attention, most violence in this country lacks an ideological motivation and involves a dispute between a single victim and offender.

This chapter reviews current knowledge about two of the most serious acts of interpersonal violence: homicide and aggravated assault. After defining these violent crimes, data is presented on (a) trends in homicide and aggravated assault over time, (b) the social correlates of these crime rates, (c) the offender profile in violent crime, (d) victim characteristics, and (e) situational elements and circumstances of the crime. Several homicide and assault syndromes are then described that combine offender, victim, and situational elements.

"Homicide and Aggravated Assault" from *Crime Profiles: The Anatomy of Dangerous Persons, Places, and Situations* by Terance D. Miethe and Richard McCleary, 1998, pp. 19–53. Reprinted with permission of Roxbury Publishing Company.

DEFINITIONS OF CRIME TYPES

Homicide and aggravated assaults are similar in that they both involve the application of force and the physical injury of one by another. The similarity of these offenses is further revealed by empirical research that shows that many murders and assaults involve "similar participants, interacting in similar ways, for similar reasons, in similar settings" (Luckenbill 1984:25; Block 1977; Pittman and Hardy 1964). The primary difference lies in whether the criminal act results in death or injury. However, situational elements in many cases—poor aim, the type of weapon, the availability of immediate medical care—may be the only practical difference between homicides and aggravated assaults.

From a legal perspective, homicide is defined as the killing of one human being by another. There are two general types. Noncriminal homicides are killings under a lawful justification or excuse, such as self-defense slayings, accidental deaths, and the execution of a death sentence by authorized state agents. Criminal homicide is comprised of murder and manslaughter. Most jurisdictions recognize several degrees of both murder and manslaughter. First-degree murders are those acts committed with deliberation, premeditation, and malice aforethought, or murders that occur in conjunction with another felony. In second-degree murders, the act is deliberate but not premeditated and the intent is only to do physical injury to the victim (Samaha 1987). Killings "in the heat of passion" and under physical provocation by the victim are often classified as voluntary manslaughters. Negligent or involuntary manslaughters are unintentional killings while acting in a wanton, reckless, or careless manner or during the commission of another unlawful act other than a felony.

Assaults are often distinguished according to whether they are simple or aggravated attacks. Aggravated assaults involve deliberate and serious physical injuries to another or cases in which a weapon is used regardless of injury. Simple assaults, in contrast, typically involve minimal or minor physical injury and no weapon.

TRENDS OVER TIME AND SOCIAL CORRELATES

Rates of homicide and aggravated assault in the U.S. have vacillated over time. Starting in the 1930s (when systematic data was first collected by the FBI), homicide rates decreased prior to and during World War II, increased in the post-war period, steadily rose throughout the 1960s and 1970s, declined briefly in the early 1980s, generally increased until 1993, and have decreased over the last two years. Assault rates followed a more dramatic and uniform increase from the mid-1950s to 1995. An estimated 21,597 homicides and 1,099,179 aggravated assaults were known to the police in 1995 (UCR 1995). These numbers represent a 5 percent increase in homicide and a 32 percent increase in aggravated assaults over the last 10 years.

Changes in homicide and assault rates over time have been attributed to several factors. First, the "legitimation of violence" hypothesis (Archer and Gartner 1984) argues that during wartime pro-violent values are reinforced and these values are carried over to post-war periods. Increases in homicide rates after World War II, the Korean War, and during and after the Vietnam War are consistent with the idea of a legitimation of violence. Second, both homicide and suicide rates have been associated with business cycles (Henry and Short 1954). Higher homicide rates occur in periods of growing economic prosperity characterized by relative deprivation for some groups. Third, rising homicide rates in the 1960s and 1970s have been linked to increased gang activity and drug trafficking in central cities. Patterns of violence in large cities strongly influence national trends. For example, about one-fourth of all murders and assaults in the U.S. in 1991 occurred in the large central cities—New York, Los Angeles, Chicago, Detroit, Houston, Philadelphia, and Washington, D.C. (UCR 1991). Fourth, increases in homicide and assault rates since World War II have been attributed to changes in people's routine activities and lifestyles that make them more visible, accessible, and exposed to risky and dangerous situations (Cohen and Felson 1979). Fifth, the unprecedented increase in the number of persons in the 15-to-24 age group has contributed to the rise in rates of murder and assault in the 1960s and 1970s. This age group has the highest risks of offending for violent crimes.

It is important to note that UCR data on homicide and assault underestimate the actual prevalence of violence in the U.S. because many individual acts of violence go unreported to the authorities, and multiple occurrences of violence in a criminal episode are often only counted once in crime reports. Nonetheless, the following estimates from UCR data and victimization surveys provide a chilling picture of the magnitude of interpersonal violence in American society over the last two decades:

- More than 450,000 Americans have been victims of homicide in the last twenty years. Between 18,000 and 25,000 homicides occurred every year since 1972 (UCR 1995; UCR Supplemental Homicide Reports 1995).
- One out of every 154 Americans will be a victim of homicide. The lifetime risks of homicide victimization are 1 out of 503 for white women, 1 in 170 for white men, 1 in 125 for African American women, and 1 in 26 for African American men (Dobrin, Wiersema, Loftin, and McDowall 1996).
- More than one million assaults each year involving serious injury are known to the police for every year since 1990 (UCR 1995).
- About 5 percent of American households experience a violent crime each year (Rand 1993).

Social Correlates of Homicide and Assault Rates

Homicide and assault rates vary widely by geographical areas and degree of urbanization. In recent UCR data, homicide rates were highest in southern states with a rate of 9.8 murders per 100,000 inhabitants, followed closely by western states at 9.0 per 100,000 (UCR 1995). Murder rates were substantially lower in the Midwest (6.9) and the Northeast (6.2). Southern and western states also had the highest rates of aggravated assault, and the lowest rates were again in the Northeast. Both homicide and assault rates were about twice as high in large urban areas as in rural counties (UCR 1995).

There was enormous variation in homicide and assault rates across U.S. cities. Many large Southern cities like New Orleans, Atlanta, and Birmingham, Alabama, far exceeded the national rate of 8.2 per 100,000 population. Some Midwestern cities such as St. Louis, Detroit, and Chicago also had high homicide rates, but other cities in the region did not. Among select cities in Western states, Los Angeles had the highest rate and Seattle was below the national average. Cities with the highest rate of aggravated assault from 1992 to 1994 were (a) Little Rock, Arkansas, (b) Atlanta, (c) Tampa, (d) St. Louis, and (e) San Bernardino, California. Even the city with the twenty-fourth highest rate (Boston) had an assault rate more than twice the national average of 438 per 100,000.

Rates of homicides and aggravated assaults in the U.S. can be explained by the presence or absence of several other social correlates. Cities and neighborhoods with high unemployment, rapid population turnover, overcrowding and housing decay, high ethnic diversity, substandard schools, high rates of single parent households, and high income inequality have the highest rates of homicide and assault. High murder and assault rates in some cities, and in the "bad side of town" within them, are often attributed to elements of social disorganization including low economic opportunity, the diversity of language and values, and the low supervision of youth (Miethe, Hughes, and McDowall 1991; Sampson and Groves 1989).

Although media reports of political terrorism and deadly civil unrest give the impression that violence is more prevalent in foreign lands, the U.S. is by far the most violent industrial nation in terms of its homicide rate. The U.S. homicide rate of 9.4 per 100,000 in 1990 was more than twice as high as India's rate, more than four times higher than Canada's and China's rate, and almost nineteen times higher than Japan's homicide rate. Explanations for the higher homicide rate in the U.S. include the wider availability of handguns, the greater social acceptance of violence as a method of conflict resolution, and the fact that violence is deeply woven into the fabric of American culture, including our street talk, prime-time television programming, "gangsta rap" and other music lyrics, and sports and leisure activities.

OFFENDER PROFILE

When compared to their distribution in the U.S. population, murder and assault offenders are disproportionately male, young (under 25), African American, and urban residents. According to UCR data, more than nine out of 10 homicide offenders are male, three-fourths are between the ages of 15 and 34, more than half are African American, and more than half of all murders occur in cities with more than 100,000 residents (UCR 1995). Police data rarely provide reliable information on offenders' employment history or social class, but the high concentration of violence in low-income areas within urban areas (see Miethe and Meier 1994; Reiss and Roth 1993) suggests that low socioeconomic status is another element in the typical offender profile. Combining these attributes reveals that poor, young, African American, urban males are the most prone to violent offending.

The demographic profile of violent offenders has changed over time in some cases but not in others. Racial and gender differences in homicide and assault offending, for example, have changed little over time. The proportion of homicide offenders under the age of 25, however, has shown a marked increase. This age group accounted for 30 percent of homicide offenders in 1970 and rose to 57 percent by 1995. Murderers under 18 years old increased from 6 percent in 1960 to about 15 percent in 1995 (UCR 1960, 1995). As described shortly, these young killers differ from other murderers in their personal characteristics, victim attributes, and situational elements of their crimes.

Court data on felony defendants provide some additional details about the typical violent offender. For example, more than half of the murder and assault suspects in the 75 largest U.S. counties have a prior felony arrest (Smith 1993). About two-thirds of these violent defendants have arrest records when misdemeanor arrests are included. Background checks during presentence investigations and clinic interviews of violent offenders often reveal a family history of abuse and neglect and the onset of antisocial behavioral patterns at an early age.

Offense Specialization and Escalation

Given that most homicide and assault offenders have prior arrest histories, one question involves whether these criminals specialize in violence and whether there is a pattern of escalation from nonviolent to violent crime. Large-scale cohort studies of the same offenders over time reveal two general trends. First, only about one out of every five persons ever arrested had an arrest for violent crime. Most arrestees for violence had long criminal careers dominated by arrests for nonviolent crime (Reiss and Roth 1993). Studies of recidivism provide limited support for specialization in that only about 7 percent of murderers released from prison are rearrested for murder within three years (BJS 1989). Second, the small number of chronic adult criminals who do specialize tend to commit either violent crime or property crime, rarely switching between the two (Blumstein et al. 1986, 1988; Farrington et al. 1988). Prediction of future violent behavior from arrest records, however, has proven to be highly inaccurate (Reiss and Roth 1993).

Planning and Spontaneity

Another aspect of the offender profile is the extent of planning and premeditation in the commission of murder and assaults. Within this context, a distinction is made between planned or premeditated acts (that is, those involving rational planning and the relative weighing of costs and benefits) and impulsive or spontaneous acts (that is, those in which the offender is guided by the "heat of passion").

Television crime dramas and the recent proliferation of stalking laws across the country give the impression that most murders and assaults are meticulously planned. They are not. Research findings indicate that most violent crimes are spontaneous, triggered by a trivial altercation or argument that quickly escalates in the heat of passion (Luckenbill 1977; Oliver 1994; Polk 1994; UCR 1995). More than half of murders in which the motive is specified involve actions that are largely spontaneous, such as mutual brawls and arguments (UCR 1995). Even for the one-third of murders that occur during the commission of other felonies (UCR 1995), it is likely that most murders happen in the heat of the moment and are not planned in advance. As situationally induced acts of violence, they exhibit little premeditation or planning.

VICTIM PROFILE

Social scientists have long recognized the similarity between the sociodemographic characteristics of violent offenders and their victims (see Miethe and Meier 1994; Singer 1981). Victims and offenders of both homicide and physical assault are generally similar in gender, race, and age, and live as neighbors, or at least in residential propinquity. Data from the UCR Supplemental Homicide Reports (1995) indicate that, like their offenders, homicide victims are disproportionately male, African American, and between 20 and 34 years old. About nine out of every 10 homicides involving one victim and one offender are intraracial. Studies of homicides in particular cities reveal that typical victims have never been married and are employed (Messner and Tardiff 1985). The most dangerous occupations for homicide victimization at work are: (a) taxicab drivers and chauffeurs, (b) police and other law enforcement officials, (c) hotel clerks, (d) garage and service-station employees, and (e) stock handlers and baggers (Castillo and Jenkins 1994).

The typical victim profile of assault is similar to that of homicide. Persons with the greatest risks of assault victimization are male, young (16–24), African American, never married, make less than $7500 per year, and live in central cities (Perkins and Klaus 1996).

The contributory role of victims in violent situations has been studied under the topic of "victim-precipitated" crime. According to Wolfgang (1958), victim-precipitated homicides are cases in which the victims are the first to resort to physical force, which then leads to their subsequent slaying. Victim-precipitated assaults occur when victims initiate physical force or insinuating language against subsequent attackers (Curtis 1974). Up to one-half of all criminal homicides may be provoked by victims who initiate physical violence (Miethe 1985). A smaller but still substantial proportion of assaults are provoked by the victims. The prevalence of victim precipitation in murder and assault cases is clearly contrary to the popular image of victims as totally innocent bystanders to predatory attacks.

Various explanations have been offered to account for the sociodemographic profile of homicide and assault victims (see Fattah 1991; Karmen 1990). According to a "routine activity" and "lifestyle" approach (Cohen and Felson 1979; Hindelang, Gottfredson, and Garofalo 1978), persons with social attributes that indicate high contact with motivated offenders (such as living in high-crime neighborhoods), potential attractiveness as crime targets (such as accessibility and visibility), and low protection or guardianship have the greatest risks of victimization. Given that those who are male, young, single, unemployed, and live in urban areas engage in a variety of public and private activities that increase their proximity and exposure to dangerous people and places, it makes sense that differences in routine activities and lifestyles may account for their differential risk of victimization. Tests of the validity of this theoretical framework for explaining violent victimization, however, have been largely inconclusive (see Miethe and Meier 1994).

SITUATIONAL ELEMENTS AND CIRCUMSTANCES

Three necessary elements for homicide and assault are the presence of (a) an offender, (b) a victim, and (c) a situational context for the crime. The situational context defines the micro-environment for crime and includes the motivation for offending, the victim-offender relationship, the physical setting, and particular situational dynamics. Each of these elements is described below.

The Motivation and Circumstances for Murder and Assault

One of the most perplexing questions in understanding crime is that of motivation. Motive is at the heart of criminal trials, television crime dramas, and our personal infatuation with crime. Inquisitive minds want to know why—not only why friends or family members are picked as crime targets, but also why offenders commit their crimes. As in law, the search for the true motive for crime is elusive. Consider the diversity of possible motivations alleged for the following high-profile violent offenses over the last decade:

- Theodore John Kaczynski, the alleged Unabomber, has been tied to letter bombs that have killed 3 people and injured 23 others over the last two decades. His anti-technology ideology is a possible motive for his actions.
- Eric and Lyle Menendez were convicted of first degree murder for the brutal shotgun slaying of their parents in Beverly Hills. Their defense was based on the "abuse excuse." The apparent motives for the murders ranged from the brothers' fear of their father's continual abuse to their desire to collect $11 million in insurance.
- Susan Smith claimed that her two young boys were abducted by a stranger in a carjacking. After a massive search, she later confessed to killing her children by driving her car into a lake with the two boys in the back seat. She may have been motivated by her belief that her lover did not like children.
- Ruth Cole, the wife of the owner of a large newspaper in Washington state, was convicted of attempted murder when she tried to hire a hitman to kill the judge who convicted her son of being the Southside (Spokane) rapist and sentenced him to multiple life terms. One possible motive was retaliation.
- Four young black males approached a white male, Bernard Goetz, on a subway at night in New York City. Goetz pulled a gun from under his jacket, fired, and shot them. Convicted in criminal court for a firearm violation, he was also ordered to pay $43 million in damages in a civil case filed by one of his attackers. Self-protection, retaliatory street justice, and "hate crimes" were some of the motives suggested in the case.
- Lorena Bobbitt cut off her husband's penis with a kitchen knife while he was sleeping and later threw the severed member out her car window. She was found not guilty of malicious wounding by reason of insanity. Alleged motives include her husband's unsatisfactory sexual performance and her desire to pay him back for a history of spousal abuse.
- Timothy McVeigh and Terry Nichols are accused of the bombing of the Federal Building in Oklahoma City that resulted in 169 deaths (including one rescue worker). Speculation is that both defendants are part of an anti-government militia and that the bombing was a warning message. McVeigh was convicted of multiple murders in Federal Court and was sentenced to death in August 1997. Nichols has not yet gone to trial.

Although the accurate determination of motive in any crime is highly subjective, social scientists have used several approaches to categorize motives. One strategy is to distinguish between instrumental and expressive motivations. Under this framework, violent acts with instrumental motivations are directed at some valued goal beyond the act itself. The Menendez brothers may have killed their parents for the instrumental goal of protecting themselves or collecting the insurance payment. Ruth Cole attempted to hire a hitman for the instrumental goal of revenge. In contrast, expressive actions are those motivated exclusively by rage, anger, frustration or, more generally, the heat of passion. Many violent acts are best considered spontaneous and unplanned outbursts of violence rather than goal-directed actions (Oliver 1994; Polk 1994). When the motivation for violence is to resolve conflict or to settle a festering argument, however, the classification of the offense as either an instrumental or expressive act becomes more difficult. Likewise, shootings by rival gang members may be driven by both instrumental and expressive motives—enhancing their own or their groups' standing, retaliating for stealing a girlfriend, defending turf, or just "kickin' around, raisin' hell."

The predominance of expressive motivations for homicides and assaults is supported by two general observations. First, convictions for voluntary manslaughter (an expressive act in the heat of passion) are far more common in criminal courts than convictions for first-degree murder (an act often judged to be instrumentally motivated). Second, although differences in convictions for murder and manslaughter may be due in large part to plea bargaining practices, independent examination of "murder rap sheets" and case narratives offers some support for the greater prevalence of expressive, nonutilitarian motivations for homicide (Luckenbill 1977; Polk 1994).

A variety of categories is used in the UCR Supplemental Homicide Reports (1995) to classify the motives or circumstances surrounding homicide. Among incidents in which a particular murder circumstance is specified, the most common motives, in decreasing order of prevalence, are arguments (49 percent), participation in other felony crimes, especially robbery and drug offenses (33 percent), youth gang activity (10 percent), brawls under the influence of drugs or alcohol (4 percent), romantic triangles (2 percent), and miscellaneous situations such as killings by babysitters, gangland slayings, and sniper attacks (2 percent). If arguments and brawls are considered spontaneous events, the majority of homicides are motivated by expressive rather than instrumental concerns. Studies of homicide in Chicago provide additional support for this conclusion. Among those killings with a clear motive, more than half of Chicago homicides are classified as expressive crimes (Block 1995).

It is widely assumed that the motives for aggravated assault are similar to those for homicide, but national data on assault are less widely available and analyzed. Evaluations of the proportion of assaults that are instrumentally motivated is somewhat problematic because assaults spurred by monetary gain are defined in police records as robberies. Nonetheless, several studies indicate that most assaults and murders result from emotional outbursts that are precipitated by trivial disputes and arguments (Curtis 1974; Luckenbill 1977; Wolfgang 1958). The variety of these disputes is clearly illustrated by the following brief police descriptions of homicides and aggravated assaults in Las Vegas:

- The victim and suspect were fighting over a drug debt and money when the suspect pulled a gun and fired four or five rounds. One round struck the victim in the chest.
- The suspect said that she had stabbed the victim during a fight over $20 he owed her.
- Two people were in a verbal argument concerning the whereabouts of a firearm. After telling the suspect he gave the gun to someone else, the suspect became irate and punched the victim. This person then ran upstairs, retrieved a firearm, and shot the other man several times.
- The victim was confronted by an ex-boyfriend at work. The suspect said, "I can't believe you've been talking about me to all your friends," as he began removing a five-inch hunting knife from its sheath. A struggle ensued with the victim suffering multiple stab wounds.
- The suspect tried to pass the victim's car and "flipped him off." Another suspect in the back seat then started shooting at the victim.
- A verbal altercation between the victim and suspect started during a pick-up basketball game. Suspect went to a gym bag and brought out two handguns. Suspect shot once in the direction of the victim but missed. Victim rushed the suspect and was then shot several times in the abdomen and upper chest. The victim was seriously injured.
- The victim was stabbed multiple times by a friend. The motive was apparently a $400 phone bill. The victim survived the attack.
- The victim was walking by a housing project when a large group of males started yelling profanities and asking him what he was doing there. The victim and suspect exchanged words which eventually led to a fistfight between the victim and four suspects. One suspect pulled out a handgun and shot the victim in the shoulder as he ran from the area.
- The victim and several friends were at a party. Rival gang members arrived and a fight broke out. Multiple shots were fired. No one was seriously injured.
- The suspect and victim were attending a party. The suspect was belligerent and was asked to leave. The victim approached and continued to argue with the suspect. The suspect pulled out a semi-automatic handgun and racked a round into the chamber. The victim said he "saw the gun and I dared him to shoot." The suspect then fired 5–7 shots, hitting the victim and another person.

These descriptions emphasize the prevalence of disputes as the underlying motivation for many homicides and assaults. They also illustrate that violent crimes derive from expressive concerns "in the heat of passion" and in situations of escalating conflict (see also Luckenbill 1977; Oliver 1994; Polk 1994).

The Victim-Offender Relationship

One major distinction between types of homicide and aggravated assault involves the relationship between victims and offenders. Three types of relationships are often identified: (a) familial (especially spouses and siblings), (b) acquaintances (including friends, neighbors, and co-workers), and (c) strangers.

Based on the analysis of national homicide data (UCR 1995), the vast majority of homicides known to the police involved acquaintances (57 percent), while fewer involve strangers (25 percent) and relatives (18 percent). Most typical acquaintances are people who are visually recognized by or had previous sporadic contact with offenders. In decreasing order of frequency, friends, girlfriends, boyfriends, neighbors, and co-workers make up the remaining "acquaintance" murders (UCR 1995). Spouse slayings are the most prevalent family murders (8 percent of all homicides), with wives nearly three times more likely to be victims than their husbands. About 4 percent of homicides involve a parent killing their offspring, 2 percent involve children killing their parents, and 2 percent are murders between siblings.

Over the last two decades, several patterns have emerged in the nature of the victim-offender relationship. First, slayings among intimates (including relatives and romantic partners) have remained fairly stable over time, ranging from 28 percent of homicides in 1976 to 25 percent in 1995 (UCR 1976, 1995). Second, the number of lethal and nonlethal attacks by strangers has increased over time. This rise in stranger violence is especially troubling because it suggests a greater sense of randomness in the selection of crime targets and less individual control over the risk of victimization. The term "mushroom," derived from the Super Mario videogame, is sometimes used to describe innocent victims caught in gang crossfire (Sherman, Steele, Laufersweiler, Hoffer, and Julian 1989).

UCR data provide limited information on the victim-offender relationship in aggravated assaults. However, given the other similarities between homicide and assault, the vast majority of aggravated assaults should also involve acquaintances and family members. Using data from multiple U.S. cities, Mulvihill, Tumin, and Curtis (1969) found that about 40 percent of aggravated assaults involved acquaintances and nearly 20 percent involved family members. Somewhat lower rates of aggravated assault by acquaintances (33 percent) and relatives (7 percent) were found in national victimization data (Perkins and Klaus 1996), but these surveys tend to underestimate crimes among intimates. These findings provide some validation for the claim that assaults, like homicides, typically involve persons with prior relationships.

Situational Dynamics

Aside from motive and the victim-offender relationship, homicides and assaults involve other situational elements that define the nature of these crimes and their social, spatial, and temporal distribution. Major situational dynamics include temporal and spatial aspects, weapon use, co-offender patterns, and the use of alcohol and drugs.

Temporal and Spatial Patterns

Violent crimes do not occur in a vacuum. Rather, they vary dramatically across time and physical space.

Most homicides and assaults occur in the evening hours and disproportionately on weekends. Nearly two-thirds of aggravated assaults happen in the evening (Perkins and Klaus 1996), and about one-half of homicides take place on weekends (National Center for Health Statistics 1993). The dangerousness of weekend nights, especially Saturday nights, is due to several factors. First, there is more public activity over the weekend, increasing one's visibility to criminals and exposure to risky and dangerous situations. Second, our cultural heritage of unwinding on the weekend and the increased use of drugs and alcohol on weekend nights further increase the dangerousness of this time period.

The most dangerous physical locations for murder and assault are victims' homes and on the street. About one-third of aggravated assaults occur on the street or in parking lots, and one-quarter in or near the victims' homes (Perkins and Klaus 1996). Victims' homes are the most common place for assaults involving known parties, but about half of stranger assaults happen in open, public places like streets and parking lots (Miethe and Meier 1994). Nearly half of the homicides in Las Vegas take place within the home. Among Philadelphia murders within the home, Wolfgang (1958) found that men tended to kill their wives in the bedroom (where male masculinity may be most commonly challenged) and women killed their husbands in the kitchen (where knives and other cutting objects are more readily available). Given the large amount of time people spend in their homes, the high risk of violent crime in this location is understandable.

Over the last decade, violent crimes that occur in the workplace have received much national attention. Disgruntled employees who shoot multiple co-workers are often noted, but such attacks make up less than 10 percent of all work-related homicides (Dobrin, Wiersema, Loftin, and McDowall 1996). Instead, about three-fourths of the homicides at work occur during robberies and other miscellaneous crimes. Because they are common locations for robberies or involve greater contact with dangerous people, the following industries have the highest risks of work-related homicides per 100,000 workers: (a) taxicab services (26.9), (b) liquor stores (8.0), (c) gasoline stations (5.6), (d) detective and protective services (5.0), (e) justice, public order, and safety industries (3.4), (f) grocery stores (3.2), and (g) jewelry stores (3.2) (Castillo and Jenkins 1994). However, only a small proportion of all homicides and assaults take place in commercial establishments and other work settings.

Weapon Use in Murder and Assault

While gun control is a hotly debated and volatile public issue, police data clearly reveal that firearms are the most common weapons in homicides and a common weapon in aggravated assaults. A firearm (primarily a handgun) is used in about two-thirds of all homicides, followed, in decreasing order of prevalence, by knives or cutting instruments, personal weapons (such as hands, fists, and feet), blunt objects, and strangulation (UCR 1995). Contrary to media images, poison and explosives are rarely used as murder weapons. Homicides by firearms have generally increased over time, whereas those involving cutting instruments have decreased in frequency. Rather than indicating an increase in gun use in homicide, however, these changes over time may reflect the greater effectiveness of paramedic units and trauma teams in reducing the lethality of nongun injuries.

Weapon use in aggravated assault cases exhibits greater parity. The most common weapons involved in assaults are those such as clubs or other blunt objects, followed closely by personal weapons, firearms, and knives or cutting instruments.

Aggravated assaults with firearms have increased over the last three decades, while assaults involving personal weapons have decreased (UCR 1970, 1995).

The type of weapon used in both homicide and assault cases, however, varies dramatically for different groups of people and violent situations. The UCR Supplemental Homicide Reports indicate that more men than women use firearms in homicides, more women use knives or sharp objects as deadly weapons. African Americans are more likely than other racial and ethnic groups to use firearms in homicides. Firearms are used in more than 90 percent of homicides involving youth gangs, but in only about two-thirds of all other murder situations (UCR Supplemental Homicide Reports 1988–1992). Similarly, a firearm is the weapon in about three-fourths of the murders committed by strangers compared to only about half of deadly attacks on family members.

Co-Offending Patterns

Although typical homicides and assaults involve single victims and single offenders, media attention tends to focus on violent acts with multiple offenders or multiple victims. Between 1988 and 1992, 63 percent of homicides involved one offender and victim, 29 percent involved multiple offenders and one victim, 5 percent were mass slayings by one offender, and only 3 percent involved multiple victims and multiple offenders.

The motivations and situational elements underlying homicides and assaults in a group context are quite different from those found in other offenses. Trivial altercations often motivate one-on-one attacks, but youth-gang attacks and hate crimes against particular groups of people are the primary circumstance surrounding violence committed against one victim by multiple offenders. When one offender commits multiple victimizations, the victims are usually family members or co-workers. Homicides and assaults involving multiple victims and multiple offenders typically occur in group melees and mass disturbances.

The presence of others may lead to either the escalation or cessation of violent situations. Given that many homicides and assaults are initiated by efforts to "save face" (Luckenbill 1977), the mere presence of others in potentially violent situations may have an audience effect that provides added pressure to resort to violence as a way of showing personal strength and enhancing status among peers. This type of group facilitation is most evident in violent offenses committed by juveniles. In contrast, others may intervene in conflict situations or be perceived as legal witnesses to the actions, thereby minimizing further violence. It is impossible to determine how many assaults or potential murders are diverted by the mere presence of or actions taken by bystanders. Nonetheless, it seems clear that the presence of others may either constrain or enhance the escalation of violent situations.

Alcohol and Drug Use

Social scientists have widely recognized the importance of drugs and alcohol as situational elements in many violent offenses (Luckenbill 1977; Parker and Rebhun 1995; Wolfgang 1958). The distribution, sale, and use of drugs has been implicated as a criminogenic factor in about 10 percent of all homicides and a substantial share of those in urban areas (Reiss and Roth 1993). Drug and alcohol use by offenders or victims is a common theme in violent crime, with about 60 percent of arrestees for violent offenses exhibiting positive drug tests (Reiss and Roth 1993). Between 20 and 40 percent of convicted murderers have been diagnosed as alcoholics (Greenberg 1981). Battles over the distribution of drugs in large metropolitan areas are a major contributory factor in increasing homicide rates in the last two decades.

Drugs and alcohol are associated with homicide and assault in three distinct ways. First, sellers of illicit drugs compete with other distributors, and violence may be an effective means of reducing this competition. Second, sellers may be violently attacked by buyers who are trying to steal their drug supplies or cash to support a habit. Third, drug and alcohol use in many cases occurs within a group context. When the presence of others similarly situated is coupled with the reduced inhibitions often associated with drug and alcohol use, a rather trivial comment or action may quickly escalate into a violent episode.

Target-Selection Strategies

From the perspective of the "reasoning criminal" (Cornish and Clarke 1986), offenders make a series of rational choices about crime commission and target-selection by weighing the relative costs and benefits of alternative courses of action. Once a decision is made to commit crime, three general characteristics are thought to underlie the selection of particular crime targets: (a) convenience and familiarity, (b) the level of protection or guardianship, and (c) the expected yield and target attractiveness.

Although most homicides and assaults involve disputes that occur in the heat of passion (suggesting that these crimes lack a rational calculus), target-selection factors are nonetheless important for explaining the nature and distribution of violent crimes. Convenience and familiarity, for example, may account for why family members and acquaintances are the most common targets for murder and assault. Non-strangers are more convenient and familiar targets because most people spend more time with friends and family than strangers and we are more aware of the particular habits and routines of intimates, allowing for better timing of the crime event and great anticipation of protective actions by the victim. Similarly, gender differences in the ability to protect oneself and physically resist an attack may explain why wives are far more likely than husbands to be victims of domestic violence. The expected yield and target attractiveness may account for the high concentration of felony homicides involving robbery and drug dealing, offenses that have the highest potential financial return. Under these conditions, target-selection factors are just as important in understanding homicide and aggravated assault as they are in accounting for more instrumentally motivated crimes like burglary, robbery, and motor vehicle theft.

QUESTIONS FOR DISCUSSION

1. Compare and contrast homicide and aggravated assault. Discuss trends of homicides and aggravated assaults from the 1930s to the 1990s. Are there any surprises? What do you expect the homicide and aggravated assault trend to look like in 2010?

2. Discuss social correlates of homicide and assault rates. What explains this phenomenon?

3. Describe the offender and victim profiles in homicides and assaults. What is interesting about your finding? Discuss situational dynamics as they relate to homicides and assaults.

Convicted Rapists' Vocabulary of Motive: Excuses and Justifications

Diana Scully and Joseph Marolla

JUSTIFYING RAPE

. . . deniers attempted to justify their behavior by presenting the victim in a light that made her appear culpable, regardless of their own actions. Five themes run through attempts to justify their rapes: (1) women as seductresses; (2) women mean "yes" when they say "no"; (3) most women eventually relax and enjoy it; (4) nice girls don't get raped; and (5) guilty of a minor wrongdoing.

1) Women as Seductresses

Men who rape need not search far for cultural language which supports the premise that women provoke or are responsible for rape. In addition to common cultural stereotypes, the fields of psychiatry and criminology (particularly the subfield of victimology) have traditionally provided justifications for rape, often by portraying raped women as the victims of their own seduction (Albin, 1977; Marolla and Scully, 1979). For example, Hollander (1924:130) argues:

> Considering the amount of illicit intercourse, rape of women is very rare indeed. Flirtation and provocative conduct, i.e., tacit (if not actual) consent is generally the prelude to intercourse.

Since women are supposed to be coy about their sexual availability, refusal to comply with a man's sexual demands lacks meaning and rape appears normal. The fact that violence and, often, a weapon are used to accomplish the rape is not considered. As an example, Abrahamsen (1960:61) writes:

> The conscious or unconscious biological or psychological attraction between man and woman does not exist only on the part of the offender toward the woman but, also, on her part toward him, which in many instances may, to some extent, be the impetus for his sexual attack. Often a women [sic] unconsciously wishes to be taken by force—consider the theft of the bride in Peer Gynt.

Like Peer Gynt, the deniers we interviewed tried to demonstrate that their victims were willing and, in some cases, enthusiastic participants. In these accounts, the rape became more dependent upon the victims' behavior than upon their own actions.

Thirty-one percent . . . of the deniers presented an extreme view of the victim. Not only willing, she was the aggressor, a seductress who lured them, unsuspecting, into sexual action. Typical was a denier convicted of his first rape and accompanying crimes of burglary, sodomy, and abduction. According to the pre-sentence reports, he had broken into the victim's house and raped her at knife point. While he admitted to the breaking and entry, which he claimed was for altruistic purposes ("to pay for the prenatal care of a friend's girlfriend"), he also argued that when the victim discovered him, he had tried to leave but she had asked him to stay. Telling him that she cheated on her husband, she had voluntarily removed her clothes and seduced him. She was, according to him, an exemplary sex partner who "enjoyed it very much and asked for oral sex. Can I have it now?" he reported her as saying. He claimed they had spent hours in bed, after which the victim had told him he was good looking and asked to see him again. "Who would believe I'd meet a fellow like this?" he reported her as saying.

In addition to this extreme group, 25 percent . . . of the deniers said the victim was willing and had made some sexual advances. An additional 9 percent . . . said the victim was willing to have sex for money or drugs. In two of these three cases, the victim had been either an acquaintance or picked up, which the rapists said led them to expect sex.

2) Women Mean 'Yes' When They Say 'No'

Thirty-four percent . . . of the deniers described their victim as unwilling, at least initially, indicating either that she had resisted or that she had said no. Despite this, and even though according to pre-sentence reports a weapon had been present in 64 percent . . . of these 11 cases, the rapists justified their behavior by arguing that either the victim had not resisted enough or that her "no" had really meant "yes." For example, one denier who was serving time for a previous rape was subsequently convicted of attempting to rape a prison hospital nurse. He insisted he had actually completed the second rape, and said of his victim: "She semi-struggled but deep down inside I think she felt it was a fantasy come true." The nurse, according to him, had asked a question about his conviction for rape, which he interpreted as teasing. "It was like she was saying, 'rape me'." Further, he stated that she had helped him along with oral sex and "from her actions, she was enjoying it." In another case, a 34-year-old man convicted of abducting and raping a 15-year-old teenager at knife point as she walked on the beach, claimed it was a pickup. This rapist said women like to be overpowered before sex, but to dominate after it begins.

> A man's body is like a coke bottle, shake it up, put your thumb over the opening and feel the tension. When you take a woman out, woo her, then she says "no, I'm a nice girl," you have to use force. All men do this. She said "no" but it was a societal no, she wanted to be coaxed. All women say "no" when they mean "yes" but it's a societal no, so they won't have to feel responsible later.

Claims that the victim didn't resist or, if she did, didn't resist enough, were also used by 24 percent . . . of admitters to explain why, during the incident, they believed the victim was willing and that they were not raping. These rapists didn't redefine their acts until some time after the crime. For example, an admitter who used a bayonet to threaten his victim, an employee of the store he had been robbing, stated:

> At the time I didn't think it was rape. I just asked her nicely and she didn't resist. I never considered prison. I just felt like I had met a friend. It took about five years of reading and going to school to change my mind about whether it was rape. I became familiar with the subtlety of violence. But at the time, I believed that as long as I didn't hurt anyone it wasn't wrong. At the time, I didn't think I would go to prison, I thought I would beat it.

Another typical case involved a gang rape in which the victim was abducted at knife point as she walked home about midnight. According to two of the rapists, both of whom were interviewed, at the time they had thought the victim had willingly accepted a ride from the third rapist (who was not interviewed). They claimed the victim didn't resist and one reported her as saying she would do anything if they would take her home. In this rapist's view, "She acted like she enjoyed it, but maybe she was just acting. She wasn't crying, she was engaging in it." He reported that she had been friendly to the rapist who abducted her and, claiming not to have a home phone, she gave him her office number—a tactic eventually used to catch the three. In retrospect, this young man had decided, "She was scared and just relaxed and enjoyed it to avoid getting hurt." Note, however, that while he had redefined the act as rape, he continued to believe she enjoyed it.

Men who claimed to have been unaware that they were raping viewed sexual aggression as a man's prerogative at the time of the rape. Thus they regarded their act as little more than a minor wrongdoing even though most possessed or used a weapon. As long as the victim survived without major physical injury, from their perspective, a rape had not taken place. Indeed, even U.S. courts have often taken the position that physical injury is a necessary ingredient for a rape conviction.

3) Most Women Eventually Relax and Enjoy It

Many of the rapists expected us to accept the image, drawn from cultural stereotype, that once the rape began, the victim relaxed and enjoyed it. Indeed, 69 percent . . . of deniers justified their behavior by claiming not only that the victim was willing, but also that she enjoyed herself, in some cases to an immense degree. Several men suggested that they had fulfilled their victims' dreams. Additionally, while most admitters used adjectives such as "dirty," "humiliated," and "disgusted," to describe how they thought rape made women feel, 20 percent . . . believed that their victim enjoyed herself. For example, one denier had posed as a salesman to gain entry to his victim's house. But he claimed he had a previous sexual relationship with the victim, that she agreed to have sex for drugs, and that the opportunity to have sex with him produced "a glow, because she was really into oral stuff and fascinated by the idea of sex with a black man. She felt satisfied, fulfilled, wanted me to stay, but I didn't want her." In another case, a denier who had broken into his victim's house but who insisted the victim was his lover and let him in voluntarily, declared "She felt good, kept kissing me and wanted me to stay the night. She felt proud after sex with me."

And another denier, who had hid in his victim's closet and later attacked her while she slept, argued that while she was scared at first, "once we got into it, she was ok." He continued to believe he hadn't committed rape because "she enjoyed it and it was like she consented."

4) Nice Girls Don't Get Raped

The belief that "nice girls don't get raped" affects perception of fault. The victim's reputation, as well as characteristics or behavior which violate normative sex role expectations, are perceived as contributing to the commission of the crime. For example, Nelson and Amir (1975) defined hitchhike rape as a victim-precipitated offense.

In our study, 69 percent . . . of deniers and 22 percent . . . of admitters referred to their victims' sexual reputation, thereby evoking the stereotype that "nice girls don't get raped." They claimed that the victim was known to have been a prostitute, or a "loose" woman, or to have had a lot of affairs, or to have given birth to a child out of wedlock. For example, a denier who claimed he had picked up his victim while she was hitchhiking stated, "To be honest, we [his family] knew she was a damn whore and whether she screwed one or 50 guys didn't matter." According to pre-sentence reports this victim didn't know her attacker and he abducted her at knife point from the street. In another case, a denier who claimed to have known his victim by reputation stated:

> If you wanted drugs or a quick piece of ass, she would do it. In court she said she was a virgin, but I could tell during sex [rape] that she was very experienced.

When other types of discrediting biographical information were added to these sexual slurs, a total of 78 percent . . . of the deniers used the victim's reputation to substantiate their accounts. Most frequently, they referred to the victim's emotional state or drug use. For example, one denier claimed his victim had been known to be loose and, additionally, had turned state's evidence against her husband to put him in prison and save herself from a burglary conviction. Further, he asserted that she had met her current boyfriend, who was himself in and out of prison, in a drug rehabilitation center where they were both clients.

Evoking the stereotype that women provoke rape by the way they dress, a description of the victim as seductively attired appeared in the accounts of 22 percent . . . of deniers and 17 percent . . . of admitters. Typically, these descriptions were used to substantiate their claims about the victim's reputation. Some men went to extremes to paint a tarnished picture of the victim, describing her as dressed in tight black clothes and without a bra; in one case, the victim was portrayed as sexually provocative in dress and carriage. Not only did she wear short skirts, but she was observed to "spread her legs while getting out of cars." Not all of the men attempted to assassinate their victim's reputation with equal vengeance. Numerous times they made subtle and offhand remarks like, "She was a waitress and you know how they are."

The intent of these discrediting statements is clear. Deniers argued that the woman was a "legitimate" victim who got what she deserved. For example, one denier stated that all of his victims had been prostitutes; pre-sentence reports indicated they were not. Several times during his interview, he referred to them as "dirty sluts," and argued "anything I did to them was justified." Deniers also claimed their victim had wrongly accused them and was the type of woman who would perjure herself in court.

5) Only a Minor Wrongdoing

The majority of deniers did not claim to be completely innocent and they also accepted some accountability for their actions. Only 16 percent . . . of deniers argued that they were totally free of blame. Instead, the majority of deniers pleaded guilty to a lesser charge. That is, they obfuscated the rape by pleading guilty to a less serious, more acceptable charge. They accepted being over-sexed, accused of poor judgment or trickery, even some violence, or guilty of adultery or contributing to the delinquency of a minor, charges that are hardly the equivalent of rape.

Typical of this reasoning is a denier who met his victim in a bar when the bartender asked him if he would try to repair her stalled car. After attempting unsuccessfully, he claimed the victim drank with him and later accepted a ride. Out riding, he pulled into a deserted area "to see how my luck would go." When the victim resisted his advances, he beat her and he stated:

> I did something stupid. I pulled a knife on her and I hit her as hard as I would hit a man. But I shouldn't be in prison for what I did. I shouldn't have all this time [sentence] for going to bed with a broad.

This rapist continued to believe that while the knife was wrong, his sexual behavior was justified.

In another case, the denier claimed he picked up his under-age victim at a party and that she voluntarily went with him to a motel. According to pre-sentence reports, the victim had been abducted at knife point from a party. He explained:

> After I paid for a motel, she would have to have sex but I wouldn't use a weapon. I would have explained. I spent money and, if she still said no, I would have forced her. If it had happened that way, it would have been rape to some people but not to my way of thinking. I've done that kind of thing before. I'm guilty of sex and contributing to the delinquency of a minor, but not rape.

In sum, deniers argued that, while their behavior may not have been completely proper, it should not have been considered rape. To accomplish this, they attempted to discredit and blame the victim while presenting their own actions as justified in the context. Not surprisingly, none of the deniers thought of himself as a rapist. A minority of the admitters attempted to lessen the impact of their crime by claiming the victim enjoyed being raped. But despite this similarity, the nature and tone of admitters' and deniers' accounts were essentially different.

EXCUSING RAPE

In stark contrast to deniers, admitters regarded their behavior as morally wrong and beyond justification. They blamed themselves rather than the victim, although some continued to cling to the belief that the victim had contributed to the crime somewhat, for example, by not resisting enough.

Several of the admitters expressed the view that rape was an act of such moral outrage that it was unforgivable. Several admitters broke into tears at intervals during their interviews. A typical sentiment was,

> I equate rape with someone throwing you up against a wall and tearing your liver and guts out of you. . . . Rape is worse than murder . . . and I'm disgusting.

Another young admitter frequently referred to himself as repulsive and confided:

> I'm in here for rape and in my own mind, it's the most disgusting crime, sickening. When people see me and know, I get sick.

Admitters tried to explain their crime in a way that allowed them to retain a semblance of moral integrity. Thus, in contrast to deniers' justifications, admitters used excuses to explain how they were compelled to rape. These excuses appealed to the existence of forces outside of the rapists' control. Through the use of excuses, they attempted to demonstrate that either intent was absent or responsibility was diminished. This allowed them to admit rape while reducing the threat to their identity as a moral person. Excuses also permitted them to view their behavior as idiosyncratic rather than typical and, thus, to believe they were not "really" rapists. Three themes run through these accounts: (1) the use of alcohol and drugs; (2) emotional problems; and (3) nice guy image.

1) The Use of Alcohol and Drugs

A number of studies have noted a high incidence of alcohol and drug consumption by convicted rapists prior to their crime (Groth, 1979; Queen's Bench Foundation, 1976). However, more recent research has tentatively concluded that the connection between substance use and crime is not as direct as previously thought (Ladouceur, 1983). Another facet of alcohol and drug use mentioned in the literature is its utility in disavowing deviance. McCaghy (1968) found that child molesters used alcohol as a technique for neutralizing their deviant identity. Marolla and Scully (1979), in a review of psychiatric literature, demonstrated how alcohol consumption is applied differently as a vocabulary of motive. Rapists can use alcohol both as an excuse for their behavior and to discredit the victim and make her more responsible. We found the former common among admitters and the latter common among deniers.

Alcohol and/or drugs were mentioned in the accounts of 77 percent . . . of admitters and 84 percent . . . of deniers and both groups were equally likely to have acknowledged consuming a substance—admitters, 77 percent . . . ; deniers, 72 percent. . . . However, admitters said they had been affected by the substance; if not the cause of their behavior, it was at least a contributing factor. For example, an admitter who estimated his consumption to have been eight beers and four "hits of acid" reported:

> Straight, I don't have the guts to rape. I could fight a man but not that. To say, "I'm going to do it to a woman," knowing it will scare and hurt her, takes guts or you have to be sick.

Another admitter believed that his alcohol and drug use,

> . . . brought out what was already there but in such intensity it was uncontrollable. Feelings of being dominant, powerful, using someone for my own gratification, all rose to the surface.

In contrast, deniers' justifications required that they not be substantially impaired. To say that they had been drunk or high would cast doubt on their ability to control themself or to remember events as they actually happened. Consistent with this, when we asked if the alcohol and/or drugs had an effect on their behavior, 69 percent . . . of admitters, but only 40 percent . . . of deniers said they had been affected.

Even more interesting were references to the victim's alcohol and/or drug use. Since admitters had already relieved themselves of responsibility through claims of being drunk or high, they had nothing to gain from the assertion that the victim had used or been affected by alcohol and/or drugs. On the other hand, it was very much in the interest of deniers to declare that their victim had been intoxicated or high: that fact lessened her credibility and made her more responsible for the act. Reflecting these observations, 72 percent . . . of deniers and 26 percent . . . of admitters maintained that alcohol or drugs had been consumed by the victim. Further, while 56 percent . . . of deniers declared she had been affected by this use, only 15 percent . . . of admitters made a similar claim. Typically, deniers argued that the alcohol and drugs had sexually aroused their victim or rendered her out of control. For example, one denier insisted that his victim had become hysterical from drugs, not from being raped, and it was because of the drugs that she had reported him to the police. In addition, 40 percent . . . of deniers argued that while the victim had been drunk or high, they themselves either hadn't ingested or weren't affected by alcohol and/or drugs. None of the admitters made this claim. In fact, in all of the 15 percent . . . of cases where an admitter said the victim was drunk or high, he also admitted to being similarly affected.

These data strongly suggest that whatever role alcohol and drugs play in sexual and other types of violent crime, rapists have learned the advantage to be gained from using alcohol and drugs as an account. Our sample were aware that their victim would be discredited and their own behavior excused or justified by referring to alcohol and/or drugs.

2) Emotional Problems

Admitters frequently attributed their acts to emotional problems. Forty percent . . . of admitters said they believed an emotional problem had been at the root of their rape behavior, and 33 percent . . . specifically related the problem to an unhappy, unstable childhood or a marital-domestic situation. Still others claimed to have been in a general state of unease. For example, one admitter said that at the time of the rape he had been depressed, feeling he couldn't do anything right, and that something had been missing from his life. But he also added, "being a rapist is not part of my personality." Even admitters who could locate no source for an emotional problem evoked the popular image of rapists as the product of disordered personalities to argue they also must have problems:

> The fact that I'm a rapist makes me different. Rapists aren't all there. They have problems. It was wrong so there must be a reason why I did it. I must have a problem.

Our data do indicate that a precipitating event, involving an upsetting problem of everyday living, appeared in the accounts of 80 percent . . . of admitters and 25 percent . . . of deniers. Of those experiencing a precipitating event, including deniers, 76 percent . . . involved a wife or girlfriend. Over and over, these men described themselves as having been in a rage because of an incident involving a woman with whom they believed they were in love.

Frequently, the upsetting event was related to a rigid and unrealistic double standard for sexual conduct and virtue which they applied to "their" woman but which they didn't expect from men, didn't apply to themselves, and, obviously, didn't honor in other women. To discover that the "pedestal" didn't apply to their wife or girlfriend sent them into a fury. One especially articulate and typical admitter described his feeling as follows. After serving a short prison term for auto theft, he married his "childhood sweetheart" and secured a well-paying job. Between his job and the volunteer work he was doing with an ex-offender group, he was spending long hours away from home, a situation that had bothered his wife. In response to her request, he gave up his volunteer work, though it was clearly meaningful to him. Then, one day, he discovered his wife with her former boyfriend "and my life fell apart." During the next several days, he said his anger had made him withdraw into himself and, after three days of drinking in a motel room, he abducted and raped a stranger. He stated:

> My parents have been married for many years and I had high expectations about marriage. I put my wife on a pedestal. When I walked in on her, I felt like my life had been destroyed, it was such a shock. I was bitter and angry about the fact that I hadn't done anything to my wife for cheating. I didn't want to hurt her [victim], only to scare and degrade her.

It is clear that many admitters, and a minority of deniers, were under stress at the time of their rapes. However, their problems were ordinary—the types of upsetting events that everyone experiences at some point in life. The overwhelming majority of the men were not clinically defined as mentally ill in court-ordered psychiatric examinations prior to their trials. Indeed, our sample is consistent with Abel *et al.* (1980) who found fewer than 5 percent of rapists were psychotic at the time of their offense.

As with alcohol and drug intoxication, a claim of emotional problems works differently depending upon whether the behavior in question is being justified or excused. It would have been counter-productive for deniers to have claimed to have had emotional problems at the time of the rape. Admitters used psychological explanations to portray themselves as having been temporarily "sick" at the time of the rape. Sick people are usually blamed for neither the cause of their illness nor for acts committed while in that state of diminished capacity. Thus, adopting the sick role removed responsibility by excusing the behavior as having been beyond the ability of the individual to control. Since the rapists were not "themselves," the rape was idiosyncratic rather than typical behavior. Admitters asserted a non-deviant identity despite their self-proclaimed disgust with what they had done. Although admitters were willing to assume the sick role, they did not view their problem as a chronic condition, nor did they believe themselves to be insane or permanently impaired. Said one admitter, who believed that he needed psychological counseling: "I have a mental disorder, but I'm not crazy." Instead, admitters viewed their "problem" as mild, transient, and curable. Indeed, part of the appeal of this excuse was that not only did it relieve responsibility, but, as with alcohol and drug addiction, it allowed the rapist to "recover." Thus, at the time of their interviews, only 31 . . . percent of admitters indicated that "being a rapist" was part of their self-concept. Twenty-eight percent . . . of admitters stated they had never thought of themselves as a rapist, 8 percent . . . said they were unsure, and 33 percent . . . asserted they had been a rapist at one time but now were recovered. A multiple "ex-rapist," who believed his "problem" was due to "something buried in my subconscious" that was triggered when his girlfriend broke up with him, expressed a typical opinion:

> I was a rapist, but not now. I've grown up, had to live with it. I've hit the bottom of the well and it can't get worse. I feel born again to deal with my problems.

3) Nice Guy Image

Admitters attempted to further neutralize their crime and negotiate a non-rapist identity by painting an image of themselves as a "nice guy." Admitters projected the image of someone who had made a serious mistake but, in every other respect, was a decent person. Fifty-seven percent . . . expressed regret and sorrow for their victim indicating that they wished there were a way to apologize for or amend their behavior. For example, a participant in a rape-murder, who insisted his partner did the murder, confided, "I wish there was something I could do besides saying 'I'm sorry, I'm sorry.' I live with it 24 hours a day and, sometimes, I wake up crying in the middle of the night because of it."

Schlenker and Darby (1981) explain the significance of apologies beyond the obvious expression of regret. An apology allows a person to admit guilt while at the same time seeking a pardon by signaling that the event should not be considered a fair representation of what the person is really like. An apology separates the bad self from the good self, and promises more acceptable behavior in the future. When apologizing, an individual is attempting to say: "I have repented and should be forgiven," thus making it appear that no further rehabilitation is required.

The "nice guy" statements of the admitters reflected an attempt to communicate a message consistent with Schlenker's and Darby's analysis of apologies. It was an attempt to convey that rape was not a representation of their "true" self. For example,

> It's different from anything else I've ever done. I feel more guilt about this. It's not consistent with me. When I talk about it, it's like being assaulted myself.
> I don't know why I did it, but once I started, I got into it. Armed robbery was a way of life for me, but not rape. I feel like I wasn't being myself.

Admitters also used "nice guy" statements to register their moral opposition to violence and harming women, even though, in some cases, they had seriously injured their victims. Such was the case of an admitter convicted of a gang rape:

> I'm against hurting women. She should have resisted. None of us were the type of person that would use force on a woman. I never positioned myself on a woman unless she showed an interest in me. They would play to me, not me to them. My weakness is to follow. I never would have stopped, let alone pick her up without the others. I never would have let anyone beat her. I never bothered women who didn't want sex; never had a problem with sex or getting it. I loved her—like all women.

Finally, a number of admitters attempted to improve their self-image by demonstrating that, while they had raped, it could have been worse if they had not been a "nice guy." For example, one admitter professed to being especially gentle with his victim after she told him she had just had a baby. Others claimed to have given the victim money to get home or make a phone call, or to have made sure the victim's children were not in the room. A multiple rapist, whose pattern was to break in and attack sleeping victims in their homes, stated:

> I never beat any of my victims and I told them I wouldn't hurt them if they cooperated. I'm a professional thief. But I never robbed the women I raped because I felt so bad about what I had already done to them.

Even a young man, who raped his five victims at gun point and then stabbed them to death, attempted to improve his image by stating:

> Physically they enjoyed the sex [rape]. Once they got involved, it would be difficult to resist. I was always gentle and kind until I started to kill them. And the killing was always sudden, so they wouldn't know it was coming.

SUMMARY AND CONCLUSIONS

Convicted rapists' accounts of their crimes include both excuses and justifications. Those who deny what they did was rape justify their actions; those who admit it was rape attempt to excuse it or themselves. This study does not address why some men admit while others deny, but future research might address this question. This paper does provide insight on how men who are sexually aggressive or violent construct reality, describing the different strategies of admitters and deniers.

Admitters expressed the belief that rape was morally reprehensible. But they explained themselves and their acts by appealing to forces beyond their control, forces which reduced their capacity to act rationally and thus compelled to rape. Two types of excuses predominated: alcohol/drug intoxication and emotional problems. Admitters used these excuses to negotiate a moral identity for themselves by viewing rape as idiosyncratic rather than typical behavior. This allowed them to reconceptualize themselves as recovered or "ex-rapists," someone who had made a serious mistake which did not represent their "true" self.

In contrast, deniers' accounts indicate that these men raped because their value system provided no compelling reason not to do so. When sex is viewed as a male entitlement, rape is no longer seen as criminal. However, the deniers had been convicted of rape, and like the admitters, they attempted to negotiate an identity. Through justifications, they constructed a "controversial" rape and attempted to demonstrate how their behavior, even if not quite right, was appropriate in the situation. Their denials, drawn from common cultural rape stereotypes, took two forms, both of which ultimately denied the existence of a victim.

The first form of denial was buttressed by the cultural view of men as sexually masterful and women as coy but seductive. Injury was denied by portraying the victim as willing, even enthusiastic, or as politely resistant at first but eventually yielding to "relax and enjoy it." In these accounts, force appeared merely as a seductive technique. Rape was disclaimed: rather than harm the woman, the rapist had fulfilled her dreams. In the second form of denial, the victim was portrayed as the type of woman who "got what she deserved." Through attacks on the victim's sexual reputation and, to a lesser degree, her emotional state, deniers attempted to demonstrate that since the victim wasn't a "nice girl," they were not rapists. Consistent with both forms of denial was the self-interested use of alcohol and drugs as a justification. Thus, in contrast to admitters, who accentuated their own use as an excuse, deniers emphasized the victim's consumption in an effort to both discredit her and make her appear more responsible for the rape. It is important to remember that deniers did not invent these justifications. Rather, they reflect a belief system which has historically victimized women by promulgating the myth that women both enjoy and are responsible for their own rape.

While admitters and deniers present an essentially contrasting view of men who rape, there were some shared characteristics. Justifications particularly, but also excuses, are buttressed by the cultural view of women as sexual commodities, dehumanized and devoid of autonomy and dignity. In this sense, the sexual objectification of women must be understood as an important factor contributing to an environment that trivializes, neutralizes, and, perhaps, facilitates rape.

Finally, we must comment on the consequences of allowing one perspective to dominate thought on a social problem. Rape, like any complex continuum of behavior, has multiple causes and is influenced by a number of social factors. Yet, dominated by psychiatry and the medical model, the underlying assumption that rapists are "sick" has pervaded research. Although methodologically unsound, conclusions have been based almost exclusively on small clinical populations of rapists—that extreme group of rapists who seek counseling in prison and are the most likely to exhibit psychopathology. From this small, atypical group of men, psychiatric findings have been generalized to all men who rape. Our research, however, based on volunteers from the entire prison population, indicates that some rapists, like deniers, viewed and understood their behavior from a popular cultural perspective. This strongly suggests that cultural perspectives, and not an idiosyncratic illness, motivated their behavior. Indeed, we can argue that the psychiatric perspective has contributed to the vocabulary of motive that rapists use to excuse and justify their behavior (Scully and Marolla, 1984).

Efforts to arrive at a general explanation for rape have been retarded by the narrow focus of the medical model and the preoccupation with clinical populations. The continued reduction of such complex behavior to a singular cause hinders, rather than enhances, our understanding of rape.

REFERENCES

Abel, Gene, Judith Becker, and Linda Skinner (1980) "Aggressive behavior and sex." *Psychiatric Clinics of North America* 3(2):133–151.
Abrahamsen, David (1960) *The Psychology of Crime.* New York: John Wiley.
Albin, Rochelle (1977) "Psychological studies of rape." *Signs* 3(2):423–435.

Athens, Lonnie (1977) "Violent crimes: A symbolic interactionist study." *Symbolic Interaction* 1(I):56–71.

Burgess, Ann Wolbert, and Lynda Lytle Holmstrom (1974) *Rape: Victims of Crisis*. Bowie: Robert J. Brady.

———— (1979) "Rape: Sexual disruption and recovery." *American Journal of Orthopsychiatry* 49(4):648–657.

Burt, Martha (1980) "Cultural myths and supports for rape." *Journal of Personality and Social Psychology* 38(2):217–230.

Burt, Martha, and Rochelle Albin (1981) "Rape myths, rape definitions, and probability of conviction." *Journal of Applied Psychology* 11(3):212–230.

Feldman-Summers, Shirley, Patricia E. Gordon, and Jeanette R. Meagher (1979) "The impact of rape on sexual satisfaction." *Journal of Abnormal Psychology* 88(I):101–105.

Glueck, Sheldon (1925) *Mental Disorders and the Criminal Law*. New York: Little Brown.

Groth, Nicholas A. (1979) *Men Who Rape*. New York: Plenum Press.

Hall, Peter M., and John P. Hewitt (1970) "The quasi-theory of communication and the management of dissent." *Social Problems* 18(1):17–27.

Hewitt, John P., and Peter M. Hall (1973) "Social problems, problematic situations, and quasi-theories." *American Journal of Sociology* 38(3):367–374.

Hewitt, John P., and Randall Stokes (1975) "Disclaimers." *American Sociological Review* 40(1):1–11.

Hollander, Bernard (1924) *The Psychology of Misconduct, Vice, and Crime*. New York: Macmillan.

Holmstrom, Lynda Lytle, and Ann Wolbert Burgess (1978) "Sexual behavior of assailant and victim during rape." Paper presented at the annual meetings of the American Sociological Association, San Francisco, September 2–8.

Kilpatrick, Dean G., Lois Veronen, and Patricia A. Resnick (1979) "The aftermath of rape: Recent empirical findings." *American Journal of Orthopsychiatry* 49(4):658–669.

Ladouceur, Patricia (1983) "The relative impact of drugs and alcohol on serious felons." Paper presented at the annual meetings of the American Society of Criminology, Denver, November 9–12.

Luckenbill, David (1977) "Criminal homicide as a situated transaction." *Social Problems* 25(2):176–187.

McCaghy, Charles (1968) "Drinking and deviance disavowal: The case of child molesters." *Social Problems* 16(l):43–49.

Marolla, Joseph, and Diana Scully (1979) "Rape and psychiatric vocabularies of motive." Pp. 301–318 in Edith S. Gomberg and Violet Franks (eds.), *Gender and Disordered Behavior: Sex Differences in Psychopathology*. New York: Brunner/Mazet.

Mills, C. Wright (1940) "Situated actions and vocabularies of motive." *American Sociological Review* 5(6):904–913.

Nelson, Steven, and Menachem Amir (1975) "The hitchhike victim of rape: A research report." Pp. 47–65 in Israel Drapkin and Emilio Viano (eds.), *Victimology: A New Focus*. Lexington, KY: Lexington Books.

Queen's Bench Foundation (1976) *Rape: Prevention and Resistance*. San Francisco: Queen's Bench Foundation.

Ruch, Libby O., Susan Meyers Chandler, and Richard A. Harter (1980) "Life change and rape impact." *Journal of Health and Social Behavior* 21(3):248–260.

Schlenker, Barry R., and Bruce W. Darby (1981) "The use of apologies in social predicaments." *Social Psychology Quarterly* 44(3):271–278.

Scott, Marvin, and Stanford Lyman (1968) "Accounts." *American Sociological Review* 33(1):46–62.

Scully, Diana, and Joseph Marolla (1984) "Rape and psychiatric vocabularies of motive: Alternative perspectives." In Ann Wolbert Burgess (ed.), *Handbook on Rape and Sexual Assault.* New York: Garland Publishing.

Shore, Barbara K. (1979) *An Examination of Critical Process and Outcome Factors in Rape.* Rockville, MD: National Institute of Mental Health.

Stokes, Randall, and John P. Hewitt (1976) "Aligning actions." *American Sociological Review* 41(5):837-849.

Sykes, Gresham M., and David Matza (1957) "Techniques of neutralization." *American Sociological Review* 22(6):664–670.

Williams, Joyce (1979) "Sex role stereotypes, women's liberation, and rape: A cross-cultural analysis of attitude." *Sociological Symposium* 25 (Winter):61–97.

QUESTIONS FOR DISCUSSION

1. Discuss the five justifications of rape the authors discovered in their research.

2. Discuss the three excuses rapists used.

3. Explain how convicted rapists' justifications and excuses for rape are enmeshed in rape myths.

Rape Myth Acceptance in College Students: How Far Have We Come?

Shelly Schaefer Hinck and Richard W. Thomas

While the acceptance of rape myths has been found to be a crucial factor in explanatory models of rape behavior (Berkowitz, 1992; Brownmiller, 1975; Burt, 1980), current researchers have found that responses to rape myth scales may reflect ceiling effects limiting the ability of the scales to differentiate those individuals who subscribe to rape supportive attitudes from those who do not (Lonsway & Fitzgerald, 1994; Schaefer & Nelson, 1993). The purpose of this study is to examine the current state of rape myth acceptance in college students and the factors which differentiate acceptance vs. nonacceptance of rape myths. This paper reviews the literature concerning rape myths, examines college students' acceptance of rape myths via two rape myth scales, and offers suggestions for future research on rape myths and rape prevention programs.

"Rape Myth Acceptance in College Students: How Far Have We Come?" by Shelly Schaefer Hinck and Richard W. Thomas from *Sex Roles,* May 1999, p. 815. Reprinted with permission of Plenum Publishing Corporation.

REVIEW OF THE LITERATURE

Research involving rape myths has found that individuals' acceptance of rape myths and rape-supportive attitudes is correlated with increased sex-role stereotyping, stronger adherence to adversarial sexual beliefs, and greater acceptance of interpersonal violence within relationships (Burt, 1980). Koss, Leonard, Beezly, & Oros (1985), in a seminal study of undetected, self-reported acquaintance rapists, found that an individual's propensity to rape (to engage in sexually assaultive, abusive, or coercive behavior in order to procure sexual intercourse) was significantly related to the degree to which they subscribed to several rape-supportive attitudes (e.g., acceptance of rape myths, adherence to traditional views of female/male sexuality, perception of sexual aggression as normal). The explanatory power of these belief systems has been replicated continually in numerous studies (c.f., Bridges, 1991; Canterbury, Grossman, & Lloyd, 1993; Harrison, Downes, & Williams, 1991; Holcomb, Holcomb, Sondag, & Williams, 1991).

Rape myths have also been linked to an individual's definition of rape; the more an individual accepts rape myths as truth, the more restrictive the definition of rape (Burt, 1981). This may help explain why few men tend to view their coercive behavior as rape. Lisak and Roth (1990) found that none of the men who had admitted to engaging in forced sexual intercourse or forced oral sex labeled their acts as rape. Moreover, research has indicated that there is a connection between acceptance of rape myths and the likelihood of using sexual force (Briere & Malamuth, 1983).

Individuals' acceptance of rape-supportive attitudes, as well as their sex-role stereotyping (traditional or liberal), has also been found to impact how they perceive date rape. For example, Coller & Resick (1987) found that women in the high sex-role stereotyping group blamed the victim more for her victimization, felt the victim led the perpetrator on, and attributed more responsibility to the victim for her situation than did women in the low sex-role stereotyping group. Check and Malamuth (1983) and Muehlenhard (1988) also found a positive correlation between sex-typed orientation (traditional or liberal) and acceptance of rape myths. In both studies, traditional subjects supported more rape myths and were more likely to perceive the rape victim as being blameworthy. This may be due to the fact that high sex-role stereotyped individuals perceive date rape in particular as being at the extreme point on a continuum of in-role sexual behaviors.

Rape justifiability is often decided by power factors inherent in a dating situation, such as who pays for the date, where the couple goes, and who initiates the dates. Unwanted sexual intercourse was rated as more justifiable if the couple went to the man's apartment, if the woman asked the man out and if the man paid all dating expenses (Muehlenhard, Friedman, & Thomas, 1985). Further support for the correlation between rape justifiability and dating practices was found in Muehlenhard's 1988 study. For both sexes, rape justifiability ratings were highest when the man paid for the date, the woman asked the man out for the date, and the date occurred in the man's apartment. Additionally, men's ratings of rape justifiability were found to be consistently higher than the women's.

Traditional persons, especially traditional men, rated the justifiability of the rape as more acceptable than nontraditional persons. Bostwick & Delucia (1992), in a replication of the Muehlenhard study, indicated that while who asked and who paid for the date impacted subjects' views as to how willing the woman was to have sex and how willing the man was to have sex, it did not affect subjects' views of rape justifiability. Subjects indicated that the man was not justified forcing the woman to have sexual intercourse without her consent. This result was inconsistent with Muehlenhard's findings. Holcomb, Holcomb, Sondag, & Williams (1991) found results similar to both Muehlenhard and Bostwick & Delucia. Sexually aggressive males and males who agreed more with rape myths were less likely to perceive the scenarios as rape, blamed the victim more, perceived the victim as desiring intercourse and viewed the assailants' behavior as less violent. The dating situation (planned date with monetary investment, a planned dutch-treat date and an unplanned pick-up date) did not affect their attitudes towards the rape situation. The dating situation did impact females' perceptions of the rape scenarios. In the pick-up scenario, females perceived the situation as rape more than the scenario depicting a monetary investment. Holcomb et al. argue that changes in the expectations and rules of the sexual script must be changed if date rape is to be prevented.

In an effort to decrease people's acceptance of rape myths, an increasing number of rape awareness and prevention programs, classes, and lectures are being offered. These programs usually have been successful in challenging rape myth acceptance. For example, Lee (1987) exposed 24 undergraduate men to a program that included: (1) a review of myths and facts about rape; (2) an experiential victim-empathy exercise; (3) a guided fantasy about responding to a potential date rape; and (4) an open discussion. He found that participants' attitudes about rape did change; after the program, participants were less accepting of rape myths. In a study focusing on improving sorority women's knowledge about the definition and components of acquaintance rape, Kamm, Mayfield, Tait, & Yonker (1991) found that the women's knowledge about acquaintance rape did increase; however, the increase was small. This was due to a ceiling effect; the scores on the pretest were so high that it left little room for improvement. Harrison et al. (1991) investigated the dimensions of students' attitudes to acquaintance rape and date rape and the effect of a program to change them. They found that men showed improvement from the pretest to the post test on both the victim blaming scale and a survey assessing their factual knowledge about acquaintance rape. Women, however, did not change as much on either the victim blaming scale or the survey assessing their factual knowledge since women's scores were so high to begin with, indicating a possible ceiling effect.

While the number of rape awareness and prevention programs, classes, and lectures being offered is increasing, researchers have still found that the acceptance of rape myths exists, and that men tend to exhibit stronger acceptance of these myths. Holcomb et al. (1991) found that in a sample of 407 males, one in four male subjects agreed with such statements as: rape was often provoked by the victim; any woman could prevent rape if she really wanted to; and women frequently cried rape falsely. Muehlenhard & Linton (1987) found that 79% of the college males in her study replied that raping a woman is justifiable if the woman was perceived as being a tease or "loose." In a study of high school students, Muehlenhard, Friedman, & Thomas (1985) found that 54% of the males thought that rape was justified if a girl "led them on." Blumberg and Lester (1991) found that high school males believed more strongly than did both high school females and college males in myths about rape. Assigning more blame to the victims of rape was associated with belief in rape myths for both high school males and females. Schaeffer and Nelson (1993) found that males who lived in co-ed dormitories were less accepting of rape myths than residents of single-sex housing or fraternities. However, fraternity-house residents did not differ significantly from single-sex residence hall residents in regard to rape myth acceptance. Schaeffer and Nelson (1993) also failed to find an effect of education concerning rape on rape myth acceptance. Exploring why there was a lack of change in attitudes towards rape and rape victims is an important area to examine. Perhaps research that explores specific factors that help justify an individual's perception that a rape has not occurred in conjunction with different types of rape awareness programs is needed.

Given the significance of the rape myth concept to the study of and intervention into rape on the college campus, we advanced the following hypotheses and research question. The first hypothesis we offer is that college students will express an overall disagreement with statements reflecting rape myths. The second hypothesis reflects the effect of gender on rape myth acceptance; specifically we suggest that women, more than men, will more strongly disagree with rape myth statements. The third hypothesis we propose is that college students who have attended a rape awareness seminar will hold a higher degree of rape "intolerant" attitudes. Finally, in order to better understand the acceptance and nonacceptance of rape myths, we ask the following research question: What factors differentiate "tolerant" vs. "intolerant" attitudes toward rape?

METHODS

Participants

Data were collected in the Spring semester of 1995 from 158 undergraduate students attending a midsize central Midwest university. Participants were students enrolled in undergraduate courses which drew populations from a variety of disciplines across campus. Participants were given credit for their voluntary participation in the study. Data were collected outside of the classroom, and all procedures were in compliance with Federal Guidelines concerning the use of Human Subjects.

Of the 158 participants, 53% (n = 85) were female and 47% (n = 73) were male. Ethnicity data revealed that 95% of the participants identified themselves as Caucasian, 2% as African-American, 1% as Hispanic, and 2% as "other." The majority of respondents were freshmen (75%); 17% were sophomores, 5% were juniors, and 3% were seniors. The median age of the group was 18 years (M = 19.25, SD = 2.35). Ten percent of the respondents reported belonging to a social fraternity or sorority, and 23% of the entire group indicated that they were members of one or more other campus group.

Procedures

A questionnaire was employed to address the research questions.[1] Relevant to the present study, respondents first were asked to provide demographic data (e.g., gender, class rank, age, race, organizational memberships) and to indicate whether or not they had ever attended a rape awareness workshop.[2] Participants then were provided with two measures of their attitudes toward/beliefs about rape. The first measure was the Harrison, Downes, and Williams' (1991) Revised Attitude Toward Rape Scale (ATR). This scale, derived in part from Field's (1978) and Barnett and Field's (1977) ATR scale, consisted of 14 rape-oriented statements reflecting two underlying factors: (1) victim blaming and denial, and (2) perceptions of factual information.[3] The second measure was Holcomb, Holcomb, Sondag, and Williams' (1991) Rape Attitude and Perception Questionnaire (RAP). This scale originally consisted of 22 statements concerning rape.[4] Two of the items from the original scale, which focused on behaviors rather than beliefs, were omitted in the present study. These 34 questions were randomly ordered and respondents were asked to indicate, on a 6-point Likert-type scale, the degree to which they agreed or disagreed with each statement.[5]

RESULTS
Acceptance of Rape-Oriented Statements

We began the analysis by examining participants' responses to each of the rape myth scales separately. Overall, participants responded to each of the rape myth scales in the hypothesized direction. Specifically, respondents reported a general disagreement with rape-supportive statements on both the Revised Attitude Toward Rape Scale (M = 4.75, SD = 0.56) and the Rape Attitude and Perception Scale (M = 4.93, SD = 0.47), providing initial support for [H.sup.1] that there is a general nonacceptance of rape-oriented statements in today's college students.

In order to empirically test [H.sup.1] we computed the overall frequencies for the agree and disagree response categories for each of the two scales.[6] Chi square analysis revealed a significant preference for disagree (nonacceptance) responses for both the ATR ([X.sup.2] = 634.36, df = 2, p [less than] .001) and the RAP ([X.sup.2] = 1423.2, df = 2, p [less than] .001), reflecting a strong tendency for college students to disagree with rape supportive beliefs and attitudes.

Next, we examined the impact of gender and attendance at a rape awareness seminar on acceptance of rape attitudes. As discussed above, these factors have been seen as critical in understanding students' adherence to rape-supportive statements; previous research has consistently found that men and individuals who have not been exposed to the existence of rape myths and rape supportive beliefs more strongly subscribe to these belief systems. MANOVA, with gender and workshop attendance as the independent variables and the two rape myth scales as the dependent variables, revealed significant main effects for both gender (F = 12.90, df = 2,145, p [less than] .001) and seminar attendance (F = 3.89, df = 2,145, p [less than] .05) in the hypothesized direction. No interaction between these variables emerged. Subsequent univariate F-tests produced significant main effects for gender on both the ATR (F = 25.94, df = 1,146, p [less than] .001) and the RAP (F = 12.03, df = 1,146, p [less than] .001). However, for seminar attendance, a significant main effect accrued only for the RAP (F = 7.82, df = 1,146, p [less than] .01). These results provide support for [H.sup.2] and [H.sup.3]; although college students tend to report disagreement with rape-oriented statements, men and individuals who have not attended a rape awareness workshop disagree less strongly with rape-oriented statements than do women and individuals who have attended some type of rape awareness workshop.

In order to further understand these results and to ascertain students' general responses to the rape-oriented statements, we combined their responses from the ATR and the RAP into a single rape belief acceptance scale (RBA).[7] As hypothesized, students' responses indicated an overall disagreement with rape belief statements (M = 4.86; SD = 0.47), with significantly more individuals rejecting these beliefs than accepting them ([X.sup.2] = 2167.68, df = 2, p [less than] .001).

ANOVA of the relationship between gender and workshop attendance and the RBA produced significant main effects for both gender (F = 21.11 (1,146) p [less than] .000) and seminar attendance (F = 6.33 (1,146) p [less than] .05); no interaction occurred between the variables.

In general, then, these results provide additional support for both [H.sup.2] and [H.sup.3]. This trend was especially strong with respect to gender; ANOVA of each of the 34 statements revealed that men and women differed significantly and in the predicted direction on 62% (21) of the rape belief statements. In contrast, participants in a rape awareness workshop significantly differed in their responses from nonparticipants on only 29% (10) of the rape belief statements.

Analysis of the specific statements produced three significant interactions. Responses to the statement, "In a woman, submissiveness equals femininity," did not differ across gender or seminar attendance. However, a significant interaction effect occurred (F = 4.23 (1,148) p = .042); attendance of a rape awareness workshop increased women's disagreement with this score (M = 4.79 vs. 5.00), but not as greatly as it did men's (M = 4.40 vs. 5.06). A second interaction effect occurred with the statement "Rapists are motivated by an overwhelming, unfulfilled sexual desire" (F = 14.58 (1,152) p = .017). Women who had not attended a rape awareness workshop disagreed significantly less with this statement (M = 3.11) than did women who had attended a workshop (M = 3.81). In contrast, men who had not attended a rape awareness workshop disagreed significantly more with this statement (M = 3.81) than did men who had attended a workshop (M = 3.06). For some reason, workshop attendance had a completely opposite effect for men and women with respect to this belief. The only other statement which produced an interaction effect was: "Many people have sex to feel close to someone, not just because they're aroused" (F = 8.86 (1,153) p = .011). Women who had participated in a workshop agreed less strongly with this statement than did women who had not (M = 4.08 vs. 3.85, respectively). However, men who had participated in a workshop agreed more strongly with this statement than did men who had not (M = 4.24 vs. 3.38, respectively).

RAPE MYTH PROFILE

In order to identify those statements which differentiated high RBA from low RBA and address [RQ.sup.1], we submitted participants' responses to all 34 rape-oriented statements, along with the variables of gender and seminar attendance, to a stepwise discriminant analysis. By using the median score of the RBA scale (Mdn = 4.87), we defined scores above the median as "high" RBA and scores below the median as "low" RBA.

The method of minimizing Wilk's lambda was used for inclusion of statements, and the criterion of p [less than] .001 was set. The results of the discriminant analysis produced a single function discriminating the two groups with respect to 19

TABLE 1   Discriminant Classification of Rape Myth Statements

Actual Group	N	Predicted Group	
		High RBA	Low RBA
High RBA	76	73 (96.1%)	3 (3.9%)
Low RBA	74	2 (2.7%)	72 (97.3%)

statements and seminar participation (see Table 1). The Wilk's lambda of .26, with approximate X2 = 188.14 (df = 20, p [less than] .001) indicated that the two groups differed significantly. The canonical of .86 suggested a strong degree of association between the two groups and the discriminant function. One hundred percent of the variance in the function was accounted for by group membership. When these 21 variables were used to reclassify subjects into the two groups, 96.1% of the high RBAs and 97.3% of the low RBAs were correctly classified, resulting in an overall correct classification of 96.67%.

Specifically, the analysis identified 19 rape-supportive statements which significantly differentiated individuals with a "strong" rejection of rape belief statements from individuals with a "less strong" rejection of rape belief statements. In addition, attendance at a rape awareness seminar emerged as a significant component in affecting these belief systems. It seems then, that even though individuals disagree with these statements, the degree of their disagreement (and possibly their adherence to rape-supportive attitudes) potentially reflects different underlying belief and attitude structures.

Inspection of the 19 statements suggested that they paralleled many of the central factors housed within rape attitude scales, including blame of the victim or denial (e.g., some women provoke rape by their appearance or behavior), adherence to sex-role stereotypes (e.g., in a woman, submissiveness equals femininity), justification for rape (e.g., if a woman is heavily intoxicated, it is OK to have sex with her), misinformation about rape ("nice" women don't get raped), and communication/relationship factors (if a woman says "no" to having sex, she means "no"). In order to empirically identify the underlying factor structure of these statements, we submitted them to a factor analysis.[8] This analysis produced seven factors accounting for 59.4% of the variance. The first factor, which accounted for 17.7% of the variance, was composed of three statements (Some women provoke rape by their appearance or behavior; Women frequently accuse men of rape to get back at them; The degree of a woman's resistance should be the major factor in determining if a rape has occurred). The second factor, which accounted for 8.8% of the variance, was composed of three statements (If a woman says "no" to having sex, she means "no"; A report of rape 2 days after the act has occurred is probably a false report; A man can control his behavior no matter how attracted he feels toward someone). The third factor, which accounted for 8.0% of the variance, was composed of three statements (Rapists are motivated by an overwhelming, unfulfilled sexual desire; Most victims of rape do not know their attacker; In a woman, submissiveness equals femininity). The fourth factor, which accounted for 6.9% of the variance, was composed of two statements (In a man, aggressiveness equals masculinity; If a woman is heavily intoxicated, it is OK to have sex with her.). The fifth factor, which accounted for 6.4% of the variance, was composed of three statements (Many people have sex to feel close to someone, not just because they're aroused; Deep down, a woman likes to be whistled at on the street; Victims of rape rarely report the crime). The sixth factor, which accounted for 5.9% of the variance, was composed of three statements (It is best that men initiate dates; Some women may enjoy being raped; Rape will never happen to me). The seventh factor, which accounted for 5.6% of the variance, was composed of two statements (When a woman says love she means love, when a man says love he means sex; A prostitute can be raped).

DISCUSSION

The central purpose of this study was to examine the current state of rape myth acceptance in college students and the factors which differentiated acceptance vs. nonacceptance of rape myths. Research tells us that individuals who engage in sexually aggressive behavior, particularly rape, subscribe to belief and attitude systems that are markedly different from those who refrain from sexually aggressive behavior. Burt (1980) offered that adversarial sexual beliefs, beliefs that "sexual relationships are fundamentally exploitative, that each party is too manipulative, sly, cheating, opaque to the other's understanding, and not to be trusted" are linked to rape myth acceptance. Additionally, negative and stereotypical attitudes toward women have been found to be associated with rape myth acceptance (see Lonsway & Fitzgerald, 1994). However, the central question remains, do the rape myth scales empirically capture the cognitive differences that exist between individuals who engage in sexually aggressive behavior and those who do not?

On the surface, the results of this study would suggest no. Respondents consistently disagreed with each of the rape-supportive statements presented to them. However, this pattern does not negate the value of rape myth scales. The analyses presented in this study suggest that significant differences accrue in the degree to which individuals disagree with these statements. This variation in responses suggests that different belief structures do exist, and manifest themselves in the strength of an individual's response. The results of the discriminant analysis suggest that specific factors such as victim blame, sex-role expectations, misinformation, and communication/relationship skills contribute to an individual's potential to subscribe to rape-supportive attitudes. The results also suggest that gender and attendance at a rape prevention workshop may impact rape myth acceptance. The findings indicate that men and individuals who have not been exposed to rape awareness information disagree less strongly with these statements than women and individuals who have been exposed to rape awareness information. However it should be noted that the type of workshop the subjects attended is not known. Research linking sex with rape myth acceptance has consistently found that men are more likely to accept rape myths than women (Lonsway & Fitzgerald, 1994). The relationship between attendance at rape workshops and rape myth acceptance however is not as clear. This inconsistency may be due to differences among the workshops. Who presents the information (male/female faculty member or male/female peer), how the information is presented (lecture, discussion, videotapes, etc.), and to whom the information is presented (all male audience, all female audience, male & female audience) may account for the variations in results. Future research linking the variations in rape prevention workshops with the nonacceptance of rape myths is needed.

The implications of these findings are two fold. First, efforts must be made to redesign rape myth scales. Definitional inconsistencies and poorly phrased statements and phrases limit our understanding of rape myths, limit our ability to explain specific results, and limit our ability to generalize beyond a particular research experiment. Changes within rape myth statements would allow researchers to access the belief and attitude structures with greater precision.

The second implication of these findings concerns the focus of rape awareness workshops. Many workshops focus exclusively on increasing individuals' awareness of the problem of rape and the different factors which precipitate sexual aggression (including belief and attitude systems). Although we applaud these efforts, in themselves, we feel they are not enough. Emphasis must be given not only to increasing the awareness of individuals about the problem of rape, but also to the internalization and enactment of the belief and attitude structures advanced in the workshops. In other words, rape prevention programs must find ways to impact the communication patterns and behavior patterns of college students in their everyday interactions. Lonsway (1996) notes that a new area for researchers lies in the exploration of "how rape education leads to desirable outcomes. For example, how do rape myth presentations lead to successful attitude change? Which myths are easiest or most beneficial to change? Which presentation method can lead to the most powerful change outcomes?" We offer that more educators need to acknowledge that different audiences may be impacted by different presentational styles. Research offers that men respond best to film and visual stimuli with sexual themes (Check & Malamuth, 1981), presentations that are less verbal and didactic (Borden, Karr, & Caldwell-Colbert, 1988) and more visually oriented workshops (Harrison, Downes, & Williams, 1991). Earle (1996) offers that the most effective program in changing attitudes is single sex, small interactive groups, with discussion rather than lecture as the primary information vehicle, while Holcomb & Schaefer (1995) indicate that by using mixed gender workshops, the likelihood of alienating men by blaming and "bashing" men as well as blaming women for miscommunication is reduced. Clearly, more research is needed; simply presenting information is useless if it cannot be used to persuade college students either to avoid situations in which rape could occur or to more effectively and competently deal with sexually aggressive behavior when faced with it.

Finally, it is important to note the limitations of the study. First, while the participants in the study reflected a fairly equal distribution of men and women, the difference in the ethnicity of the subjects was limited. Approximately 95% of the participants were Caucasian. Lonsway (1996) offers that socioeconomic and cultural variables may impact rape vulnerability, which in turn may be linked to rape myth acceptance. Research has indicated that African-American students and Hispanic students were more accepting of rape myths than Caucasian students. However, it should be noted that currently the research has not consistently established a relationship between racial and ethnic identity and the acceptance of rape myths. Lonsway and Fitzgerald (1994) suggest that cultural history, religious tradition, and sex-role expectations may be better predictors of rape myth acceptance than race. Additionally, the fact that all subjects are college students also limits the generalizations that can be made concerning the research. Variables such as occupation, knowledge about and awareness of rape, and age are all variables that have been found to influence rape myth acceptance, although in some instances indirectly. While student populations are certainly important to study in terms of rape myth acceptance, examining other populations may yield valuable information about rape, rape myth acceptance, and ultimately rape prevention programs and their effectiveness.

NOTES

1. The questionnaire addressed several areas relevant to participants' dating behaviors and perceptions of rape. The data reported here reflect only a portion of this larger study.

2. Only participants' attendance at a workshop was elicited. Participants were not asked about the content or format of this workshop. Although content and formats of workshops vary considerably, the literature indicates that rape-oriented beliefs and rape myths are a consistent component of such workshops.

3. Original psychometric analysis of these scales reported alpha reliability estimates of .77 for the victim blaming or denial factor and .64 for the perceptions of factual information factor.

4. Psychometric analysis of these scales reported an alpha reliability coefficient of .86. A multidisciplinary panel of experts reviewed the survey instrument to estimate its content validity, face validity, and readability. The instrument was tested a second time and the Chronbach alpha reliability coefficient was .77.

5. In the original studies, a 5-point Likert-type scale was used. We employed a 6-point scale to clearly differentiate agreement from disagreement.

6. Specifically, for each scale, we determined the number of responses which indicated agreement and disagreement and summed these across statements. With 158 participants, 2212 responses were possible for the ATR and 3160 responses were possible for the RAP. These frequencies were then submitted to Chi-square analysis.

7. Characteristics of the two scales suggested this move. Many of the statements included in each of the scales appeared quite similar, and the two scales exhibited only a moderate degree of association ($r = .59$). In addition, we were unable to replicate the underlying factor structure of the ATR. Finally, the combination of the ATR and the RAP increased the internal reliability, as computed by Chronbach's Alpha, from .70 and .73 (respectively) to .83.

8. Factors were created through varimax rotation. The Scree criterion was employed to define item inclusion.

REFERENCES

Barnett, N.J., & Field, H. S. (1977). Sex differences in university students' attitudes toward rape. *Journal of College Student Personnel, 18,* 93–96.

Berkowitz, A. (1992). College men as perpetrators of acquaintance rape and sexual assault: A review of recent research. *Journal of American College Health, 40,* 175–181.

Blumberg, M. L., & Lester, D. (1991). High school and college students' attitudes toward rape. *Adolescence, 26,* 727–729.

Bordon, L. A., Karr, S. K., & Caldwell-Colbert, A. T. (1988). Effects of a university rape prevention program on attitudes and empathy toward rape. *Journal of College Student Development, 29,* 132–138.

Bostwick, T. D., & Delucia, J. L. (1992). Effects of gender and specific eating behavior on perceptions of sex willingness and data rape. *Journal of Social and Clinical Psychology, 11,* 14–25.

Bridges, J. S. (1991). Perceptions of date and stranger rape: A difference in sex role expectations and rape-supportive beliefs. *Sex Roles, 24,* 291–307.

Briere, J., & Malamuth, N. M. (1983). Self-reported likelihood of sexually aggressive behavior: Attitudinal versus sexual explanations. *Journal of Research in Personality, 17,* 315–323.

Brownmiller, S. (1975). *Against our will: Men, women and rape.* New York: Bantam Books.

Buchwald, E., Fletcher, P., & Roth, M. (1993). *Transforming a rape culture.* Minneapolis, MN: Milkweed Editions.

Bureau of Justice Statistics. (1995). *Criminal victimization in the United States.* Washington, DC: U.S. Department of Justice.

Burt, M. R. (1980). Cultural myths and supports for rape. *Journal of Personality and Social Psychology, 38,* 217–230.

Burt, M. R. (1991). Rape myths and acquaintance rape. In A. Parrot & L. Bechhofer (Eds.), *Acquaintance rape: The hidden crime.* New York: Wiley.

Cantebury, R. J., Grossman, S. J., & Lloyd, E. (1993). Drinking behaviors and lifetime incidents of date rape among high school graduates upon entering college. *College Student Development, 27,* 75–84.

Check, J. P., & Malamuth, M. M. (1983). Sex-role stereotyping and reactions to depictions of stranger versus acquaintance rape. *Journal of Personality and Social Psychology, 45,* 344–356.

Coller, S. A., & Resick, P. A. (1987). Women's attributions of responsibility for date rape: The influence of empathy and sex-role stereotyping. *Violence and Victims, 2,* 115–125.

Earle, J. P. (1996). Acquaintance rape workshops: Their effectiveness in changing the attitudes of first year college men. *NASPA Journal, 34,* 2–15.

Field, H. S. (1978). Attitudes toward rape: A comparative analysis of police, rapists, crisis counselors, and citizens. *Journal of Personality and Social Psychology, 36,* 156–179.

Gray, M. D., Lesser, D., Quinn, & Bounds, C. (1987). The effectiveness of personalizing acquaintance rape prevention: Programs on perception of vulnerability and on reducing risk-taking behavior. *Journal of College Student Development, 31,* 217–220.

Harrison, P. J., Downes, J., & Williams, M. D. (1991). Date and acquaintance rape: Perceptions and attitude change strategies. *Journal of College Student Development, 32,* 131–139.

Holcomb, D. R., & Schaefer, R. W. (1995). Enhancing dating attitudes through peer education as a date rape prevention strategy. *The Peer Facilitator Quarterly, 12,* 16–20.

Holcomb, D. R., Holcomb, L. C., Sondag, K. A., & Williams, N. (1991). Attitudes about date rape: Gender differences among college students. *College Student Journal, 25,* 434–439.

Kahn, A. S., Mathie, V. A., & Torgler, C. (1994). Rape scripts and rape acknowledgment. *Psychology of Women Quarterly, 18,* 53–66.

Koss, M. P., Leonard, K. E., Beezley, D. A., & Oros, C. J. (1985). Nonstranger sexual aggression: A discriminant analysis of the psychological characteristics of undetected offenders. *Sex Roles, 12,* 981–992.

Lee, L. (1987). Rape prevention: Experimental training for men. *Journal of Counseling and Development, 66,* 100–101.

Lisak, D., & Roth, S. (1990). Motives and psychodynamics of self-reported, unincarcerated rapists. *American Journal of Orthopsychiatric, 60,* 268.

Lively, K. (1996, April 26). "Drug arrests rise again," *The Chronicle of Higher Education,* pp. A37–A49.

Lonsway, L. A., & Fitzgerald, L. F. (1994). Rape myths: In review. *Psychology of Women Quarterly, 18,* 133–164.

Muehlenhard, C. L. (1988). Misinterpreted dating behaviors and the risk of date rape. *Journal of Social and Clinical Psychology, 6,* 20–37.

Muehlenhard, C. L., & Linton, M. A. (1987). Date rape and sexual aggression in dating situations: Incidence and risk factors. *Journal of Counseling Psychology, 34,* 186–196.

Muehlenhard, C. L., Friedman, D. E., & Thomas, C. M. (1985). Is date rape justifiable? The effects of dating activity, who initiated, who paid, and men's attitudes toward women. *Psychology of Women Quarterly, 9,* 287–309.

National Crime Victimization Survey Report. (1992). Criminal victimization in the United States, NCJ-139563.

Schaeffer, A. M., & Nelson, E. S. (1993). Rape-supportive attitudes: Effects of on-campus residence and education. *Journal of College Student Development, 34,* 175–179.

QUESTIONS FOR DISCUSSION

1. Discuss some of the rape justifications the authors found in their study. Do you agree or disagree with the authors? Explain your response.

2. What are the results of the study? What are the two implications of the study? According to the authors, what are the two limitations of the study?

3. How could replicating the Lee (1987) study help your college deal with rape? Discuss each part of the study in your response.

CHAPTER 10

WHITE-COLLAR CRIME

Denying the Guilty Mind: Accounting for Involvement in a White-Collar Crime

Michael L. Benson

Adjudication as a criminal is, to use Garfinkel's (1956) classic term, a degradation ceremony. The focus of this article is on how offenders attempt to defeat the success of this ceremony and deny their own criminality through the use of accounts. However, in the interest of showing in as much detail as possible all sides of the experience undergone by these offenders, it is necessary to treat first the guilt and inner anguish that is felt by many white-collar offenders even though they deny being criminals. This is best accomplished by beginning with a description of a unique feature of the prosecution of white-collar crimes.

In white-collar criminal cases, the issue is likely to be *why* something was done, rather than *who* did it (Edelhertz, 1970:47). There is often relatively little disagreement as to what happened. In the words of one Assistant U.S. Attorney interviewed for the study:

> If you actually had a movie playing, neither side would dispute that a person moved in this way and handled this piece of paper, etc. What it comes down to is, did they have the criminal intent?

"Denying the Guilty Mind: Accounting for Involvement in a White-Collar Crime" by Michael L. Benson from *Criminology*, vol. 23 (November 1985), pp. 589–599. Reprinted with permission of American Society of Criminology.

If the prosecution is to proceed past the investigatory stages, the prosecutor must infer from the pattern of events that conscious criminal intent was present and believe that sufficient evidence exists to convince a jury of this interpretation of the situation. As Katz (1979:445–446) has noted, making this inference can be difficult because of the way in which white-collar illegalities are integrated into ordinary occupational routines. Thus, prosecutors in conducting trials, grand jury hearings, or plea negotiations spend a great deal of effort establishing that the defendant did indeed have the necessary criminal intent. By concentrating on the offender's motives, the prosecutor attacks the very essence of the white-collar offender's public and personal image as an upstanding member of the community. The offender is portrayed as someone with a guilty mind.

Not surprisingly, therefore, the most consistent and recurrent pattern in the interviews, though not present in all of them, was denial of criminal intent, as opposed to the outright denial of any criminal behavior whatsoever. Most offenders acknowledged that their behavior probably could be construed as falling within the conduct proscribed by statute, but they uniformly denied that their actions were motivated by a guilty mind. This is not to say, however, that offenders *felt* no guilt or shame as a result of conviction. On the contrary, indictment, prosecution, and conviction provoke a variety of emotions among offenders.

The enormous reality of the offender's lived emotion (Denzin, 1984) in admitting guilt is perhaps best illustrated by one offender's description of his feelings during the hearing at which he pled guilty.

> You know (the plea's) what really hurt. I didn't even know I had feet. I felt numb. My head was just floating. There was no feeling, except a state of suspended animation. . . . For a brief moment, I almost hesitated. I almost said not guilty. If I had been alone, I would have fought, but my family. . . .

The traumatic nature of this moment lies, in part, in the offender's feeling that only one aspect of his life is being considered. From the offender's point of view his crime represents only one small part of his life. It does not typify his inner self, and to judge him solely on the basis of this one event seems an atrocious injustice to the offender.

For some the memory of the event is so painful that they want to obliterate it entirely, as the two following quotations illustrate.

> I want quiet. I want to forget. I want to cut with the past.

> I've already divorced myself from the problem. I don't even want to hear the names of certain people ever again. It brings me pain.

For others, rage rather than embarrassment seemed to be the dominant emotion.

> I never really felt any embarrassment over the whole thing. I felt rage and it wasn't false or self-serving. It was really (something) to see this thing in action and recognize what the whole legal system has come to through its development, and the abuse of the grand jury system and the abuse of the indictment system. . . .

The role of the news media in the process of punishment and stigmatization should not be overlooked. All offenders whose cases were reported on by the news media were either embarrassed or embittered or both by the public exposure.

> The only one I am bitter at is the newspapers, as many people are. They are unfair because you can't get even. They can say things that are untrue, and let me say this to you. They wrote an article on me that was so blasphemous, that was so horrible. They painted me as an insidious, miserable creature, wringing out the last penny. . . .

Offenders whose cases were not reported on by the news media expressed relief at having avoided that kind of embarrassment, sometimes saying that greater publicity would have been worse than any sentence they could have received.

In court, defense lawyers are fond of presenting white-collar offenders as having suffered enough by virtue of the humiliation of public adjudication as criminals. On the other hand, prosecutors present them as cavalier individuals who arrogantly ignore the law and brush off its weak efforts to stigmatize them as criminals. Neither of these stereotypes is entirely accurate. The subjective effects of conviction on white-collar offenders are varied and complex. One suspects that this is true of all offenders, not only white-collar offenders.

The emotional responses of offenders to conviction have not been the subject of extensive research. However, insofar as an individual's emotional response to adjudication may influence the deterrent or crime-reinforcing impact of punishment on him or her, further study might reveal why some offenders stop their criminal behavior while others go on to careers in crime (Casper, 1978:80).

Although the offenders displayed a variety of different emotions with respect to their experiences, they were nearly unanimous in denying basic criminality. To see how white-collar offenders justify and excuse their crimes, we turn to their accounts. The small number of cases rules out the use of any elaborate classification techniques. Nonetheless, it is useful to group offenders by offense when presenting their interpretations.

ANTITRUST VIOLATORS

Four of the offenders have been convicted of antitrust violations, all in the same case involving the building and contracting industry. Four major themes characterized their accounts. First, antitrust offenders focused on the everyday character and historical continuity of their offenses.

> It was a way of doing business before we even got into the business. So it was like why do you brush your teeth in the morning or something. . . . It was part of the everyday. . . . It was a method of survival.

The offenders argued that they were merely following established and necessary industry practices. These practices were presented as being necessary for the well-being of the industry as a whole, not to mention their own companies. Further, they argued that cooperation among competitors was either allowed or actively promoted by the government in other industries and professions.

The second theme emphasized by the offenders was the characterization of their actions as blameless. They admitted talking to competitors and admitted submitting intentionally noncompetitive bids. However, they presented these practices as being done not for the purpose of rigging prices nor to make exorbitant profits. Rather, the everyday practices of the industry required them to occasionally submit bids on projects they really did not want to have. To avoid the effort and expense of preparing full-fledged bids, they would call a competitor to get a price to use. Such a situation might arise, for example, when a company already had enough work for the time being, but was asked by a valued customer to submit a bid anyway.

> All you want to do is show a bid, so that in some cases it was for as small a reason as getting your deposit back on the plans and specs. So you just simply have no interest in getting the job and just call to see if you can find someone to give you a price to use, so that you didn't have to go through the expense of an entire bid preparation. Now that is looked on very unfavorably, and it is a technical violation, but it was strictly an opportunity to keep your name in front of a desired customer. Or you may find yourself in a situation where somebody is doing work for a customer, has done work for many, many years and is totally acceptable, totally fair. There is no problem. But suddenly they (the customer) get an idea that they ought to have a few tentative figures, and you're called in, and you are in a moral dilemma. There's really no reason for you to attempt to compete in the circumstance. And so there was a way to back out.

Managed in this way, an action that appears on the surface to be a straightforward and conscious violation of antitrust regulations becomes merely a harmless business practice that happens to be a "technical violation." The offender can then refer to his personal history to verify his claim that, despite technical violations, he is in reality a law-abiding person. In the words of one offender, "Having been in the business for 33 years, you don't just automatically become a criminal overnight."

Third, offenders were very critical of the motives and tactics of prosecutors. Prosecutors were accused of being motivated solely by the opportunity for personal advancement presented by winning a big case. Further, they were accused of employing prosecution selectively and using tactics that allowed the most culpable offenders to go free. The Department of Justice was painted as using antitrust prosecutions for political purposes.

The fourth theme emphasized by the antitrust offenders involved a comparison between their crimes and the crimes of street criminals. Antitrust offenses differ in their mechanics from street crimes in that they are not committed in one place and at one time. Rather, they are spatially and temporally diffuse and are intermingled with legitimate behavior. In addition, the victims of antitrust offenses tend not to be identifiable individuals, as is the case with most street crimes. These characteristics are used by antitrust violators to contrast their own behavior with that of common stereotypes of criminality. Real crimes are pictured as discrete events that have beginnings and ends and involve individuals who directly and purposely victimize someone else in a particular place and a particular time.

> It certainly wasn't a premeditated type of thing in our cases as far as I can see. . . . To me it's different than _____ and I sitting down and we plan, well, we're going to rob this bank tomorrow and premeditatedly go in there. . . . That wasn't the case at all. . . . It wasn't like sitting down and planning I'm going to rob this bank type of thing. . . . It was just a common everyday way of doing business and surviving.

A consistent thread running through all of the interviews was the necessity for antitrust-like practices, given the realities of the business world. Offenders seemed to define the situation in such a manner that two sets of rules could be seen to apply. On the one hand, there are the legislatively determined rules—laws—which govern how one is to conduct one's business affairs. On the other hand, there is a higher set of rules based on the concepts of profit and survival, which are taken to define what it means to be in business in a capitalistic society. These rules do not just regulate behavior; rather, they constitute or create the behavior in question. If one is not trying to make a profit or trying to keep one's business going, then one is not really "in business." Following Searle (1969:33–41), the former type of rule can be called a regulative rule and the latter type a constitutive rule. In certain situations, one may have to violate a regulative rule in order to conform to the more basic constitutive rule of the activity in which one is engaged.

This point can best be illustrated through the use of an analogy involving competitive games. Trying to win is a constitutive rule of competitive games in the sense that if one is not trying to win, one is not really playing the game. In competitive games, situations may arise where a player deliberately breaks the rules even though he knows or expects he will be caught. In the game of basketball, for example, a player may deliberately foul an opponent to prevent him from making a sure basket. In this instance, one would understand that the fouler was trying to win by gambling that the opponent would not make the free throws. The player violates the rule against fouling in order to follow the higher rule of trying to win.

Trying to make a profit or survive in business can be thought of as a constitutive rule of capitalist economies. The laws that govern *how* one is allowed to make a profit are regulative rules, which can understandably be subordinated to the rules of trying to survive and profit. From the offender's point of view, he is doing what businessmen in our society are supposed to do—that is, stay in business and make a profit. Thus, an individual who violates society's laws or regulations in certain situations may actually conceive of himself as thereby acting more in accord with the central ethos of his society than if he had been a strict observer of its law. One might suggest, following Denzin (1977), that for businessmen in the building and contracting industry, an informal structure exists below the articulated legal structure, one which frequently supersedes the legal structure. The informal structure may define as moral and "legal" certain actions that the formal legal structure defines as immoral and "illegal."

TAX VIOLATORS

Six of the offenders interviewed were convicted of income tax violations. Like antitrust violators, tax violators can rely upon the complexity of the tax laws and an historical tradition in which cheating on taxes is not really criminal. Tax offenders would claim that everybody cheats somehow on their taxes and present themselves as victims of an unlucky break, because they got caught.

> Everybody cheats on their income tax, 95% of the people. Even if it's for ten
> dollars it's the same principle. I didn't cheat. I just didn't know how to report it.

The widespread belief that cheating on taxes is endemic helps to lend credence to the offender's claim to have been singled out and to be no more guilty than most people.

Tax offenders were more likely to have acted as individuals rather than as part of a group and, as a result, were more prone to account for their offenses by referring to them as either mistakes or the product of special circumstances. Violations were presented as simple errors which resulted from ignorance and poor record-keeping. Deliberate intention to steal from the government for personal benefit was denied.

> I didn't take the money. I have no bank account to show for all this money, where all this money is at that I was supposed to have. They never found the money, ever. There is no Swiss bank account, believe me.
>
> My records were strictly one big mess. That's all it was. If only I had an accountant, this wouldn't even of happened. No way in God's creation would this ever have happened.

Other offenders would justify their actions by admitting that they were wrong while painting their motives as altruistic rather than criminal. Criminality was denied because they did not set out to deliberately cheat the government for their own personal gain. Like the antitrust offenders discussed above, one tax violator distinguished between his own crime and the crimes of real criminals.

> I'm not a criminal. That is, I'm not a criminal from the standpoint of taking a gun and doing this and that. I'm a criminal from the standpoint of making a mistake, a serious mistake. . . . The thing that really got me involved in it is my feeling for the employees here, certain employees that are my right hand. In order to save them a certain amount of taxes and things like that, I'd extend money to them in cash, and the money came from these sources that I took it from. You know, cash sales and things of that nature, but practically all of it was turned over to the employees, because of my feeling for them.

All of the tax violators pointed out that they had no intention of deliberately victimizing the government. None of them denied the legitimacy of the tax laws, nor did they claim that they cheated because the government is not representative of the people (Conklin, 1977:99). Rather, as a result of ignorance or for altruistic reasons, they made decisions which turned out to be criminal when viewed from the perspective of the law. While they acknowledged the technical criminality of their actions, they tried to show that what they did was not criminally motivated.

VIOLATIONS OF FINANCIAL TRUST

Four offenders were involved in violations of financial trust. Three were bank officers who embezzled or misapplied funds, and the fourth was a union official who embezzled from a union pension fund. Perhaps because embezzlement is one crime in this sample that can be considered *mala in se,* these offenders were much more forthright about their crimes. Like the other offenders, the embezzlers would not go so far as to say "I am a criminal," but they did say "What I did was wrong, was criminal, and I knew it was." Thus, the embezzlers were unusual in that they explicitly admitted responsibility for their crimes. Two of the offenders clearly fit Cressey's scheme as persons with financial problems who used their positions to convert other people's money to their own use.

Unlike tax evasion, which can be excused by reference to the complex nature of tax regulations or antitrust violations, which can be justified as for the good of the organization as a whole, embezzlement requires deliberate action on the part of the offender and is almost inevitably committed for personal reasons. The crime of embezzlement, therefore, cannot be accounted for by using the same techniques that tax violators or antitrust violators do. The act itself can only be explained by showing that one was under extraordinary circumstances which explain one's uncharacteristic behavior. Three of the offenders referred explicitly to extraordinary circumstances and presented the offense as an aberration in their life history. For example, one offender described his situation in this manner:

> As a kid, I never even—you know kids will sometimes shoplift from the dime store—I never even did that. I had never stolen a thing in my life and that was what was so unbelievable about the whole thing, but there were some psychological and personal questions that I wasn't dealing with very well. I wasn't terribly happily married. I was married to a very strong-willed woman and it just wasn't working out.

The offender in this instance goes on to explain how, in an effort to impress his wife, he lived beyond his means and fell into debt.

A structural characteristic of embezzlement also helps the offender demonstrate his essential lack of criminality. Embezzlement is integrated into ordinary occupational routines. The illegal action does not stand out clearly against the surrounding set of legal actions. Rather, there is a high degree of surface correspondence between legal and illegal behavior. To maintain this correspondence, the offender must exercise some restraint when committing his crime. The embezzler must be discrete in his stealing; he cannot take all of the money available to him without at the same time revealing the crime. Once exposed, the offender can point to this restraint on his part as evidence that he is not really a criminal. That is, he can compare what happened with what could have happened in order to show how much more serious the offense could have been if he was really a criminal at heart.

> What I could have done if I had truly had a devious criminal mind and perhaps if I had been a little smarter—and I am not saying that with any degree of pride or any degree of modesty whatever, [as] it's being smarter in a bad, an evil way— I could have pulled this off on a grander scale and I might still be doing it.

Even though the offender is forthright about admitting his guilt, he makes a distinction between himself and someone with a truly "devious criminal mind."

Contrary to Cressey's (1953:57–66) findings, none of the embezzlers claimed that their offenses were justified because they were underpaid or badly treated by their employers. Rather, attention was focused on the unusual circumstances surrounding the offense and its atypical character when compared to the rest of the offender's life. This strategy is for the most part determined by the mechanics and organizational format of the offense itself. Embezzlement occurs within the organization but not for the organization. It cannot be committed accidentally or out of ignorance. It can be accounted for only by showing that the actor "was not himself" at the time of the offense or was under such extraordinary circumstances that embezzlement was an understandable response to an unfortunate situation. This may explain the finding that embezzlers tend to produce accounts that are viewed as more sufficient by the justice system than those produced by other offenders (Rothman and Gandossy, 1982). The only plausible option open to a convicted embezzler trying to explain his offense is to admit responsibility while justifying the action, an approach that apparently strikes a responsive chord with judges.

FRAUD AND FALSE STATEMENTS

Ten offenders were convicted of some form of fraud or false statements charge. Unlike embezzlers, tax violators, or antitrust violators, these offenders were much more likely to deny committing any crime at all. Seven of the ten claimed that they, personally, were innocent of any crime, although each admitted that fraud had occurred. Typically, they claimed to have been set up by associates and to have been wrongfully convicted by the U.S. Attorney handling the case. One might call this the scapegoat strategy. Rather than admitting technical wrongdoing and then justifying or excusing it, the offender attempts to paint himself as a victim by shifting the blame entirely to another party. Prosecutors were presented as being either ignorant or politically motivated.

The outright denial of any crime whatsoever is unusual compared to the other types of offenders studied here. It may result from the nature of the crime of fraud. By definition, fraud involves a conscious attempt on the part of one or more persons to mislead others. While it is theoretically possible to accidentally violate the antitrust and tax laws, or to violate them for altruistic reasons, it is difficult to imagine how one could accidentally mislead someone else for his or her own good. Furthermore, in many instances, fraud is an aggressively acquisitive crime. The offender develops a scheme to bilk other people out of money or property, and does this not because of some personal problem but because the scheme is an easy way to get rich. Stock swindles, fraudulent loan scams, and so on are often so large and complicated that they cannot possibly be excused as foolish and desperate solutions to personal problems. Thus, those involved in large-scale frauds do not have the option open to most embezzlers of presenting themselves as persons responding defensively to difficult personal circumstances.

Furthermore, because fraud involves a deliberate attempt to mislead another, the offender who fails to remove himself from the scheme runs the risk of being shown to have a guilty mind. That is, he is shown to possess the most essential element of modern conceptions of criminality: an intent to harm another. His inner self would in this case be exposed as something other than what it has been presented as, and all of his previous actions would be subject to reinterpretation in light of this new perspective. For this reason, defrauders are most prone to denying any crime at all. The cooperative and conspiratorial nature of many fraudulent schemes makes it possible to put the blame on someone else and to present oneself as a scapegoat. Typically, this is done by claiming to have been duped by others.

Two illustrations of this strategy are presented below.

> I figured I wasn't guilty, so it wouldn't be that hard to disprove it, until, as I say, I went to court and all of a sudden they start bringing in these guys out

of the woodwork implicating me that I never saw. Lot of it could be proved
that I never saw.

Inwardly, I personally felt that the only crime that I committed was not telling
on these guys. Not that I deliberately, intentionally committed a crime against
the system. My only crime was that I should have had the guts to tell on these
guys, what they were doing, rather than putting up with it and then trying to
gradually get out of the system without hurting them or without them thinking
I was going to snitch on them.

Of the three offenders who admitted committing crimes, two acted alone and
the third acted with only one other person. Their accounts were similar to others
presented earlier and tended to focus on either the harmless nature of their violations
or on the unusual circumstances that drove them to commit their crimes. One
claimed that his violations were only technical and that no one besides himself had
been harmed.

First of all, no money was stolen or anything of that nature. The bank didn't
lose any money. . . . What I did was a technical violation. I made a mistake.
There's no question about that, but the bank lost no money.

Another offender who directly admitted his guilt was involved in a check-kiting
scheme. In a manner similar to embezzlers, he argued that his actions were motivated
by exceptional circumstances.

I was faced with the choice of all of a sudden, and I mean now, closing the
doors or doing something else to keep that business open. . . . I'm not going
to tell you that this wouldn't have happened if I'd had time to think it over, because I think it probably would have. You're sitting there with a dying patient.
You are going to try to keep him alive.

In the other fraud cases more individuals were involved, and it was possible and perhaps necessary for each offender to claim that he was not really the culprit.

DISCUSSION: OFFENSES, ACCOUNTS, AND DEGRADATION CEREMONIES

The investigation, prosecution, and conviction of a white-collar offender involves him
in a very undesirable status passage (Glaser and Strauss, 1971). The entire process
can be viewed as a long and drawn-out degradation ceremony with the prosecutor
as the chief denouncer and the offender's family and friends as the chief witnesses.
The offender is moved from the status of law-abiding citizen to that of convicted felon.
Accounts are developed to defeat the process of identity transformation that is the
object of a degradation ceremony. They represent the offender's attempt to diminish
the effect of his legal transformation and to prevent its becoming a publicly validated
label. It can be suggested that the accounts developed by white-collar offenders take
the forms that they do for two reasons: (1) the forms are required to defeat the success of the degradation ceremony, and (2) the specific forms used are the ones available given the mechanics, history, and organizational context of the offenses.

Three general patterns in accounting strategies stand out in the data. Each can be characterized by the subject matter on which it focuses: the event (offense), the perpetrator (offender), or the denouncer (prosecutor). These are the natural subjects of accounts in that to be successful, a degradation ceremony requires each of these elements to be presented in a particular manner (Garfinkel, 1956). If an account giver can undermine the presentation of one or more of the elements, then the effect of the ceremony can be reduced. Although there are overlaps in the accounting strategies used by the various types of offenders, and while any given offender may use more than one strategy, it appears that accounting strategies and offenses correlate.

REFERENCES

Casper, Jonathan D. 1978. *Criminal Courts: The Defendant's Perspective*. Washington, D.C.: U.S. Department of Justice.

Conklin, John E. 1977. *Illegal But Not Criminal: Business Crime in America*. Englewood Cliffs, N.J.: Prentice-Hall.

Cressey, Donald. 1953. *Other People's Money*. New York: Free Press.

Denzin, Norman K. 1977. Notes on the criminogenic hypothesis: A case study of the American liquor industry. *American Sociological Review* 42:905–920.

———. 1984. *On Understanding Emotion*. San Francisco: Jossey-Bass.

Edelhertz, Herbert. 1970. *The Nature, Impact, and Prosecution of White Collar Crime*. Washington, D.C.: U.S. Government Printing Office.

Garfinkel, Harold, 1956. Conditions of successful degradation ceremonies. *American Journal of Sociology* 61:420–424.

Glaser, Barney G. and Anselm L. Strauss. 1971. *Status Passage*. Chicago: Aldine.

Katz, Jack. 1979. Legality and equality: Plea bargaining in the prosecution of white-collar crimes. *Law and Society Review* 13:431–460.

Rothman, Martin and Robert F. Gandossy. 1982. Sad tales: The accounts of white-collar defendants and the decision to sanction. *Pacific Sociological Review* 4:449–473.

Searle, John R. 1969. *Speech Acts*. Cambridge: Cambridge University Press.

QUESTIONS FOR DISCUSSION

1. Explain the emotional responses of white-collar offenders. How would you compare their responses to why people commit street crimes like rape, robbery, murder, and assault?

2. According to the author, how do white-collar offenders justify and excuse their crimes?

3. Give a detailed explanation of how an antitrust violator, tax violator, or financial trust violator commits the crime. Which of the violations is the most frequently committed? Do you know anyone who has committed this violation? How did you react to the person(s)?

A Sociology of Hackers [1]

Tim Jordan and Paul Taylor

INTRODUCTION

The growth of a world-wide computer network and its increasing use both for the construction of online communities and for the reconstruction of existing societies means that unauthorised computer intrusion, or hacking, has wide significance. The 1996 report of a computer raid on Citibank that netted around $10 million indicates the potential seriousness of computer intrusion. Other, perhaps more whimsical, examples are the attacks on the CIA world-wide web site, in which its title was changed from Central Intelligence Agency to Central Stupidity Agency, or the attack on the British Labour Party's web-site, in which titles like "Road to the Manifesto" were changed to "Road to Nowhere." These hacks indicate the vulnerability of increasingly important computer networks and the anarchistic, or perhaps destructive, world-view of computer intruders (Miller, 1996; Gow and Norton-Taylor, 1996). It is correct to talk of a world-view because computer intrusions come not from random, obsessed individuals but from a community that offers networks and support, such as the long running magazines *Phrack* and *2600*. At present there is no detailed sociological investigation of this community, despite a growing number of racy accounts of hacker adventures.[2] To delineate a sociology of hackers, an introduction is needed to the nature of computer-mediated communication and of the act of computer intrusion, the hack. Following this the hacking community will be explored in three sections: first, a profile of the number of hackers and hacks; second, an outline of its culture through the discussion of six different aspects of the hacking community; and third, an exploration of the community's construction of a boundary, albeit fluid, between itself and its other, the computer security industry.[3] Finally, a conclusion that briefly considers the significance of our analysis will be offered.

In the early 1970s, technologies that allowed people to use decentred, distributed networks of computers to communicate with each other globally were developed.[4] By the early 1990s a new means of organising and accessing information contained on computer networks was developed that utilised multi-media 'point and click' methods, the World-Wide Web. The Web made using computer networks intuitive and underpinned their entry into mass use. The size of this global community of computer communicators is difficult to measure[5] but in January 1998 there were at least 40 million (Hafner and Lyons, 1996; Quarterman, 1990; Jordan, 1998a; Rickard, 1995; Quarterman, 1993). Computer communication has also become key to many industries, not just through the Internet but also through private networks, such as those that underpin automated teller services. The financial industry is the clearest

"A Sociology of Hackers" by Tim Jordan and Paul Taylor from *The Sociological Review,* November, 1998, p. 757. Reprinted by permission of Blackwell Publishers, Ltd., Oxford, England.

example of this, as John Perry Barlow says, "cyberspace is where your money is." Taken together, all the different computer networks that currently exist control and tie together vital institutions of modern societies; including telecommunications, finance, globally distributed production and the media (Castells, 1996; Jordan, 1998a). Analysis of the community which attempts to illicitly use these networks can begin with a definition of the 'hack'.

Means of gaining unauthorised access to computer networks including guessing, randomly generating or stealing a password. For example, in the Prestel hack, which resulted in the Duke of Edinburgh's mail-box becoming vulnerable, the hacker simply guessed an all too obvious password (222222 1234) (Schifreen, hacker, interview). Alternatively, some computers and software programmes have known flaws that can be exploited. One of the most complex of these is 'IP spoofing' in which a computer connected to the Internet can be tricked about the identity of another computer during the process of receiving data from that computer (Felten et al., 1996; Shimomura, 1996; Littman, 1996). Perhaps most important of all is the ability to 'social engineer'. This can be as simple as talking people into giving out their passwords by impersonating someone, stealing garbage in the hope of gaining illicit information (trashing) or looking over someone's shoulder as they use their password (shoulder surfing). However, what makes an intrusion a hack or an intruder a hacker is not the fact of gaining illegitimate access to computers by any of these means but a set of principles about the nature of such intrusions. Turkle identifies three tenets that define a good hack: simplicity, the act has to be simple but impressive; mastery, however simple it is the act must derive from a sophisticated technical expertise; and, illicit, the act must be against some legal, institutional or even just perceived rules (Turkle, 1984: 232).[6] Dutch hacker Ralph used the example of stealing free telephone time to explain the hack:

> It depends on how you do it, the thing is that you've got your guys that think up these things, they consider the technological elements of a phone-booth, and they think, 'hey wait a minute, if I do this, this could work', so as an experiment, they cut the wire and it works, now they're hackers. Okay, so it's been published, so Joe Bloggs reads this and says "hey, great, I have to phone my folks up in Australia," so he goes out, cuts the wire, makes phone calls. He's a stupid ignoramus, yeah? (Ralph, hacker, interview)

A second example would be the Citibank hack. In this hack, the expertise to gain unauthorised control of a bank was developed by a group of Russian hackers who were uninterested in taking financial advantage. The hacker ethic to these intruders was one of exploration and not robbery. But, drunk and depressed, one of the hackers sold the secret for $100 and two bottles of vodka, allowing organised criminals to gain the expertise to steal $10 million (Gow and Norton-Taylor, 1996). Here the difference between hacking and criminality lay in the communally held ethic that glorified being able to hack Citibank but stigmatised using that knowledge to steal. A hack is an event that has an original moment and, though it can be copied, it loses its status as a hack the more it is copied. Further, the good hack is the object in-itself that hackers desire, not the result of the hack (Cornwall, 1985: vii).

The key to understanding computer intrusion in a world increasingly reliant on computer-mediated communication lies in understanding a community whose aim is the hack. It is this community that makes complex computer intrusion possible and a never ending threat, through the limitless search for a good hack. It is this community that stands forever intentionally poised both at the forefront of computer communications and on the wrong side of what hackers see as dominant social and cultural norms.

COMPUTER UNDERGROUND: DEMOGRAPHICS

Analysing any intentionally illicit community poses difficulties for the researcher. The global and anonymous nature of computer-mediated communication exacerbates such problems because generating a research population from the computer underground necessitates self-selection by subjects and it will be difficult to check the credentials of each subject. Further methodological difficulties involved in examining a self-styled 'outlaw' community that exists in cyberspace are indicated by the Prestel hacker.

> There used to be a hacking community in the UK, the hackers I used to deal with 8 or 9 years ago were all based in North London where I used to live and there were 12 of us around the table at the local Chinese restaurant on a Friday night . . . within about 20 minutes of me and my colleague Steve Gold being arrested: end of hacking community. An awful lot of phone calls went around, a lot of discs got buried in the garden, and a lot of people became ex-hackers and there's really no-one who'll talk now (Schifreen, hacker, interview).

Demographic data is particularly difficult to collect from an underground community.[7] However, some statistics are available. Following presentation of these, an in-depth exploration of the hacking community on the basis of qualitative research will be presented. After investigating the US police force's crackdown on the computer underground in the early 1990s, Sterling estimated there were 5,000 active hackers with only around 100 in the elite who would be "skilled enough to penetrate sophisticated systems" (Sterling, 1992: 76–77). For the same period, Clough and Mungo estimated there were 2,000 of 'the really dedicated, experienced, probably obsessed computer freaks' and possibly 10,000 others aspiring to this status (Clough and Mungo, 1992: 218).[8] Though no more than an indication, the best, indeed only, estimates for the size of the hacking community or computer underground are given by these figures.

Another means of measuring the size of the computer underground is by its effects. Though this cannot hope to indicate the actual number of hackers, as one hacker can be responsible for extensive illicit adventures, measuring the extent of hacking allows one indication of the underground's level of activity. Three surveys are available that generate evidence from the 'hacked' rather than hackers: the 1990 UK Audit Commission's survey, the 1993 survey conducted as part of this research project, and the 1996 War Room Research, information systems security survey.[9] Results from all three sources will be presented, focusing on the amount of hacking.

The 1990 UK Audit Commission surveyed 1,500 academic, commercial and public service organisations in the United Kingdom. This survey found 5% of academic, 14% of commercial and 11.5% of public service organisations had suffered computer intrusion (Audit Commission, 1990). A survey was conducted as part of this research project (hereafter referred to as the Taylor survey) and received 200[10] responses, of which 64.5% had experienced a hack, 18.5% a virus only and 17% no detected illicit activity (Taylor, 1993). The 1996 WarRoom survey received 236 responses from commercial USA firms (Fortune 1,000 companies) of which 58% reported attempts by outsiders to gain computer access in the 12 months prior to July 1996, 29.8% did not know and 12.2% reported no such attempts. The types of intrusions can be categorised as 38.3% malicious, 46.5% unidentifiable as malicious or benign and 15.1% benign [11] (WarRoom, 1996).

The level of hacking activity reported in these surveys varies greatly between the Audit Commission on the one hand and the Taylor and WarRoom surveys on the other. A number of possibilities explain this. The lower level of hacking comes from a survey of UK organisations, while Taylor was over half from the USA and a third UK and WarRoom was solely USA. This might suggest a higher level of hacking into USA organisations, though this says nothing about the national source of a hack. Second, the Audit Commission survey has a much larger sample population and consequently should be more reliable. However, third, the WarRoom and Taylor surveys stressed the confidentiality of respondents. This is a key issue as organisations show a consistently high level of caution in reporting hacks. The WarRoom survey found that 37% of organisations would only report computer intrusion if required by law, that 22% would report only if "everybody else did," that 30% would only report if they could do so anonymously and only 7% would report anytime intrusion was detected (WarRoom, 1996). From this perspective the Audit Commission survey may have under-reported hacking because it did not place sufficient emphasis on the confidentiality of responses. Fourth, the Taylor and WarRoom surveys were conducted later than the Audit Commission survey and may reflect rising levels of or rising awareness of hacking. Unfortunately, there is no way of deciding which of these factors explain the differences in reported levels of hacking.

The available statistics suggest the computer underground may not be very large, particularly in the number of elite hackers, but may be having a significant effect on a range of organisations. If the Taylor and WarRoom surveys are accurate nearly two-thirds of organisations are suffering hacks. To grasp the nature of hackers requires turning to the qualitative fieldwork conducted in this project.

INTERNAL FACTORS: TECHNOLOGY, SECRECY, ANONYMITY, MEMBERSHIP FLUIDITY, MALE DOMINANCE, AND MOTIVATIONS

> To find 'hacker culture' you have to take a very wide view of the cyberspace terrain and watch the interactions among physically diversified people who have in common a mania for machines and software. What you will find will be a gossamer framework of culture. (Marotta, hacker, interview)

The 'imagined community' that hackers create and maintain can be outlined through the following elements: technology, secrecy, anonymity, boundary fluidity, male dominance and motivations. Community is here understood as the collective identity that members of a social group construct or, in a related way, as the 'collective imagination' of a social group. Both a collective identity and imagination allow individuals to recognise in each other membership of the same community. The computer underground, or at least the hacking part of it, can be in this way understood as a community that offers certain forms of identity through which membership and social norms are negotiated. Even though some of these forms are externally imposed, the nature of Internet technology for example, the way these forms are understood allows individuals to recognise in each other a common commitment to an ethic, community or way of life. This theorisation draws on Anderson's concept of the imagined community and on social movement theories that see movements as dispersed networks of individuals, groups and organisations that combine through a collectively articulated identity. Anderson names the power of an imagined identity to bind people, who may never meet each other, together in allegiance to a common cause. Social movement theories grasp the way movements rely on divergent networks that are not hierarchically or bureaucratically unified but are negotiated between actors through an identity that is itself the subject of much of the negotiation (Jordan, 1995; Diani, 1992; Anderson, 1991). These perspectives allow us to grasp a hacking community that can use computer mediated communication to exist world-wide and in which individuals often never physically meet.[12]

Technology

The hacking community is characterised by an easy, if not all-consuming, relationship with technology, in particular with computer and communications technology.

> We are confronted with . . . a generation that has lived with computers virtually from the cradle, and therefore have no trace of fear, not even a trace of reverence. (Professor Herschberg, academic, interview)

Hackers share a certain appreciation of or attitude to technology in the assumption that technology can be turned to new and unexpected uses. This attitude need not be confined to computer mediated communication. Dutch hacker Dell claimed to have explored the subterranean tunnels and elevator shafts of Amsterdam, including government fall-out shelters (Dell, hacker, interview), while Utrecht hacker Ralph argued hacking "pertains to any field of technology. Like, if you haven't got a kettle to boil water with and you use your coffee machine to boil water with, then that in my mind is a hack, because you are using technology in a way that it's not supposed to be used" (Ralph, hacker, interview). It is the belief that technology can be bent to new, unanticipated purposes that underpins hackers' collective imagination.

Secrecy

Hackers demonstrate an ambivalent relationship to secrecy. A hack demands secrecy, because it is illicit, but the need to share information and gain recognition demands publicity. Sharing information is key in the development of hackers, though it makes keeping illicit acts hidden from law enforcement difficult. Hackers often hack in groups, both in the sense of physically being in the same room while hacking and of hacking separately but being in a group that physically meets, that frequents bulletin boards, on-line places to talk, and exchanges information. It is a rare story of a hacker's education that does not include being trained by more experienced hackers or drawing on the collective wisdom of the hacking community through on-line information. Gaining recognition is also important to hackers. A member of the Zoetermeer hacking group noted "Hacking can be rewarding in itself, because it can give you a real kick sometimes. But it can give you a lot more satisfaction and recognition if you share your experiences with others. . . . Without this group I would never have spent so much time behind the terminals digging into the operating system" (Zoetermeer, hackers, interview). A good hack is a bigger thrill when shared and can contribute to a hacker gaining status and access to more communal expertise. For example, access to certain bulletin boards is only given to those proven worthy.

A tension between the need to keep illicit acts away from the eyes of police and other authority figures but in front of the eyes of peers or even the general public defines hackers' relationship to secrecy. No hack exemplifies this more than a WorldWide Web hack where the object is to alter an internationally accessible form of public communication but at the same time not be caught. In the case of the Labour Party hack, the hacker managed to be quoted on the front page of UK national newspapers, by ringing up the newspapers to tell them to look at the hack before it was removed, but also kept his/her identity secret. A further example is that many hackers take trophies in the form of copied documents or pieces of software because a trophy proves to the hacking community that the hacker 'was there'. The problem is that a trophy is one of the few solid bases for prosecuting hackers. Ambivalence toward secrecy is also the source of the often-noted fact that hackers are odd criminals, seeking publicity. As Gail Thackeray, one-time police nemesis of hackers, noted "What other group of criminals . . . publishes newsletters and holds conventions?" (Thackeray, cited in Sterling, 1992:181).[13]

Anonymity

The third component of the hacking community is anonymity. As with technology what is distinctive is not so much the fact of online anonymity, as this is a widely remarked aspect of computer-mediated communication (Dery, 1993:561), but the particular understanding of anonymity that hackers take up. Anonymity is closely related to secrecy but is also distinct. Secrecy relates to the secrecy of the hack, whereas anonymity relates to the secrecy of a hacker's offline identity. Netta Gilboa notes one complex version of this interplay of named and hidden identity on an on-line chat channel for hackers.

> Hackers can log into the #hack channel using software . . . that allows them
> to come in from several sites and be on as many separate connections, appear
> ing to be different people. One of these identities might then message you pri
> vately as a friend while another is being cruel to you in public. (Gilboa, 1996:
> 102–103)

Gilboa experienced the construction of a number of public identities all intended to mask the 'real' identity of a hacker. A second example of this interplay of anonymity and publicity is the names or 'handles' hackers give themselves and their groups. These are some of the handles encountered in this research: Hack-Tic (group), Zoetermeer (group), Altenkirch (German), Eric Bloodaxe, Faustus, Maelstrom, Mercury, Mofo. Sterling notes a long list of group names—such as Kaos Inc., Knights of Shadow, Master Hackers, MAD!, Legion of Doom, Farmers of Doom, the Phirm,

Inner Circle I and Inner Circle II. Hackers use names to sign their hacks (sometimes even leaving messages for the hacked computer's usual users), to meet on-line and to bolster their self-image as masters of the hack, all the while keeping their offline identity secret.[14]

Membership Fluidity

The fourth quality of the hacking community is the speed at which membership changes. Hacking shares the characteristics ascribed to many social movements of being an informal network rather than a formally constituted organisation and, as such, its boundaries are highly permeable (Jordan, 1995; Diani, 1992). There are no formal ceremonies to pass or ruling bodies to satisfy to become a hacker. The informal and networked nature of the hacking community, combined with its illicit and sometimes obsessional nature means that a higher turnover of hackers occurs (Clough and Mungo, 1992:18). Hackers form groups within the loose overall structure of the hacking community and these may aspire to be formally organised, however the pressures of law enforcement means that any successful hacking group is likely to attract sustained attention at some point (Quittner and Slatalla, 1995).

> People come and go pretty often and if you lay off for a few months and then come back, almost everyone is new. There are always those who have been around for years . . . I would consider the hacking community a very informal one. It is pretty much anarchy as far as rule-making goes. . . . The community was structured only within the framework of different hacking 'groups'. Legion of Doom would be one example of this. A group creates its own rules and usually doesn't have a leader . . . The groups I've been in have voted on accepting new members, kicking people out, etc. (Eric Bloodaxe, hacker, member of Legion of Doom, interview)

Gilboa claims that the future of hacking will be a split between lifelong hackers, often unable to quit because of police records and suspicion, and 90% of hackers who will move on "when they get a job they care about or a girlfriend who sucks up their time" (Gilboa, 1996:111). A more prosaic, but equally potent, reason why the hacking community's membership is fluid is given by hacker Mike: "if you stop, if you don't do it for one week then things change, the network always changes. It changes very quickly and you have to keep up and you have to learn all the tricks by heart, the default passwords, the bugs you need" (Mike, hacker, interview). The sheer speed at which computer communications technology changes requires a powerful commitment from hackers.

Male Dominance

The fifth component of hacking culture is male dominance and an associated misogyny. Research for this project and literature on hackers fails to uncover any significant evidence of female hackers (Taylor, 1993:92). Gilboa states "I have met more than a thousand male hackers in person but less than a dozen of them women" (Gilboa, 1996:106). This imbalance is disproportionate even in the field of computer mediated communication (Spertus, 1991:i). A number of factors explain the paucity of women generally in the computer sciences: childhood socialisation, where boys are taught to relate to technology more easily than girls; education in computers occurs in a masculine environment; and, a gender bias towards men in the language used in computer science (Spertus, 1991; Turkle, 1984; Taylor, 1993:91–103). With these factors producing a general bias towards males in relation to computers, the drive towards the good hack exacerbates this as it involves a macho, competitive attitude (Keller, 1988:58). Hackers construct a more intensely masculine version of the already existing male bias in the computer sciences.

> When Adam delved and Eve span . . . who was then the gentleman? Well, we see that Adam delves into the workings of computers and networks and meanwhile Eve spins, what? Programmes? Again, my wife programmes and she has the skills of a hacker. She has had to crack security in order to do her job. But she does it as her job, not for the abstract thrill of discovering the unknown. Even spins. Females who compute would rather spend their time building a good system, than breaking into someone else's system. (Mercury, hacker, interview)

Whether Mercury's understanding of differences between men and women is accurate or not, the fact that he, and many other hackers, have such attitudes means the hacking community will almost certainly feel hostile to women. Added to these assumptions of, at best, separate spheres of male and female expertise in computing is the problem that anonymity often fuels sexual harassment. "The fact that many networks allow a user to hide his real name . . . seems to cause many males to drop all semblance of civilisation. Sexual harassment by email is not uncommon" (Freiss, hacker, interview). Gilboa, a woman, recounts an epic tale of harassment that included hackers using her on-line magazine as a 'tutorial' example of how to charge phone calls to someone else, taking over her magazine entirely and launching a fake version, being called a prostitute, child molester, and drug dealer, having her phone calls listened to, her phone re-routed or made to sound constantly engaged and having her email read. One answer to Gilboa's puzzlement at her treatment lies in the collective identity hackers share and construct that is in part misogynist.

Motivations

Finally, hackers often discuss their motivations for hacking. They are aware of, and often glory in, the fact that the life of a dedicated hacker seems alien to those outside the hacking community. One result of this is that hackers discuss their motivations. These are sometimes couched as self-justifications, sometimes as explanations and sometimes as agonised struggles with personal obsessions and failures. However, whatever the content of such discussions, it is the fact of an ongoing discourse around the motivation to hack that builds the hacking community. These discussions are one more way that hackers can recognise in each other a common identity that provides a collective basis for their community. A number of recurring elements to these discussions can be identified.

First, hackers often confess to an addiction to computers and/or to computer networks, a feeling that they are compelled to hack. Second, curiosity as to what can be found on the world-wide network is also a frequent topic of discussion. Third, hackers often claim their offline life is boring compared to the thrill of illicit searches in online life. Fourth, the ability to gain power over computer systems, such as NASA, Citibank or the CIA web site, is an attraction. Fifth, peer recognition from other hackers or friends is a reward and goal for many hackers, signifying acceptance into the community and offering places in a hierarchy of more advanced hackers. Finally, hackers often discuss the service to future computer users or to society they are offering because they identify security loopholes in computer networks. Hackers articulate their collective identity, and construct a sense of community, by discussing this array of different motivations.

> I just do it because it makes me feel good, as in better than anything else that I've ever experienced . . . the adrenaline rush I get when I'm trying to evade authority, the thrill I get from having written a program that does something that was supposed to be impossible to do, and the ability to have social relations with other hackers are all very addictive . . . For a long time, I was extremely shy around others, and I am able to let my thoughts run free when I am alone with my computer and a modem hooked up to it. I consider myself addicted to hacking . . . I will have no moral or ethical qualms about system hacking until accounts are available to the general public for free . . . Peer recognition was very important, when you were recognised you had access to more. (Maelstrom, hacker, interview)

Maelstrom explores almost the whole range of motivations including curiosity, the thrill of the illicit, boredom, peer recognition and the social need for free or cheap access. By developing his own interpretation out of the themes of motivation, he can simultaneously define his own drives and develop a sense of community. It is this double movement in which individual motivations express the nature of a community, that makes the discussions of motivations important for hackers. Finally, the motivations offered by perhaps the most famous of all hackers, Kevin Mitnick, provides another common articulation of reasons for hacking.

> You get a better understanding of cyberspace, the computer systems, the operating systems, how the computer systems interact with one another, that basically was my motivation behind my hacking activity in the past. It was just from the gain of knowledge and the thrill of adventure, nothing that was well and truly sinister as trying to get any type of monetary gain or anything. (Mitnick, hacker, interviewer)

INTERNAL FACTORS: CONCLUSION

These six factors all function largely between hackers, allowing them a common language and a number of resources through which they can recognise each other as hackers and through which newcomers can become hackers. These are resources internal to the hacking community, not because they do not affect or include nonhackers but because their significance is largely for other hackers. Put another way, these are the resources hackers use to discuss their status as hackers with other hackers, they are collectively negotiated within the boundaries of the hacker community. This raises the issue of how an external boundary is constructed and maintained. How do hackers recognise a distinction between inside and outside? How do hackers adjust, reinvent and maintain such a distinction? This is the subject of the third and final section of this definition of the hacker community. External factors: the boundary between computer underground and the computer security industry.

Hackers negotiate a boundary around their community by relating to other social groups. For example, hackers have an often spectacular relationship to the media. Undoubtedly the most important relationship to another community or group is their intimate and antagonistic bond to the computer security industry (CSI). This relationship is constitutive of the hacking community in a way that no other is. Put another way, there is no other social group whose existence is necessary to the existence of the hacking community. Here is a sample of views of hackers from members of CSI.

> Hackers are like kids putting a 10 pence piece on a railway line to see if the train can bend it, not realising that they risk derailing the whole train. (Mike Jones, security awareness division, Department of Trade and Industry, UK, interview)

> Electronic vandalism. (Warman, London Business School, interview)

> Somewhere near vermin. (Zmudsinski, system engineer/manager, USA, interview)

Naturally, hackers often voice a similar appreciation of members of CSI. For example, while admitting psychotic tendencies exist in the hacking community Mofo notes:

> my experience has shown me that the actions of 'those in charge' of computer systems and networks have similar 'power trips' which need to be fulfilled. Whether this psychotic need is developed or entrenched before one's association with computers is irrelevant. (Mofo, hacker, interview)

However, the boundary between these two communities is not as clear as such attitudes might suggest. This can be seen in relation to membership of the communities and the actions members take.

Hackers often suggest the dream that their skills should be used by CSI to explore security faults, thereby giving hackers jobs and legitimacy to pursue the hack by making them members of CSI. The example of a leading member of one of the most famous hacker groups, the Legion of Doom, is instructive. Eric Bloodaxe, aka Chris Goggans, became a leading member of the hacking community before helping to set up a computer security firm, Comsec, and later moving to become senior network security engineer for WheelGroup, a network security company (Quittner and Slatalla, 1995:145–147 and 160). On the CSI side, there have been fierce debates over whether hackers might be useful because they identify security problems (Spafford, 1990; Denning, 1990). Most striking, a number of CSI agencies conduct hacking attacks to test security. IBM employ a group of hackers who can be hired to attack computer systems and the UK government has asked 'intelligence agents' to hack its secure email system for government ministers (Lohr, 1997; Hencke, 1998).[15] In the IBM case, an attempt at differentiating the hired hackers from criminal hackers is made by hiring only hackers without criminal records (a practice akin to turning criminals who have not been caught into police) (Lohr, 1997). Both sides try to assure themselves of radical differences because they undertake similar actions. For example, Bernie Cosell was a USA commercial computer systems manager and one of the most vehement anti-hackers encountered in this study, yet he admitted he hacked

> once or twice over the years. I recall one incident where I was working over the weekend and the master source hierarchy was left read-protected, and I really needed to look at it to finish what I was doing, and this on a system where I was not a privileged user, so I 'broke into' the system enough to give myself enough privileges to be able to override the file protections and get done what I needed . . . at which point I put it all back and told the systems administrator about the security hole. (Cosell, USA systems manager, interview)

More famous is the catalogue of hacks Clifford Stoll had to perpetrate in his pursuit of a hacker, which included borrowing other people's computers without permission and monitoring other people's electronic communications without permission (Stoll, 1989; Thomas, 1990). Such examples mean that differences between the two communities cannot be expressed through differences in what they do but must focus on the meaning of actions. Delineating these meanings is chiefly done through ethical debates about the nature of hacking conducted through analogies drawn between cyberspace and non-virtual or real space.

CSI professionals often draw analogies between computer intrusion and a range of widely understood crimes. These analogies drawn on the claim that a computer is something like a bank, car or house that can be 'got into'. Using this analogy makes it easy to understand the danger of hackers, people who break into banks, schools or houses usually do so for nefarious purposes. The ethical differences between hackers and the CSI become clearly drawn. The problem with such analogies is that, on further reflection, hackers seem strange burglars. How often does a burglar leave behind an exact copy of the video recorder they have stolen? But this unreal situation is a more accurate description of theft in cyberspace because taking in cyberspace overwhelmingly means copying. Further, hacker culture leads hackers to publicise their break-ins, sometimes even stressing the utility of their break-ins for identifying system weaknesses. What bank robbers ring up a bank to complain of lax security? The simple analogy of theft breaks down when it is examined and must be complicated to begin to make sense of what hackers do.

> There is a great difference between trespassing on my property and breaking into my computer. A better analogy might be finding a trespasser in your high-rise office building at 3 am and learning that his back-pack contained some tools, some wire, a timer and a couple of detonation caps. He could claim that he wasn't planting a bomb, but how can you be sure? (Cosell, USA systems manager, interview)

Cosell's analogy continues to draw on real world or physically based images of buildings being entered but tries to come closer to the reality of how hackers operate. However, the ethical component of the analogy has been weakened because the damage hackers cause becomes implied, where is the bomb?[16] Cosell cannot claim there will definitely be a bomb, only that it is possible. If all possible illegal actions were prohibited then many things would become illegal, such as driving because it is possible to speed and then hurt someone in an accident. The analogy of breaking and entering is now strong on implied dangers but weak on the certainty of danger. The analogies CSI professionals use continue to change if they try to be accurate. "My analogy is walking into an office building, asking a secretary which way it is to the records room and making some Xerox copies of them. Far different than breaking and entering someone's home" (Cohen, CSI, interview). Clearly there is some ethical content here, some notion of theft of information, but it is ethically far muddier than the analogy burglar offers. At this point, the analogy breaks down entirely because the ethical content can be reversed to one that supports hackers as "whistle-blowers" of secret abuses everyone should know about.

> The concept of privacy is something that is very important to a hacker. This is so because hackers know how fragile privacy is in today's world. . . . In 1984 hackers were instrumental in showing the world how TRW kept credit files on millions of Americans. Most people had not even heard of a credit file until this happened . . . More recently, hackers found that MCI's 'Friends and Family' programme allowed anybody to call an 800 number and find out the numbers of everyone in a customer's 'calling circle'. As a bonus, you could also find out how these numbers were related to the customer . . . In both the TRW and MCI cases, hackers were ironically accused of being the ones to invade privacy. What they really did was help to educate the American consumer. (Goldstein, 1993)

The central analogy of CSI has now lost its ethical content. Goldstein reverses the good and bad to argue that the correct principled action is to broadcast hidden information. If there is some greater social good to be served by broadcasting secrets, then perhaps hackers are no longer robbers and burglars but socially responsible whistle blowers. In the face of such complexities, CSI professionals sometimes abandon the analogy of breaking and entering altogether; "it is no more a valid justification to attack systems because they are vulnerable than it is valid to beat up babies because they can't defend themselves" (Cohen, CSI, interview). Here many people's instinctive reaction would be to side with the babies, but a moment's thought reveals that in substance Cohen's analogy changes little. A computer system is not human and if information in it is needed by wider society, perhaps it should be attacked.

The twists and turns of these analogies show that CSI professionals use them not so much to clearly define hacking and its problems, but to establish clear ethical differences between themselves and hackers. The analogies of baby-bashing and robbery all try to establish hacking as wrong. The key point is that while these analogies work in an ethical and community building sense, they do not work in clearly grasping the nature of hacking because analogies between real and virtual space cannot be made as simply as CSI professionals would like to assume.

> Physical (and biological) analogies are often misleading as they appeal to an understanding from an area in which different laws hold. . . . Many users (and even 'experts') think of a password as a 'key' despite the fact that you can easily guess the password, while it is difficult to do the equivalent for a key. (Brunnstein, academic, Hamburg University, interview)

The process of boundary formation between the hacking and CSI communities occurs in the creation of analogies by CSI professionals to establish ethical differences between the communities and their reinterpretation by hackers. However, this does not exclude hackers from making their own analogies.

> Computer security is like a chess-game, and all these people that say breaking into my computer systems is like breaking into my house: bull-shit, because securing your house is a very simple thing, you just put locks on the doors and bars on the windows and then only brute force can get into your house, like smashing a window. But a computer has a hundred thousand intricate ways to get in, and it's a chess game with the people that secure a computer. (Gongrijp, Dutch hacker, interview)

Other hackers offer similar analogies that stress hacking is an intellectual pursuit. "I was bored if I didn't do anything . . . I mean why do people do crosswords? It's the same thing with hackers" (J. C. van Winkel, hacker, interview). Gongrijp and van Winkel also form boundaries through ethical analogy. Of course, it is an odd game of chess or crossword that results in the winner receiving thousands of people's credit records or access to their letters. Hackers' elision of the fact that a game of chess has no result but a winner and a loser at a game of chess whereas hacking often results in access to privileged information, means their analogies are both inaccurate and present hacking as a harmless, intellectual pursuit. It is on the basis of such analogies and discussions that the famed 'hacker ethic' is often invoked by hackers. Rather than hackers learning the tenets of the hacker ethic, as seminally defined by Steven Levy, they negotiate a common understanding of the meaning of hacking of which the hacker ethic provides a ready articulation.[17] Many see the hacker ethic as a foundation of the hacker community, whereas we see the hacker ethic as the result of the complex construction of a collective identity.

The social process here is the use of analogies to physical space by CSI and hackers to establish a clear distinction between the two groups. In these processes can be seen the construction by both sides of boundaries between communities that are based on different ethical interpretations of computer intrusion, in a situation where other boundaries, such as typical actions or membership, are highly fluid.

CONCLUSION

The nature of the hacking community needs to be explored in order to grasp the social basis that produces hacking as a facet of computer networks. The figures given previously and the rise of the World-Wide Web hack, offering as it does both spectacular publicity and anonymity, point to the endemic nature of hackers now that world-wide computer networks are an inescapable reality. Hackers show that living in a networked world means living in a risky world. The community found by this research articulates itself in two key directions. First there are a number of components that are the subject of ongoing discussion and negotiation by hackers with other hackers. In defining and redefining their attitudes to technology, secrecy, anonymity, membership change, male dominance and personal motivations, hackers create an imagined community. Second, hackers define the boundaries of their community primarily in relation to the Computer Security Industry. These boundaries stress an ethical interpretation of hacking because it can be difficult to clearly distinguish the activities or membership of the two communities. Such ethics emerge most clearly through analogies used by members of each community to explain hacking.

Hackers are often pathologised as obsessed, isolated young men. The alien nature of online life allows people to believe hackers more easily communicate with machines than humans, despite hackers' constant use of computers to communicate with other humans. Fear of the power of computers over our own lives underpins this terror. The very anonymity that makes their community difficult to study, equally makes hackers an easy target for pathologising. For example, Gilboa's experience of harassment outlined earlier led her to pathologise hackers, suggesting work must be done exploring the characteristics of hackers she identified—such as lack of fathers or parental figures, severe depression and admittance to mental institutions (Gilboa, 1996:112). Similar interpretations of hackers are offered from within their community: "All the hackers I know in France have (or have had) serious problems with their parents" (Condat, hacker, interview). Our research strongly suggests that psychological interpretations of hackers that individualise hackers as mentally unstable are severely limited because they miss the social basis of hacking. Gilboa's experience is no less unpleasant but all the more understandable when the male dominance of the hacking community is grasped.

The fear many have of the power of computers over their lives easily translates into the demonisation of those who manipulate computers outside of society's legitimate institutions. Journalist Jon Littman once asked hacker Kevin Mitnick if he thought he was being demonised because new and different fears had arisen with society becoming increasingly dependent on computers and communications. Mitnick replied "Yeah . . . That's why they're instilling fear of the unknown. That's why they're scared of me. Not because of what I've done, but because I have the capability to wreak havoc" (Mitnick, cited in Littman, 1996:205). The pathological interpretation of hackers is attractive because it is based on the fear of computers controlling our lives. What else could someone be but mad, if s/he is willing to play for fun on computer systems that control air traffic, dams or emergency phones? The interpretation of hackers as members of an outlaw community that negotiates its collective identity through a range of clearly recognisable resources does not submit to the fear of computers. It gains a clearer view of hackers, who have become the nightmare of information societies despite very few documented cases of upheaval caused by hackers. Hacking cannot be clearly grasped unless fears are put aside to try and understand the community of hackers, the digital underground. From within this community, hackers begin to lose their pathological features in favour of collective principles, allegiances and identities.

NOTES

1. Thanks to Sally Wyatt, Alan White, Ian Taylor and two anonymous referees for comments on this piece.

2. Meyer and Thomas (1989) and Sterling (1992) provide useful outlines of the computer underground, while Rosteck (1994) provides an interesting interpretation of hackers as a social movement. Previous accounts lack detailed survey work.

3. This analysis draws on extensive fieldwork consisting of both a quantitative questionnaire (200 respondents) that outlines the extent and nature of hacking and 80 semi-structured interviews with hackers (30), computer security professionals (30) and other interested parties (20). A full methodology and list of interviewees is available in Taylor (1993). All notes of the following form (Schifreen, hacker, interview indicate that Schifreen was a hacker interviewed for this project.

4. It is of course impossible to provide an adequate history of computer networking here and would distract from the main purpose of present arguments. A summary and full references for such a history can be found in Jordan, (1998a).

5. See Jordan, (1998a) for a full discussion of methodologies for counting Internet users.

6. The concept of a 'hacker' has had several manifestations, with at least four other possibilities than a computer intruder. This paper is concerned solely with hacker in the sense of a computer intruder, though see Taylor, (1993) for further discussion (Levy, 1984; Coupland, 1995). It should also be noted that hacking makes most sense within a society in which knowledge has become extensively commodified and is subject to a process in which it can be extensively copied (Mosco and Wasco, 1988).

7. One indication of these difficulties is that the passage of the Computer Misuse Act 1990 in the UK meant it was difficult to persuade UK hackers to discuss their activities but a lack of comparable legislation in the Netherlands removed one barrier to several Dutch hackers allowing interviews to go ahead. For an extensive discussion of the difficulties and advantages of this research methodology, see Taylor, (1993: chapter 2). For a general discussion of such difficulties see Jupp (1989).

8. Professional security consultants, whose interests are best served by a large underground, have placed the number of hackers as high as 50,000 or 35,000 (Sterling, 1992: 77; Gilboa, 1996: 98).

9. A fourth survey exists, the 1991 UK National Computing Centre Survey, but investigates 'logical breaches' (disruption to computer systems) and only provides tangential evidence of hacking. We became aware of John Howard's work too late for inclusion in this analysis (Howard, 1997).

10. Academic (39.5%), commercial (41%), public service organisations (2.5%), other (14%) and some combination of the above (3%).

11. The following categories from the WarRoom survey were joined to create categories of clearly malicious, neither malicious nor benign, and clearly benign: malicious—manipulated data integrity (6.8), introduced virus (10.6), denied use of service (6.3), compromised trade secrets (9.8), stole/diverted money (0.3), harassed personnel (4.5); neither—installed sniffer (6.6), stole password files (5.6), trojan logons (5.8), IP spoofing (4.8), downloaded data (8.1), compromised email/documents (12.6), other (3.0); and, benign—probing/scanning of system (14.6), publicised intrusion (0.5). It is of course possible to argue that any intrusion is malicious and to dispute the division given above.

12. Much more, of course, could be said about the nature of community and the theories referred to here. To prevent this paper becoming a theoretical exposition of well-known work, the understanding of community will be left here.

13. Hackers do indeed hold conferences, such as HoHoCon, SummerCon, PumpCon and DefCon (Rosteck, 1994). See Littman, (1996: 41–44) for a description of such a conference.

14. Anonymity also enables some of the darker fears that emerge about hackers. Finding fearsomely named gangs of hackers running amok in supposedly secure systems can give rise to exaggerated fears, which hackers are often happy to live up to, at least rhetorically (Barlow, 1990).

15. Our research also leads us to believe that CSI uses teams of hackers to test security far more often than CSI professionals publicly admit.

16. Other CSI professionals offered similar analogies, such as finding someone looking at a car or aeroplane engine.

17. Steven Levy distilled a hacker ethic from the early, non-computer intruder, hackers. This ethic is often invoked by all types of hackers and Levy defines the tenets as: all information should be free; mistrust authority, promote decentralisation; hackers should be judged by their hacking, not by bogus criteria such as degrees, age, race or position; you can create art and beauty on a computer; and, computers can change your life for the better (Levy, 1984: 40–45).

References

Anderson, B., (1991), *Imagined Communities,* second edition, London: Verso.

Audit Commission, (1990), "Survey of Computer Fraud and Abuse," Audit Commission.

Barlow, J. P., (1990), "Crime and Puzzlement," *Whole Earth Review,* Fall 1990, 44–57.

Castells, M., (1996), *The Rise of the Network Society: The information age,* volume 1, Oxford: Blackwell.

Cherny, L. and Weise, E., (eds), (1996), *Wired Women: Gender and new realities in cyberspace,* Seattle: Seal Press.

Clough, B. and Mungo, P., (1992), *Approaching Zero: Data crime and the computer underworld,* London: Faber and Faber.

Coupland, D., (1995), *Microserfs,* London: HarperCollins.

deamon9/route/infinity, (1996), "IP-Spoofing Demystified," Phrack, 7 (48), also available at http://www.gcocitics.com/CapeCanaveral/3498/.

Denning, P., (ed.), (1990), *Computers Under Attack: Intruders, worms and viruses,* New York: Addison-Wesley.

Dery, M., (ed.), (1993), *Flame Wars,* London: Duke University Press.

Diani, M., (1992), "The Concept of a Social Movement," *The Sociological Review, 40* (1): 1–25.

Dreyfus, S., (1997), *Underground: Tales of hacking, madness and obsession on the electronic frontier,* Kew: Mandarin.

Felten, E., Balfanz, D., Dean, D. and Wallack, D., (1996), "Web-Spoofing: An Internet con game," Technical Report 540–96, Department of Computer Science, Princeton University, also at http://www.cs.princeton.edu/sip.

Gilboa, N., (1996), "Elites, Lamers, Narcs and Whores: Exploring the computer underground," in Cherny, L. and Weise, E. (eds), (1996), 98–113.

Goldstein, E., (1993), "Hacker Testimony to House Sub-committee Largely Unheard," *Computer Underground Digest,* 5.43.

Godell, J., (1996), *The Cyberthief and the Samurai: The true story of Kevin Mitnick and the man who hunted him down,* New York: Dell.

Gow, D. and Norton-Taylor, R., (1996), "Surfing Superhighwaymen," *The Guardian* newspaper, 7/12/1996, 28.

Hafner, K. and Lyons, M., (1996), *Where Wizards Stay Up Late: The origins of the Internet,* New York: Simon and Schuster.

Hafner, K. and Markoff, J., (1991), *Cyberpunk: Outlaws and hackers on the computer frontier,* London: Corgi.

Harasim, L., (ed.), (1993), *Global Networks: Computers and international communication,* Cambridge: MIT Press.

Hencke, D., (1998), "Whitehall Attempts to Foil Net Hackers," *Guardian Weekly*, 26 April, 8.

Howard, J., (1997), "Information Security," unpublished PhD dissertation, Carnegie Mellon University, available at http://www.cert.org.

Jordan, T., (1995), "The Unity of Social Movements," *The Sociological Review*, 43 (4): 675–692.

Jordan, T., (1998a), *Cyberpower: A sociology and politics of cyberspace and the Internet*, London: Routledge.

Jordan, T., (1998b), "New Space? New Politics: Cyberpolitics and the Electronic Frontier Foundation," in Jordan, T. and Lent, A. (eds), (1998).

Jordan, T. and Lent, A., (eds), (1998), *Storming the Millennium: The new politics of change*, London: Lawrence and Wishart.

Jupp, C., (1989), *Methods of Criminological Research*, London: Unwin Hyman.

Keller, L., (1988), "Machismo and the Hacker Mentality: Some personal observations and speculations," paper presented to WiC (Women in Computing) Conference.

Levy, S., (1984), *Hackers: Heroes of the computer revolution*, Harmondsworth: Penguin.

Littman, J., (1996), *The Fugitive Game: Online with Kevin Mitnick, the inside story of the great cyberchase*, Boston: Little, Brown and Co.

Ludlow, P., (ed.), (1996), *High Noon on the Electronic Frontier*, Cambridge: MIT Press. NCC, (1991), "Survey of Security Breaches," National Computing Centre.

Meyer, G. and Thomas, J., (1989), "The Baudy World of the Byte: A post-modernist interpretation of the Computer Underground," paper presented at the American Society of Criminology annual meeting, Reno, November 1989.

Miller, S., (1996), "Hacker takes over Labour's cyberspace," *The Guardian* newspaper, 10/12/1996, 1.

Mosco, V. and Wasko, M., (eds), (1988), *The Political Economy of Information*, Wisconsin: University of Wisconsin Press.

Quarterman, J., (1990), *The Matrix: Computer networks and conferencing systems worldwide*, Bedford: Digital Press.

Quarterman, J., (1993), "The Global Matrix of Minds," in Harasim, L. (ed.), 1993, 35–56.

Quittner, J. and Slattalla, M., (1995), *Masters of Deception: The gang that ruled cyberspace*, London: Vintage.

Ross, A., (1991), *Strange Weather*, London: Verso.

Rosteck, T., (1994), Computer Hackers: rebels with a cause," honours thesis. Sociology and Anthropology, Concordia University, Montreal, also at http://www.geocities.com /CapeCanaveral/3498/.

Shimomura, R., (1996), *Takedown: the pursuit and capture of Kevin Mitnick, the world's most notorious cybercriminal—by the man who did it*, with John Markoff, London: Secker and Warburg.

Spafford, E., (1990), "Are Computer Hacker Break-Ins Ethical?," Princeton University Technical Report, CSD-TR-994, Princeton.

Spertus, E., (1991), "Why are there so few female computer scientists?," unpublished paper, MIT.

Sterling, B., (1992), *The Hacker Crackdown: Law and disorder on the electronic frontier*, London: Viking.

Sterling, B., (1994), "The Hacker Crackdown three years later," only published electronically, available at http://www.uel.ac.uk/research/nprg.

Stoll, C., (1989), *The Cuckoo's Egg: Tracking a spy through the maze of counterespionage*, New York: Simon and Schuster.

Taylor, P., (1993), "Hackers: a case-study of the social shaping of computing," unpublished Ph.D. dissertation, University of Edinburgh.

Thomas, J., (1990), "Review of The Cuckoo's Egg," *Computer Underground Digest,* 1.06.
Turkle, S., (1984), *The Second Self: Computers and the human spirit,* London: Granada.
WarRoom, (1996), "1996 Information Systems Security Survey," WarRoom Research, LLC,
    available at http://www.infowar.com/.

QUESTIONS FOR DISCUSSION

1. Define hacking. Discuss how people hack into computers and telephone lines.

2. Discuss three of the six different aspects of the hacking community as presented by the authors.

3. How do the authors discuss pathologisation of hackers? What evidence is presented to reject the pathologisation of hackers?

DRUGS AND DEVIANT BEHAVIOR

Becoming a Marihuana User

Howard S. Becker

The use of marihuana is and has been the focus of a good deal of attention on the part of both scientists and laymen. One of the major problems students of the practice have addressed themselves to has been the identification of those individual psychological traits which differentiate marihuana users from nonusers and which are assumed to account for the use of the drug. That approach, common in the study of behavior categorized as deviant, is based on the premise that the presence of a given kind of behavior in an individual can best be explained as the result of some trait which predisposes or motivates him to engage in the behavior (Marcovitz and Meyers, 1944; Gaskill, 1945; Charen and Perelman, 1946).

This study is likewise concerned with accounting for the presence or absence of marihuana use in an individual's behavior. It starts, however, from a different premise: that the presence of a given kind of behavior is the result of a sequence of social experiences during which the person acquires a conception of the meaning of the behavior, and perceptions and judgments of objects and situations, all of which make the activity possible and desirable. Thus, the motivation or disposition to engage in the activity is built up in the course of learning to engage in it and does not antedate this learning process. For such a view it is not necessary to identify those "traits" which "cause" the behavior. Instead, the problem becomes one of describing the set of changes in the person's conception of the activity and of the experience it provides for him (Mead, 1934:277–80).

This paper seeks to describe the sequence of changes in attitudes and experience which lead[s] to *the use of marihuana for pleasure*. Marihuana does not produce addiction, as do alcohol and the opiate drugs; there is no withdrawal sickness and no ineradicable craving for the drug (cf. Adams, 1942). The most frequent pattern of use might be termed "recreational." The drug is used occasionally for the pleasure the user finds in it, a relatively casual kind of behavior in comparison with that connected with the use of addicting drugs. The term "use for pleasure" is meant to emphasize the noncompulsive and casual character of the behavior. It is also meant to eliminate from consideration here those few cases in which marihuana is used for its prestige value only, as a symbol that one is a certain kind of person, with no pleasure at all being derived from its use.

The analysis presented here is conceived of as demonstrating the greater explanatory usefulness of the kind of theory outlined above as opposed to the predispositional theories now current. This may be seen in two ways: (1) Predispositional theories cannot account for that group of users (whose existence is admitted) (cf. Kolb, 1938; Bromberg, 1939) who do not exhibit the trait or traits considered to cause the behavior and (2) such theories cannot account for the great variability over time of a given individual's behavior with reference to the drug. The same person will at one stage be unable to use the drug for pleasure, at a later stage be able and willing to do so, and, still later, again be unable to use it in this way. These changes, difficult to explain from a predispositional or motivational theory, are readily understandable in terms of changes in the individual's conception of the drug, as is the existence of "normal" users.

The study attempted to arrived at a general statement of the sequence of changes in individual attitude and experience which have always occurred when the individual has become willing and able to use marihuana for pleasures and which have not occurred or not been permanently maintained when this is not the case. This generalization is stated in universal terms in order that negative cases may be discovered and used to revise the explanatory hypothesis (Lindesmith, 1947: chap. 1).

Fifty interviews with marihuana users from a variety of social backgrounds and present positions in society constitute the data from which the generalization was constructed and against which it was tested. The interviews focused on the history of the person's experience with the drug, seeking major changes in his attitude toward it and in his actual use of it and the reasons for these changes. The final generalization is a statement of that sequence of changes in attitude which occurred in every case known to me in which the person came to use marihuana for pleasure. Until a negative case is found, it may be considered as an explanation of all cases of marihuana use for pleasure. In addition, changes from use to nonuse are shown to be related to similar changes in conception, and in each case it is possible to explain variations in the individual's behavior in these terms.

This paper covers only a portion of the natural history of an individual's use of marihuana, starting with the person having arrived at the point of willingness to try marihuana. He knows that others use it to "get high," but he does not know what this means in concrete terms. He is curious about the experience, ignorant of what it may turn out to be, and afraid that it may be more than he has bargained for. The steps outlined below, if he undergoes them all and maintains the attitudes developed in them, leave him willing and able to use the drug for pleasure when the opportunity presents itself.

I

The novice does not ordinarily get high the first time he smokes marihuana, and several attempts are usually necessary to induce this state. One explanation of this may be that the drug is not smoked "properly," that is, in a way that ensures sufficient dosage to produce real symptoms of intoxication. Most users agree that it cannot be smoked like tobacco if one is to get high:

> Take in a lot of air, you know, and . . . I don't know how to describe it, you don't smoke it like a cigarette, you draw in a lot of air and get it down in your system and then keep it there. Keep it there as long as you can.

Without the use of some such technique,[1] the drug will produce no effects, and the user will be unable to get high:

> The trouble with people like that [who are not able to get high] is that they're just not smoking it right, that's all there is to it. Either they're not holding it down long enough, or they're getting too much air and not enough smoke, or the other way around or something like that. A lot of people just don't smoke it right, so naturally nothing's gonna happen.

If nothing happens, it is manifestly impossible for the user to develop a conception of the drug as an object which can be used for pleasure, and use will therefore not continue. The first step in the sequence of events that must occur if the person is to become a user is that he must learn to use the proper smoking technique in order that his use of the drug will produce some effects in terms of which his conception of it can change.

Such a change is, as might be expected, a result of the individual's participation in groups in which marihuana is used. In them the individual learns the proper way to smoke the drug. This may occur through direct teaching:

> I was smoking like I did an ordinary cigarette. He said, "No, don't do it like that." He said, "Suck it, you know, draw in and hold it in your lungs . . . for a period of time."
> I said, "Is there any limit of time to hold it?"
> He said, "No, just till you feel that you want to let it out." So I did that three or four times.

Many new users are ashamed to admit ignorance and, pretending to know already, must learn through the more indirect means of observation and imitation:

> I came on like I had turned on [smoked marihuana] many times before, you know. I didn't want to seem like a punk to this cat. See, like I didn't know the first thing about it—how to smoke it, or what was going to happen, or what. I just watched him like a hawk—I didn't take my eyes off him for a second, because I wanted to do everything just as he did it. I watched how he held it, how he smoked it, and everything. Then when he gave it to me I just came on cool,

Howard S. Becker   303

as though I knew exactly what the score was. I held it like he did and took a
poke just the way he did.

No person continued marihuana use for pleasure without learning a technique
that supplied sufficient dosage for the effects of the drug to appear. Only when this
was learned was it possible for a conception of the drug as an object which could be
used for pleasure to emerge. Without such a conception, marihuana use was considered meaningless and did not continue.

II

Even after he learns the proper smoking technique, the new user may not get high
and thus not form a conception of the drug as something which can be used for pleasure. A remark made by a user suggested the reason for this difficulty in getting high
and pointed to the next necessary step on the road to being a user:

> I was told during an interview, "As a matter of fact, I've seen a guy who was
> high out of his mind and didn't know it."
> I expressed disbelief: "How can that be, man?"
> The interviewee said, "Well, it's pretty strange, I'll grant you that, but I've
> seen it. This guy got on with me, claiming that he'd never got high, one of those
> guys, and he got completely stoned. And he kept insisting that he wasn't high.
> So I had to prove to him that he was."

What does this mean? It suggests that being high consists of two elements: the
presence of symptoms caused by marihuana use and the recognition of these symptoms and their connection by the user with his use of the drug. It is not enough, that
is, that the effects be present; they alone do not automatically provide the experience
of being high. The user must be able to point them out to himself and consciously
connect with his having smoked marihuana before he can have this experience.
Otherwise, regardless of the actual effects produced, he considers that the drug has
had no effect on him: "I figured it either had no effect on me or other people were
exaggerating its effect on them, you know. I thought it was probably psychological,
see." Such persons believe that the whole thing is an illusion and that the wish to be
high leads the user to deceive himself into believing that something is happening
when, in fact, nothing is. They do not continue marihuana use, feeling that "it does
nothing" for them.

Typically, however, the novice has faith (developed from his observation of users
who do get high) that the drug actually will produce some new experience and continues to experiment with it until it does. His failure to get high worries him, and he
is likely to ask more experienced users or provoke comments from them about it. In
such conversations he is made aware of specific details of his experience which he
may not have noticed or may have noticed but failed to identify as symptoms of being high:

> I didn't get high the first time. . . . I don't think I held it in long enough. I proba-
> bly let it out, you know, you're a little afraid. The second time I wasn't sure, and
> he [smoking companion] told me, like I asked him for some of the symptoms or

something, how would I know, you know. . . . So he told me to sit on a stool. I sat on—I think I sat on a bar stool—and he said, "Let your feet hang," and then when I got down my feet were real cold, you know.

And I started feeling it, you know. That was the first time. And then about a week after that, sometime pretty close to it, I really got on. That was the first time I got on a big laughing kick, you know. Then I really knew I was on.

One symptom of being high is an intense hunger. In the next case the novice becomes aware of this and gets high for the first time:

They were laughing the hell out of me because like I was eating so much. I just scoffed [ate] so much food, and they were just laughing at me, you know. Sometimes I'd be looking at them, you know, wondering why they're laughing, you know, not knowing what I was doing. [Well, did they tell you why they were laughing eventually?] Yeah, yeah, I come back, "Hey, man, what's happening?" Like, you know, like I'd ask, "What's happening?" and all of a sudden I feel weird, you know. "Man, you're on, you know. You're on pot [high on marihuana]." I said, "No, am I?" Like I don't know what's happening.

The learning may occur in more indirect ways:

I heard little remarks that were made by other people. Somebody said, "My legs are rubbery," and I can't remember all the remarks that were made because I was very attentively listening for all these cues for what I was supposed to feel like.

The novice, then, eager to have this feeling, picks up from other users some concrete referents of the term "high" and applies these notions to his own experience. The new concepts make it possible for him to locate these symptoms among his own sensations and to point out to himself a "something different" in his experience that he connects with drug use. It is only when he can do this that he is high. In the next case, the contrast between two successive experiences of a user makes clear the crucial importance of the awareness of the symptoms in being high and reemphasizes the important role of interaction with other users in acquiring the concepts that make this awareness possible:

[Did you get high the first time you turned on?] Yeah, sure. Although, come to think of it, I guess I really didn't. I mean, like that first time it was more or less of a mild drunk. I was happy, I guess, you know what I mean. But I didn't really know I was high, you know what I mean. It was only after the second time I got high that I realized I was high the first time. Then I knew that something different was happening.

[How did you know that?] How did I know? If what happened to me that night would of happened to you, you would've known, believe me. We played the first tune for almost two hours—one tune! Imagine, man! We got on the stand and played this one tune, we started at nine o'clock. When we got finished I looked at my watch, it's a quarter to eleven. Almost two hours on one tune. And it didn't seem like anything.

I mean, you know, it does that to you. It's like you have much more time or something. Anyway, when I saw that, man, it was too much. I knew I must really be high or something if anything like that could happen. See, and then they explained to me that that's what it did to you, you had a different sense of time and everything. So I realized that's what it was. I knew then. Like the first time, I probably felt that way, you know, but I didn't know what's happening.

It is only when the novice becomes able to get high in this sense that he will continue to use marihuana for pleasure. In every case in which use continued, the user had acquired the necessary concepts with which to express to himself the fact that he was experiencing new sensations caused by the drug. That is, for use to continue, it is necessary not only to use the drug so as to produce effects but also to learn to perceive these effects when they occur. In this way marihuana acquires meaning for the user as an object which can be used for pleasure.

With increasing experience the user develops a greater appreciation of the drug's effects; he continues to learn to get high. He examines succeeding experiences closely, looking for new effects, making sure the old ones are still there. Out of this there grows a stable set of categories for experiencing the drug's effects whose presence enables the user to get high with ease.

The ability to perceive the drug's effects must be maintained if use is to continue; if it is lost, marihuana use ceases. Two kinds of evidence support this statement. First, people who become heavy users of alcohol, barbiturates, or opiates do not continue to smoke marihuana, largely because they lose the ability to distinguish between its effects and those of the other drugs.[2] They no longer know whether the marihuana gets them high. Second, in those few cases in which an individual uses marihuana in such quantities that he is always high, he is apt to get this same feeling that the drug has no effect on him, since the essential element of a noticeable difference between feeling high and feeling normal is missing. In such a situation, use is likely to be given up completely, but temporarily, in order that the user may once again be able to perceive the difference.

III

One more step is necessary if the user who has now learned to get high is to continue use. He must learn to enjoy the effects he has just learned to experience. Marihuana-produced sensations are not automatically or necessarily pleasurable. The taste for such experience is a socially acquired one, not different in kind from acquiring tastes for oysters or dry martinis. The user feels dizzy; thirsty; his scalp tingles; he misjudges time and distances; and so on. Are these things pleasurable? He isn't sure. If he is to continue marihuana use, he must decide that they are. Otherwise, getting high, while a real enough experience, will be an unpleasant one he would rather avoid.

The effects of the drug, when first perceived, may be physically unpleasant or at least ambiguous:

> It started taking effect, and I didn't know what was happening, you know, what it was, and I was very sick. I walked around the room, walking around the room trying to get off, you know; it just scared me at first, you know. I wasn't used to that kind of feeling.

In addition, the novice's naïve interpretation of what is happening to him may further confuse and frighten him, particularly if he decides, as many do, that he is going insane:

> I felt I was insane, you know. Everything people done to me just wigged me. I couldn't hold a conversation, and my mind would be wandering, and I was

always thinking, oh, I don't know, weird things, like hearing music different. . . . I get the feeling that I can't talk to anyone. I'll goof completely.

Given these typically frightening and unpleasant first experiences, the beginner will not continue use unless he learns to redefine the sensations as pleasurable:

It was offered to me, and I tried it. I'll tell you one thing. I never did enjoy it at all. I mean it was just nothing that I could enjoy. [Well, did you get high when you turned on?] Oh, yeah, I got definite feelings from it. But I didn't enjoy them. I mean I got plenty of reactions, but they were mostly reactions of fear. [You were frightened?] Yes. I didn't enjoy it. I couldn't seem to relax with it, you know. If you can't relax with a thing, you can't enjoy it, I don't think.

In other cases the first experiences were also definitely unpleasant, but the person did become a marihuana user. This occurred, however, only after a later experience enabled him to redefine the sensations as pleasurable:

[This man's first experience was extremely unpleasant, involving distortion of spatial relationships and sounds, violent thirst, and panic produced by these symptoms.] After the first time I didn't turn on for about, I'd say, ten months to a year. . . . It wasn't a moral thing; it was because I'd gotten so frightened, bein' so high. An' I didn't want to go through that again, I mean, my reaction was, "Well, if this is what they call bein' high, I don't dig [like] it." . . . So I didn't turn on for a year almost, accounta that. . . .

Well, my friends started, an' consequently I started again. But I didn't have any more, I didn't have that same initial reaction, after I started turning on again. [In interaction with his friends he became able to find pleasure in the effects of the drug and eventually became a regular user.]

In no case will use continue without such a redefinition of the effects as enjoyable.

This redefinition occurs, typically, in interaction with more experienced users who, in a number of ways, teach the novice to find pleasure in this experience which is at first so frightening (Charen and Perelman, 1946:679). They may reassure him as to the temporary character of the unpleasant sensations and minimize their seriousness, at the same time calling attention to the more enjoyable aspects. An experienced user describes how he handles newcomers to marihuana use:

Well, they get pretty high sometimes. The average person isn't ready for that, and it is a little frightening to them sometimes. I mean, they've been high on lush [alcohol], and they get higher that way than they've ever been before, and they don't know what's happening to them. Because they think they're going to keep going up, up, up till they lose their minds or begin doing weird things or something. You have to like reassure them, explain to them that they're not really flipping or anything, that they're gonna be all right. You have to just talk them out of being afraid. Keep talking to them, reassuring, telling them it's all right. And come on with your own story, you know: "The same thing happened to me. You'll get to like that after awhile." Keep coming on like that; pretty soon you talk them out of being scared. And besides they see you doing it and nothing horrible is happening to you, so that gives them more confidence.

The more experienced user may also teach the novice to regulate the amount he smokes more carefully, so as to avoid any severely uncomfortable symptoms while retaining the pleasant ones. Finally, he teaches the new user that he can "get to like it after awhile." He teaches them to regard those ambiguous experiences formerly defined as unpleasant as enjoyable. The older user in the following incident is a person whose tastes have shifted in this way, and his remarks have the effect of helping others to make a similar redefinition:

> A new user had her first experience of the effects of marihuana and became frightened and hysterical. She "felt like she was half in and half out of the room" and experienced a number of alarming physical symptoms. One of the more experienced users present said, "She's dragged because she's high like that. I'd give anything to get that high myself. I haven't been that high in years."

In short, what was once frightening and distasteful becomes, after a taste for it is built up, pleasant, desired, and sought after. Enjoyment is introduced by the favorable definition of the experience that one acquires from others. Without this, use will not continue, for marihuana will not be for the user an object he can use for pleasure.

In addition to being a necessary step in becoming a user, this represents an important condition for continued use. It is quite common for experienced users suddenly to have an unpleasant or frightening experience, which they cannot define as pleasurable, either because they have used a larger amount of marihuana than usual or because it turns out to be a higher-quality marihuana than they expected. The user has sensations which go beyond any conception he has of what being high is and is in much the same situation as the novice, uncomfortable and frightened. He may blame it on an overdose and simply be more careful in the future. But he may make this the occasion for a rethinking of his attitude toward the drug and decide that it no longer can give him pleasure. When this occurs and is not followed by a redefinition of the drug as capable of producing pleasure, use will cease.

The likelihood of such a redefinition occurring depends on the degree of the individual's participation with other users. Where this participation is intensive, the individual is quickly talked out of his feeling against marihuana use. In the next case, on the other hand, the experience was very disturbing, and the aftermath of the incident cut the person's participation with other users to almost zero. Use stopped for three years and began again only when a combination of circumstances, important among which was a resumption of ties with users, made possible a redefinition of the nature of the drug:

> It was too much, like I only made about four pokes, and I couldn't even get it out of my mouth, I was so high, and I got real flipped. In the basement, you know, I just couldn't stay in there anymore. My heart was pounding real hard, you know, and I was going out of my mind; I thought I was losing my mind completely. So I cut out of this basement, and this other guy, he's out of his mind, told me, "Don't, don't leave me, man. Stay here." And I couldn't.
>
> I walked outside, and it was five below zero, and I thought I was dying, and I had my coat open; I was sweating, I was perspiring. My whole insides were all . . . , and I walked about two blocks away, and I fainted behind a bush. I don't know how long I laid there. I woke up, and I was feeling the worst, I can't describe it at all, so I made it to a bowling alley, man, and I was trying to act normal, I was trying to shoot pool, you know, trying to act real normal, and I couldn't lay and I couldn't stand up and I couldn't sit down, and I went up and

laid down where some guys that spot pins lay down, and that didn't help me, and I went down to a doctor's office. I was going to go in there and tell the doctor to put me out of my misery . . . because my heart was pounding so hard, you know. . . . So then all weekend I started flipping, seeing things there and going through hell, you know, all kinds of abnormal things. . . . I just quit for a long time then.

[He went to a doctor who defined the symptoms for him as those of a nervous breakdown caused by "nerves" and "worries." Although he was no longer using marihuana, he had some recurrences of the symptoms which led him to suspect that "it was all his nerves."] So I just stopped worrying, you know; so it was about thirty-six months later I started making it again. I'd just take a few pokes, you know. [He first resumed use in the company of the same user-friend with whom he had been involved in the original incident.]

A person, then, cannot begin to use marihuana for pleasure, or continue its use for pleasure, unless he learns to define its effects as enjoyable, unless it becomes and remains an object which he conceives of as capable of producing pleasure.

IV

In summary, an individual will be able to use marihuana for pleasure only when he goes through a process of learning to conceive of it as an object which can be used in this way. No one becomes a user without (1) learning to smoke the drug in a way which will produce real effects; (2) learning to recognize the effects and connect them with drug use (learning, in other words, to get high); and (3) learning to enjoy the sensations he perceives. In the course of this process he develops a disposition or motivation to use marihuana which was not and could not have been present when he began use, for it involves and depends on conceptions of the drug which could only grow out of the kind of actual experience detailed above. On completion of this process he is willing and able to use marihuana for pleasure.

He has learned, in short, to answer "Yes" to the question: "Is it fun?" The direction his further use of the drug takes depends on his being able to continue to answer "Yes" to this question and, in addition, on his being able to answer "Yes" to other questions which arise as he becomes aware of the implications of the fact that the society as a whole disapproves of the practice: "Is it expedient?" "Is it moral?"[3] Once he has acquired the ability to get enjoyment out of the drug, use will continue to be possible for him. Considerations of morality and expediency, occasioned by the reactions of society, may interfere and inhibit use, but use continues to be a possibility in terms of his conception of the drug. The act becomes impossible only when the ability to enjoy the experience of being high is lost, through a change in the user's conception of the drug occasioned by certain kinds of experience with it.

In comparing this theory with those which ascribe marihuana use to motives or predispositions rooted deep in individual behavior, the evidence makes it clear that marihuana use for pleasure can occur only when the process described above is undergone and cannot occur without it. This is apparently so without reference to the nature of the individual's personal makeup or psychic problems. Such theories assume that people have stable modes of response which predetermine the way they will act in relation to any particular situation or object and that, when they come in contact with the given object or situation, they act in the way in which their makeup predisposes them.

This analysis of the genesis of marihuana use shows that the individuals who come in contact with a given object may respond to it at first in a great variety of ways. If a stable form of new behavior toward the object is to emerge, a transformation of meanings must occur, in which the person develops a new conception of the nature of the object (cf. Strauss, 1952). This happens in a series of communicative acts in which others point out new aspects of his experience to him, present him with new interpretations of events, and help him achieve a new conceptual organization of his world, without which the new behavior is not possible. Persons who do not achieve the proper kind of conceptualization are unable to engage in the given behavior and turn off in the direction of some other relationship to the object or activity.

This suggests that behavior of any kind might fruitfully be studied developmentally, in terms of changes in meanings and concepts, their organization and reorganization, and the way they channel behavior, making some acts possible while excluding others.

NOTES

1. A pharmacologist notes that this ritual is in fact an extremely efficient way of getting the drug into the bloodstream (Walton, 1938:48).

2. "Smokers have repeatedly stated that the consumption of whiskey while smoking negates the potency of the drug. They find it very difficult to get 'high' while drinking whiskey and because of that smokers will not drink while using the 'weed'" (cf. New York City Mayor's Committee on Marihuana, 1944:13).

3. Another paper will discuss the series of developments in attitude that occurs as the individual begins to take account of these matters and adjust his use to them.

REFERENCES

Adams, Roger. "Marihuana." *Bulletin of the New York Academy of Medicine* 18 (November 1942):705–30.

Bromberg, Walter. "Marihuana: A Psychiatric Study." *Journal of the American Medical Association* 113 (July 1939):11.

Charen, Sol, and Perelman, Luis. "Personality Studies of Marihuana Addicts." *American Journal of Psychiatry* 102 (March 1946):674–82.

Gaskill, Herbert S. "Marihuana, an Intoxicant." *American Journal of Psychiatry* 102 (September 1945):202–4.

Kolb, Lawrence. "Marihuana." *Federal Probation* 11 (July 1938):22–25.

Lindesmith, Alfred R. *Opiate Addiction.* Bloomington: Principle Press, 1947.

Marcovitz, Eli, and Meyers, Henry J. "The Marihuana Addict in the Army." *War Medicine* 4 (December 1944):382–91.

Mead, George Herbert. *Mind, Self and Society:* Chicago: University of Chicago Press, 1934.

New York City Mayor's Committee on Marihuana. *The Marihuana Problem in the City of New York.* Lancaster, Pennsylvania: Jacques Cattrell Press, 1944.

Strauss, Anselm. "The Development and Transformation of Monetary Meanings in the Child." *American Sociological Review* 17 (June 1952):275–86.

Walton, R. P. *Marihuana: America's New Drug Problem*. Philadelphia: J. B. Lippincott, 1938.

QUESTIONS FOR DISCUSSION

1. The author describes the sequence of changes in attitudes and experience that leads to the use of marijuana for pleasure. What is the sequence? Give examples from the reading. Can this sequence explain other deviant acts? Explain your response.

2. According to the author, until a negative case is found people only use marijuana for pleasure. Do you agree or disagree with the statement? Explain your response.

3. The author suggests that being high consists of two elements. What two elements is the author referring to and can you have one element without the other when getting high? Explain your responses.

Drifting into Cocaine Dealing

Sheigla Murphy, Dan Waldorf, and Craig Reinarman

No American who watched television news in the 1980s could have avoided images of violent drug dealers who brandished bullets while driving BMW's before being hauled off in handcuffs. This new stereotype of a drug dealer has become a staple of popular culture, the very embodiment of evil. He works for the still more vile villains of the "[Columbian] cartel," who make billions on the suffering of millions. Such men are portrayed as driven by greed and utterly indifferent to the pain from which they profit.

We have no doubt that some such characters exist. Nor do we doubt that there may be a new viciousness among some of the crack cocaine dealers who have emerged in ghettos and barrios already savaged by rising social problems and falling social programs: We have grave doubts, however, that such characterizations tell us anything about cocaine sellers more generally. If our interviews are any guide, beneath every big-time dealer who may approximate the stereotype there are hundreds of smaller sellers who do not.

This paper describes such sellers, not so much as a way of debunking a new devil but rather as a way of illuminating how deviant careers develop and how the identities of the individuals who move into this work are transformed. Along with the many routine normative strictures against drug use in our culture, there has been a mobilization in recent years for a "war on drugs" which targets cocaine dealers in particular. Many armaments in the arsenal of social control from propaganda to prisons have been employed in efforts to dissuade people from using/selling such substances. In such a context it is curious that ostensibly ordinary people not only

"Drifting into Cocaine Dealing" by Sheigla Murphy, Dan Waldorf, and Craig Reinarman from *Qualitative Sociology*, vol. 13, no. 4 (1990), pp. 487–507. Plenum Publishing Co. Reprinted by permission of Kluwer Academic/Plenum Publishing Co.

continue to use illicit drugs but also take the significant additional step of becoming drug sellers. To explore how this happens, we offer an analysis of eighty depth interviews with former cocaine sellers. We sought to learn something about how it is that otherwise conventional people—some legally employed, many well educated—end up engaging in a sustained pattern of behavior that their neighbors might think of as very deviant indeed.

DEVIANT CAREERS AND DRIFT

Our reading of this data was informed by two classic theoretical works in the deviance literature. First, in *Outsiders,* Howard Becker observed that, "The career lines characteristic of an occupation take their shape from the problems peculiar to that occupation. These, in turn, are a function of the occupation's position vis-à-vis other groups in society" (1963:102). He illustrated the point with the dance musician, caught between the jazz artist's desire to maintain creative control and a structure of opportunities for earning a living that demanded the subordination of this desire to mainstream musical tastes. Musicians' careers were largely a function of how they managed this problem. When the need to make a living predominated, the basis of their self conceptions shifted from art to craft.

Of course, Becker applied the same proposition to more deviant occupations. In the next section, we describe five discrete modes of becoming a cocaine seller which center on "the problems peculiar to" the world of illicit drug use and which entail a similar shift in self conception. For example, when a drug such as cocaine is criminalized, its cost is often greatly increased while its availability and quality are somewhat limited. Users are thus faced with the problems of avoiding detection, reducing costs, and improving availability and quality. By becoming involved in sales, users solve many of these problems and may also find that they can make some money in the bargain. As we will show, the type of entree and the level at which it occurs are functions of the individual's relationship to networks of other users and suppliers. At the point where one has moved from being a person who *has* a good connection for cocaine to a person who *is* a good connection for cocaine, a subtle shift in self conception and identity occurs.

Becker's model of deviant careers entails four basic steps, three of which our cocaine sellers took. First, the deviant must somehow avoid the impact of conventional commitments that keep most people away from intentional nonconformity. Our cocaine sellers passed this stage by ingesting illegal substances with enough regularity that the practice became normalized in their social world. Second, deviant motives and interests must develop. These are usually learned in the process of the deviant activity and from interaction with other deviants. Here too our cocaine sellers had learned the pleasures of cocaine by using it, and typically were moved toward involvement in distribution to solve one or more problems entailed in such use. Once involved, they discovered additional motivations which we will describe in detail below.

Becker's third step in the development of deviant careers entails public labeling. The person is caught, the rule is enforced, and his or her public identity is transformed. The new master status of "deviant," Becker argues, can be self fulfilling when it shapes others' perceptions of the person and limits his or her possibilities for resuming conventional roles and activities. Few of our respondents had been publicly labeled deviant, but they did describe a gradual change in identity that may be likened to self-labeling. This typically occurred when they deepened their deviance by dealing on top of using cocaine. This shift in self conception for our subjects was more closely linked to Becker's fourth step—movement into an organized deviant group in which people with a common fate and similar problems form subcultures. There they learn more about solving problems and ideologies which provide rationales for continuing the behavior, thus further weakening the hold of conventional norms and institutions and solidifying deviant identities. In the case of our subjects, becoming sellers further immersed them into deviant groups and practices to the point where many came to face the problems of, and to see themselves as, "dealers."

The fact that these processes of deeper immersion into deviant worlds and shifts in self conception were typically gradual and subtle brought us to a second set of theoretical reference points in the work of David Matza (1964; 1969).[1] In his research on delinquency, Matza discovered that most so-called delinquents were not self-consciously committed to deviant values or lifestyles, but on the contrary continued to hold conventional beliefs. Most of the time they were law abiding, but because the situation of "youth" left them free from various restraints, they often *drifted* in and out of deviance. Matza found that even when caught being delinquent, young people tended to justify or rationalize their acts through "techniques of neutralization" (Sykes and Matza, 1957) rooted in conventional codes of morality. Although we focus on *entering* selling careers, we found that Matza's concept of drift (1964) provided us with a useful sensibility for making sense of our respondents' accounts. The modes of entree they described were as fluid and noncommittal as the drift into and out of delinquency that he described.

None of the career paths recounted by our subjects bear much resemblance to stereotypes of "drug dealers."[2] For decades the predominant image of the illicit drug dealer was an older male reprobate sporting a long, shabby overcoat within which he had secreted a cornucopia of dangerous consciousness-altering substances. This proverbial "pusher" worked school yards, targeting innocent children who would soon be chemically enslaved repeat customers. The newer villains have been depicted as equally vile but more violent. Old or new, the ideal-typical "drug dealer" is motivated by perverse greed and/or his own addiction, and has crossed a clearly marked moral boundary, severing most ties to the conventional world.

The cocaine sellers we interviewed, on the other hand, had more varied and complex motives for selling cocaine. Moreover, at least within their subcultures, the moral boundaries were both rather blurry and as often wandered along as actually crossed. Their life histories reminded us of Matza's later but related discussion of the *overlap* between deviance and conventionality:

> Overlap refers to . . . the marginal rather than gross differentiation between deviant and conventional folk and the considerable though variable interpenetration of deviant and conventional culture. Both themes sensitize us to the regular exchange, traffic, and flow—of persons as well as styles and precepts—that occur among deviant and conventional worlds. (1969:68)

Our subjects were already seasoned users of illicit drugs. For years their drug use coexisted comfortably with their conventional roles and activities; having a deviant dimension to their identities appeared to cause them little strain. In fact, because their use of illicit drugs had gone on for so long, was so common in their social worlds, and had not significantly affected their otherwise normal lives, they hardly considered it deviant at all.

Thus, when they began to sell cocaine as well as use it, they did not consider it a major leap down an unknown road but rather a series of short steps down a familiar path. It was not as if ministers had become mobsters; no sharp break in values, motives, world views, or identities was required. Indeed, few woke up one morning and made a conscious decision to become sellers. They did not break sharply with the conventional world and actively choose a deviant career path; most simply drifted into dealing by virtue of their strategies for solving the problems entailed in using a criminalized substance, and only then developed additional deviant motives centering on money.

To judge from our respondents, then, dealers are not from a different gene pool. Since the substances they enjoy are illegal, most regular users of such drugs become involved in some aspect of distribution. There is also a growing body of research on cocaine selling and distribution that has replaced the simplistic stereotype of the pusher with complex empirical evidence about underground economies and deviant careers (e.g., Langer, 1977; Waldorf et al., 1977, 1991; Adler, 1985; Plasket and Quillen, 1985; Morales, 1986a, 1986b; Sanchez and Johnson, 1987; Sanabria, 1988; and Williams, 1989). Several features of underground economies or black markets in drugs contribute to widespread user participation in distribution. For example, some users who could obtain cocaine had other user-friends who wanted it. Moreover, the idea of keeping such traffic among friends offered both sociability and safety. For others, cocaine's high cost inspired many users to become involved in purchasing larger amounts to take advantage of volume discounts. They then sold part of their supply to friends in order to reduce the cost of personal use. The limited supply of cocaine in the late seventies and early eighties made for a sellers' market, providing possibilities for profits along with steady supplies. For most of our subjects, it was not so much that they learned they could make money and thus decided to become dealers but rather, being involved in distribution anyway, they learned they could make money from it. As Becker's model suggests, deviant motives are learned in the course of deviant activities; motivation follows behavior, not the other way around.

After summarizing our sampling and interviewing procedures, we describe in more detail (1) the various modes and levels of entree into cocaine sales; (2) some of the practices, rights and responsibilities entailed in dealing; and (3) the subtle transformation of identity that occurred when people who consider themselves rather conventional moved into careers considered rather deviant.

SAMPLE AND METHODS

The sample consists of 80 ex-sellers who sold cocaine in the San Francisco Bay Area. We interviewed them in 1987 and 1988. Most had stopped selling before crack sales peaked in this area. Only five of the eighty had sold crack or rock. Of these five, two had sold on the street and two had sold in "rock party houses"[3] as early as 1978. It is important to note, therefore, that the sellers we describe are very likely to be different from street crack dealers in terms of the product type, selling styles, visibility, and thus the risks of arrest and attendant violence.

The modes and levels of entree we describe should not be considered exhaustive. They are likely to vary by region, subculture, and level of dealing. For example, our sample and focus differed from those of Adler (1985), who studied one community of *professional* cocaine dealers at the *highest levels* of the distribution system. Her ethnographic account is rich in insights about the lifestyles and career contingencies of such high-level dealers and smugglers. These subjects decided to enter into importing and/or dealing and to move up the ranks in this deviant occupation in order to obtain wealth and to live the sorts of lives that such wealth made possible. Adler's dealers were torn, however, between the lures of fast money and the good life and the stress and paranoia inherent in the scene. Thus, she reported "oscillations" wherein her dealers moved in and out of the business, usually to be lured back in by the possibility of high profits. Our dealers tended to have different motivations, career trajectories, and occupational exigencies. Most were lower in the hierarchy and non-professional (some maintained "straight" jobs); few set out to achieve success in an explicitly deviant career, to amass wealth, or to live as "high rollers." Moreover, our study was cross-sectional rather than longitudinal, so our focus was on how a wide variety of cocaine sellers entered careers rather than on the full career trajectories of a network of smugglers and sellers.

To be eligible for the study our respondents had to have sold cocaine steadily for at least a year and to have stopped selling for at least 6 months. We designed the study to include only *former* sellers so that respondents would feel free to describe all their activities in detail without fear that their accounts could somehow be utilized by law enforcement authorities.

They spoke of six different levels or types of sellers: smugglers, big dealers, dealers, sellers (unspecified), bar dealers, and street dealers. The social organization of cocaine sales probably varies in other areas. We located and interviewed ex-sellers from the full range of these dealer-identified sales levels, but we have added two categories in order to provide a more detailed typology. Our eight levels of sales were defined according to the units sold rather than the units bought. So, for example, if a seller bought quarters or eighths of ounces and regularly sold grams, we categorized him or her as a gram dealer rather than a part-ounce dealer.

Levels of Sales	Number of Interviews
Smugglers	2
Kilograms/pounds	13
Parts of kilos and pounds	6
Ounce dealers	18
Part-ounce dealers	13
Gram dealers	12
Part-gram dealers	11
Crack dealers	5
Total	80

Unlike most other studies of dealing and the now infamous street crack dealers, the majority of our respondents sold cocaine hydrochloride (powder) in private places. There are a number of styles of selling drugs—selling out of homes, selling out of rock houses and shooting galleries, selling out of party houses, selling out of rented "safe houses" and apartments, delivery services (using telephone answering, answering machines, voice mail and telephone beepers), car meets,[4] selling in bars, selling in parks, and selling in the street. Within each type there are various styles. For example, in some African-American communities in San Francisco a number of sellers set up business on a street and respond to customers who come by on foot and in automobiles. Very often a number of sellers will approach a car that slows down or stops to solicit customers; drugs and money are exchanged then and there. Such sales activities are obvious to the most casual observers; even television camera crews often capture such transactions for the nightly news. On certain streets in the Mission District, a Latino community in San Francisco, street drug sales are less blatant. Buyers usually walk up to sellers who stand on the street among numerous other people who are neither buyers nor sellers. There, specific transactions rarely take place on the street itself; the participants generally retreat to a variety of shops and restaurants. Buyers seldom use cars for transactions and sellers tend not to approach a car to solicit customers.

Despite the ubiquity of street sales in media accounts and the preponderance of street sellers in arrest records, we set out to sample the more hidden and more numerous sellers who operate in private. Most users of cocaine hydrochloride are working- or middle-class. They generally avoid street sellers both because they want to avoid being observed and because they believe that most street sellers sell inferior quality drugs (Waldorf et al., 1991). Further, we found that people engaged in such illegal and furtive transactions tend to prefer dealing with people like themselves, people they know.

We located our respondents by means of chain referral sampling techniques (Biernacki and Waldorf, 1981; Watters and Biernacki, 1989). This is a method commonly used by sociologists and ethnographers to locate hard-to-find groups and has been used extensively in qualitative research on drug use (Lindesmith, 1947; Becker, 1953; Feldman, 1968; Preble and Casey, 1969; Rosenbaum, 1981; Biernacki, 1986). We initiated the first of our location chains in 1974–1975 in the course of a short-term ethnography of cocaine use and sales among a small friendship network (Waldorf et al., 1977). Other chains were developed during a second study of cocaine cessation conducted during 1986–1987 (Reinarman et al., 1988; Macdonald et al., 1988; Murphy et al., 1989; Waldorf et al., 1991). Another three chains were developed during the present study. We located the majority of our respondents via referral chains developed by former sellers among their previous customers and suppliers. Initial interviewees referred us to other potential respondents whom we had not previously known. In this way we were able to direct our chains into groups of ex-sellers from a variety of backgrounds.

We employed two interview instruments: an open-ended, exploratory interview guide designed to maximize discovery of new and unique types of data, and a more structured survey designed to gather basic quantifiable data on all respondents. The open-ended interviews were tape-recorded, transcribed, and content-analyzed. These interviews usually took from 2 to 4 hours to complete, but when necessary we conducted longer and/or follow-up interviews (e.g., one woman was interviewed for 10 hours over three sessions). The data analyzed for this paper was drawn primarily from the tape-recorded depth interviews.

There is no way to ascertain if this (or any similar) sample is representative of all cocaine sellers. Because the parameters of the population are unknowable, random samples on which systematic generalizations might be based cannot be drawn. We do know that, unlike other studies of drug sellers, we placed less emphasis on street sellers and included dealers at all levels. We also attempted to get a better gender and ethnic mix than studies based on captive samples from jails or treatment programs. Roughly one in three (32.5%) of our dealers are female and two of five (41.2%) are persons of color.

Our respondents ranged in age from 18 to 60, with a mean age of 37.1 years. Their education level was generally high, presumably an indication of the relatively large numbers of middle-class people in the sample (see Table 1).

DEALERS

Dealers are people who are "fronted" (given drugs on consignment to be paid for upon sale) and/or who buy quantities of drugs for sale. Further, in order to be considered a dealer by users or other sellers a person must: (1) have one or more reliable connections (suppliers); (2) make regular cocaine purchases in amounts greater than a single gram (usually an eighth of an ounce or greater) to be sold in smaller units; (3) maintain some consistent supplies for sale; and (4) have a network of customers who make purchases on a regular basis. Although the stereotype of a dealer holds that illicit drug sales are a full-time occupation, many dealers, including members of our sample, operate part-time and supplement income from a legal job.

As we noted in the introduction, the rather average, ordinary character of the respondents who fit this definition was striking. In general, without prior knowledge or direct observation of drug sales, one would be unable to distinguish our respondents from other, non-dealer citizens. When telling their career histories, many of our respondents invoked very conventional, middle-class American values to explain their involvement in dealing (e.g., having children to support, mortgages or rent to pay in a high-cost urban area, difficulty finding jobs which paid enough to support a family). Similarly, their profits from drug sales were used in "normal" ways—to buy children's clothes, to make house or car payments, to remodel a room. Moreover, like Matza's delinquents, most of our respondents were quite law-abiding, with the obvious exception of their use and sales of an illicit substance.

When they were not dealing, our respondents engaged in activities that can only be described as mainstream American. For example, one of our dealers, a single mother of two, found herself with a number of friends who used cocaine and a good connection. She needed extra income to pay her mortgage and to support her children, so she sold small amounts of cocaine within her friendship network. Yet while she sold cocaine, she worked at a full-time job, led a Girl Scout troop, volunteered as a teacher of cardiopulmonary resuscitation (CPR) classes for young people, and went to Jazzercise classes. Although she may have been a bit more civic-minded than

Demographics (N = 80)   **TABLE 1**

Age: Range = 18–60
       Mean 37.1
       Median = 35.4

	Number	Percent
Sex:		
Male	54	67.5
Female	26	32.5
Ethnicity:		
African-American	28	35.0
White	44	58.8
Latino(a)	4	5.0
Asian	1	1.2
Education:		
Less than high school grad	11	13.8
High school graduate	18	22.5
Some college	31	38.8
B.A. or B.S. degree	12	15.0
Some graduate	3	3.8
Graduate degree	5	6.3

Percentages may not equal 100% due to rounding.

many others, her case served to remind us that cocaine sellers do not come from another planet.

MODES OF ENTREE INTO DEALING

Once they began selling cocaine, many of our respondents moved back and forth between levels in the distribution hierarchy. Some people dealt for short periods of time and then quit, only to return several months later at another level of sales.[5] The same person may act as a broker on one deal, sell a quarter gram at a profit to a friend on another, and then pick up an ounce from an associate and pass it on to another dealer in return for some marijuana in a third transaction. In a few instances each of these roles [was] played by the same person within the same twenty-four hour period.

But whether or not a dealer/respondent moved back and forth in this way, s/he usually began selling in one of five distinct ways. All five of these modes of entree presuppose an existing demand for cocaine from people known to the potential dealers. A person selling any line of products needs two things, a group of customers and a product these customers are interested in purchasing. Cocaine sellers are no different. In addition to being able and willing to pay, however, cocaine customers must also be trustworthy because these transactions are illegal.

The first mode of entree, *the go-between,* is fairly straightforward. The potential seller has a good cocaine connection and a group of friends who place orders for cocaine with him/her. If the go-between's friends use cocaine regularly enough and do not develop their own connections, then a period of months or even years might go by when the go-between begins to spend more and more time and energy purchasing for them. Such sellers generally do not make formal decisions to begin dealing; rather, opportunities regularly present themselves and go-betweens gradually take advantage of them. For example, one 30-year-old African-American who became a gram dealer offered this simple account of his passage from go-between to seller:

> Basically, I first started because friends pressured me to get the good coke I could get. I wasn't even making any money off of it. They'd come to me and I'd call up my friend who had gotten pretty big selling a lot of coke. (Case # E-5)

This went on for six months before he began to charge his friends money for it. Then his connection started fronting him eighths of ounces at a time, and he gradually became an official dealer, regularly selling drugs for a profit. Others who began in this way often took only commissions-in-kind (a free snort) for some months before beginning to charge customers a cash markup.

Another African-American male began selling powdered cocaine to snorters in 1978 and by the mid-eighties had begun selling rock cocaine (crack) to smokers. He described his move from go-between to dealer as follows:

> Around the time I started indulging [in cocaine] myself, people would come up and say, "God, do you know where I can get some myself?" I would just say, "Sure, just give me your money," I would come back and either indulge with them or just give it to them depending on my mood. I think that's how I originally set up my clientele. I just had a certain group of people who would come to me because they felt that I knew the type of people who could get them a real quality product.
>
> And pretty soon I just got tired of, you know, being taken out of situations or being imposed upon. . . . I said that it would be a lot easier to just do it myself. And one time in particular, and I didn't consider myself a dealer or anything, but I had a situation one night where 5 different people called me to try to get cocaine . . . not from me but it was like, "Do you know where I can get some good cocaine from?" (Case # E-11)

Not all go-betweens-cum-dealers start out so altruistically. Some astute businessmen and women spot the profit potential early on and immediately realize a profit, either in-kind (a share of the drugs purchased) or by tacking on a surcharge to the purchase price. The following respondent, a 39-year-old African-American male, described this more profit-motivated move from go-between to formal seller:

> Well, the first time that I started it was like I knew where to get good stuff . . . and I had friends that didn't know where to get good stuff. And I knew where to get them really good stuff and so I would always put a couple of dollars on it, you know, if got it for $20 I would sell it to, them for $25 or $30 or whatever.
>
> It got to be where more and more people were coming to me and I was going to my man more and I would be there 5 or 6 times a day, you know. So he would tell me, "Here, why don't you take this, you know, and bring me x-amount of dollars for it." So that's how it really started. I got fronted and I was doing all the business instead of going to his house all the time, because he had other people that were coming to his house and he didn't want the traffic. (Case # E-13)

The second mode of entree is the *stash dealer,* or a person who becomes involved in distribution and/or sales simply to support or subsidize personal use. The name is taken from the term "stash," meaning a personal supply of marijuana (see Fields, 1985, on stash dealers in the marijuana trade). This 41-year-old white woman who sold along with her husband described her start as a stash dealer this way:

Q: So what was your motivation for the sales?
A: To help pay for my use, because the stuff wasn't cheap and I had the means and the money at the time in order to purchase it, where our friends didn't have that amount of money without having to sell something. . . . Yeah, friendship, it wasn't anything to make money off of, I mean we made a few dollars. . . . (Case # E-7)

The respondents who entered the dealing world as stash dealers typically started out small (selling quarter and half grams) and taking their profits in product. However, this motivation contributed to the undoing of some stash dealers in that it led to greater use, which led to the need for greater selling, and so on. Unless they then developed a high-volume business that allowed them to escalate their cocaine use and still make profits, the reinforcing nature of cocaine tempted many of them to use more product than was good for business.

Many stash dealers were forced out of business fairly early on in their careers because they spent so much money on their own use they were financially unable to "re-cop" (buy new supplies). Stash dealers often want to keep only a small number of customers in order to minimize both the "hassle" of late-night phone calls and the risk of police detection, and they do not need many customers since they only want to sell enough to earn free cocaine. Problems arise, however, when their small group of customers do not buy the product promptly. The longer stash dealers had cocaine in their possession, the more opportunities they had for their own use (i.e., for profits to "go up your nose"). One stash dealer had an axiom about avoiding this: "It ain't good to get high on your own supply" (Case # E-57). The predicament of using rather than selling their product often afflicts high-level "weight dealers" as well, but they are better able to manage for longer periods of time due to larger volumes and profit margins.

The third mode of entry into cocaine selling had to do with users' desire for high-quality, unadulterated cocaine. We call this type the *connoisseur.* Ironically, the motivation for moving toward dealing in this way is often health-related. People who described this mode of entree described their concerns, as users, about the possible dangers of ingesting the various adulterants or "cuts" commonly used by dealers to increase profits. User folklore holds that the larger the quantity purchased, the purer the product. This has been substantiated by laboratory analysis of the quality of small amounts of street drugs (typically lower) as opposed to larger police seizures (typically higher).

The connoisseur type of entry, then, begins with the purchase of larger quantities of cocaine than they intend to use in order to maximize purity. Then they give portions of the cocaine to close friends at a good price. If the members of the network start to use more cocaine, the connoisseurs begin to make bigger purchases with greater regularity. At some point they begin to feel that all this takes effort and that it makes sense to buy large quantities not only to get purer cocaine but to make some money for their efforts. The following 51-year-old, white business executive illustrated the connoisseur route as follows:

> I think the first reason I started to sell was not to make money or even to pay for my coke, because I could afford it. It was to get good coke and not to be snorting a lot of impurities and junk that people were putting into it by cutting it so much. So I really think that I started to sell it or to get it wholesale so that I would get the good stuff. And I guess my first, . . . what I did with it in the beginning, because I couldn't use all that I had to buy to get good stuff, I sold it to some of my friends for them to sell it, retail it. (Case # E-16)

Connoisseurs, who begin by selling unneeded quantities, often found they unlearned certain attitudes when they moved from being volume buyers looking for quality toward becoming dealers looking for profit. It was often a subtle shift, but once their primary motivation gradually changed from buying-for-purity to buying-to-sell they found themselves beginning to think and act like dealers. The shift usually occurred when connoisseurs realized that the friends with whom they had shared were in fact customers who were eager for their high-quality cocaine and who often made demands on their time (e.g., friends seeking supplies not merely for themselves, but for other friends a step or two removed from the original connoisseur). Some connoisseurs also became aware of the amount of money that could be made by becoming businesslike about what had been formerly friendly favors. At such points in the process they began to buy-to-sell, for a profit, as well as for the purpose of obtaining high-quality cocaine for personal use. This often meant that, rather than buying sporadically, they had to make more regular buys; for a successful businessperson must have supplies when customers want to buy or they will seek another supplier.

The fourth mode of entree into cocaine selling is an *apprenticeship*. Like the other types, apprentices typically were users who already had loosened conventional normative strictures and learned deviant motives by interacting with other users and with dealers; and they, too, drifted into dealing. However, in contrast to the first three types, apprentices moved toward dealing less to solve problems inherent in using a criminalized substance than to solve the problems of the master dealer. Apprenticeships begin in a personal relationship where, for example, the potential seller is the lover or intimate of a dealer. This mode was most often the route of entry for women, although one young man we interviewed learned to deal from his father. Couples often start out with the man doing the dealing—picking up the product, handling the money, weighing and packaging, etc. The woman gradually finds herself acting as an unofficial assistant—taking telephone messages, sometimes giving people prepackaged cocaine and collecting money. Apprentices frequently benefit from being involved with the experienced dealer in that they enjoy both supplies of high-quality cocaine and indirect financial rewards of dealing.

Some of our apprentices moved into official roles or deepened their involvement when the experienced dealer began to use too much cocaine to function effectively as a seller. In some such cases the abuse of the product led to an end of the relationship. Some apprentices then left dealing altogether while others began dealing on their own. One 32-year-old African-American woman lived with a pound dealer in Los Angeles in 1982. Both were freebasers (cocaine smokers) who sold to other basers. She described her evolution from apprentice to dealer this way:

> I was helping him with like weighing stuff and packaging it and I sort of got to know some of the people that were buying because his own use kept going up. He was getting more out of it, so I just fell into taking care of it partly because I like having the money and it also gave me more control over the situation, too, for awhile, you know, until we both got too out of it. (Case # E-54)

The fifth mode of entree into cocaine selling entailed the *expansion of an existing product line*. A number of the sellers we interviewed started out as marijuana salespersons and learned many aspects of the dealers' craft before they ever moved to cocaine. Unlike in the other modes, in this one an existing marijuana seller already had developed selling skills and established a network of active customers for illicit drugs. Expansion of product line (in business jargon, horizontal integration) was the route of entry for many of the multiple-ounce and kilo cocaine dealers we interviewed. The combination of the availability of cocaine through their marijuana connection and their marijuana customers' interest in purchasing cocaine, led many marijuana sellers to add cocaine to their product line.

Others who entered dealing this way also found that expanding from marijuana to cocaine solved some problems inherent in marijuana dealing. For example, cocaine is far less bulky and odoriferous than marijuana and thus did not present the risky and costly shipping and storage problems of multiple pounds of marijuana. Those who entered cocaine selling via this product line expansion route also recognized, of course, that there was the potential for higher profits with cocaine. They seemed to suggest that as long as they were already taking the risk, why shouldn't they maximize the reward? Some such dealers discontinued marijuana sales altogether and others merely added cocaine to their line. One white, 47-year-old mother of three grown children described how she came to expand her product line:

Q: How did you folks [she and her husband] get started dealing?
A: The opportunity just fell into our lap. We were already dealing weed and one of our customers got this great coke connection and started us onto dealing his product. We were selling him marijuana and he was selling us cocaine.
Q: So you had a network of weed buyers, right? So you could sell to those . . . ?
A: There was a shift in the market. Yeah, because weed was becoming harder [to find] and more expensive and a bulkier product. The economics of doing a smaller, less bulkier product and more financially rewarding product like cocaine had a certain financial appeal to the merchant mentality. (Case # E-1)

CONSCIOUS DECISION TO SELL

As noted earlier, the majority of our sample were middle-class wholesalers who, in the various ways just described, drifted into dealing careers. The few street sellers we interviewed did not drift into sales in the same way. We are obliged to note again that the five modes of entry into cocaine selling we have identified should not be taken as exhaustive. We have every reason to believe that for groups and settings other than those we have studied there are other types of entree and career trajectories. The five cases of street sellers we did examine suggest that entree into street-level sales was more of a conscious decision of a poor person who decided to enter an underground economy, not an effort to solve a user's problems. Our interviews with street sellers suggest that they choose to participate in an illicit profit-generating activity largely because licit economic opportunities were scarce or nonexistent. Unlike our other types, such sellers sold to strangers as well as friends, and their place of business was more likely to be the street corner rather than homes, bars, or nightclubs. For example, one 30-year-old Native American ex-prostitute described how she became a street crack dealer this way:

> I had seen in the past friends that were selling and stuff and I needed extra money so I just one day told one of my friends, you know, if he could help me, you know, show me more or less how it goes. So I just went by what I seen. So I just started selling it. (Case # E-AC 1)

A few higher-level dealers also made conscious decisions to sell (see Adler, 1985), particularly when faced with limited opportunity structures. Cocaine selling, as an occupation, offers the promise of lavish lifestyles otherwise unattainable to most ghetto youth and other impoverished groups. Dealing also provides an alternative to the low-paying, dead-end jobs typically available to those with little education and few skills. A 55-year-old African-American man who made his way up from grams to ounce sales described his motivation succinctly: "The chance presented itself to avoid the 9 to 5" (Case # E-22).

Street sellers and even some higher-level dealers are often already participating in quasi-criminal lifestyles; drug sales are simply added to their repertoire of illicit activities. The perceived opportunity to earn enormous profits, live "the good life," and set your own work schedule are powerful enticements to sell. From the perspective of people with few life chances, dealing cocaine may be seen as their only real chance to achieve the "American Dream" (i.e., financial security and disposable income). Most of our sample were not ghetto dwellers and/or economically disadvantaged. But for those who were, there were different motivations and conscious decisions regarding beginning sales. Popular press descriptions of cocaine sellers predominantly portray just such street sellers. Although street sellers are the most visible, our data suggest that they represent what might be called the tip of the cocaine dealing iceberg.

LEVELS OF ENTRY

The levels at which a potential dealer's friends/connections were selling helped determine the level at which the new dealer entered the business. If the novitiate was moving in social scenes where "big dealers" [were] found, then s/he [was] likely to begin by selling grams and parts of grams. When supplies were not fronted, new dealers' personal finances, i.e., available capital, also influenced how much they could buy at one time.

Sellers move up and down the cocaine sales ladder as well as in and out of the occupation (see Adler, 1985). Some of our sellers were content to remain part-ounce dealers selling between a quarter and a half an ounce a week. Other sellers were more ambitious and eventually sought to become bigger dealers in order to increase profits. One interviewee reported that her unusually well organized suppliers had sales quotas, price fixing, and minimum purchase expectations which pushed her toward expansion. The levels of sales and selling styles of the new dealer's suppliers, then, interacted with personal ambitions to influence eventual sales careers.

Another important aspect of beginning to sell cocaine is whether the connection is willing to "front" the cocaine (risk a consignment arrangement), rather than requiring the beginner to pay in full. Having to pay "up front" for one's inventory sometimes slowed sales by tying up capital, or even deterred some potential dealers from entering the business. Fronted cocaine allowed people with limited resources to enter the occupation. Decisions to front or not to front were based primarily on the connection's evaluation of the new seller's ability to "move" the product. This was seen as a function of the potential volume of business the beginning seller could generate among his/her networks of friends and/or customers. The connection/fronter also evaluates the trustworthiness of the potential dealer, as well as their own capability of absorbing the loss should the deal "go bad" and the frontee be unable to pay. The judgment of the fronter is crucial, for a mistake can be very costly and there is no legal recourse.

LEARNING TO DEAL

In the go-between, stash and connoisseur modes of entree, novices gradually learn the tricks of the trade by observing the selling styles of active dealers, and ultimately by doing. Weighing, packaging, and pricing the product are basic techniques. A scale, preferably a triple-beam type . . . accurate to the tenth of a gram, is a necessary tool. In the last ten years answering machines, beepers, and even cellular phones have become important tools as well. Learning how to manage customers and to establish selling routines and rules of procedure are all essential skills that successful dealers must master.

The dealers who enter sales through the apprenticeship and product line expansion modes have the advantage of their own or their partner/seller's experience. Active marijuana sellers already have a network of customers, scales, familiarity with metric measures, and, most important, a connection to help them move into a new product line. Apprentices have lived with and/or observed the selling styles of their dealer/mentors and have access to their equipment, connections and customers. Both apprentices and marijuana dealers who have expanded into cocaine also know how to "maintain a low profile" and avoid any kind of attention that might culminate in arrest. In this way they were able to reduce or manage the paranoia that often inheres in drug dealing circles.

Many sellers learn by making mistakes, often expensive mistakes. These include: using too much cocaine themselves, fronting drugs to people who do not pay for them, and adding too much "cut" (usually an inactive adulterant such as vitamin B) to their product so they develop a reputation for selling inferior cocaine and sometimes have difficulty selling the diluted product. One 32-year-old African-American male made one such error in judgment by fronting to too many people who did not "come through." It ended up costing him $15,000:

> It was because of my own recklessness that I allowed myself to get into that position. There was a period where I had a lot of weight that I just took it and just shipped it out to people I shouldn't have shipped it out to. . . . I did this with 10 people and a lot of them were women to be exact. I had a lot of women coming over to my house and I just gave them an ounce apiece one time. . . . So when maybe 6 of those people didn't come through . . . there was a severe cramp in my cash flow. This made me go to one of the family members to get the money to re-cop. (Case # E-11)

Business Sense/People Sense

Many people have a connection, the money to make the initial buy, a reputation for being reliable, and a group of friends interested in buying drugs, but still lack the business sense to be a successful dealer. Just because a person drifts into dealing does not mean that he or she will prosper and stay in dealing. We found a variety of ways in which people initially became dealers, few of which hinged on profits. But what determined whether they continued dealing was their business sense. Thus even though a profit orientation had little to do with becoming a dealer, the ability to consistently realize profits had a major influence over who remained a dealer. In this sense, cocaine selling was like any other capitalist endeavor.

According to our respondents, one's ability to be a competent dealer depended on being able to separate business from pleasure. Success or failure at making this separation over time determined whether a profit was realized. Certain business practices were adopted by prosperous dealers to assist them in making this important distinction. For example, prepackaging both improves quality control and helps keep inventory straight; establishing rules for customers concerning when they can purchase and at what prices reduces the level of hassle; limiting the amount of fronting can reduce gross sales volume, but it also reduces financial risk and minimizes the amount of debt collection work; and limiting their own personal use keeps profits from disappearing up one's nose or in one's pipe.

Being a keen judge of character was seen as another important component of being a skilled dealer. Having the "people skills" to judge whether a person could be trusted to return with the money for fronted supplies, to convince people to pay debts that the dealer had no legal mechanisms for collecting, and to engender the trust of a connection when considerable amounts of money were at stake, are just a few of the sophisticated interpersonal skills required of a competent dealer.

Adler also discusses the importance of a "good personal reputation" among upper level dealers and smugglers:

> One of the first requirements for success, whether in drug trafficking, business enterprise broadly, or any life undertaking, is the establishment of a good personal reputation. To make it in the drug world, dealers and smugglers had to generate trust and likability. (1985:100)

Adler's general point applies to our respondents as well, although the experiences of some of our middle and lower level dealers suggested a slight amendment: A likable person with a good reputation could sell a less than high quality product, but an unlikable person, even one with a bad reputation, could still do a considerable amount of business if s/he had an excellent product. One 47-year-old white woman described her "difficult" husband/partner, "powder keg Paul":

> He would be so difficult, you couldn't believe it. Somebody [this difficult] better have a super primo product to make all this worthwhile. . . . He's the kind of guy you don't mind buying from because you know you'll get a good product, but he's the kind of guy you never want to sell to . . . he was that difficult. (Case # E-1)

High-quality cocaine, in other words, is always at a premium in this subculture, so even without good people skills a dealer or connection with "good product" was tolerated.

From User to Dealer:
The Transformation of Identity

In each of our respondents' deviant careers there occurred what Becker referred to as a change in self conception. Among our respondents, this took the form of a subtle shift in identity from a person who *has* a good connection for cocaine to a person who *is* a good connection for cocaine. There is a corresponding change in the meaning of, and the motives for, selling. The relationship between the seller and the customer undergoes a related transformation, from "picking up something for a friend" to conducting a commercial transaction. In essence, dealing becomes a business quite like most others, and the dealer gradually takes on the professional identity of a business person. Everett Hughes, writing on the sociology of work, urged social scientists to remember that when we look at work,

> We need to rid ourselves of any concepts which keep us from seeing that the essential problems of men at work are the same whether they do their work in the laboratories of some famous institution or in the messiest vat of a pickle factory. (1951:313)

When they had fully entered the dealer role, our respondents came to see selling cocaine as a job—work, just like other kinds of work save for its illegality. For most, selling cocaine did not mean throwing out conventional values and norms. In fact, many of our respondents actively maintained their conventional identities (see Broadhead, 1983). Such identities included those of parents, legally employed workers, neighbors, church-goers and softball players, to list just a few. Dealer identities tended not to replace former, "legitimate" identities but were added to a person's repertoire of more conventional identities.

Like everyone else in modern life, sellers emphasized one or another dimension of their identities as appropriate to the situation. In his study of heroin addicts Biernacki notes that, "The arrangement of identities must continuously be managed in such a way as to stress some identities at certain points in particular social worlds and situations, and at the same time to de-emphasize others" (1986:23). Our sellers, too, had to become adept at articulating the proper identity at the proper time. By day, one woman dealer was a concerned mother at her daughter's kindergarten field trip, and that same evening she was an astute judge of cocaine quality when picking up an ounce from her connection. At least for our interviewees, selling cocaine rarely entailed entirely terminating other social roles and obligations.

Yet, at some point in all of our sellers' careers, they found themselves transformed from someone who has a good connection to someone who is a good connection, and they gradually came to accept the identity of dealer as a part of their selves. Customers began to treat them like a salesperson, expecting them to be available to take calls and do business and even for services such as special off-hour pickups and deliveries or reduced rates for volume purchases. When dealers found themselves faced with such demands, they typically began to feel *entitled* to receive profits from selling. They came to be seen as dealers by others, and in part for this reason, came to see themselves as dealers. As Becker's (1963) model suggests, selling *behavior* usually preceded not only motivation but also changes in attitude and identity. As one 38-year-old white woman put it,

> I took over the business and paid all my husband's debts and started to make some money. One day I realized I was a coke dealer. . . . It was scary, but the money was good. (Case # E-75)

Acceptance of the dealer identity brings with it some expectations and values shared by dealers and customers alike. Customers have the expectation that the dealer will have a consistent supply of cocaine for sale. Customers also expect that the dealer will report in a fairly accurate manner the quality of his/her present batch of drugs within the confines of the *caveat emptor* philosophy that informs virtually all commercial activities in market societies. Buyers do not expect sellers to denigrate their product, but they do not expect the dealer to claim that their product is "excellent" if it is merely "good." Customers assume the dealer will make a profit, but dealers should not be "too greedy." A greedy dealer is one who makes what is estimated by the buyer to be excessive profits. Such estimations of excessiveness vary widely among customers and between sellers and buyers. But the fact that virtually all respondents spoke of some unwritten code of fairness suggests that there is, in E. P. Thompson's (1971) phrase, a "moral economy" of drug dealing that constrains the drive for profit maximization even within an illicit market.[6]

For their part, dealers expect that customers will act in a fashion that will minimize their chances of being arrested by being circumspect about revealing their dealer status. One simply did not, for example, bring to a dealer's house friends whom the dealer had not met. Dealers want customers to appreciate the risks undertaken to provide them with cocaine. And dealers come to feel that such risks deserve profits. After all, the seller is the one who takes the greatest risks; s/he could conceivably receive a stiff jail sentence for a sales conviction. While drifting into dealing and selling mostly to friends and acquaintances mitigated the risks of arrest and reduced their paranoia, such risks remained omnipresent.

In fact, the growing realization of such risks—and the rationalization it provided for dealing on a for-profit basis—was an integral part of becoming a cocaine seller. As our 38-year-old white woman dealer put it, "When it's all said and done, I'm the one behind bars, and I had better have made some money while I was selling or why in the hell take the risk?" (Case # E-75)

NOTES

1. Adler also refers briefly to Matza's formulations within her discussion of becoming a dealer (pp. 127–128, 1985).

2. It must be noted at the outset that the predominantly white, working- and middle-class cocaine sellers we interviewed are very likely to differ from inner-city crack dealers depicted in the media. While there is now good reason to believe that both the profits and the violence reported to be endemic in the crack trade have been exaggerated (e.g., Reuter, 1990, and Goldstein et al., 1989, respectively), our data are drawn from a different population, selling a different form of the drug, who were typically drawn to selling for different reasons. Thus the exigencies they faced and their responses to them are also likely to differ from those of inner-city crack sellers.

3. Rock party houses are distinct from "rock houses" or "crack houses." In the former, sellers invite only selected customers to their homes to smoke rock and "party." Unlike crack houses, where crack is sold to all comers, outsiders are never invited to rock party houses, and the arrangement is social and informal. Proprietors of both types, however, charge participants for the cocaine.

4. Car meets are transactions that take place in cars. Arrangements are made over the telephone in advance and both buyer and seller arrange to meet at parking lots, usually at busy shopping centers, and exchange drugs and money. Each arrives in his or her own car and leaves separately.

5. These movements back and forth among different levels of involvement in dealing were different from the "shifts and oscillations" found among the cocaine dealers studied by Adler (1985:133–141). She studied a circle of high-level dealers over an extended period of field work and found that the stresses and strains of dealing at the top of the pyramid often led her participants to attempt to get out of the business. While many of our interviewees felt similar pressures later in their careers and subsequently quit, our focus here is on becoming a cocaine seller.

6. In addition to lore about "righteous" and "rip off" dealers, there were present other norms that suggested the existence of such an unwritten code or moral economy, e.g., refusing to sell to children or to adults who "couldn't handle it" (i.e., had physical, financial, familial, or work-related problems because of cocaine use).

REFERENCES

Adler, P. (1985). *Wheeling and Dealing: An Ethnography of an Upper-Level Drug Dealing Community*. New York: Columbia University Press.

Becker, H. S. (1953). "Becoming a marijuana user." *American Journal of Sociology* 59: 235–242.

Becker, H. S. (1963). *Outsiders*. New York: Free Press.

Biernacki, P. (1986). *Pathways from Heroin Addiction*. Philadelphia: Temple University Press.

Biernacki, P., and Waldorf, D. (1981). "Snowball sampling: problems and techniques of chain referral sampling." *Sociological Methods and Research* 10:141–163.

Broadhead, R. (1983). *The Private Lives and Professional Identity of Medical Students*. New Brunswick, NJ: Transaction Books.

Feldman, H. W. (1968). "Ideological supports to becoming and remaining a heroin addict." *Journal of Health and Social Behavior* 9:131–139.

Fields, A. (1985). "Weedslingers: a study of young black marijuana dealers." *Urban Life* 13: 247–270.

Goldstein, P., Brownstein, H., Ryan, P., and Belucci, P. (1989). "Crack and homicide in New York City, 1988." *Contemporary Drug Problems* 16:651–687.

Hughes, E. (1951). "Work and the self." In John Rohrer and Muzafer Sherif (eds.), *Social Work at the Crossroads*. New York: Harper and Brothers, 313–323.

Langer, J. (1977). "Drug entrepreneurs and dealing culture." *Social Problems* 24:377–386.

Lindesmith, A. (1947). *Addiction and Opiates*. Chicago: Aldine Press.

Macdonald, P., Waldorf, D., Reinarman, C., and Murphy, S. (1988). "Heavy cocaine use and sexual behavior." *Journal of Drug Issues* 18:437–455.

Matza, D. (1964). *Delinquency and Drift*. New York: Wiley.

Matza, D. (1969). *Becoming Deviant*. Englewood Cliffs. NJ: Prentice Hall.

Morales, E. (1986a). "Coca culture: the white gold of Peru." *Graduate School Magazine of City University of New York* 1:4–11.

Morales, E. (1986b). "Coca and cocaine economy and social change in the Andes in Peru." *Economic Development and Social Change* 35:143–161.

Murphy, S., Reinarman, C., and Waldorf, D. (1989). "An eleven year follow-up of a network of cocaine users." *British Journal of the Addictions* 84:427–436.

Plasket, B., and Quillen, E. (1985). *The White Stuff*. New York: Dell Publishing Company.

Preble, E., and Casey, J. H., Jr. (1969). "Taking care of business: the heroin user's life on the streets." *The International Journal of the Addictions* 4:1–24.

Reinarman, C., Waldorf, D., and Murphy, S. (1988). "Scapegoating and social control in the construction of a public problem: empirical and critical findings on cocaine and work." *Research in Law, Deviance and Social Control* 9:37–62.

Reuter, P. (1990). *Money from Crime: The Economics of Drug Dealing*. Santa Monica, CA: Rand Corporation.

Rosenbaum, M. (1981). *Women on Heroin*. New Brunswick, NJ: Rutgers University Press.

Sanabria, H. (1988). *Coca, Migration and Socio-Economic Change in a Bolivian Highland Peasant Community*. Ph.D. thesis, University of Wisconsin.

Sanchez, J., and Johnson, B. (1987). "Women and the drug crime connection: crime rates among drug abusing women at Riker's Island." *Journal of Psychoactive Drugs* 19:205–215.

Sykes, G., and Matza, D. (1957). "Techniques of neutralization." *American Sociological Review* 22:664–670.

Thompson, E. P. (1971). "The moral economy of the English crowd in the eighteenth century." *Past and Present* 50:76–136.

Waldorf, D., Reinarman, C., Murphy, S., and Joyce, B.

(1977). *Doing Coke: An Ethnography of Cocaine Snorters and Sellers*. Washington, DC: Drug Abuse Council.

Waldorf, D., Reinarman, C., and Murphy, S. (1991). *Cocaine Changes*. Philadelphia: Temple University Press.

Watters, J. K., and Biernacki, P. (1989). "Targeted sampling: options for the study of hidden populations." *Social Problems* 36:416–430.

Williams, T. (1989). *The Cocaine Kids*. New York: Addison-Wesley.

QUESTIONS FOR DISCUSSION

1. Describe the general process by which people can "drift" into becoming a drug dealer.

2. One of the conclusions of this study is that when they were not dealing drugs, the dealers were leading pretty conventional lives. What are the implications of this finding for the difference between "deviance" and "conventionality"?

3. There are different modes of entry into drug dealing. What are the five modes described by the authors of this study?

The Origins of Crack

James A. Inciardi, Hilary L. Surratt, Dale D. Chitwood, and Clyde B. McCoy

The first mention of crack cocaine in the major media occurred on November 17, 1985.[1] Buried within the pages of that Monday edition of the prestigious *New York Times,* journalist Donna Boundy, in writing about a local drug abuse treatment program, unceremoniously commented, "Three teenagers have sought this treatment already this year . . . for cocaine dependence resulting from the use of a new form of the drug called 'crack' or rock-like pieces of prepared 'freebase' (concentrated) cocaine" (*New York Times,* 17 Nov. 1985, p. B12).

Although Boundy, like so many after her, had erred in describing crack as "freebase" or "concentrated" cocaine,[2] her mere mentioning of what was ostensibly an old drug initiated a major media event. Crack suddenly took on a life of its own, and in less than 11 months the *New York Times,*[3] the *Washington Post,* the *Los Angeles Times,* the wire services, *Time, Newsweek,* and *U.S. News & World Report* collectively had served the nation with more than 1,000 stories in which crack had figured prominently. Or as social critic Malcolm Gladwell (1986, p. 11) recalled the episode, " . . . coverage feeding coverage, stories of addiction and squalor multiplying across the land."

And then CBS capped their reporting with "48 Hours on Crack Street," a prime-time presentation that reached 15 million viewers and became one of the highest rated documentaries in the history of television. Not to be outdone, NBC offered "Cocaine Country," culminating a 6-month stretch in which the network had broadcast more than 400 reports on drug abuse.

As the crack frenzy was mounting during the summer of 1986, a number of researchers in the drug community were somewhat perplexed. Although *Newsweek* (Smith, 1986) claimed crack was the biggest story since Vietnam and the fall of the Nixon presidency, and other media giants compared the spread of crack with the plagues of medieval Europe, researchers were finding crack to be not a national epidemic, but a phenomenon isolated to but a few inner-city neighborhoods in less than a dozen urban areas. By late August, crack hysteria had reached such proportions that the Drug Enforcement Administration (DEA) felt compelled to respond. Based on reports from its field agents and informants in cities throughout the country, a DEA (1986) report concluded,

> Crack is readily available in Atlanta, Boston, Detroit, Houston, Kansas City, Miami, New York City, Newark, San Diego, San Francisco, Seattle, and St. Louis.

Availability at some level has also been reported in Dallas, Denver, Los Angeles, Minneapolis, Phoenix, and Washington, D.C. Crack generally is not available in Chicago, New Orleans, and Philadelphia.

Crack is currently the subject of considerable media attention. The result has been a distortion of the public perception of the extent of crack use as compared to the use of other drugs. With multi-kilogram quantities of cocaine hydrochloride available and with snorting continuing to be the primary route of cocaine administration, crack presently appears to be a secondary rather than primary problem in most areas.

Curiously, most of the major newspapers, networks, and weekly magazines ignored the DEA report. It was not until the revelations about Lt. Col. Oliver L. North and the Iran-contra connection toward the close of 1986 that the media coverage concerning crack declined significantly.

In contrast with the media contention that crack was "everywhere," observations during the summer of 1986 tended to concur with DEA's position (see Inciardi, 1987). Additional support in this regard came from New York City. Throughout 1986, crack seemed to be concentrated in the city's Washington Heights section, an inner-city neighborhood at the northern end of Manhattan. Many of the streets in Washington Heights had been transformed into outdoor drug marketplaces. One of the more curious aspects of the situation was that the streets frequently were clogged with cars from other parts of New York City, its suburbs, its neighboring states, and other locations (*New York Times,* 21 Jan. 1987, p. B1). But outside of Washington Heights at that time, crack was generally unavailable. Elsewhere, as Dr. Sidney H. Schnoll, at the time affiliated with the Northwestern University Medical School, commented about Chicago late in the summer of 1986: "It's a hoax! There's just no crack in Chicago!" (personal communication, August 1986).

Nevertheless, *Newsweek* (Morganthou, Greenberg-Fink, Murr, Miller, & Raine, 1986) described the crack scene as "an inferno of craving and despair." *Time* (Lamar, 1986, p. 16) stated it somewhat differently: "In minutes the flash high is followed by a crashing low that can leave a user craving another hit."

On the same day, this story appeared in *USA Today:* "Katrina Linton was 17 when she first walked into a crack house in the Bronx. By then she was selling her body to crack dealers just to support her $900-a-day habit" (16 June, 1986, p. 1A).

In these and other media stories the implication was clear: Crack plunged the user almost immediately into the nightmare worlds of Charles Adams, Stephen King, and Rod Serling, from which there was little chance of return. But to researchers and clinicians in the drug field who remembered the media's portrayal of the so-called PCP epidemic a decade earlier, reports of the pervasiveness of crack were regarded with skepticism. Interestingly, the media and the drug professionals were at the same time both right and wrong about what they were saying. During the summer and fall of 1986, contrary to media claims, crack indeed had *not* been an epidemic drug problem in the United States. Crack was there, but it was not until the beginning of 1987 that it really began to assert itself, eventually becoming perhaps the most degrading drug of the century.

UNRAVELING CRACK COCAINE

The history of crack dates back to at least the early 1970s, but to fully understand its evolution, a short diversion into a few other products of the coca leaf is warranted. More specifically, "freebase" and "coca paste" are prominent players in the story of crack.

Powder Cocaine and Freebase Cocaine

During the late 1960s, when cocaine had begun its contemporary trek from the underground to mainstream society, most users viewed it as a relatively "safe" drug. They snorted it in relatively small quantities, and use typically occurred within a social-recreational context (Siegel, 1977). But as the availability of cocaine increased in subsequent years, so too did the number of users and the ways of ingesting it. Some began sprinkling street cocaine on tobacco or marijuana and smoking it as a cigarette or in a pipe, but this method did not produce effects distinctly different from snorting (Grinspoon & Bakalar, 1985). But a new alternative soon became available called *freebasing,* the smoking of "freebase cocaine."

Freebase cocaine is actually a different chemical product than cocaine itself. In the process of freebasing, cocaine hydrochloride is transformed to the base state in a crystalline form. The crystals are then crushed and heated in a special glass pipe. By 1977 some 4 million people were estimated to be users of cocaine in the United States (Abelson, Cohen, Schrayer, & Rappaport, 1978), with as many as 10% of these freebasing the drug exclusively (Siegel, 1982). Yet few outside of the drug-using and drug research and treatment communities were even aware of the existence of the freebase culture, and even fewer had an understanding of the additional complications that freebasing had introduced to the cocaine scene.

The complications are several. First, cocaine in any of its forms is highly seductive. With freebasing, the euphoria is more intense than when the drug is snorted. Moreover, this profound euphoria subsides into intense craving after only a few minutes, thus influencing many users to continue freebasing for days at a time—until either they, or their drug supplies, are fully exhausted. Second, freebasing is expensive. When snorting cocaine, a single gram can last the social user an entire weekend or longer. With street cocaine ranging in price anywhere from $40 to $120 a gram depending on availability and purity, even this method of ingestion can be an expensive recreational pursuit. Yet with freebasing, the cost can undergo a geometric increase. Habitual users have been known to freebase continuously for 3 or 4 days without sleep, using up to 150 grams of cocaine in a 72-hour period. Third, a special danger of freebasing is the proximity of highly flammable ether (or rum when it is used instead of water as a coolant in the pipe) to an open flame. This problem is enhanced because the user generally is suffering from a loss of coordination produced by cocaine or a combination of cocaine and alcohol. As such, there have been many freebasing situations where the volatile concoction has exploded in the face of the user.

By 1980 reports of major problems associated with freebasing had begun to reach a national audience, crystallized by the near death by explosion of comedian-actor Richard Pryor, presumably the result of freebasing.[4]

Coca Paste—Pasta Basica de Cocaina

Common in the drug-using communities of Colombia, Bolivia, Venezuela, Ecuador, Peru, and Brazil is the use of coca paste, known to most South Americans as *basuco, susuko, pasta basica de cocaina,* or just simply *pasta* (Jeri, 1984). Perhaps best known as *basuco,* coca paste is one of the intermediate products in the processing of the coca leaf into cocaine. It is typically smoked straight or in cigarettes mixed with either tobacco or marijuana.

The smoking of coca paste became popular in South America beginning in the early 1970s. It was readily available, inexpensive, had a high cocaine content, and was absorbed quickly. As researchers studied the phenomenon, however, they quickly realized that paste smoking was far more dangerous than any other form of cocaine use. In addition to cocaine, paste contains traces of all the chemicals used initially to process the coca leaves—kerosene, sulfuric acid, methanol, benzoic acid, and the oxidized products of these solvents, plus any number of other alkaloids that are present in the coca leaf (Almeida, 1978). One analysis undertaken in Colombia in 1986 found all of these chemicals, plus traces of brick dust, leaded gasoline, ether, and various talcs (Bogota *El Tiempo,* 19 June 1986, p. 2-D).

When the smoking of paste was first noted in South America, it seemed to be restricted to the coca-processing regions of Bolivia, Colombia, Ecuador, and Peru, appealing primarily to low-income groups because of its cheap price when compared with that of refined cocaine (Jeri, Sanchez, & Del Pozo, 1976). By the early 1980s, however, it had spread to other South American nations, to various segments of the social strata, and throughout the decade paste smoking further expanded to become a major drug problem for much of South America.[5] At the same time, coca paste made its way to the United States—first to Miami, its initial smuggling port of entry, and then elsewhere.[6] Interestingly, the paste quickly became known to young North American users as "bubble gum," likely due to the phonetic association of the South American *basuco* with the American Bazooka brand bubble gum.[7]

The Coca Paste/Crack Cocaine Connection

Contrary to popular belief, crack is not a new substance, having been first reported in the literature during the early 1970s (*The Gourmet Cokebook,* 1972). At that time, however, knowledge of crack, known then as "base" or "rock" (not to be confused with "rock cocaine," a cocaine hydrochloride product for intranasal snorting), seemed to be restricted to segments of cocaine's freebasing subculture. Crack is processed from cocaine hydrochloride by adding ammonia or baking soda and water and heating this mixture to remove the hydrochloride. The result is a pebble-sized crystalline form of cocaine base.

Contrary to another popular belief, crack is neither "freebase cocaine" nor "purified cocaine." Part of the confusion about what crack actually is comes from the different ways the word *freebase* is used in the drug community. Freebase (the noun) is a drug, a cocaine product converted to the base state from cocaine hydrochloride after adulterants have been chemically removed. Crack is converted to the base state *without* removing the adulterants. Freebasing (the act) means to inhale vapors of cocaine base, of which crack is but one form. Finally, crack is not purified cocaine because during its processing, the baking soda remains as a salt, thus reducing its homogeneity somewhat. Informants in the Miami drug subculture indicate that the purity of crack ranges as high as 80%, but generally contains much of the filler and impurities found in the original cocaine hydrochloride, along with some of the baking soda (sodium bicarbonate) and cuts (expanders, for increasing bulk) from the processing. And interestingly, crack gets its name from the fact that the residue of sodium bicarbonate often causes a crackling sound when the substance is smoked.[8]

As to the presence of crack in the drug communities of the early 1970s, it was available for only a short period of time before it was discarded by freebase cocaine aficionados as an inferior product. Many of them referred to it as "garbage freebase" because of the many impurities it contained. In this regard, a 42-year-old Miami cocaine user commented in 1986,

> Of course crack is nothing new. The only thing that's new is the name. Years ago it was called *rock, base,* or *freebase,* although it really isn't *true* "freebase." It was just an easier way to get something that gave a more potent rush, done the same way as now with baking soda. It never got too popular among the 1970s cokeheads because it was just not as pure a product as conventional freebase.

The rediscovery of crack seemed to occur simultaneously on the East and West coasts early in the 1980s. As a result of the Colombian government's attempts to reduce the amount of illicit cocaine production within its borders, it apparently, at least for a time, successfully restricted the amount of ether available for transforming coca paste into cocaine hydrochloride. The result was the diversion of coca paste into cocaine hydrochloride. The result was the diversion of coca paste from Colombia, through Central America and the Caribbean, into South Florida for conversion into cocaine. Spillage from shipments through the Caribbean corridor introduced the smoking of coca paste to local island populations, who in turn developed the forerunner of crack cocaine in 1980 (Hall, 1986). Known as *baking-soda base, base-rock, gravel,* and *roxanne,* the prototype was a smokable product composed of coca paste, baking soda, water, and rum. Migrants from Jamaica, Haiti, Trinidad, and locations along the Leeward and Windward islands chain introduced the crack prototype to Caribbean inner-city populations in Miami's immigrant undergrounds, where it was ultimately produced from powder cocaine rather than paste. As a Miami-based immigrant from Barbados commented in 1986 about the diffusion of what he referred to as "baking-soda paste,"

> Basuco and baking-soda paste seemed to come both at the same time. There was always a little cocaine here and there in the islands, but not too much, and it wasn't cheap. Then 'bout five, maybe six, years ago, the paste hit all of the islands. It seemed to happen overnight—Barbados, Saint Lucia, Dominica, and [Saint] Vincent and [Saint] Kitts—all at the same time.[9]
>   . . . Then I guess someone started to experiment, and we got the rum-soda-paste concoction. We brought it to Miami when we came in '82, and we saw that the Haitians too were into the same combination.

Apparently, at about the same time, a Los Angeles basement chemist rediscovered the rock variety of baking soda cocaine, and it was initially referred to as "cocaine rock" (Inciardi, 1988, p. 470). It was an immediate success, as was the East Coast type, for a variety of reasons. First, it could be "smoked" rather than snorted.

When cocaine is smoked, it is more rapidly absorbed and crosses the blood-brain barrier within 6 seconds. Hence it creates an almost instantaneous high. Second, it was cheap. Although a gram of cocaine for snorting may cost $60 or more depending on its purity, the same gram can be transformed into anywhere from 5 to 30 "rocks." For the user, this meant that individual rocks could be purchased for as little as $2, $5 (*nickel rocks*), $10 (*dime rocks*), or $20. For the seller, $60 worth of cocaine hydrochloride (purchased wholesale for $30) could generate as much as $100 to $150 when sold as rocks. Third, it was easily hidden and transportable, and when hawked in small glass vials, it could be readily scrutinized by potential buyers. As a South Miami narcotics detective described it during the summer of 1986,[9]

> Crack has been a real boon to both buyer and seller. It's cheap, real cheap. Anybody can come up with $5 or $10 for a trip to the stars. But most important, it's easy to get rid of in a pinch. Drop it on the ground and it's almost impossible to find; step on it and the damn thing is history. All of a sudden your evidence ceases to exist.

By the close of 1985 when crack had finally come to the attention of the national media, it was predicted to be the "wave of the future" among the users of illegal drugs (*New York Times,* 29 Nov. 1985, p. A1). Media stories also reported that crack was responsible for rising rates of street crime. As a cover story in *USA Today* put it,

> Addicts spend thousands of dollars on binges, smoking the contents of vial after vial in crack or "base" houses—modern-day opium dens—for days at a time without food or sleep. They will do anything to repeat the high, including robbing their families and friends, selling their possessions and bodies. (*USA Today,* 16 June 1986, p. 1A).

As the media blitzed the American people with lurid stories depicting the hazards of crack, Congress and the White House began drawing plans for a more concerted war on crack and other drugs. At the same time, crack use was reported in Canada, most European nations, Hong Kong, South Africa, Egypt, India, Mexico, Belize, Bermuda, Barbados, Colombia, Brazil, and the Philippines.[10]

CRACK IN THE MIAMI INNER CITY

Although the use of crack cocaine became evident in most major cities across the United States during the latter half of the 1980s, cocaine and crack tend to be associated more often with Miami than most other urban areas. In part, this is due to Miami's association with the cocaine wars of the late 1970s and early 1980s, and with South Florida's reputation for cocaine importation and distribution (see Allman, 1987; Buchanan, 1987; Carr, 1990; Rieff, 1987; Rothchild, 1985). No doubt the image of the city as America's Casablanca, or the Casablanca on the Caribbean, presented in TV's *Miami Vice* contributed as well. In reality, crack is indeed a significant facet of Miami street life, and the Miami experience is targeted in much of this article to illustrate the players, the situations, the adventures, the degradation, and the tragedies associated with crack use.

Crack was already part of the street scene in Miami by 1980. A report in 1980 to the National Institute on Drug Abuse noted that many of the 55 cocaine "freebasers" who were enrolled in an exploratory study of patterns of cocaine use in Miami, Florida, were already using baking soda to transform cocaine hydrochloride into smokable cocaine (Chitwood, 1980; Martinez, 1980). According to several informants, the drug could be purchased at inner-city *get-off houses* (shooting galleries)[11] by 1981. A longtime heroin user and resident of Miami's Overtown community recalled:

> I remember it clear like it was yesterday, 'cause I remember my brother Freddie and me were out celebrating. He had just finished doin' eighteen months for a B & E [breaking and entering] and this was the first I had seen him since he was out. That was the last time he done any hard time, and he got out in May of '81.
>
> Anyway, there was this place on 17th Street, near a little park. It was a get-off house, you know, a shooting gallery. Freddie wanted some white boy [heroin], so since he was just out an' all that I told him it was on me. So we go to this place on 17th. After we're there a few minutes the house man [shooting gallery owner] shows me these small cocaine rocks. I forget exactly what he called them, but later on we know'd it as crack. Said they were comin' down every day from Little Haiti and he'd been dealin' them out of his place for three months for the smokin' cokeys [cocaine users]. I remember Freddie laughin' about it, that with there bein' coke there and the kind of people it attracted the place wouldn't be respectable much longer.

The use of crack and the existence of crack houses proliferated in Miami and elsewhere throughout the 1980s. Subsequent to the initial media sensationalism, press coverage targeted the involvement of youths in crack distribution, the violence associated with struggles to control the crack marketplace in inner-city neighborhoods, and the child abuse, child neglect, and child abandonment by crack-addicted mothers.[12] In Miami, although the violence associated with crack distribution never reached the proportions apparent in other urban centers (Inciardi, 1990), crack use was nevertheless a major drug problem. By 1989 the Drug Enforcement Administration had estimated there were no fewer than 700 operating crack houses in the greater Miami area (DEA, 1989). As in other urban locales, the production, sale, and use of crack (Bourgois, 1989; Fagen & Chin, 1989; Hall, 1988/1989; Massing, 1989; Mieczkowski, 1990; Wallace, 1989), as well as prostitution and sex for drugs exchanges,[13] became prominent features of the Miami crack scene.

In addition to the media coverage of crack in Miami, empirical data also document its presence in Miami well before the drug's so-called discovery by the *New York Times*. A large-scale study of cocaine use and related criminality was conducted between April 1988 and March 1990 in the Miami (Dade County), Florida, metropolitan area (Inciardi & Pottieger, 1994). Of the 699 cocaine users interviewed, 94.1% had used crack. Twenty-six (4.0%) of the crack-using respondents had used crack prior to 1980, with the earliest report of crack use in 1973. Of the remaining respondents, most (66.9%) first used crack between 1984 and 1986.

A subsequent study of seriously delinquent youths in Miami observed that crack use existed in the early 1980s. By the mid-1980s, it was not uncommon for delinquent youths to be established users of crack since their early teens (Inciardi & Pottieger, 1991).

DOING CRACK

Crack is known by many pseudonyms. Most commonly, it might be *cracks, hard white, white,* or *flavor.* Furthermore, there are *bricks, boulders,* and *eight-ball* (large rocks or slabs of crack), *doo-wap* (two rocks), as well as *crumbs, shake,* and *kibbles and bits* (small rocks or crack shavings). The *dope man* or *bond man* (crack dealer) can deliver a *cookie* (a large quantity of crack, sometimes as much as 90 rocks), which he carries in his *bomb bag* (any bag in which drugs are conveyed for delivery) to any crack house in his neighborhood territory. The *dope man* may also *deal* (sell) or *juggle* (sell, for double what it is worth) his crack on the street.

In many crack houses the drug might be displayed on *boards* (tables, mirrors, or bulletin boards), whereas in the street crack is hawked in small glass vials or plastic bags. In a few locales, these bags are sealed or stamped with a brand name. Such a practice affords the illusion of quality control and gives the buyer a specific name to ask for. In New York City, crack labeling has included such brands as "White Cloud," "Conan," and "Handball," and in Miami, the better known labels of the early 1990s included "Cigarette" (named after the high-performance racing boat), "Biscayne Babe" (an epithet for prostitutes who *stroll* Miami's Biscayne Boulevard), "Olympus" (perhaps from Greek mythology), "Bogey" (of *Key Largo* fame), "Noriega's Holiday" (after the former Panamanian dictator), "Bush" (after the former president, who escalated a "war on drugs in Miami"), and "Pablo" (after the late head of the Medellin cartel, Pablo Escobar).

"Beaming Up"

Many crack users are uncertain about what crack actually is. Some know it as cocaine that has been "cooked" into a hard solid form called a "rock." Exactly what is in crack, in addition to street cocaine, is also debated: Some say baking soda; others say ether. Actually, most users don't know or care. A few argue that crack is the purest form of cocaine, but others hold that "freebase" is. Still others believe crack *is* freebase.

The *rocking up* (preparation) of crack is done in a variety of ways, all of which require baking soda that binds with cocaine and frees it from hydrochloride salt. In most instances a *cut* (expander) of some sort is also added to increase the volume and weight of the crack, and hence, the profits. Typical in this behalf are *comeback, swell up, blow up,* and *rush,* all of which are cocaine analogues (novocaine, lidocaine, and benzocaine) that bind with the cocaine when cooked. Crack is easy to cook, regardless of whether the recipe calls for a serving for 2 or 200 or the heat source is a handheld cigarette lighter or a microwave.

Crack is "smoked" in a variety of ways—special glass pipes, makeshift smoking devices fabricated from beer and soda cans, jars, bottles, and other containers known as *stems, straight shooters, skillets, tools, ouzies,* or more directly, the *devil's dick.* A *beam* (from "Beam me up, Scotty" of TV's original *Star Trek*) is a hit of crack, as is a *bubb, backs* (a single hit), and *back up* (a second hit). Crack is also smoked with marijuana in cigarettes, called *geek joints, lace joints,* and *pin joints.* Some users get high from a *shotgun*—secondary smoke exhaled from one crack user into the mouth of another.

Users typically smoke for as long as they have crack or the means to purchase it—money or sex, stolen goods, furniture, or other drugs. It is rare that smokers have but a single hit. More likely they spend $50 to $500 during a *mission*—a 3- or 4-day binge, smoking almost constantly, 3 to 50 rocks per day. During these cycles, crack users rarely eat or sleep. And once crack is tried, for many users it is not long before it becomes a daily habit. For example, a recovering crack user commented,

> I smoked it Thursday, Friday, Saturday, Monday, Tuesday, Wednesday, Thursday, Friday, Saturday on that cycle. I was working at that time. I would spend my whole $300 check. Everyday was a crack day for me. My day was not made without a hit. I could smoke it before breakfast, don't even have breakfast or I don't eat for three days.

And a crack user/dealer reported:

> For the past five months I've been wearing the same pants. And sneakers are new but with all the money you make a day at least $500/$600 a day you don't want to spend a $100 in clothes. Everything is rocks, rocks, rocks, rocks, rocks. And to tell you the truth I don't even eat well for having all that money. You don't even want to have patience to sit down and have a good dinner. I could tell you rock is . . . I don't know what to say. I just feel sorry for anyone who falls into it.

As a final note here, crack has been called the fast-food variety of cocaine. It is cheap, easy to conceal, vaporizes with practically no odor, and the gratification is swift: an intense, almost sexual euphoria that lasts less than 5 minutes. Given these attributes, it would appear that crack cocaine might be a safer alternative to powder cocaine. But such a conclusion is far from accurate. There are many problems associated with cocaine use, including a high addiction potential, hyperstimulation, digestive disorders, nausea, loss of appetite, weight loss, tooth erosion, brain abscess, stroke, cardiac irregularities, occasional convulsions, and sometimes paranoid psychoses and delusions of persecution. Smoking crack as opposed to snorting cocaine results in a more immediate and direct absorption of the drug. Smoking produces a quicker and more compelling high that greatly increases the potential for dependence. Moreover, there is increased risk of acute toxic reactions, including brain seizure, cardiac irregularities, respiratory paralysis, paranoid psychosis, and pulmonary dysfunction.

The tendency to binge on crack for days at a time, neglecting food, sleep, and basic hygiene, severely compromises physical health. As such, crack users often appear emaciated. They lose interest in their physical appearance. Many have scabs on their faces, arms, and legs, the results of burns and picking on the skin (to remove bugs and other insects believed to be crawling *under* the skin). Crack users may have burned facial hair from carelessly lighting their smoking paraphernalia, they may have burned lips and tongues from the hot stems of their pipes, and many seem to cough constantly. The tendency of both male and female crack users to engage in high-frequency unprotected sex with numerous anonymous partners increases their risk for sexually transmitted diseases, including AIDS.

According to national surveys, crack has not become prevalent in the general population. But for reasons difficult to understand, crack's appeal in the majority of the nation's inner cities has endured. Perhaps the best explanation of crack's appeal in the inner city comes from anthropologist Philippe Bourgois (1989):

> Substance abuse in general, and crack in particular, offers . . . [a] metamorphosis. Users are instantaneously transformed from being unemployed, depressed high school dropouts, despised by the world . . . into being a mass of heart-palpitating pleasure, followed . . . by a jaw-gnashing crash and wide-awake alertness that provides their lives with concrete purpose: Get more crack—fast! (642–643).

NOTES

1. The media story of crack can actually be traced to the latter part of 1984 when the Los Angeles dailies began reporting on local "rock houses" where small pellets of cocaine could be had for as little as $25 (for example, *Los Angeles Times,* 25 Nov. 1984, pp. CC1, CC8. *Newsweek* (11 Feb. 1985, p. 33) later gave a half page to the Los Angeles item, but the term *crack* was never used, and little attention was given to the matter.

2. Rather than "concentrated" or "purified" cocaine, crack might be better described as the "fast-food" analogue of cocaine.

3. Following Boundy's brief mention of crack earlier in November, it was likely that the front-page story in the November 29, 1985, issue of the *Times,* headlined "A New Purified Form of Cocaine Causes Alarm as Abuse Increases," represented the beginning of the drug's concentrated media attention.

4. Although Pryor denied he had been using cocaine at the time of the June 1980 explosion, he later admitted he had been freebasing for three days prior to the event (see *Time,* July 6, 1981, p. 63). In a 1986 interview with Barbara Walters, Pryor once again changed his story, claiming the fire was the result of a suicide attempt because he couldn't overcome his dependence on freebase.

5. See Caracas (Venezuela) *El Universal,* October 4, 1985, pp. 4, 30; Caracus *Zeta,* September 12–23, 1985, pp. 39–46; Manaus (Brazil) *Jornal Do Comercio,* May 20, 1986, p. 16; Bogota *El Tiempo,* June 1, 1986, p. 3-A; Medellin *El Colombiano,* July 22, 1986, p. 16-A; Bogota *El Tiempo,* October 6, 1986, p. 7-A; Lima (Peru) *El Nacional,* November 14, 1986, p. 13; La Paz (Bolivia) *Presencia,* March 3, 1988, Sec. 2, p. 1; Sao Paulo (Brazil) *Folba de Sao Paulo,* June 11, 1987, p. A29; Buenos Aires (Argentina) *La Prensa,* June 20, 1987, p. 9; Sao Paulo *O Estado de Sao Paulo,* March 8, 1988, p. 18; Bogota *El Espectador,* April 2, 1988, pp. 1A, 10A; La Paz *El Diario,* October 21, 1988, p. 3; Cochabamba (Bolivia) *Los Tiempos,* June 13, 1989, p. B5; Sao Paulo *O Estado de Sao Paulo,* June 18, 1989, p. 32; Rio de Janiero (Brazil) *Manchete,* October 28, 1989, pp. 20–29; Philadelphia *Inquirer,* September 21, 1986, p. 25A; Timothy Ross, "Bolivian Paste Fuels Basuco Boom," *WorldAIDS,* September 1989, p. 9.

6. Curiously, coca paste was reportedly available in Italy during 1987. See Milan *Corriere Della Sera,* October 26, 1987, p. 8.

7. During the latter part of the 1980s, a new form of coca paste smoking was noticed, principally in Brazil. For years in the nations of Peru, Bolivia, Ecuador, Argentina, and Brazil, the term *pitillo* (also "petilho," and "pitilio" in Portuguese) had referred to a marijuana cigarette, or marijuana laced with coca paste. The new "pitillo," however, also referred to in parts of Brazil as "Bolivian crack," was marijuana and coca paste residue—the dregs left in the processing drum after coca paste precipitate had been removed. Although no analyses of this residue have been reported in the literature, it is suspected the product has even higher concentrations of sulfuric acid and petroleum products than does coca paste. See Sao Paulo *Folba*

de Sao Paulo, July 30, 1986, p. 15; Cochabamba *Los Tiempos,* October 13, 1986, p. 5; Sao Paulo *O Estado de Sao Paulo,* November 10, 1990, p. 5, Rio de Janiero *O Globo,* November 30, 1986, p. 18; Buenos Aires *La Prensa,* June 20, 1987, p. 9; Sao Paulo *Veja,* December 12, 1990, pp. 22–23.

8. Some comment seems warranted on the practice of referring to crack as "smokable co-caine." Technically, crack is not really smoked. "Smoking" implies combustion, burning, and the inhalation of smoke. Tobacco is smoked. Marijuana is smoked. Crack, on the contrary, is actually inhaled. The small pebbles or rocks, having a relatively low melting point, are placed in a special glass pipe or other smoking device and heated. Rather than burning, crack vapor-izes and the fumes are inhaled.

9. For those unfamiliar with the geography of the Caribbean, the locations spoken of by this informant are part of the Leeward and Windward islands. The Leeward Islands are the northern segment of the Lesser Antilles and stretch some 400 miles in a southerly arc from the Virgin Islands to Dominica. The Windward Islands are the southern part of the Lesser Antilles, stretching some 200 miles from Martinique south to Grenada. Barbados is located just west of the southern half of the Windward chain, but is not geographically part of it.

10. Windsor (Canada) *Windsor Star,* June 26, 1986, p. A13; Toronto *Globe and Mail,* September 2, 1987, p. A5; Ottawa *Citizen,* February 13, 1988, p. A15; Belfast (Ireland) *News Letter,* July 9, 1986, p. 3; Helsinki (Finland) *Uusi Suomi,* July 28, 1986, p. 8; Rio de Janeiro *O Globo,* May 24, 1986, p. 6; Hong Kong *South China Morning Post,* August 2, 1986, p. 16; Johannesburg (South Africa) *Star,* September 23, 1986, p. 1M; Cape Town (South Africa) *Argus,* March 10, 1987, p. 13; Johannesburg *City Press,* January 7, 1990, p. 5; Milan (Italy) *Panorama,* May 3, 1987, pp. 58–59; Oslo (Norway) *Arbeiderbladet,* June 4, 1987, p. 13; Madrid (Spain) *El Alcazar,* September 14, 1986, p. 11; Nuevo Laredo (Mexico) *El Diario de Nuevo Laredo,* October 12, 1986, Sec. 4, p. 1; Calcutta (India) *Statesman,* October 16, 1986, p. 1; Belize City (Belize) *Beacon,* October 25, 1986, pp. 1, 14; London *Al-Fursan,* September 13, 1986, pp. 51–53; London *Sunday Telegraph,* April 12, 1987, p. 1; Geneva (Switzerland) *Journal de Geneve,* December 26, 1986, p. 1; Brussels (Belgium) *Le Soir,* November 10/11, 1986, p. 3; Munich (West Germany) *Sueddeutsche Zeitung,* October 18/19, 1986, p. 12; Lisbon (Portugal) *O Jornal,* January 30–February 5, 1987, p. 40; Hamburg (West Germany) *Die Zeit,* April 24, 1987, p. 77; Hamilton (Bermuda) *Royale Gazette,* December 3, 1987, p. 3; Bridge-town (Barbados) *Weekend Nation,* January 15–16, 1988, p. 32; Stockholm (Sweden) *Dagens Nyheter,* January 2, 1988, p. 6; Grand Cayman (Bahamas) *Caymanian Compass,* January 20, 1988, pp. 1–2; Paris *Liberation,* January 12, 1990, p. 33; Bangkok (Thailand) *Siam Rat,* August 13, 1988, p. 12.

11. Get-off houses and shooting galleries are places where many injection drug users go to use drugs and rent injection equipment.

12. See Ron Rosenbaum, "Crack Murder: A Detective Story," *New York Times Magazine,* February 15, 1987, pp. 29–33, 57, 60; *Newsweek,* February 22, 1988, pp. 24–25; *Time,* March 7, 1988, p. 24; *Newsweek,* March 28, 1988, pp. 20–29; *Newsweek,* April 27, 1988, pp. 35–36; *Time,* May 9, 1988, pp. 20–33; *New York Times,* June 23, 1988, pp. A1, B4; *Time,* December 5, 1988, p. 32; *New York Doctor,* April 10, 1989, pp. 1, 22; *U.S. News & World Report,* April 10, 1989, pp. 20–32; *New York Times,* June 1, 1989, pp. A1, B4; *New York Times* (National Edition), August 11, 1989, pp. 1, 10; Andrew C. Revkin, "Crack in the Cradle," *Discover,* September 1989, pp. 62–69.

13. See "Sex for Crack: How the New Prostitution Affects Drug Abuse Treatment," *Substance Abuse Report,* 19 (November 15, 1988), pp. 1–4; "Syphilis and Gonorrhea on the Rise Among Inner-City Drug Addicts," *Substance Abuse Report,* 20 (June 1, 1989), pp. 1–2; "Syphilis and Crack Linked in Connecticut," *Substance Abuse Report,* 20 (August 1, 1989), pp. 1–2; *New York Times,* August 20, 1989, pp. 1, 36; *New York Times,* October 9, 1989,

pp. A1, A30; *U.S. News & World Report,* October 23, 1989, pp. 29–30; *Newsweek,* September 25, 1989, p. 59; *Miami Herald,* October 22, 1989, pp. 1G, 6G.

REFERENCES

Abelson, H., Cohen, R., Schrayer, R., & Rappaport, M. (1978). Drug experience, attitudes and related behavior among adolescents and adults. In *Annual Report.* Washington, DC: Office of Drug Abuse Policy.

Allman, T. D. (1987). *Miami: City of the future.* New York: Atlantic Monthly Press.

Almeida, M. (1978). Contrabucion al estudio de la historia natural de la dependencia a la pasta basica de cocaina. *Revista de Neuro-Psiquiatria, 41,* 44–45.

Bourgois, P. (1989). In search of Horatio Alger: Culture and ideology in the crack economy. *Contemporary Drug Problems, 16,* 619–649.

Bogota El Tiempo, (1986, June 19). p. 2D.

Buchanan, E. (1987). *The corpse had a familiar face: Covering Miami, America's hottest beat.* New York: Random House.

Carr, P. (1990). *Sunshine states.* New York: Doubleday.

Chitwood, D. D. (1980). Patterns of cocaine use: Preliminary Observations. *Local Drug Abuse: Trends, Patterns and Issues* (pp. 156–166). Rockville, MD: National Institute on Drug Abuse.

Drug Enforcement Administration. (1986, August 22). *Special report. The crack situation in the United States.* Unpublished release from the Strategic Intelligence Section, Drug Enforcement Administration, Washington, DC.

Drug Enforcement Administration. (1989). *Crack/cocaine: Overview 1989.* Washington, DC: Author.

Fagen, J., & Chin, K. (1989). Initiation into crack and cocaine: A tale of two epidemics. *Contemporary Drug Problems, 16,* 579–617.

Gladwell, M. (1986, October 27). A new addiction to an old story. *Insight,* pp. 8–12.

The gourmet cokebook: A complete guide to cocaine. (1972). Sonoma, CA: White Mountain Press.

Grinspoon, L., & Bakalar, J. B. (1985). *Cocaine: A drug and its social evolution.* New York: Basic Books.

Hall, J. N. (1986). Hurricane crack. *Street Pharmacologist, 10,* 1–2.

Hall, J. N. (1988/1989, Fall/Winter). Cocaine smoking ignites America. *Street Pharmacologist,* pp. 28–30.

Inciardi, J. A. (1987). Beyond cocaine: Basuco, crack, and other coca products. *Contemporary Drug Problems, 14,* 461–492.

Inciardi, J. A. (1990). The crack/violence connection within a population of hard-core adolescent offenders. In M. DeLaRosa, E. Y. Lambert, & B. Gropper (Eds.), *Drugs and violence: Causes, correlates, and consequences* (pp. 92–111). Rockville, MD: National Institute on Drug Abuse.

Inciardi, J. A., & Pottieger, A. E. (1991). Kids, crack and crime. *Journal of Drug Issues 21,* 257–270.

Inciardi, J. A., & Pottieger, A. E. (1994). Crack cocaine use and street crime. *Journal of Drug Issues, 24,* 273–292.

Jeri, F. R. (1984, April–June). Coca-paste smoking in some Latin American countries: A severe and unabated form of addiction. *Bulletin on Narcotics,* pp. 15–31.

Jeri, F. R., Sanchez, C., & Del Pozo, T. (1976). Consumo de drogas peligrosas por miembros familiares de la fuerza armada y fuerza policial peruana. *Revista de la Sanidad de las Fuerzas Policiales, 37,* 104–112.

Lamar, J. V., Jr. (1986). Crack, a cheap and deadly cocaine is spreading menace. *Time,* June 2, pp. 16–18.

Martinez, R. (1980). Freebase cocaine. *Local Drug Abuse; Trends, Patterns, and Issues* (pp. I-67–I-70). Rockville, MD: National Institute on Drug Abuse.

Massing, M. (1989, October 1). Crack's destructive sprint across America. *New York Times Magazine,* pp. 38–41, 58–59, 62.

Mieczkowski, T. (1990). Crack distribution in Detroit. *Contemporary Drug Problems, 18,* 9–30.

Morganthou, T., Greenberg-Fink, V., Murr, A., Miller, M., & Raine, G. (1986). *Newsweek,* June 16, pp. 16–22.

New York Times, (1985). Nov. 17, p. B12.

New York Times, (1985). Nov. 29, p. A1

New York Times, (1987). Jan. 21, p. B1.

Rieff, D. (1987). *Going to Miami: Exiles, tourists, and refugees in the new America.* Boston: Little, Brown.

Rothchild, J. (1985). *Up for grabs: A trip through time and space in the sunshine state.* New York: Viking Press.

Siegel, R. K. (1977). Cocaine: Recreational use and intoxication. In R. C. Petersen & R. Stillman (Eds.), *Cocaine: 1977* (pp. 119–136). Rockville, MD: National Institute on Drug Abuse.

Siegel, R. K. (1982). Cocaine smoking. *Journal of Psychoactive Drugs, 14,* 271–359.

Smith, R. M. (1986, June 16). The drug crisis. *Newsweek,* p. 15.

Smith, R. M. (1986, June 16). *USA Today,* p. 1A.

Wallace, B. (1989). Psychological and environmental determinants of relapse in crack cocaine smokers. *Journal of Substance Treatment, 6,* 95–106.

QUESTIONS FOR DISCUSSION

1. Discuss the history of crack cocaine.

2. According to the authors, why was crack cocaine an immediate success on the West and East Coasts?

3. According to the authors, "crack has been called the fast-food variety of cocaine." What does this mean? Are there any other fast-food drugs? Explain your response.

CHAPTER 12

ALCOHOL AND DEVIANT BEHAVIOR

The Alcoholically Divided Self

Norman K. Denzin

The alcoholically divided self (and its other) lives two modes of existence, referenced by the terms *sober* and *intoxicated*. These two modes of existence contradict one another, producing deep divisions within the inner and outer structures of the subject's self and the self of the other. Alcohol thickens these divisions, leading the subject to live an emotionally divided self. The subject, like his or her other, is in the grip of negative emotions, including ressentiment, anger, fear, self-loathing, self-pity, self-hatred, despair, anguish, remorse, guilt, and shame (see Denzin, 1984a: 283 on ressentiment and resentment). The alcoholic's self is disembodied. He or she experiences a separation between an alcoholically distorted inner stream of consciousness and a painful, often bruised, bloated, and diseased body he or she lives from within. His or her alcoholic self-pride mobilizes the negative feelings held toward the other.

The following argument organizes my analysis. *The alcoholic and his or her other are trapped within an interactional circuit of progressively differentiated alcoholic and nonalcoholic conduct (schismogenesis) that transforms their relationship into a painful field of negative, contrasting emotional experience. If unchecked, this relationship moves slowly toward self-destruction.* Like the violent family (Denzin,

"The Alcoholically Divided Self" from *The Alcoholic Self* by Norman K. Denzin, 1987, pp. 135–149. Sage Publications. Reprinted by permission of the author.

1984b: 490–491), the alcoholic relationship will move through nine interactional stages: (1) denial of alcoholism and violence; (2) pleasure derived from alcoholism and violence; (3) the building up of mutual hostility; (4) the development of misunderstandings; (5) jealousy (especially sexual); (6) increased alcoholic violence; and either (7) the eventual collapse of the system; or (8) the resolution of the situation into an unsteady, yet somewhat stable state of recurring alcoholic violence; or (9) the transformation of the relationship into a "recovering" alcoholic situation.

ACT ONE: THE ALCOHOLIC SITUATION

Four categories of persons are present in Act One: the alcoholic, his family, friends, and coworkers (Kellerman, 1969: 2). In Act One the alcoholic has passed through the stage of being a heavy social drinker into the phases of crucial or chronic alcoholism. The alcoholic still believes that he is in control of his own destiny. The alcoholic lives in a culture that sanctions drinking in a wide variety of contexts, yet he seems to be unable to drink like normal drinkers. He has learned that alcohol makes him feel better, yet he drinks hard and fast, often secretly (Kellerman, 1969: 3). The power to choose whether to drink or not is lost. When drinking, the alcoholic ignores rules of social conduct, often becomes emotionally uncontrollable, and continually embarrasses his significant others. He often creates a crisis that requires that they intercede on his behalf. In the process the drinker becomes increasingly more dependent on them, yet persists in believing that he is independent and in control of the situation. He adopts a reactive stance toward the problems he creates, waiting for others to react to him, while the alcoholic reacts to their reactions.

The alcoholic and his other have built up, through years of heavy drinking, an alcoholic-centered relationship. His spouse, or lover, has supported his drinking. The spouse or lover has denied the problems alcohol has created, and has attempted to control his drinking, all the time maintaining the myth that he is a heavy social drinker.

The following interaction is typical of the experiences the alcoholic and his other confront in Act One. The speaker recounts an experience before he came to A.A.

> I'd gotten mad at my family. They'd come home and found me drunk. They left and went to Burger King for dinner and left me alone. I called a cab, got a bottle of gin, and checked into a hotel under another name. Late that night I called home and said I was going to kill myself. Did she want to do anything? She asked where I was and I wouldn't tell her. I hung up, passed out, and woke up the next morning and called home and asked for a ride to work. They came and picked me up and dropped me off. They looked at me and said "What's going on?" I said, "Nothing, I just needed to get away." Then I got mad. I screamed at them. I said, "Get off my case. You're driving me crazy," and I slammed the door and left them there. That night I took them out for dinner and brought home roses. Everything was "lovey dovey" [field conversation, April 3, 1981].

In Act One the alcoholic and his other coexist in a field of contrasting emotional experiences. Emotional violence, sexual intimacy, physical abuse, gifts offered out of guilt, DUIs, suicide threats, and trips to counselors chart the emotional roller coaster the alcoholic and his other ride in Act One (see following). This act will continue until the alcoholic produces a situation that neither he nor alcohol can handle. That is, he requires the massive assistance of others, including financial aid, legal counsel, medical attention, and help from his coworkers. However, until this occurs, the alcoholic and his others remain within the uneasy interactional space they have created for one another.

Together they have produced *the alcoholic situation* that represents four interactional patterns organized around alcohol and its consumption by the alcoholic. The *open drinking context* displays the drinker with a drink in his hands. The *closed drinking context* references those situations in which the drinker attempts to hide the fact that he has been drinking. The *sober context* is a nondrinking situation in which it is evident that the alcoholic probably is sober. The *normally intoxicated but in control context* references the setting in which the alcoholic is maintaining a level of alcohol intake that others regard as normal and acceptable (see Glaser and Strauss, 1967).

Each time the alcoholic interacts with his other, the other must determine if he has been drinking, and if so, how much. Two interpretive frameworks (Goffman, 1974: 10–11; Bateson, 1972d: 187) compete for attention in the alcoholic situation. The "sober" framework produces one set of definitions regarding accountable and nonaccountable violent, emotional conduct. The "intoxicated," or "he has been drinking" framework produces another set of meanings and interpretations. These two frameworks may exist side by side in the same interactional situation. Drunk, the alcoholic confronts his sober other, attempting to move her into the "intoxicated" framework, or the alcoholic denies his intoxication and attempts to speak to the other from the "sober" framework. The clash of these two interpretive points of view leads to hostility, feelings of anger, and ressentiment. Witness the following account, offered by Merryman (1984: 6). The interactants are an alcoholic wife Abby and her husband Martin. Abby has just returned from a weekend escape to the family's cottage on the New England coast. Martin has picked her up from the airport. They are driving home. Afraid of what her husband will say about her drinking, Abby has taken a drink of vodka in the women's restroom at the airport.

> Abby sat silent, waiting, watching, intensely aware of Martin. . . . She could see right now from his hurt mad expression that the next words from his mouth would be sarcastic hints for self-improvement—unconnected to any tenderness or sensitivity. And then she would defend herself, and they would be at it again, and there would be no real hope.
>
> Presently, Martin's eyes flicked away from the highway and glanced over his glasses toward Abby. "You look completely burnt out," he said, "Had your nose in the sauce?"
>
> Abby twisted toward him, hands clenched in her lap, voice controlled. "It so happens I almost missed the plane and haven't had any breakfast. I feel like I've been in a Waring blender set on chop. . . ."
>
> Martin's eyes remained unswervingly on the road, his face set and grim. When he spoke again, his voice was cutting, punishing. "How much did you drink?"

Reading the Alcoholic

Fearful of the effects alcohol may have upon the alcoholic's conduct, family members, like Martin above, are, as Jackson (1962: 475–477) observes, always alert to what phase of the drinking cycle the alcoholic is in when he or she comes before them. The interaction patterns in the alcoholic-dependent relationship vary, as Steinglass and Robertson (1983: 272) suggest, by whether or not the alcoholic is in a stable or non-stable, sober or "wet" state. The stability of the alcoholic's emotionality, whether predictable or nonpredictable, intentional or nonintentional, whether it is charming or cruel, harsh and aloof, self-pitying and withdrawn, fawning and conciliatory, hostile and angry, will be seen to vary by real and imputed states of intoxication. Confronted with the self-fulfilling definition that she has been drinking before she comes into their presence, the alcoholic may, as did Abby above, attempt to mask the effects of alcohol upon her conduct. Or the alcoholic may react with anger and produce the negative experiences family members wish to avoid.

The underlying premise of all interactions in the alcohol-dependent relationship questions the presence or absence of alcohol in the drinker's consciousness. This premise makes problematic the intentionality of the alcoholic's conduct, for her actions and meanings can always be interpreted from within the "she has been drinking" framework. Such a premise diminishes the alcoholic's standing in the relationship, making her dependent on others for the valued self-definitions she seeks. That is, the alcoholic knows and they know—and she knows that they know—that she could have been drinking. Doubt and anger thus cling to every interaction between the alcoholic and her other. Because alcohol has become the central social object in the alcohol-dependent relationship, it—not emotionality or the mutual exchange of selves—becomes the dominant, if not the hidden, focus of all interactions.

Violent Emotionality

The alcoholic and her other live in emotional violence, which I define as negative emotionality turned into active, embodied, hostile interactions with herself and with others. If violence is understood to reference the attempt to regain through force something that has been lost (Denzin, 1984a: 169), then the alcoholic attempts to use emotional and physical force to regain the sense of self-pride she has lost to alcohol and to the other.

The violence the alcoholic engages in may be verbal or physical, or both. It may be sporadic or frequent, and it may involve such actions as the sexual abuse of a spouse or child, the threat to use a weapon (including knives and guns), the actual hitting or beating of another, and the driving of an automobile while intoxicated and killing another. Violent alcoholic emotionality ranges from physical violence, as just indicated, to inflicted emotionality, as well as spurious, playful, real, and paradoxical violent emotional acts (Denzin, 1984a: 185–190). Violent emotionality, in all its forms, is embedded in the daily life of the alcoholic and his or her emotional associates. The interiority of alcoholic existence frightfully illuminates the emotional violence that has been observed in nonalcoholic, violent relationships (Denzin, 1984a: 190–197; 1984b: 483–513). The alcoholic relationship is an emotionally violent relationship.

The Five Forms of Alcoholic Violence

There are five forms of alcoholic violence that are woven through the four interactional-drinking situations the alcoholic and his or her other produce. As indicated above, every time the other interacts with the alcoholic he or she must determine if the alcoholic has been drinking. This master definition then structures the meanings that are applied to the alcoholic's conduct.

The first form of alcoholic violence is *emotional violence*. In emotional violence the alcoholic inflicts his or her emotionality on the other. He or she does this with emotional outbursts and in emotional scenes that get out of control. The following interaction describes alcoholic, emotional violence.

> I came home tired and beat. The house was a mess, the dogs were loose, and she was in the bedroom taking a nap. Supper wasn't even started yet. I fixed a drink, turned on the news. She came out, yelled at me for having that drink. I'd heard it a thousand times before. I couldn't take it anymore. I threw the drink in her face, grabbed her arm and yelled, "Where's my God damned supper! You never do anything around here." She hit back at me, called me no good. She ran and got my wood carving that I'd been making her for Christmas. She laughed at it, called it stupid and dumb. She threw it against the wall. That's when I lost it. I ran at her. She called the police. How can she ever trust me again? How can we start over? I don't know what's wrong with me. I ain't been the same since I got home from Nam. (Denzin, 1984b: 501).

The second form of alcoholic violence is *playful, alcoholic violence*. Here the alcoholic, usually in the open drinking context, plays at being violent, but serious violence is not intended. He playfully slaps his child on the shoulder, or smacks his wife on the cheek, for example. He takes a step away from "literal" reality and makes a play at being violent, but in a nonviolent way. Playful violence is conveyed through winks, smiles, voice intonations, shrugs, hand and shoulder movements, a toss of the head, and so on (Lynch; 1982: 29).

The third form of negative, violent emotionality is *spurious* and/or *accidental alcoholic violence*. As with playful violence, deliberate nonviolence is intended. However, the alcoholic's actions carry the meaning of real, intended violence. His hand slips as he reaches out to touch the other and he slaps her, rather than caressing her. He stumbles, falls, and crashes through a window. He passes out while driving and runs the car through the front door of his house. Though inwardly the alcoholic does not intend to be violent, his outward actions convey violence to the other. Unlike playful violence in which the alcoholic communicates through playful actions that his actions could be interpreted both as violence and nonviolence, in accidental violence, real violent consequences follow from his conduct.

Real alcoholic violence is the fourth form of negative, violent emotionality. Here the alcoholic intends to be violent, is violent, and his or her violent intentions are felt by the other. He or she may hit, hurl an insult, throw a knife, or fire a gun at the other. In real violence the alcoholic embodies a violent line of action from which he or she cannot willfully walk away. Real, alcoholic violence is felt in the bodies of both the alcoholic and his or her other who becomes a victim. It is "naked emotion," often raw and brutal.

The following account illuminates the differences between playful, accidental, and real violence.

> When my husband drank he would do crazy things that he didn't mean. But it got to the point where it didn't matter if he meant it or not. The first time he threw my daughter's teddy bear at me and missed and knocked over the vase of flowers on the dining room table and we both laughed at how absurd it was. We

thought something crazy was going on. How could he be mad at me if he threw a teddy bear at me? But it got worse. The next time he knocked me down the stairs and said it was an accident. I knew he was mad at me and he was trying to hit me. He said it was accidental. I never believed him. Finally he got out of hand. He came home slightly drunk. I asked him "How much have you been drinking?" He swore at me. Slammed the kitchen door. I guess he went to the basement and got a drink. He came back up with the drink in his hands and threw it at me. He missed and it went through the front window. Then he came at me, grabbed me by the throat, and started to shake me. He was screaming all the time. He said he hated me and wanted to kill me. I got loose. Ran out of the house. Called the police. They came and arrested him. I filed for separation after that. We're divorced now [field conversation, September 1, 1983, 42-year-old female, occupational nurse].

If violent interactions are to be interpreted as being present, then the other must be able to distinguish between playful, accidental, and real alcoholic mood states. He or she must also be able to understand those moments when *real violence* of a non-alcoholic nature is intended. In this situation the alcoholic is held, without doubt, fully accountable for these actions.

More often, however, alcohol is present in the situation when the alcoholic's violent emotionality erupts. His or her actions occur within the following message frame:

> Because I have been drinking all messages here are untrue.
> I don't want to be violent.
> I want to be violent.

The alcoholic and his other are trapped within this message system. *Paradoxical alcoholic violence* of a "real" order is produced and experienced, but seemingly it is non-intended. This is the fifth form of alcoholic violence. Hence the other must choose whether or not to listen to the alcohol speaking. He or she may discount the alcohol and listen to the words that were not spoken, but would have been spoken, if the alcoholic had not been drinking. If he or she is not willing to suspend disbelief in the words that were in fact spoken, there is no recourse other than to hear what in fact was said. He or she is trapped. If the other knows the alcoholic didn't mean what was said, then how can the other discount the effects of the emotional and physical violence just inflicted upon him or her? Further, if the other feels that the alcoholic has self-control over his or her drinking, then he or she believes that the alcoholic drank in order to be violent. In which case the alcoholic is doubly accountable for his or her actions.

The confluence of playful, accidental, real, and paradoxical violence within the sober and intoxicated interactional contexts that the alcoholic and his or her other produces causes recurring emotional chaos. This situation serves to locate emotional violence, in all its forms, at the center of the alcohol-dependent relationship.

Alcoholic Identities

The alcoholic relationship solidifies into a set of reciprocally expected, alcoholic identities that center on alcohol and drinking. Children may become scapegoats for the family's problems. In reaction to the alcoholic situation, they may become rebellious, withdrawn, or delinquent. They may attempt to become pseudo-parents, taking over the mothering or fathering responsibilities of the alcoholic. They may become family stars, or heroes, and they may become overachievers. What they will become, to use the currently employed phrase, are adult children of alcoholics (Woititz, 1983).

Spouses, family members, friends, lovers, coworkers, employers, physicians, and children become *enablers* and *coalcoholic dependents* in the relationship. These others assist the alcoholic when she gets in trouble. They buy the alcoholic drinks and bring alcohol home for her. They give the alcoholic money when she needs it. They become dependent upon the alcoholic's dependency and mold identities that place them in a "helping" relationship with her. Enablers often become *victims,* for they are victimized by the alcoholic's inability to meet the ordinary demands that have been placed upon her. Victims do the alcoholic's work for her. Victims may also become *provokers.* The provoker becomes the other who feeds back into the alcoholic relationship all the bitterness, resentment, fear, anger, and hostility that is felt when the alcoholic turns against him or her and attacks the provoker for attempting to control the alcoholic's drinking (Kellerman, 1969). Resenting such control, the alcoholic flails out. The victim in turn becomes a *martyr,* bearing the cross of the alcoholic relationship upon his or her shoulders. The victim resents the alcoholic's resentment.

Resenting this attitude, the alcoholic builds a deep hatred toward the victim. He or she provokes the alcoholic's anger. The victim is blamed for everything that goes wrong in the relationship. The alcoholic may drink in order to blot out the painful loss of control experienced from the victim's hands. A recovering alcoholic reported the following angry thoughts he held toward his mother while he was drinking:

> She was domineering and tried to control me. We lived together in a little house trailer she bought. She'd say "Now, Harry, don't drink today." I'd get mad as hell, storm out, go to the bar on the corner and have me a drink. I'd look at her face in the glass and drink to it. I'd say, "Set em up. I'm getting drunk today. She can't tell me what to do!" And I would [field conversation, 36-year-old male, in treatment for the third time, June 22, 1984].

THE ALCOHOLIC SITUATION AS A STRUCTURE OF CONTRASTING EMOTIONS

Contrasting emotional experiences, as first indicated, are layered through the alcoholic relationship. First charming and loving, then cruel and hostile, the alcoholic generates a field of experience that is negative, positive, alienating, ambiguous, ambivalent, and ultimately self-destructive. The alcoholic and his or her others stand at the center of this field of shifting, contrasting emotions. The following account speaks to this feature of their shared life. The speaker is at his first A.A. meeting. He offers the labels Jekyll and Hyde for the two sides of his emotionally divided self. In his mid-forties, he is married, has two children and works for a large accounting firm. His father and grandfather were alcoholics. Before he stopped drinking, seven days earlier, he had been told to leave work because he was drunk. Not believing this, he had checked into a local hospital to have his blood alcohol level tested. The test confirmed that he was, in fact, intoxicated. He speaks the following:

> When I drink I become another person. Like a Dr. Jekyll and a Mr. Hyde (or whatever they're called). I get violent. I swear, I throw things. Last Saturday, a

week ago, I threw the kitchen table at my father-in-law. I grabbed my wife (she only weighs 98 pounds) by the throat 'cause she said I was drunk when I came home. My little girls were hanging on my leg, telling me not to hurt Mommy! Christ! What's wrong with me? I'm not violent. I don't swear. I'm quiet. I always wear a smile. I'm easy going. Even when things are going bad I smile and say it'll work out. But I stop and have that first beer and the next thing you know I'm drunk and there till the bar closes. Then the wife's mad. Screaming at me when I come in the door. I feel guilty, mad. Mad at myself. Mad at her. Hell, I know I'm drunk. She don't have to tell me. Why'd she throw it up at me like that? I don't want to be like this any more than she wants me to be drunk. I get crazy, like last Saturday, last week. Then we don't talk. Now she's gone! Took the girls. Told me to get professional help. Are you people professional? I guess you must be cause you're not drinking. You must have something I don't have. Maybe I can get it. I'll be back [field conversation, September 25, 1984].

Self-pride lies behind this alcoholic's anger at his wife. He knows that he is drunk when he comes home. He knows that he has failed to control his drinking once again. His violence represents an attempt to regain a sense of self-worth, or self-pride, in the face of this humiliating situation. He resents her failure to acknowledge this fact. The structures of contrastive experience that are confronted by the alcoholic self and his other revolve around the negative emotions of self-hatred, fear, anxiety, anger, and violence, but most centrally the master emotions of *ressentiment* (Nietzsche, 1897: 35–39, 63–65, 77–78, 90, 110; Scheler, 1961: 39–44, 224–228, 283), and *self-pride* (see Denzin, 1986c).

Negative Emotionality

Restless and uneasy, the alcoholically divided self is always on the move, always seeking new experiences to fill, if only momentarily, the loneliness of self that is felt.

The emotions that divide the self alternate between momentary feelings of positive self-worth, and deep, underlying feelings of doubt, self-hatred, despair, and anguish. Alcoholic drinking, fueled by self-pride and risk-taking, is pursued in a futile attempt to overcome the deep inadequacies that are felt. The alcoholic is unable to sustain over any length of time a positive definition of self.

The following account addresses this aspect of the alcoholic self. The speaker is a 45-year-old advertising executive. He has been in A.A. for over 12 years, but has never had more than 11 months of continuous sobriety. He has been married and divorced. He drives expensive cars, wears expensive, yet casual clothes, and once owned an expensive home. He has been in over five treatment centers and has been exposed to reality therapy, EST, hypnosis, and Individualized Alcohol Behavioral Therapy. He has been addicted to valium and has smoked marijuana on a regular basis. When he starts drinking he begins with very dry Beefeater martinis, but usually ends with sweet liquors, including Southern Comfort and Irish Mist. He has been diagnosed manic-depressive, depressive, an alcohol addict, and a sociopath. He speaks of himself in the following words:

I have two parts to me. One part wants to take credit for what I accomplish. Like the magazine. It's beautiful. It's one of the best in the country. Best photos. Best layout, best writers. [He had published a monthly magazine for six months, before he lost it during his last binge.] It's just like me. Pretty on the outside, empty, nothing on the inside. Pretty boy, call me. Nothing in here [points to his chest].

You see, that's why the other part of me wants to destroy what the good part creates. I don't deserve for good things to happen to me. The sick side of me says destroy it all. If that side gets too strong I drink. I want to drink. And I do, and once I have the first drink the alcohol starts to talk to me and it says have another drink. You're rotten. Run, you don't deserve what you have. Blow it all. Run and leave. Kill yourself. You won't live to be 50 anyway. Remember your old man killed himself with this stuff when he was 45.

These thoughts fill me up. I can't get away from them. I feel guilty if I do well. I'm so damned sick, even when I'm well I'm sick. Pretty Boy on the outside [looks vacantly outside through the windows]. Pretty Boy's sick. Ready to go? [field conversation, April 1, 1983].

Uniqueness and Fear

The alcoholic self cultivates a particular moral individualism that rests on the felt uniqueness of its own alienation. The alcoholic cuts herself off from the rest of society, feeling a profound alienation from all with whom she comes in contact. The alcoholic self nurtures an inner goodness, seeking to hide within a solitude that is solely hers. Rejecting objective experiences with others, the alcoholically divided self lives within the private world of insulated madness that alcohol produces. She assigns a sense of moral superiority to this inner world that is uniquely her own (see Laing, 1965: 94–95). From the unsteady vantage point of this alcoholically produced position, the alcoholic subject directs a world that threatens to come apart at any instant. All the while she lives an inner fear, knowing that the false walls she has built will crumble at any moment.

Consider the following account given by a recovering 44-year-old alcoholic, four years in A.A.

I felt that nobody else knew how to live their life. I felt that my way was superior; that I knew something other people didn't. It was impossible for me to work or deal with others because I felt this way about them. I lived my life in my studio and in my tiny office where I worked. I stayed at home as much as I could. When I drank I felt that I knew who I was and what was right. In those feelings I gained a strength and a way of thinking that was unique to me. But then I collapsed. Everything came in on me. I went crazy. I came to A.A. and I found out that my way, that all the certainties I had held onto, that all my thoughts were empty too. I didn't know how to live either! I'm still learning [field conversation, June 17, 1984].

ALCOHOLIC RESSENTIMENT

Ressentiment (Scheler, 1961: 39–40) toward the past, toward others, toward the present and the future is felt. The repeated experiencing and reliving of this temporal and emotional attitude toward others and toward time itself characterizes the alcoholic self. The emotional attitude of ressentiment is negative, hostile, and includes the interrelated feelings of anger, wrath, envy, intense self-pride, and the desire for revenge

(Denzin, 1984a: 283). Behind alcoholic ressentiment lies alcoholic pride (see below). This emotional frame of reference is fueled by alcohol that elevates the alcoholic's feelings of self-strength and power. However, the power that is felt is illusive and fleeting, always leaving in its wake the underlying ressentiment that the alcoholic drank to annihilate.

The centrality of resentment in the alcoholic relationship is revealed in the following interaction between J, and A. A, the male companion of J, has recently started attending Al-Anon meetings. He has been sponsored by a woman who has been in Al-Anon since 1968. Her name is Mary. J is speaking to C and K, two recovering alcoholics who have come to see him. J has been drinking for 2 days, after having been sober for 30 days. He found a full bottle of vodka in a closet when he was cleaning his and A's house.

> Come in, I'm drunk again. Here, come meet Mary [pointing to A who is seated at the dining room table]. She's been going to Al-Anon. Haven't you, Mary! Oh my, we're so much better now. Aren't we, Mary? Mary knows how to handle me now. Mary taught her "Tough Love." Well, YOU can take your TOUGH LOVE, A, and shove it. Oh, look at him, I hurt his little feelings. I'm so sorry, A, you see, I'm a sick man. You can't get mad at me. You have to love me [field conversation, as reported, September 16, 1984].

The satire, sarcasm, and double meanings that are evident in J's monologue, including his referencing of A with the feminine Mary, while he, J fills the feminine, housewife position in the relationship, speaks to the inability of the alcoholic to form "a true partnership with another human being" (A.A., 1953: 53). Using alcohol to reverse his position in the relationship, J attempts to dominate A through his words and his actions. As he expresses his hurt feelings through retaliatory vindictiveness he drives A further away from him. Resentful over A's attendance at Al-Anon meetings, he blocks and rebuffs any efforts by A to move forward in the relationship.

The return to drinking by J brings to the surface latent ressentiment that exists in his alcoholic relationship with A. Unable, or unwilling, to free himself of the anger, fear, wrath, and envy he feels toward A, his desire for revenge surfaces as he drinks.

Alcoholically induced and experienced emotionality lies at the heart of J's alcoholism. The euphoric and depressive physiological and neurological effects of the drug alcohol magnify the emotional divisions that exist within his self. His relations with A are similarly distorted when he drinks. Alcohol has become a sign, which when brought into the relationship, signifies J's alienation from A.

John Berryman (1973: 102–103) describes the telephone interactions Wilbur, an approximately 60-year-old recovering patient, had with his parents, whom he called two to five times daily from the treatment center:

> "They need me," he said many times. His father, drunk from morning to night was urging him to come home, and Wilbur was anxious to go although they fought like madmen right through every call and it was perfectly clear to everybody but Wilbur that he got a *bang* out of holding his own against his frightful Dad from a safe telephone distance (Father's ax, kept in the kitchen where *he* drank: "I'll chop you, Wilbur, late one night, I'll chop you"), reveling in the fact that he himself was not only sober at the moment but being *treated* for the disease that was killing them both. "They need me," he said stubbornly.

The fuel for stubbornly felt ressentiment is present in this account, including Wilbur's desire for revenge. The key to ressentiment is the repeated experiencing and reliving of the emotional feelings that draw the subject to another. Each occasion of interaction between Wilbur and his father becomes an occasion for the production of new feelings of anger, wrath, and perhaps hatred. These new feelings become part of the emotional repertoire that binds the father and son together. They experience the new feelings of ressentiment against the backdrop of all previous negative interactions. Drawn to one another through negativity, they both know that the self-feelings they derive from this relationship can be found nowhere else. The father and son are participants in the classic alcoholic relationship in which both partners are alcoholic. The complexity of their interaction is increased by the fact that they stand in a father-son relationship to one another. They both exist as alcoholically divided selves. The pivotal emotions that join their emotionally divided selves are lodged in the ressentiment they share and produce together.

The anger that the alcoholic feels toward his or her other turns into hatred. That hatred, when reciprocated, as in the case of Wilbur and his father, is increased. It destroys the affection that perhaps once existed between them. Thus it is with alcoholic ressentiment, for hatred and anger are layered upon one another, until the interactants in the social relationship are bound together only through negativity and mutual self-disdain.

Self-Pride

The threats of self-pride that ressentiment magnifies are evidenced in the following statement. The speaker is J, the male homosexual quoted above. He is intoxicated as he speaks. His remarks are directed, as before, to A:

> I hate you. You're cruel to me. You're a pig! You're not half the man W is. You can't hold a candle to him. You take away my respect. You don't tell me my flowers are nice. You've been mean to me. You undercut me. You make me feel ugly. My sister loves me. My aunt loves me. I don't care if you do hate me. You make me hate myself. When I'm sober you can't do this to me. I feel good about myself. I feel proud to be a man. Why do you make me drink like this? I want to feel good about myself and have some pride again in what I do [field conversation, July 15, 1984].

Here the speaker blames the other for making him drunk, but he also blames his other for undercutting his self-pride. The ressentiment that flows through his talk masks a desire to regain a sense of self he feels he has lost to A, but also to alcohol. Pride in self, then, organizes and structures the alcoholic's deep-seated feelings of anger and wrath toward the other. The desire for revenge is embedded in this emotional attitude, which is best interpreted within the larger framework of ressentiment toward the other and toward self.

End of Act One

The drinking alcoholic and his or her other(s), to summarize Act One, live inside a recurring structure of negative, contrastive emotional experiences that become self-addictive. Each member of the relationship is trapped in this cycle of destructive emotionality, which becomes, for the alcoholic, nearly as addictive as the alcohol that he or she consumes. The alcoholic and his or her other live hatred, fear, anxiety, anger, and ressentiment. Never far from the next drink, or the next emotional battle with his or her significant other, each day becomes a repetition of the day before. They have stabilized their relationship at this brittle level. Act Two begins with either (1) the eruption of violence that goes beyond emotional attacks on the other; or (2) the alcoholic producing a problematic situation that traps him or her and exposes his or her alcoholism to others. Both of these sequences of action provoke a radical restructuring in the alcoholic's relationship with himself or herself and the significant other.

QUESTIONS FOR DISCUSSION

1. The author refers to the "divided self." What does he mean by this term?

2. The author suggests that becoming an alcoholic is like being in a play. Discuss the alcoholic situation.

3. Alcoholism is often associated with violence. What are the five forms of violence described in this article?

Some People Can Really Hold Their Liquor

Craig MacAndrew and Robert B. Edgerton

The conventional understanding of alcohol's workings has it that once alcohol is inside us, its toxic action produces changes in two fundamentally different kinds of behavior. First, its pernicious action on our innards is held to result in a marked impairment in our ability to perform at least certain of our sensorimotor skills. And with this aspect of the conventional understanding we have no quarrel. In the face of the available evidence, it is impossible to imagine how one could seriously doubt that the presence of alcohol in the body does, in fact, produce various sensorimotor performance decrements. Nor can one dispute the fact that alcohol in sufficiently high concentrations produces grave and even fatal bodily malfunctioning. In this profound sense, then, there can be no question that alcohol *is* the potent change-producer that everyone claims it to be.

The conventional understanding also insists, however, that alcohol depresses the activity of "the higher centers of the brain," thereby producing a state of affairs in which neither man's reason nor his conscience is any longer capable of performing its customary directive and inhibitory functions. It is with this aspect of the conventional understanding that we now propose to take issue.

"Some People Can Really Hold Their Liquor" from *Drunken Comportment: A Social Explanation* by Craig MacAndrew and Robert B. Edgerton, 1969, pp. 13–36. Reprinted with permission of the authors.

We have already noted that, despite its near unanimous acceptance, the formal status of this supposition that alcohol is a "moral" as well as a sensorimotor incapacitator is only and at best that of an hypothesis. Just what, after all, is the empirical warrant for the contention that alcohol reduces man to the status of a mere creature of his now unrestrained impulses? Well, the argument goes, one has but to open his eyes and look around, for the evidence is everywhere. Haven't we all seen people do things when they were drunk that they would not *think* of doing (would not think *seriously* of doing) when they were sober? To this we would reply that of course we have. But we would hasten to add that this is not *all* we have seen; and as John Dewey never tired of reminding us, we must accept consequences impartially.

In point of fact, while changes in comportment of the sort that we customarily construe as disinhibited are certainly a sometime corollary of drunkenness, they are anything but an *inevitable* corollary. While the sheer occurrence of changes between one's "sober" and one's "drunken" comportment is beyond question, it is an equally incontestable fact that these changes are of a most incredible diversity. Relative to our comportment when sober, we may, for instance, become boisterous or solemn, depressed or euphoric, repugnantly gregarious or totally withdrawn, vicious or saintly, ready at last to say our say or stoically noncommittal, energetic or lackadaisical, amorous or hostile . . . but the list could be continued for pages. The point is that, with alcohol inside us, our comportment may change in any of a wondrously profuse variety of ways. Indeed, it is precisely this variability that constitutes the problem. For how can the conventional understanding of alcohol *qua* disinhibitor possibly accommodate the fact that even within our own culture people who have made their bodies alcoholled differ so drastically both within and among themselves in their subsequent doings? Thus, while we are all aware, for instance, of the connection between drunkenness and such things as promiscuity and crime, we also know that everyone who gets drunk does not, *ipso facto,* become promiscuous and/or criminal. Furthermore, even in regard to those who do, we know that they do not comport themselves in such a fashion on *every* occasion that they lift a few. How is it, we would ask (and the question is not an idle one), that the same man, in the same bar, drinking approximately the same amount of alcohol, may, on three nights running, be, say, surly and belligerent on the first evening, the spirit of amiability on the second, and morose and withdrawn on the third? Are our impulses—the very wellsprings of conduct, by most accounts—really so frivolous? And what, finally, of the person about whom we say, "He can really hold his liquor?" In the absence of anything observably untoward in such a one's drunken comportment are we seriously to presume that he is devoid of inhibitions?

But there is little to be gained by laboring this point any further, if only because everyone—including even the most vociferous spokesman for the conventional wisdom—is already perfectly familiar with these sorts of puzzles as they pertain to drunken comportment in our own society. Familiarity, in this case, however, would seem to have given rise to nothing; for despite their patent relevance to the evaluation of the conventional understanding, the puzzles have been and continue to be ignored.

How, then, to proceed? Perhaps by turning from the seen-but-unnoticed puzzles with which the variability of drunken comportment confronts us in our everyday lives and looking instead to "the same" phenomena as they naturally occur in cultures which are foreign to our own, we shall stand a better chance of creating that necessary "shock of recognition" which apparently must precede a willingness to take these everyday puzzles seriously. This, at any rate, is our hope. For whatever may be the final judgment as to the merit of the alternative formulation that we shall advance in later chapters, of this much we are firmly convinced: the conventional explanation cannot possibly account for what is actually going on.

One or another form of alcoholic beverage has long been consumed in the vast majority of the world's societies. Although the descriptive literature on these hundreds of societies usually fails to provide specific accounts of how people actually conduct themselves when drunk—and where mention is made, it typically takes the form of but a passing remark—the ethnographic literature does contain sufficient documentation to support the following generalization: When people are drunk, not only are various of their sensorimotor capabilities impaired (there are no exceptions in this regard), their comportment often changes as well. And to this we would add as a secondary generalization that when changes in comportment *are* reported, more often than not they take the form of "changes-for-the-*worse*."

The Abipone Indians are a good example of a people whose comportment undergoes such a metamorphosis once they have alcohol inside them. When first described in the early 1800's, the Abipone were warlike, tent-dwelling horsemen who traversed the great plains of the Paraguayan Chaco. Martin Dobrizhoffer (1822) describes their natural state as follows:

> The Abipones, in their whole deportment, preserve a decorum scarce credible to Europeans. Their countenance and gait display a modest cheerfulness, and manly gravity tempered with gentleness and kindness. Nothing licentious, indecent, or uncourteous, is discoverable in their actions. In their daily meetings, all is quiet and orderly. Confused vociferations, quarrels, or sharp words, have no place there. They love jokes in conversation, but are adverse to indecency and ill-nature. If any dispute arises, each declares his opinion with a calm countenance and unruffled speech: they never break out into clamours, threats and reproaches, as is usual to certain people of Europe. These praises are justly due to the Abipones as long as they remain sober: but when intoxicated, they shake off the bridle of reason, become distracted, and quite unlike themselves (Vol. 3, p. 136).

"Quite unlike themselves" indeed, as Dobrizhoffer's sharply contrasting account of Abipone comportment during a drinking party vividly illustrates:

> Disputes are frequent among them concerning preeminence in valour, which produce confused clamours, fighting, wounds and slaughter. . . . It often happens that a contention between two implicates and incites them all, so that snatching up arms, and taking the part, some of one, some of the other, they furiously rush to attack and slay one another. This is no uncommon occurrence in drinking parties and is sometimes carried on for many hours with much vociferation of the combatants, and no less effusion of blood (Vol. 2, pp. 436–37).

Clearly, this transformation in Abipone comportment from a taciturn civility to outright savagery is admirably in accord with what one would expect if alcohol were in fact the disinhibitor that the conventional wisdom takes it to be.[1] Nor are similar examples hard to come by; societies in which such Jekyll-to-Hyde transformations occur are to be found in virtually all parts of the world.

But, and here is the rub, it is not *always* so. Consider, for example, the anthropologist C. Nimuendajú's (1948) account of the Yuruna Indians, a warlike, headhunting tribe living in the Xingu region of South America's tropical forest. While the Yuruna drink substantial quantities of *malicha* (which they make from fermented manioc root), not only do they fail to become "disinhibited," they withdraw entirely into themselves and behave much as though no one else existed (p. 238). Nor can it be said that the Yuruna constitute "the exception which proves the rule" (whatever that may mean). The Yuruna are but one of many societies in which persons consume appreciable, and in some cases prodigious, quantities of alcohol without displaying anything like the wholesale changes-for-the-worse that characterize Abipone drunkenness. In fact, in many of these societies there is scant evidence of *anything* that might reasonably be termed "disinhibited." Over the course of this chapter, we shall want to examine several such societies which have recently been studied in some detail.

What we actually find when we examine the phenomenon of drunkenness as it occurs throughout the world is a series of infinite gradations in the degree of "disinhibition" that is manifested in drunken comportment. Because of this, it might be well to begin by taking a closer look at a society in which the inhabitants' comportment falls somewhere *between* that of the Abipone and that of the Yuruna. For an example of such a society we move to the Peruvian Andes, to the Indians of Vicos as described by the anthropologist William Mangin (1957). The *hacienda* of Vicos is a largely self-contained community inhabited by approximately 1,800 Indians who, being in large measure both geographically and culturally isolated from the ongoing "life of the nation," have retained much of their traditional way of life. What of drinking and drunkenness in this *hacienda?* Ceremonial drinking has existed in Peru since pre-Conquest times, and to this day drinking remains an integral feature of both formal and informal community life in Vicos. Here is Mangin's depiction of the present-day drinking practices of these Indians:

> In Vicos, small children are given corn beer and everyone over 16 years of age drinks (*aguardiente*). Drinking by most adults, particularly adult males, is usually followed by drunkenness, and in many instances a man or woman may be drunk for several days in succession. The incidence and frequency of drinking and the amounts consumed seem to be very high. Drinking is a social activity, however, and drinking customs are integrated with the most basic and powerful institutions in the community. Drinking and drunkenness do not seem to lead to any breakdown in interpersonal relations, nor do they seem to interfere with the performance of social roles by individuals (p. 58).

Mangin's overall evaluation is that for the Vicosino the role of alcohol is "prevailingly integrative" in nature. While he does suggest that the frequency of violence seems to increase when people are drunk, his statement is a cautious one, for he goes on to add that "drunk or sober, there is actually very little violence in Vicos. Few fights were noted during the field investigation. And of those noted, the same two men seemed to be involved most of the time" (p. 63). Regarding other manifestations of what we usually consider to be "drunken impulsivity," he writes:

> Sexual activity of a premarital variety appears to increase during fiestas, but this may not be a function of drinking. Several male informants told the author that they purposely stayed sober at times during fiestas so that they could "escape" with a girl. Extramarital sexual activity, which is a disruptive force in Vicos culture, seems to occur mostly when individuals are quite sober. . . . Most of the crimes committed by Vicosinos during the field study were carried out while sober. In only two cases (one of which occurred before the [21 month] field study and one during it) could it be documented that drunkenness was associated with criminality (p. 63).

Thus, while something or other of an apparently "disinhibited" sort may occasionally occur when the inhabitants of Vicos are drunk, the paucity of specifically drunken transgressions contained in the above roster places Vicosino drunken comportment in sharp contrast to that of the Abipone.

We turn now to an examination of five societies in which not even *this* degree of drunken "disinhibition" is to be found. For our first example, we select the anthropologist Dwight Heath's (1958) account of drinking among the Camba of Eastern Bolivia. The Camba, who number approximately 80,000, are a mestizo people, the descendants of colonial Spaniards and indigenous Indians. Having rejected traditional tribal ways of life, a few Camba live as independent land-owning agriculturalists or as squatters, but most work as tenants on one or another of the vast *haciendas* into which the area is divided. From the owners or managers of these *haciendas* they receive food, housing, clothing, and a small wage in exchange for their labor. Because of the primitiveness of the infrastructure of the Bolivian economy and the nature barriers that limit even nonmechanized transport, those Camba whom Heath studied have been effectively isolated from extensive contact with other population centers. As a result, their way of life has undergone only slight modification since the Spanish colonial period.

As for their drinking, the Camba drink a distillate of sugar cane which, with good reason, they call *alcohol*. Chemical analysis has shown this beverage to be 89 percent ethyl alcohol—approximately twice as potent as a good scotch or bourbon. And the Camba drink this beverage *undiluted!* Furthermore, with the exception of a few who have joined one or another fundamentalist Protestant sect, all Camba drink and "most of them become intoxicated at least twice each month" (p. 498). Because Heath's (1958) description of Camba drinking practices is unusually rich in detail, it warrants extended quotation:

> The behavioral patterns associated with drinking are so formalized as to constitute a secular ritual. Members of the group are seated in chairs in an approximate circle in a yard or, occasionally, in a hut. A bottle of *alcohol* and a single water glass rest on a tiny table which forms part of the circle. The "sponsor" of the party pours a glassful (about 300 cc.) at the table, turns and walks to stand in front of whomever he wishes, nods and raises the glass slightly. The person addressed smiles and nods while still seated; the "sponsor" toasts with "Salud" (health), or "A su salud" (to your health), drinks half of the glassful in a single quick draught, and hands it to the person he has toasted, who then repeats the toast and finishes the glass in one gulp. While the "sponsor" returns to his seat, the recipient of the toast goes to the table to refill the glass and to repeat the ritual. There are no apparent rules concerning whom one may toast, and in this

> sense toasts proceed in no discernible sequence. A newcomer is likely to receive a barrage of toasts when he first joins a drinking group, and sometimes an attractive girl may be frequently addressed, but there tends to be a fairly equal distribution of toasts over a period of several hours. To decline a toast is unthinkable to the Camba, although as the party wears on and the inflammation of mouth and throat makes drinking increasingly painful, participants resort to a variety of ruses in order to avoid having to swallow an entire glassful of *alcohol* each time. These ruses are quite transparent (such as turning one's head aside and spitting out a fair portion) and are met with cajoling remonstrances to "Drink it all!" Such behavior is not an affront to the toaster, however, and the other members of the group are teasing more than admonishing the deviant. After the first three or four hours virtually everyone "cheats" this way and almost as much *alcohol* is wasted as is consumed. Also, as the fiesta wears on, the rate of toasting decreases markedly: during the first hour a single toast is completed in less than two minutes; during the third hour it slows to five minutes or more. A regular cycle of activity can be discerned, with a party being revived about every six hours. When a bottle is emptied, one of the children standing quietly nearby takes it away and brings a replacement from the hut. When the supply is exhausted, members of the group pool their funds to buy more; they send a child to bring it from the nearest seller.
>
> The ritual sequence described above is the only way in which the Camba drink, except at wakes where a different but equally formalized pattern of behavior is followed (pp. 499–500).

Heath (1958) also provides an hour-by-hour account of the first of the repetitive cycles into which these extended drinking bouts are divided:

> The Camba usually begin drinking shortly after breakfast. . . . as a party wears on, the effects of intoxication become apparent. After two or three hours of fairly voluble and warm social intercourse, people tend to become thick-lipped and intervals of silence lengthen. By the fourth hour there is little conversation; many people stare dumbly at the ground except when toasted, and a few who may have fallen asleep or "passed out" are left undisturbed. Once a band or guitarist starts playing, the music interminable and others take over as individual players pass out. The sixth hour sees a renewed exhilaration as sleepers waken and give the party a "second wind." This cycle is repeated every five, six, or seven hours, day and night, until the *alcohol* gives out or the call to work is sounded (pp. 500–501).

Clearly, there can be no question that the Camba get very drunk indeed; gross sensorimotor incapacitation could hardly be more evident. But what of their comportment, and what, specifically, of drunken "disinhibition"?

> Among the Camba drinking does not lead to expressions of aggression in verbal or physical form. . . . Neither is there a heightening of sexual activity: obscene joking and sexual overtures are rarely associated with drinking. Even when drunk, the Camba are not given to maudlin sentimentality, clowning, boasting or "baring of souls" (p. 501).

Actually the only change Heath noted between the Cambas' sober and drunken comportment was an increase in "volubility" and "self-confidence." And even this he observed not in the acute stages of intoxication, but only in the early stages of drinking and upon recovery of consciousness. In sum, the Camba drink often and to a point of intoxication that is extreme by anyone's standards, yet the expression of impulsivity is quite unaffected thereby. So much, then, for the "disinhibiting" effects of alcohol upon the Camba.

For a second example of a society in which disinhibition fails to occur during drunkenness, we turn to Gerardo and Alicia Reichel-Dolmatoff's (1961) account of life in the small mestizo village of Aritama, located in northern Colombia. Situated in the tropical foothills of the Sierra Nevada de Santa Marta, Aritama lies in a region that is both geographically and culturally intermediate between that of the more advanced Creoles of the urban and rural lowlands and the autochthonous Indian tribes of the higher mountains. While the Creole lowlanders consider Aritama a backward Indian village, the majority of its inhabitants manifest a tri-ethnic mixture of Indian, Caucasian, and Negro traits. Although their culture is not so untouched as that of the Indians, neither is it so developed as the Creoles, being now at a stage through which the latter have already passed. As contacts with the lowland towns have increased, the inhabitants of Aritama have become acutely aware that many of their customs and ways are "unprogressive"; and change is everywhere in evidence.

As for the character of life in present-day Aritama, we would begin by noting that the people are inordinately self-conscious. Whether with neighbor or with stranger, "They are always afraid of giving themselves away somehow, of being ridiculed because of the things they say or do, or of being taken advantage of by persons in authority" (1961, p. xvii). Not only do they avoid close personal relationships, they have firm cultural support for so doing. Thus, the Reichel-Dolmatoffs (1961) inform us that merely "to ask personal questions and to show interest in other people's lives is, according to local standards, one of the worst breaches of proper conduct" (p. xvii). While they live their lives behind a mask of formal politeness, hostility is endemic to Aritama—as witness, for instance, the fact that immortality is deemed desirable "only insofar as it offers the opportunity for the spirit to take revenge that he was unable to take during his lifetime" (p. 281). So controlled are the adult males that it was only during cock fights—which, interestingly enough, are the sole sport in which any interest is shown—that they were observed to throw off their deep-seated reserve and, in their excitement, become truly spontaneous.

As for the nature of the relations between the sexes, Aritama women look to their unions (whether they be marriage, free union, or concubinage) almost exclusively in terms of economic security, while men see above all a chance to demonstrate their virility by having as many children as possible. Not only are such relationships devoid of mutual trust, the Reichel-Dolmatoffs (1961) state flatly that in all homes, whether or not they are sanctioned by Catholic marriage, "the dominant impression is one of open hostility" (p. 186). Nor is any attempt made to hide this state of affairs. Quite the contrary, while fighting typically takes the form of verbal abuse and violent gestures, physical assaults also occur. Where relative tranquility does prevail between the partners, it is a subject of ridicule; neighbors or friends "try in every way to disturb the apparent harmony by gossip, false accusations, and open insinuations of sorcery" (1961, p. 190). Indeed, hostility between the sexes is so much a part of things that the Aritamans find it quite impossible to accept the fact that in other places couples lives together in peace and happiness.

The children of such unions are not desired in and of themselves, but only for their "asset-value." Even in earliest infancy, the child is seldom, if ever, fondled. When he is handled, he is handled roughly—"like a dead weight." Mothers consider suckling to be both tiresome and physically debilitating. When weaning occurs, it is abrupt, being accomplished by rubbing lemon juice or chili peppers on the nipples. The whole affair is taken as a joke and no one pays any serious attention to the infants' consequent rage. The fathers are, if anything, even less supporting; at best they are indifferent to their newborn and at worst they openly verbalize a profound loathing. Throughout childhood a father will rarely even touch his child save to administer punishment. Nor is the death of one's baby likely to evoke deeply felt sorrow. The Reichel-Domatoffs (1961) report four recent cases in which small babies died "simply because their mothers were unwilling to make the effort to feed them properly . . ." (p. 89). They report that public opinion took scarcely any notice of these deaths.

Child rearing in Aritama is predicated upon the notion that both "good" and "evil" character traits are inherited; and because of this, a child's "bad" behavior can never be blamed upon his parents' *doings*. Each parent, often in the presence of the other parent, points out to the child the undesirable traits of the other as examples of how one ought *not* to behave. In consequence, once the child understands the nature of these sundry accusations, his respect for his parents is all but obliterated. Thus, while obedience and respect are taught from infancy on, hostility between the generations is the rule: "Both boys and girls throw stones at adults, strike them when angry, or insult them with obscene words. Adults with physical handicaps or old people are mocked and insulted; animals are often beaten mercilessly . . ." (1961, p. 103). Nor is there affection between siblings: "The older ones take away the youngers' food or playthings, beat and push them whenever they can, and try to make them cry" (p. 102).

In adolescence, however, all of this suddenly changes:

> His former mobility of facial expression and gesture is transformed. His face becomes a rigid mask of "seriousness" which betrays no emotion. The spontaneity of play gives way to silent deliberation. All conversation is dominated by extreme caution so that no word may betray inner feelings. Routine questions as to health, family, or work are answered with monotonous formulas, and any other inquiry is answered with a stereotyped "I don't know, maybe" (*no sé, quizás*). . . . A tremendous mechanism is set into motion completely hiding the individual behind a wall of control and formality. To smile, to laugh, to talk, to ask questions, to joke about people and things, these are now improper. All the former manifestations of aggressiveness are gone; no anger, no fit of sudden rage, no obscene language are now indulged in. In their place are aloofness and apparent indifference. But there is also a certain troubled alertness, seen in the quick furtive glance, in the nervous twitching of the hands, in the halting walk. An exaggerated self-consciousness makes the youth behave as if he were continuously watched, criticized or, worse still, ridiculed. This inner tension which characterizes the adolescent continues into adulthood, no balance ever being achieved, it seems (1961, pp. 112–13).

The stage thus set, we now ask what happens to this "rigid mask of seriousness" when the Aritamans get drunk? Although illegal, some thirty to forty primitive distilleries are operating in the general area of Aritama, and the rum they produce enjoys a reputation for quality that extends far beyond the immediate area. In Aritama, if anywhere, we would seem to have every reason to expect that this rum would be consumed in great quantity by the inhabitants and that it would play merry hell with the veneer of formal politeness and controlled "seriousness" with which they confront the world. This is not, however, what we find. Instead, the Reichel-Dolmatoffs (1961) report that although boisterous drunkenness is very prevalent in the Creole lowlands, in Aritama "Even during the fiesta season, at marriages, baptisms, or wakes, one hardly ever sees an intoxicated person, and *those who can be seen are unobtrusive and silent*" (pp. 196–97; emphasis ours).

Although by prevailing Latin American peasant standards the people of Aritama are neither heavy nor regular drinkers, drinking to the point of drunkenness does occur. But the resulting comportment is scarcely what one would expect from the members of so inhibited a society:

> The various stages of intoxication are quite characteristic. After the first euphoria accompanied by small talk and a few jokes there may be some singing, and someone may go and bring a drum and play it for a while, but soon all conversation stops and gloominess sets in. There is never open physical aggressiveness of serious proportions, nor is there merry socializing, romantic serenading, obscene talk, or political discussion of any kind. One man will sing, perhaps, another play the drum or rattle, while the others sit and listen, drinking in silence and only rarely making physical contacts or attempts at conversation (1961, p. 197).

Neither is the drinking bout the occasion for the youth of Aritama to "sow their wild oats":

> Boys of fifteen or sixteen years will occasionally spend a night singing, drinking, and playing music in the company of older men who invite them on such sprees. However, the youth rarely seems to enjoy drinking and takes part in such nightly adventures mainly to demonstrate his new manliness. But as soon as the new status is achieved, i.e., when the boy has left his home, such sprees become rare and are marked by increased seriousness. A man might drink and drum all night long without once losing his composure, without becoming aggressive, sentimental, verbose, or amorous (1961, p. 113).

Even among the gravediggers, a group whose occupational task constitutes one of Aritama's few institutionalized occasions for drinking, the resulting drunkenness is not counted as a reward, but an occupational hazard. That it is totally devoid of enjoyment is obvious from the following account:

> They work and drink in silence with tense, harassed faces, digging, carrying stones, drinking again and again until they fall to the ground. Some lie in a stupor in the grass covering the graves of strangers or close relatives; others stumble between the mounds offering their bottles to each other. But there is no joking or storytelling. Only once in a while a man will say, "So we are digging a grave! Doing a good job at it, too. Working hard, drinking hard. Now we are drunk and the job is done" (1961, p. 383).

It is apparent, then, that in Aritama, regardless of the occasion and regardless of the degree of intoxication that is achieved, the "rigid mask of seriousness" remains firmly in place. In Aritama, as with the Camba, it is evident that alcohol lacks the power to "disinhibit."

For our third example, we turn to Micronesia, specifically to the small atoll of Ifaluk, which is located about midway between Yap and Truk in the heart of the Caroline Islands. This half square mile of coral rock and sand, with a total population of approximately 250, has recently been studied by the anthropologists Burrows and Spiro (1953) and by a team of natural scientists led by Marston Bates (Bates and Abbott, 1958).

Aside from an occasional trading ship, Ifaluk's contact with the West was virtually nonexistent through the nineteenth century. Under the nominal control of first Spain and then, briefly, Germany, the Caroline Islands were taken over by Japan in 1914; following World War I they were mandated to Japan by the League of Nations. The Japanese retained control until World War II, in the aftermath of which the Carolines became an American trust territory. Through it all, however, everyday life on this small atoll has been little affected. Perhaps by good fortune, Ifaluk was gifted with neither the population, the strategic position, nor the raw materials to warrant more than passing stabs at "civilizing" by any of this array of controlling powers.

The people of Ifaluk hold that their population was once much larger than it is now, and there is a good deal of evidence to suggest that this is so. For example, while there are 60 married couples in the present population, there are only 81 inhabitants under 20 years of age (Burrows and Spiro, 1953, p. 5). Since the married population is not reproducing itself, it is small wonder that in their regard for the infant there is a total reversal of the Aritama pattern. Here infants are highly prized and given the most profuse attention and indulgence. Thus, Spiro (Burrows and Spiro, 1953) writes:

> The infant is idealized and indulged to a degree that is unthinkable according to Western standards. . . . no infant is left alone, day or night, asleep or awake, until it can walk. To isolate a baby would be to commit a major atrocity, for if a baby is left alone, "by and by dies, no more people" (p. 257).

And again:

> Babies are constantly being handled, kissed, hugged and played with. The people have an inordinate desire to fondle babies, and babies are always the center of attention and of attraction in every home and in every gathering. . . . A baby knows only smiling and laughing faces, soft arms and soft words. No baby is ever handled roughly, no baby is scowled at or spoken to with bitterness (p. 245).

Verily, on Ifaluk the infant is king! But by the age of four or five, innocence is lost and so, too, is this affectual paradise. Overt signs of affection are abruptly withdrawn, and the young child is left largely to his own devices. Spiro (Burrows and Spiro, 1953) sums up the starkness of this contrast between age levels as follows: "The minutest frustration of the infant is attended to immediately, whereas much greater frustrations of the child are not only ignored, but often provoke amusement" (p. 275). It does not take great powers of insight to anticipate the consequences of so abrupt a transition: temper tantrums, rivalry for the attention of adults, generalized negativism, fighting, etc.

But on so small an atoll, aggression cannot be allowed to go unchecked. The people of Ifaluk wisely regard cooperation to be the primary social value, and aggression "the most heinous of offenses" (Burrows and Spiro, 1953, p. 276). Thus, from the time that children can speak, their training is directed with an awesome single-mindedness, and with no little harshness, to stamping out all manifestations of discord. And, "Despite the high incidence of aggression in children, this goal of the socialization process is attained with unqualified success, for there are no overt expressions of hostility in the interpersonal relations of Ifaluk adults" (p. 278).

That the resulting harmony is not all that it might be, however, can be seen in their treatment of dogs, for to be a dog on Ifaluk is truly to "lead a dog's life." Spiro (Burrows and Spiro, 1953) notes that "The children, as well as the adults, mistreat the dogs. . . . Children often kick the dogs, pull their tails, and maltreat them in other ways" (p. 278). In a word, Ifaluk, like Aritama, would seem to present a paradigm case of a society in which aggressiveness, inhibited under normal conditions, ought to be manifested with a vengeance under conditions of drunkenness.

Once again, however, this is not what we find. Life on Ifaluk depends on the abundant presence of the coconut palm; its fruit is both the people's basic dietary staple and the source of their alcoholic beverage. The juice, left unstrained, ferments in four days, producing an intoxicant that has been described (Bates and Abbott, 1958) as having "about the strength of good beer" (p. 77). That the ingestion of this intoxicant is often accompanied by dramatic changes-for-the-worse is evidenced by accounts from many parts of Micronesia. Speaking of the consequences of its use on the island of Butaritari in the Gilberts, Robert Lewis Stevenson (published 1912), not without reason, termed it "a devilish intoxicant, the counsellor of crime" (p. 238). That Stevenson's remark was more than an example of mere poetic license is indicated in many other early accounts of life in this area. Consider, for instance, F. W. Christian's (1899) sarcastic account of the aftermath of toddy-drinking festivals in the Gilberts:

> These merry meetings invariably terminated in a fierce free-fight, where men
> and women joined in the melee with ironwood clubs and wooden swords, thickly
> studded with sharks' teeth, with which they inflicted ghastly lacerations (p. 165).

And on the islands even closer to Ifaluk, there are many accounts that document a history of violence and discord in the wake of toddy-drinking.[2] Indeed, it was because the chiefs of Ifaluk had observed the deleterious consequences of toddy-drinking on some of the neighboring (and more Westernized) atolls that "they had decided that for the peace and security of the island it would be best if even fermentation were not allowed" (Bates and Abbott, 1958, p. 77). However, "their prohibition works about as well as that tried for some years by the United States. That is to say, those who enjoy drinking fermented toddy . . . make it quietly and drink it without interference" (Burrows and Spiro, 1953, p. 44).

And what happens to their comportment when they do so? According to Burrows and Spiro (1953), essentially nothing: "Some of the men drink glass after glass in the course of an evening. A slightly bleary look about the eyes, and a tendency to be jovial or sentimentally friendly, were the only effects we noticed" (p. 44). Elsewhere, Burrows (1952) wrote that "indulgence in coconut toddy seemed to have a mellowing effect. *We never saw or heard of a 'fighting drunk'*" (emphasis ours, p. 24). In confirmation of these observations we have Bates and Abbott's (1958) account of two farewell parties held in their honor, one given by the young men of Ifaluk and the other by the island's chiefs. Of the first party they report that the toddy "spread a warm feeling of good fellowship through everyone. . . . reality lost all its hard corners, and every man became a brother and the world a paradise" (pp. 121–22). Of the latter party they report only that "we were all very gay, because we liked each other so much; and very sad because we would have to part so soon" (p. 122). As with the Camba and the people of Aritama, again the question arises: How is it that on Ifaluk, where one would expect that alcohol would have a field day in releasing pent-up hostility, it has mysteriously lost all force?

For a fourth example of a society in which alcohol proves itself to be strangely incapable of producing disinhibition, we turn to the rural Japanese fishing community of Takashima, as observed in 1950–1951 by the anthropologist Edward Norbeck (1954). A community of 33 households and 188 inhabitants, Takashima is located on a small island in the Inland Sea. It lies about a quarter of a mile off the Japanese coast, and is almost directly opposite the mainland town of Shionasu, where the children go to school and where the adults market, attend meetings of the area's fishing cooperative, etc. Aside from such contacts as these with Shionasu, most of Takashima's inhabitants have little to do with any outside communities, their attention being directed in large measure simply toward making ends meet. Although almost all Takashima males are fishermen, most households do have small plots of land that the women keep under intensive cultivation. Land suitable for agricultural production is very limited on Takashima, however, and since none is ever put up for sale, it is come by only through inheritance. Still, with the exception of purchased rice, the diet of the typical household is largely confined to what members of the household have themselves raised or secured. Through the sale of their excess fish, however, they are very much a part of the money economy, and the amount of money a household possesses is the chief factor in determining its social prestige within the community.

The basic social unit of Takashima society is the household which usually embraces at least three generations and often four. Child rearing, economic affairs, entertainment, and traditional religious observances all revolve around the household. How do the members get along? Norbeck (1954) states that although there is a strong feeling of attachment between most couples, quarreling, although shameful, is considered a normal part of domestic life: "The usual course of such quarrels [being] that the wife eventually gives in, but often not until she has had her say" (p. 51). In so circumscribed a community as Takashima, little occurs that does not almost immediately become common knowledge. Not only is the state of one's domestic relations an open book, but "the habits, reactions, capabilities, and failings of each person are well-known to all other persons" (Norbeck, 1954, p. 115).

Since Takashima's inhabitants are thrown into almost inescapable contact with each other, it is not surprising that the ideal man should be one with whom all relations are smooth. Although he should not be affable to the point of self-effacement, neither should he be "pushing."

> He must never seem to contradict others or openly express opinions at variance
> with others. He adjusts his actions in accordance with the status of those with
> whom he has intercourse; elders must be respected and all shades of age and sex
> distinctions and their attendant behavior patterns are taken into account. . . .
> Actions or transactions without precedent leave the average person at a loss
> and are avoided as much as possible (Norbeck, 1954, p. 115).

Still, frictions do arise, and in situations that might become discordant, the usual and preferred course of action is that of avoidance. If avoidance is impossible (which is not infrequently the case), self-control is relied upon to sustain the tranquility of the community. And to good effect, for "beyond the initial situation when tempers and voices may rise briefly, overt expression of dislike or anger toward other persons is not made beyond the confines of one's own household" (p. 117).

So much, then, for the expression—or more correctly, the suppression—of aggression in Takashima. We would now note that in the area of sexuality things are kept similarly "under wraps."

> Rules of sex morality for women are strict, and an illegitimate child is a calamity which brings disgrace to the household and near ruin to the luckless girl. The average young man demands a virginal bride, and in the rare case when a groom fails to regard this as a matter of major importance, his parents will nevertheless demand it. Even the most wayward girl is well aware of the premium placed on chastity and knows that even one slip, if it becomes known, will almost surely prevent her from making a good marriage (Norbeck, 1954, p. 161).

As with aggression, so too with things sexual—the expectation determines the reality. Norbeck writes that "the Takashima girl is now almost invariably a virgin at the time of her marriage" (1954, p. 162). Furthermore, there has not been a single case of illegitimate birth within the *buraku* of Takashima for twenty years, and there has been only one case of even suspected adultery in approximately the same period. Nor does this puritanical orientation relate only to sexual activity; outright obscenity is considered *yaban* ("uncivilized"), the custom of the distant past. Indeed, even the non-prurient discussion of things sexual is considered a *hazu-kashii tokoro* ("an occasion for embarrassment"), and the subject is usually avoided. Even the custom of "dating," a custom of which the inhabitants of Takashima are well aware, is considered "too bold and daring" (p. 173). So, too, is Western-style dancing, which, although attempted by some, is also considered "a little new and bold for the *buraku*" (p. 85).

So much, then, for the constrictions placed upon everyday affairs in Takashima. It, too, would seem to be an ideal example of a community in which alcohol's purported ability to disinhibit would produce waves of truly tidal proportions. Let us now look at what actually happens. Norbeck (1954) describes a conventional drinking occasion as follows:

> Feasts, and particularly wedding feasts, are the most pleasurable social occasions of life, for even during difficult times strong effort is made to have food and liquor available in liberal quantities at these times. . . . When stomachs are comfortably full of banquet food and the liquor has begun to evince itself, the gathering becomes highly informal. . . . A common form of conviviality is to call upon everyone to render a song. There is hesitation at first until someone bolder than average . . . has sung a song. From that point there is no difficulty, and even the most unaccomplished singers take their turns. Applause follows every turn, and as the party spirit waxes, someone rises and dances to his own singing or to that of others. The party has now reached its height, and several dancers may perform at once. Men, usually careful to avoid bodily contact with other persons, throw their arms affectionately about one another in an excess of friendliness (pp. 87–88).

Such celebrations commence before nightfall and continue until midnight or later. That the participants do, in fact, become quite drunk is evidenced by Norbeck's (1954) observation that those drunken guests who must return to the mainland are "carefully shepherded by their wives and the more clear-headed men for fear of falling from the small boats and drowning during the crossing to Shionasu" (p. 88).

Or consider Norbeck's description of the "slambang drunk" (1954, p. 156) that occurs during the Autumn Festival:

> "[It] is the one occasion of the year when young unmarried men (from about sixteen years of age upward) may, without censure, become thoroughly drunk. . . .
>
> Many of the youths are already drunk by the time the *sendairoku* (portable shrine) is assembled, although, traditionally, serious drinking does not occur until later. It is customary to carry the shrine, weaving and staggering, back and forth along the main path centering about the community hall for an interval of perhaps an hour while children and adults watch. . . . [Following this] the young men, many of them weaving from drink and all sweating profusely, convene at someone's house where they continue drinking until [all] the *sake* is consumed (p. 155).

Although more of the participants remain sober today than formerly, the Autumn Festival and its aftermath is still an occasion for boisterous drunkenness. The participants do not, however, become quarrelsome, for the day of the Festival is formally defined as a day in which personal animosities are forgotten.

Not all drinking in Takashima is confined to such formal or quasi-formal occasions, however. Norbeck (1954) writes, for instance, that in the aftermath of a session of gambling—and gambling, while somewhat cyclical, is endemic among Takashima males—"winners are expected to be extravagant with their easily gained money and are chided by their opponents into spending if they are not liberal in the purchase of *shochu* and *sake*" (p. 85). Given the fact that "money is the prime topic of conversation for most men [and that] their thoughts appear to revolve constantly around the cost of material objects, and the amount of money possessed by other (and richer) individuals" (p. 206), one would expect the resulting drunkenness to be particularly acrimonious. This is not the case, however, for Norbeck gives the following as his summary depiction of the effects of alcohol upon the Takashima drinker:

> Reactions to alcohol are usually rapid. For many persons the first obvious reaction is a flushing of the face; often only a few of the tiny cups of *sake* or *shochu* will produce a vivid glow. This quick and intense flushing is considered both embarrassing and ugly, but it is also amusing and the subject of friendly jokes. Continued drinking soon results in a good-natured drunkenness, camaraderie, laughter, jokes, songs and dances, which are considered the inevitable result if not the objective of continued drinking (p. 72).

In summary, Takashima—with its insistence upon interpersonal harmony and its rigorously inforced puritanical approach to all things sexual—would seem to be still another society that should be ideally susceptible to alcohol's toxic effects. But not only does alcohol fail to produce tidal waves of aggression and sexuality, it does not even produce ripples. Thus, Takashima joins the list of societies in which alcohol has somehow lost its sting.

We turn now to a fifth and final example of a society in which drunken "disinhibition" *ought* to occur, but does not. This time we examine the town of Juxtlahuaca, in the state of Oaxaca, Mexico, as described by the anthropologists Kimball and Romaine Romney (1963). Juxtlahuaca has a population of approximately 3,600 people, 3,000 of whom are of mixed Spanish and Indian descent, speak Spanish, live in the central town, and differ little in their way of life from the inhabitants of the hundreds of other towns that make up village Mexico. Our interest in Juxtlahuaca, however, lies not in these townsmen, but in the remaining inhabitants—some 600 Mixtec Indians who live as agriculturalists in the *barrio* of Santo Domingo, a "neighborhood" set off from the rest of the town by a deep ravine. These Indians (all members of the *barrio* are Mixtec Indians) speak Mixteco first and Spanish second, if at all, for their heritage derives not from village Mexico, but from their indigenous Indian culture, which has existed in the area for at least the past 2,000 years.

Juxtlahuaca, then, is a town divided. While most of its inhabitants—the Spanish-speaking townsmen—live and work much as do villagers all over Mexico, the Mixtecan minority maintain a distinct language and culture within the confines of their relatively isolated *barrio*. Separate they are, but equal they are not, for the townspeople look down upon the Indians, and they avoid interaction with them wherever possible. The Romneys (1963) report that when interaction does occur, "derogation of Indians by townspeople and deference to townspeople by Indians seem to be the accepted pattern of behavior" (p. 563). Nor are the Mixtecans any better off economically: industry is nonexistent, animal husbandry is of little economic importance, land is scarce, and price levels—both of the goods the Indians must purchase and of the agricultural products they must sell—are set by the townspeople. In short, "it is difficult to make ends meet in the *barrio,* and it is virtually impossible to become rich" (p. 585).

In light of all this, it is small wonder that the Indians' main identifications center on the *barrio* itself and that "none identify much with the larger town of Juxtlahuaca, let alone with the state" (Romney and Romney, 1963, p. 561). Since their experience provides them scant cause to expect much from the outside world, their primary efforts are directed to the maintenance of the reputable character of their membership in the *barrio* proper. What sort of conduct does this entail? The Romneys summarize the character of Mixtec interaction as follows:

> In his relations with other men in the barrio the Indian pattern is adjustive and permissive, while within the town the Spanish pattern is one of ordering and dominating. The statuses of leadership in the barrio . . . are thought of as obligations rather than something to be striven for competitively. . . . [Indeed] envy and competitiveness are regarded as a minor crime. An Indian in a position of prominence never gives orders to his fellows. He may point out the pattern to be followed in a ritual or suggest practical modes of action, but this is in the manner of dispensing knowledge, not of dominating others either by force of personality or by authority of position. . . . Group decisions are made by consensus rather than by majority rule or dictatorial fiat. . . . [In a word] adjustment without friction is the goal, and, if this proves to be impossible, withdrawal rather than domination is the answer (1963, p. 565).

Thus, the Mixtecans place a high value on tranquility; and in the face of abundant reason to expect the contrary, their everyday life is, in fact, remarkably serene. In addition to the threat of ostracism from the ongoing life of the *barrio,* two additional factors to account for the tranquility of normal life are suggested by the Romneys.

First, throughout childhood the parents place great emphasis upon training in the control of aggression. For example (and the examples are numerous), most mothers reported that the sole reason for which they would physically punish their children would be to prevent them from fighting back. Secondly, the Mixtecans share a deeply held pattern of beliefs concerning the causal role of jealousy, anger, and aggression in the production of illness, which acts as a strong deterrent even to remaining in situations that *might* give rise to such emotions. The result of this three-pronged attack appears to be highly successful, for while Indian informants were able to recall specific acts of aggression that had occurred within their lifetimes, all were agreed that such instances were extremely rare. Furthermore, during their full year of residence among the Mixtecans, the Romneys neither witnessed nor heard of a single instance of sober aggression.

Clearly, then, the Mixtecan is a person who is rarely if ever overtly aggressive when he is sober. But what happens when alcohol assaults the higher centers of the Mixtecan brain? Once again we would seem to have legitimate grounds for expecting all hell to break loose. In fact, however, although these Indians drink truly prodigious amounts of alcohol—frequently to the point of passing out—*nothing of the kind occurs.* While the drunken comportment of the Spanish-speaking townspeople follows a pattern quite in keeping with what the disinhibition theory would predict (among them, drunken violence is commonplace), and while the Indian inhabitants of the *barrio* are aware of such drunken changes-for-the-worse in the townspeople, the Romneys (1963, p. 611) report that the Indians specifically deny that alcohol is capable of producing aggression in themselves. And indeed, although the Romneys witnessed numerous instances in which the Mixtecans drank themselves into states of gross intoxication, even at the fiestas during which "the men generally drank a great deal [they] were never observed to become loud or aggressive" (p. 611).

The Romneys (1963) did observe one instance of Mixtecan aggression during their year long study. When they asked the cause of this outburst they were informed that although Pedro was drunk at the time, "[he] had gone to the city to work in the past and there had picked up the habit of smoking marijuana. . . . [Prior to the outburst] he had been smoking marijuana; and, in their opinion, it was only a combination of the marijuana and drinking that could give rise to and account for such aggressive behavior" (p. 609). The Romneys (p. 686) also report a second variant case—that of an adult Mixtecan male whose drinking patterns and drunken aggressivity resembled that of the townspeople. But this person was variant in *most* crucial respects. His wife was Mexican (not Mixtecan), he was the only adult male in the *barrio* whose primary means of support was other than agricultural (he worked in the central part of town as a secretary in the mayor's office, being one of but three *barrio* adults who was considered literate), and, because his identification was with the townspeople, he participated in none of the ceremonies of the *barrio*. In a word, whether drunk or sober, he was very un-Mixtecan both in outlook and in action. So much, then, for drunkenness in Juxtlahuaca.

We have now presented five societies—the Camba, Aritama, Ifaluk, Takashima, and a Mixtec Indian *barrio*—in which the "disinhibiting" effects of alcohol are nowhere to be seen. Even during periods of extreme intoxication, the inhibitions that are normally in effect *remain* in effect. Drunken persons in these societies (and in others like them which we could have selected in their stead) may stagger, speak thickly, and become stuporous, without any corresponding display of changes-for-the-worse. Alcohol *can* be consumed—and, in many societies it *is* consumed—in immense quantities without producing any appreciable changes in behavior save for a progressive impairment in the exercise of certain of one's sensorimotor capabilities. Indeed, the only significant change in comportment reported for any of these societies is an increased volubility or sociability. But it is difficult to attribute even this to the direct action of alcohol, for such changes often begin quite early in the drinking process—they often begin, that is, *before* any appreciable degree of actual bodily intoxication could possibly have taken place.

That such societies should exist argues tellingly against the conventionally accepted notion that alcohol is a substance whose toxic action so impairs man's normally operative controls that he becomes a mere creature of impulse, inexorably doing things that he would not do under normal conditions. In a word, if alcohol were a "superego solvent" for one group of people due to its toxic action, then this same disinhibiting effect *ought* to be evident in *all* people. In point of fact, however, *it is not.* This is the first puzzle with which we confront the conventional wisdom.

NOTES

1. But even with the Abipone, these changes are selective in character. See Chapter 4, pp. 76–77.

2. For an extensive statement of the current situation on these islands, see P. R. and P. M. Toomin (1963, pp. 88–94 and *passim*).

QUESTIONS FOR DISCUSSION

1. Discuss the drinking patterns found among Indians as presented in the reading.

2. Discuss the drinking patterns found among the Japanese.

3. What would be a multicultural perspective of drinking?

College Alcohol Use: A Full or Empty Glass?

Henry Wechsler, Beth E. Molnar,
Andrea E. Davenport, and John S. Baer

Data from the Harvard School of Public Health College Alcohol Study (1993) were used to describe weekly alcohol consumption and its associated problems among a representative national sample of college students. The median number of drinks consumed/week by all students, regardless of drinking status, was 1.5. When students were divided by drinking pattern, the median number of drinks/week was 0.7 for those who did not binge drink and 3.7 for those who did so infrequently. For frequent binge drinkers, the median was considerably higher: 14.5 drinks/week. Nationally, 1 in 5 five college students is a frequent binge drinker. Binge drinkers consumed 68% of all the alcohol that students reported drinking, and they accounted for the majority of alcohol related problems. The data indicate that behavioral norms for alcohol consumption vary widely among students and across colleges. Therefore, it may not be possible to design an effective "one size fits all" approach to address college alcohol use.

College students' drinking constitutes a major challenge to public health. Numerous studies document heavy patterns of use and problems related to use. These studies stress how frequently students drink, how many students drink in a heavy episodic pattern or are binge drinkers, and the associated costs to the drinkers and to those around them.[1–3] Studies of this nature frame their results as "bad news" and are typically used to raise awareness and bring action by administrators and public officials.

Curiously, raising awareness about extreme behaviors of college students could be viewed as counter to the purposes of the public health messages being conveyed. Despite the well-recognized patterns of heavy drinking noted in the studies listed above, other studies suggest that students often tend to overestimate both the acceptability and the actual drinking behavior of their peers.[4–7] Students' beliefs that such extreme norms exist may serve to justify and explain extreme behavior and, thus, may influence students to engage in heavy drinking.

An intervention technique following from this line of inquiry is to challenge and attempt to correct these inaccurate perceptions. Prevention programs that use norm-correction strategies, ranging from campus-wide information campaigns [8,9] to providing personalized feedback and advice, have produced encouraging results.[10–12]

But what is normative for college students? What kinds of data can and should be used to challenge normative beliefs? For example, when considered in terms of average standard drinks consumed per week, Meilman et al. [13] reported that 51% of students at 4-year colleges reported consuming 1 drink or less in a typical week.

These norms present quite a different picture of college student life from those mentioned above. However, the nature of the way these data are presented (when all students are grouped together, regardless of drinking status) provides especially low estimates. Although statistically accurate, such norms can be dismissed by students as false.

In this article, we use the data from the Harvard School of Public Health College Alcohol Study (CAS) to present seemingly disparate measures of college student drinking. We seek to provide administrators and those designing prevention programs, as well as students, with clear estimates (calculated in different ways) to show what is normative use of alcohol on college campuses. We also show that, when we divide students into groups according to drinking patterns, some groups of students consume much more alcohol than others and account for the bulk of alcohol consumed and the majority of drinking-related problems on campus.

METHOD

Procedure

This report is based on data from the CAS survey of 17,592 students at 140 participating colleges in 1993. The colleges represent a cross-section of American higher education. Details of the study design and other findings have been previously published in the Journal of American College Health and other professional journals. [3,14–17]

A 20-page mailed questionnaire asked students about their drinking attitudes and behaviors, including an assessment of rates of binge drinking and total alcohol consumed. Students were also asked about campus and personal problems resulting from their drinking. We defined a drink for the respondents as either a 12-oz (360 mL) can/bottle of beer, a 4-oz (120 mL) glass of wine, a 12-oz (360 mL) bottle/can of wine cooler, or a 1.25 oz (37 mL) shot of liquor straight or in a mixed drink. In this study, we defined binge drinking as the consumption of 5 or more drinks in a row for men and 4 or more drinks in a row for women during the 2 weeks immediately preceding the survey.[15]

We combined four questions to assess binge drinking status for each student: (a) respondent's sex, (b) recency of the last drink, (c) how many times they had 5 or more drinks in a row over the last 2 weeks (for men), and (d) how many times they had 4 drinks in a row (for women). (If the responses were missing for any of these items, the student was excluded from the binge drinking analysis.)

We used three categories of drinkers in the current analysis: (a) nonbinge drinkers were those who consumed alcohol in the past 30 days but did not binge in the 2 weeks preceding the survey, (b) infrequent binge drinkers were those who binged one or two times in the past 2 weeks, and (c) frequent binge drinkers were defined as those who binged three or more times in the past 2 weeks.

We calculated the volume of alcohol consumed among students who reported they drank in the past 30 days. The 12,140 respondents who indicated that they had alcohol during that time period were asked two questions that we used to determine the volume of alcohol consumed per week. The first question asked on how many occasions the respondent had a drink of alcohol in the past 30 days. Response choices, in number of occasions, were 1–2, 3–5, 6–9, 10–19, 20–39, or 40 or more occasions. For analysis, we used the midpoint of each of the response categories to define how often the students drank in the past 30 days; we calculated the maximum answer of 40 or more as 40.

We used the second question to determine the volume of alcohol consumed: "In the past 30 days, on those occasions when you drank alcohol, how many drinks did you usually have?" The response choices for "usual number of drinks" were 1, 2, 3, 4, 5, 6, 7, 8, 9+. We calculated the maximum response choice of 9+ as 9 drinks.

We determined the number of drinks the students consumed per month by multiplying the number of occasions of drinking by the usual number of drinks in a 30-day period. The possible number of drinks in the past 30 days therefore ranged from 1.5 to 360 for this analysis. Dividing this number by 4 resulted in the number of drinks consumed by the students per week. Thus, the number of drinks per week ranged from 0 (.375 was rounded down to 0) to 90 drinks. We computed both medians and means (see Tables 1 and 2) to make the results presented in this study comparable to other published studies that report either one or both of these measures of central tendency.

We included questions to determine the extent of alcohol-related problems in the college population. If the students reported any drinking in the previous year (n = 14,588), we asked them to indicate whether they experienced any of 12 problems attributable to their own drinking (see Table 3 for specific problems assessed). The students were given the response choices of (a) not at all, (b) once, and (c) twice or more. At least one problem was reported by 70.7% of the students who drank in the past year for whom we have complete data for drinking-related problems.

RESULTS

The data in Table 1 show weekly alcohol consumption for all students, with men and women listed separately. Figures for nonbinge drinkers, drinkers, and frequent binge drinkers are shown. Consumption is reported as percentiles, or cumulative percentages. Data presented in this way make evident the percentages of students reporting consumption up to different numbers of drinks. We did not give students who completely abstained from drinking a separate column in this table because all of them drank zero drinks per week. They are, however, included in the column showing percentiles for the total sample. For all students, the median weekly consumption is 1 drink for women, 2 drinks for men, and 1.5 for the sample overall. These medians are quite similar to those from 1994–1996 reported by Meilman et al. [13] for 4-year institutions.

The mean number of drinks per week for all students in the current study was found to be 5.1 (Table 2). This is somewhat higher than the 4.5 mean number of drinks per week reported by Meilman et al. [13(p202)] for 1992–1994. However, the latter is based on students attending both 2-year and 4-year institutions.

"All students" represents a quite heterogeneous population. For this study, we divided the sample according to patterns of drinking to illustrate how the volume consumed differs between groups. Frequent binge drinkers consumed an average (mean) of 17.9 drinks per week, infrequent binge drinkers consumed an average of 4.8 drinks per week, and the combination of drinkers who did not binge drink as well as those who did not drink at all consumed an average of 0.8 drinks per week.

Looking instead at medians for the sample divided by drinking patterns, the median number of drinks per week for frequent binge drinkers was 14.5, the median number of drinks for infrequent binge drinkers was 3.7, and the median number of drinks per week for nonbinge drinkers and nondrinkers was very close to zero.

The data in Table 2 clearly show that the frequent binge drinkers consumed the majority of the alcohol—68% of the 87,008 drinks consumed, although they constituted only 19% of the total sample. The infrequent binge drinkers, who represented 24% of the total sample, consumed close to their share, or 23% of the alcohol.

Binge drinkers as a whole represent less than half of the college population (44%), but they account for almost all (91%) of the alcohol consumed by college students. The nonbinge drinkers and nondrinkers represented 56% of the sample, yet they consumed only 9% of the alcohol.

At least one problem attributable to the students' drinking was reported by 70.7% (n = 9,781) of the respondents who drank in the past year. The frequent binge drinkers, however, accounted for approximately half of the students who reported they had experienced any of the 12 problems shown on the list (Table 3). For example, of those students who drove after drinking or binge drinking, 41% were classified as frequent binge drinkers; almost 59% of those who damaged property or were injured were in this binge category. For most of the problems we surveyed, the frequent binge drinkers accounted for more than double their 24% representation among drinkers in the sample.

These patterns of binge drinking vary widely across college campuses. Among the 140 campuses surveyed, the percentages of respondents who met binge drinking criteria ranged from 1% on the campus with the lowest percentage to 70% at the school with the highest percentage. At almost one third of the campuses (44 out of 140), more than 50% of the students reached or exceeded the binge drinking definition we established for this study.

COMMENT

Norms for alcohol consumption on college campuses can be presented in many different formats. The CAS data, consistent with other national estimates,[1] suggests that drinking on college campuses is both extreme and not extreme. Distinct audiences will take away different meanings, depending on which presentation of the data they choose to examine.

College students' drinking is risky and dangerous when one considers that many young people are heavy episodic or binge drinkers. In interpreting our previous studies, using the above criterion of 5 drinks in a row for men and 4 for women, we suggested that 25% of the students binged occasionally and 19% did so frequently.

An intensive look at students' drinking patterns indicates that binge drinkers account for the lion's share of alcohol consumed and the problems encountered on campuses. Administrators and others concerned with public health should use this information in formulating campus prevention programs and policies.

TABLE 1 Weekly Alcohol Consumption in Percentiles, by Binge Drinking Status and Gender

	Binge Drinking			
	Total Sample		Frequent	
	% Women	% Men	% Women	% Men
Drinks per Week	(n = 10,020)	(n = 7,340)	(n = 1,684)	(n = 1,631)
0	32.0	26.1	0.4	0.3
1	54.2	40.7	1.2	0.7
2	66.3	51.0	4.0	1.2
3	72.9	57.1	8.4	1.7
4	78.6	62.7	16.5	3.2
5	81.0	65.3	22.5	4.5
6	84.5	69.4	29.8	7.1
7	86.0	71.6	32.1	8.3
8	88.4	74.8	42.5	12.6
9	90.6	78.2	52.6	19.4
10	90.6	78.2	52.6	19.4
11	92.8	81.8	61.3	28.8
12	92.8	81.8	61.3	28.8
13	93.3	83.1	63.7	33.4
14	93.3	83.1	63.7	33.4
15	94.9	85.7	71.6	41.0
16	94.9	85.7	71.6	41.0
17	95.2	87.4	73.0	47.6
18	96.6	89.1	81.1	54.3
19	96.6	89.1	81.1	54.3
20	96.6	89.1	81.1	54.3
21	96.6	89.1	81.1	54.3
22	97.7	91.6	87.1	64.0
23	97.7	91.6	87.1	64.0
24	97.7	91.6	87.1	64.0
25	98.5	92.9	91.4	69.4
26	98.5	92.9	91.4	69.4
27	98.5	92.9	91.4	69.4
28	98.5	92.9	91.4	69.4
29	98.8	94.2	93.2	74.6
30	98.9	94.4	93.8	75.5
31	98.9	94.4	93.8	75.5
32	98.9	94.4	93.8	75.5
33	99.3	97.3	95.9	88.4
34	99.3	97.3	95.9	88.4
35	99.3	97.3	95.9	88.4
36	99.3	97.3	95.9	88.4
37	99.4	97.8	96.6	90.3
38	99.4	97.8	96.6	90.3
39	99.4	97.8	96.6	90.3
40	99.4	97.8	96.7	90.5
41–45	99.5	98.3	97.4	92.5
46–50	99.6	98.4	97.4	92.7
51–55	99.6	98.5	98.0	93.4
56–60	99.8	98.9	98.6	95.2
61+100	100	100	100	100

	Binge drinking			
	Infrequent		Did Not Binge	
Drinks per Week	% Women (n = 2,128)	% Men (n = 1,965)	% Women (n = 4,392)	% Men (n = 2,539)
0	0.8	0.4	35.1	30.9
1	14.3	6.0	76.9	66.8
2	40.6	23.6	90.0	82.4
3	58.2	36.9	94.5	89.0
4	72.2	50.6	97.4	93.5
5	77.8	58.3	97.8	94.2
6	86.3	68.3	98.8	96.3
7	89.7	73.7	99.5	98.0
8	92.5	81.1	99.6	98.6
9	94.5	87.5	99.7	98.9
10	94.5	87.6	99.7	98.9
11	97.1	93.0	99.9	99.2
12	97.1	93.0	99.9	99.2
13	97.7	93.7	99.9	99.2
14	97.7	93.7	99.9	99.2
15	98.8	96.3	99.9	99.6
16	98.8	96.3	99.9	99.6
17	98.9	97.1	99.9	99.7
18	99.4	97.9	100	99.7
19	99.4	97.9	100	99.7
20	99.5	97.9	100	99.7
21	99.5	97.9	100	99.7
22	99.6	99.1	100	99.8
23	99.6	99.1	100	99.8
24	99.6	99.1	100	99.8
25	99.7	99.5	100	99.8
26	99.7	99.5	100	99.8
27	99.7	99.5	100	99.8
28	99.7	99.5	100	99.8
29	99.7	99.6	100	99.9
30	99.9	99.6	100	100
31	99.9	99.6	100	100
32	99.9	99.6	100	100
33	99.9	99.7	100	100
34	99.9	99.7	100	100
35	99.9	99.7	100	100
36	99.9	99.7	100	100
37	100	99.8	100	100
38	100	99.8	100	100
39	100	99.8	100	100
40	100	99.9	100	100
41–45	100	100	100	100
46–50	100	100	100	100
51–55	100	100	100	100
56–60	100	100	100	100
61	100	100	100	100

Note: Students who did not drink were included in the total sample, but an abstainer category as a fourth column is not included.

< 0.375 drinks/wk were rounded to zero.

TABLE 2   Percentage, Mean, and Median Numbers of Drinks Consumed by College Students in Each Drinking Category

Drinking Category	N	% of Sample	n	M	SD	% of Mdn	Total
				Drinks per Week			
				All Students			
All	17,046	100	87,008	5.1	9.5	1.5	100
Frequent bingers	3,317	19	59,388	17.9	15.0	14.5	68
Infrequent bingers	4,099	24	19,591	4.8	4.3	3.7	23
Nondrinkers/nonbingers	9,630	57	8,029	0.8	1.7	0.7	9
				Women (n = 9,810)			
Frequent bingers	1,685	17	22,439	13.3	11.8	9.4	64
Infrequent bingers	2,130	22	8,319	3.9	3.6	3.0	24
Nondrinkers/nonbingers	5,995	61	4,284	0.7	1.4	0.0	12
				Men (n = 7,236)			
Frequent bingers	1,632	23	36,949	22.6	16.3	18.1	71
Infrequent bingers	1,969	27	11,273	5.7	4.7	4.0	22
Nondrinkers/nonbingers	3,635	50	3,744	1.0	2.2	0.0	7

One would be mistaken to assume that such heavy episodic drinking is the norm for all (or even most) students. More than half of the students surveyed did not binge drink at all. When norms are considered across all students in all colleges, the median number of drinks (1.5) consumed per student during a week is very small. The frequent binge drinkers who consume 68% of the alcohol raise the mean, or average number of drinks per week, to 5.1, which seems high.

Patterns of drinking also vary between college campuses. These facts are critically important for students to know. Our data also show that, for some subgroups and some colleges, "norms" about alcohol consumption, that is, the patterns, volume, and problems associated with drinking that are normative on a campus, can indeed be quite extreme.

Our conclusion from the present examination of the CAS data is that it is not possible to design an effective "one size fits all" approach to address college alcohol use. The data indicate that, rather than being uniformly high, behavioral norms for alcohol consumption vary widely among students and across colleges.

Some of the difficulty in changing norms may be that heavy drinking is highly visible and may be assumed to be the norm, even at schools where our data suggest that this is not the case. Colleges where relatively less heavy drinking takes place may benefit most from marketing campaigns informing students about the actual rates of drinking to challenge false assumptions that heavy drinking is "typical" of life at their college.[9]

In many settings, although highly visible, frequent heavy drinking is not normative. However, in settings or schools with a high proportion of heavy drinkers, information or marketing campaigns become less straightforward. In some heavy drinking settings, actual norms may be so high that presentation of data unfortunately

Percentage of Frequent Binge Drinkers (n = 3,315)   **TABLE 3**
Who Encountered Various Problems Related to Their Drinking

Problem	%
Missed class	53.8
Fell behind	53.9
Did something later regretted	45.4
Experienced blackouts	52.3
Argued with friends	49.7
Engaged in	
Unplanned sex	49.7
Unprotected sex	52.3
Damaged property	58.8
Had trouble with police	58.4
Was injured	58.9
Overdosed on alcohol	41.1
Drove after drinking or binging	40.6
Had 5 or more of above problems	53.9

Data Analysis: We used the Statistical Analysis System (SAS), Version 6.07[18] for all analyses, and we used contingency table analysis to determine frequencies.

confirms assumptions or beliefs about others' risky behavior. Different intervention approaches should be employed in such settings. Yet, when working with students individually [11] or in small or well-defined groups, norms can be selected (i.e., by gender, living settings, Greek affiliation, or, as in this article, by drinking patterns) to create contrasts with expectations and beliefs. Presentations of actual norms for other students' drinking can demonstrate that not everyone uses alcohol in the same way.

REFERENCES

[1.] Presley CA, Meilman PW, Cashin JR. *Alcohol and Drugs on American College Campuses: Use, Consequences, and Perceptions of the Campus Environment,* Vol 4: 1992–94. Carbondale, IL: Southern Illinois University; 1996.

[2.] Johnston LD, O'Malley PM, Bachman JG. *National Survey Results on Drug Use From the Monitoring the Future Study 1975–1994;* Vol 2, College Students and Young Adults. US Department of Health and Human Services; 1996. NIH Publication No 96-4027.

[3.] Wechsler H, Davenport A, Dowdall G, Moeykens B, Castillo S. Health and behavioral consequences of binge drinking in college: A national survey of students at 140 campuses. *JAMA.* 1994;272:1672–1677.

[4.] Perkins HW, Berkowitz AD. Perceiving the community norms of alcohol use among students: Some research implications for campus alcohol education programming. *Int J Addict.* 1986;21:961–976.

[5.] Perkins HW, Wechsler H. Variation in perceived college drinking norms and its impact on alcohol abuse: A nationwide study. *J Drug Issues.* 1996;26:961–974.

[6.] Baer JS, Stacy A, Larimer M. Biases in the perception of drinking norms among college students. *J Stud Alcohol.* 1991; 52(6):580–586.

[7.] Baer JS, Carney MM. Biases in the perceptions of the consequences of alcohol use among college students. *J Stud Alcohol.* 1993;54:54–60.

[8.] Haines M, Spear SF. Changing the perception of the norm: A strategy to decrease binge drinking among college students. *J Am Coll Health.* 1996;45(3): 134–140.

[9.] Haines MP. A Social Norms Approach to Preventing Binge Drinking at Colleges and Universities. The Higher Education Center for Alcohol and Other Drug Prevention. US Dept of Education, Publication No. ED/OPE/96-18. Education Development Center; 1996.

[10.] Agostinelli G, Brown JM, Miller WR. Effects of normative feedback on consumption among heavy drinking college students. *J Alcohol Drug Educ.* 1995;25:31–40.

[11.] Marlatt GA, Baer JS, Kivlahan DR, et al. Screening and brief intervention for high-risk college student drinkers: Results from a two-year follow-up assessment. *J Consult Clin Psychol.* In press.

[12.] Baer JS. Etiology and secondary prevention of alcohol problems with young adults. In: Baer JS, Marlatt GA, McMahon RJ, eds. *Addictive Behaviors Across the Lifespan.* Newbury Park, CA: Sage; 1996:111–137.

[13.] Meilman PW, Presley CA, Cashin JR. Average weekly alcohol consumption: Drinking percentiles for American college students. *J Am Coll Health.* 1997;45:201–204.

[14.] Wechsler H, Dowdall GW, Davenport A, Castillo S. Correlates of college student binge drinking. *Am J Public Health.* 1995;85:921–926.

[15.] Wechsler H, Dowdall GW, Davenport A, Rimm EB. A gender-specific measure of binge drinking among college students. *Am J Public Health.* 1995;85:982–985.

[16.] Wechsler H, Davenport AE, Dowdall GW, Grossman SJ, Zanakos SI. Binge drinking, tobacco, and illicit drug use and involvement in college athletics. A survey of students at 140 American colleges. *J Am Coll Health.* 1997;45(5): 195–200.

[17.] Meilman PW, Cashin JR, Mckillip J, Presley CA. Understanding the three national databases on collegiate alcohol and drug use. *J Am Coll Health.* 1998;46(4): 159–162.

[18.] *SAS/SAT User's Guide.* Version 6, Fourth Edition. Cary, NC: SAS Institute Inc. 1994.

QUESTIONS FOR DISCUSSION

1. Discuss the finding in Table 1. How does your discussion of Table 1 relate to drinking on your campus? Compare the results in Table 1 to you and your peer group.

2. Discuss the finding in Table 3. How does your discussion of Table 3 relate to drinking on your campus? Compare the results in Table 3 to you and your peer group.

3. Discuss the results of the study. How are the results similar to your college?

SEXUAL DEVIANT BEHAVIOR

The Madam as Teacher:
The Training of House Prostitutes

Barbara Sherman Heyl

Although the day of the elaborate and conspicuous high-class house of prostitution is gone, houses still operate throughout the United States in a variety of altered forms. The business may be run out of trailers and motels along major highways, luxury apartments in the center of a metropolis or rundown houses in similar, industrialized cities. (Recent discussions of various aspects of house prostitution include: Gagnon and Simon, 1973:226–7; Hall, 1973:115–95; Heyl, 1974; Jackson, 1969:185–92; Sheehy, 1974:185–204; Stewart, 1972; and Voglotti, 1975:25–80.) Madams sometimes find themselves teaching young women how to become professional prostitutes. This paper focuses on one madam who trains novices to work at the house level. I compare the training to Bryan's (1965) account of the apprenticeship of call girls and relate the madam's role to the social organization of house prostitution.

Bryan's study of thirty-three Los Angeles call girls is one of the earliest interactionist treatments of prostitution. His data focus on the process of entry into the occupation of the call girl and permit an analysis of the structure and content of a woman's apprenticeship. He concluded that the apprenticeship of call girls is mainly directed toward developing a clientele, rather than sexual skills (1965:288, 296–7).

But while Bryan notes that pimps seldom train women directly, approximately half of his field evidence in fact derives from pimp–call girl apprenticeships. Thus, in Bryan's study (as well as in subsequent work on entry into prostitution as an occupation) there was a missing set of data on the more typical female trainer-trainee relationship and on the content and the process of training at other levels of the business in nonmetropolitan settings. This paper attempts to fill this gap.

I. ANN'S TURN-OUT ESTABLISHMENT

A professional prostitute, whether she works as a streetwalker, house prostitute, or call girl, can usually pick out one person in her past who "turned her out," that is, who taught her the basic techniques and rules of the prostitute's occupation.[1] For women who begin working at the house level, that person may be a pimp, another "working girl," or a madam. Most madams and managers of prostitution establishments, however, prefer not to take on novice prostitutes, and they may even have a specific policy against hiring turn-outs (see Erwin [1960:204–5] and Lewis [1942: 222]). The turn-out's inexperience may cost the madam clients and money; to train the novice, on the other hand, costs her time and energy. Most madams and managers simply do not want the additional burden.

It was precisely the madam's typical disdain for turn-outs that led to the emergence of the house discussed in this paper—a house specifically devoted to training new prostitutes. The madam of this operation, whom we shall call Ann, is forty-one years old and has been in the prostitute world twenty-three years, working primarily at the house level. Ann knew that pimps who manage women at this level have difficulty placing novices in houses. After operating several houses staffed by professional prostitutes, she decided to run a school for turn-outs partly as a strategy for acquiring a continually changing staff of young women for her house. Pimps are the active recruiters of new prostitutes, and Ann found that, upon demonstrating that she could transform the pimps' new, square women into trained prostitutes easily placed in professional houses, pimps would help keep her business staffed.[2] Ann's house is a small operation in a middle-sized industrial city (population 300,000), with a limited clientele of primarily working-class men retained as customers for ten to fifteen years and offered low rates to maintain their patronage.

Although Ann insists that every turn-out is different, her group of novices is remarkably homogeneous in some ways. Ann has turned out approximately twenty women a year over the six years while she has operated the training school. Except for one Chicano, one black and one American Indian, the women were all white. They ranged in age from eighteen to twenty-seven. Until three years ago, all the women she hired had pimps. Since then, more women are independent (so-called "outlaws"), although many come to Ann sponsored by a pimp. That is, in return for being placed with Ann, a turn-out gives the pimp a percentage of her earnings for a specific length of time. At present eighty percent of the turn-outs come to Ann without a long-term commitment to a pimp. The turn-outs stay at Ann's on the average of two to three months. This is the same average length of time Bryan (1965:290) finds for the apprenticeship in his call-girl study. Ann seldom has more than two or three women in training at any one time. Most turn-outs live at the house, often just a large apartment near the older business section of the city.

II. THE CONTENT OF THE TRAINING

The data for the following analysis are of three kinds. First, tape recordings from actual training sessions with fourteen novices helped specify the structure and content of the training provided. Second, lengthy interviews with three of the novices and multiple interviews with Ann were conducted to obtain data on the training during the novice's first few days at the house before the first group training sessions were conducted and recorded by Ann. And third, visits to the house on ten occasions and observations of Ann's interaction with the novices during teaching periods extended the data on training techniques used and the relationship between madam and novice. In addition, weekly contact with Ann over a four-year period allowed repeated review of current problems and strategies in training turn-outs.

Ann's training of the novice begins soon after the woman arrives at the house. The woman first chooses an alias. Ann then asks her whether she has ever "Frenched a guy all the way," that is, whether she has brought a man to orgasm during the act of fellatio. Few of the women say they have. By admitting her lack of competence in a specialized area, the novice has permitted Ann to assume the role of teacher. Ann then launches into instruction on performing fellatio. Such instruction is important to her business. Approximately eighty percent of her customers are what Ann calls "French tricks." Many men visit prostitutes to receive sexual services, including fellatio, their wives or lovers seldom perform. This may be particularly true of the lower- and working-class clientele of the houses and hotels of prostitution (Gagnon and Simon, 1973:230). Yet the request for fellatio may come from clients at all social levels; consequently, it is a sexual skill today's prostitute must possess and one she may not have prior to entry into the business (Bryan, 1965:293; Winick and Kinsie, 1971:180, 207; Gray, 1973:413).

Although Ann devotes much more time to teaching the physical and psychological techniques of performing fellatio than she does to any other sexual skill, she also provides strategies for coitus and giving a "half and half"—fellatio followed by coitus. The sexual strategies taught are frequently a mixture of ways for stimulating the client sexually and techniques of self-protection during the sexual acts. For example, during coitus, the woman is to move her hips "like a go-go dancer's" while keeping her feet on the bed and tightening her inner thigh muscles to protect herself from the customer's thrust and full penetration. Ann allows turn-outs to perform coitus on their backs only, and the woman is taught to keep one of her arms across her chest as a measure of self-defense in this vulnerable position.

After Ann has described the rudimentary techniques for the three basic sexual acts—fellatio, coitus, and "half and half"—she begins to explain the rules of the house operation. The first set of rules concerns what acts the client may receive for specific sums of money. Time limits are imposed on the clients, roughly at the rate of $1 per minute; the minimum rate in this house is $15 for any of the three basic positions. Ann describes in detail what will occur when the first client arrives: he will be admitted by either Ann or the maid; the women are to stand and smile at him, but not speak at him (considered "dirty hustling"); he will choose one of the women and go to the bedroom with her. Ann accompanies the turn-out and the client to the bedroom and begins teaching the woman how to check the man for any cuts or open sores on the genitals and for any signs of old or active venereal disease. Ann usually rechecks each client herself during the turn-out's first two weeks of work. For the first few days Ann remains in the room while the turn-out and client negotiate the sexual contract. In ensuing days Ann spends time helping the woman develop verbal skills to "hustle" the customer for more expensive sexual activities.

The following analysis of the instruction Ann provides is based on tape recordings made by Ann during actual training sessions in 1971 and 1975. These sessions took place after the turn-outs had worked several days but usually during their first two weeks of work. The tapes contain ten hours of group discussion with fourteen different novices. The teaching tapes were analyzed according to topics covered in the discussions, using the method outlined in Barker (1963) for making such divisions in the flow of conversation and using Bryan's analysis of the call girl's apprenticeship as a guide in grouping the topics. Bryan divides the content of the training of call girls into two broad dimensions, one philosophical and one interpersonal (1965:291–4). The first emphasizes a subcultural value system and sets down guidelines for how the novice *should* treat her clients and her colleagues in the business. The second dimension follows from the first but emphasizes actual behavioral techniques and skills.

The content analysis of the taped training sessions produced three major topics of discussion and revealed the relative amount of time Ann devoted to each. The first two most frequently discussed topics can be categorized under Bryan's dimension of interpersonal skills; they were devoted to teaching situational strategies for managing clients. The third topic resembles Bryan's value dimension (1965:291–2).

The first topic stressed physical skills and strategies. Included in this category were instruction on how to perform certain sexual acts and specification of their prices, discussion of particular clients, and instruction in techniques for dealing with certain categories of clients, such as "older men" or "kinky" tricks. This topic of physical skills also included discussion of, and Ann's demonstration of, positions designed to provide the woman maximum comfort and protection from the man during different sexual acts. Defense tactics, such as ways to get out of a sexual position and out of the bedroom quickly, were practiced by the novices. Much time was devoted to analyzing past encounters with particular clients. Bryan finds similar discussions of individual tricks among novice call girls and their trainers (1965:293). In the case of Ann's turn-outs these discussions were often initiated by a novice's complaint or question about a certain client and his requests or behavior in the bedroom. The novice always received tips and advice from Ann and the other women present on how to manage that type of bedroom encounter. Such sharing of tactics allows the turn-out to learn what Gagnon and Simon call "patterns of client management" (1973:2321).

Ann typically used these discussions of bedroom difficulties to further the training in specific sexual skills she had begun during the turn-out's first few days at work. It is possible that the addition of such follow-up sexual training to that provided during the turn-out's first days at the house results in a more extensive teaching of actual sexual skills than that obtained either by call girls or streetwalkers. Bryan finds that in the call-girl training—except for fellatio—"There seems to be little instruction concerning sexual techniques as such, even though the previous sexual experience of the trainee may have been quite limited" (1965:293). Gray (1973:413) notes that her sample of streetwalker turn-outs were rarely taught specific work strategies:

They learned these things by trial and error on the job. Nor were they schooled in specific sexual techniques: usually they were taught by customers who made the specific requests.

House prostitution may require more extensive sexual instruction than other forms of the business. The dissatisfied customer of a house may mean loss of business and therefore loss of income to the madam and the prostitutes who work there. The sexually inept streetwalker or call girl does not hurt business for anyone but herself; she may actually increase business for those women in the area should dissatisfied clients choose to avoid her. But the house depends on a stable clientele of satisfied customers.

The second most frequently discussed topic could be labeled: client management/ verbal skills. Ann's primary concern was teaching what she calls "hustling." "Hustling" is similar to what Bryan terms a "sales pitch" for call girls (1965:292), but in the house setting it takes place in the bedroom while the client is deciding how much to spend and what sexual acts he wishes performed. "Hustling" is designed to encourage the client to spend more than the minimum rate.[3] The prominence on the teaching tapes of instruction in this verbal skill shows its importance in Ann's training of novices.

On one of the tapes Ann uses her own turning-out experience to explain to two novices (both with pimps) why she always teaches hustling skills as an integral part of working in a house.

Ann as a Turn-out[4]

ANN: Of course, I can remember a time when I didn't know that I was supposed to hustle. So that's why I understand that it's difficult to *learn* to hustle. When I turned out it was $2 a throw. They came in. They gave me their $2. They got a hell of a fuck. And that was it. Then one Saturday night I turned *forty-four* tricks! And Penny [the madam] used to put the number of tricks at the top of the page and the amount of money at the bottom of the page—she used these big ledger books. Lloyd [Ann's pimp] came in at six o'clock and he looked at that book and he just *knew* I had made all kinds of money. Would you believe I had turned forty-two $2 tricks and two $3 tricks—because two of 'em got generous and gave me an extra buck! [Laughs] I got my ass whipped. And I was so tired—I thought I was going to die—I was 15 years old. And I got my ass whipped for it. [Ann imitates an angry Lloyd:] "Don't you know you're supposed to ask for more money?!" No, I didn't. Nobody told me that. All they told me was it was $2. So that is learning it the *hard* way. I'm trying to help you learn it the *easy* way, if there is an easy way to do it.

In the same session Ann asks one of the turn-outs (Linda, age eighteen) to practice her hustling rap.

Learning the Hustling Rap

ANN: I'm going to be a trick. You've checked me. I want you to carry it from there.
[Ann begins role-playing: she plays the client; Linda, the hustler.]

LINDA: [mechanically] What kind of party would you like to have?

ANN: That had all the enthusiasm of a wet noodle. I really wouldn't *want* any party with that because you evidently don't want to give me one.

LINDA: What kind of party would you *like* to have?

ANN: I usually take a half and half.

LINDA: Uh, the money?

ANN: What money?

LINDA: The money you're supposed to have! [loudly] 'Cause you ain't gettin' it for free!

ANN: [Upset] Linda, if you *ever,* ever say that in my joint . . . Because that's fine for street hustling. In street hustling, you're going to *have* to hard-hustle those guys or they're not going to come up with anything. Because they are going to *try* and get it for free. But when they walk in here, they *know* they're not going to get it for free to begin with. So try another tack—just a little more friendly, not quite so hard-nosed. [Returning to role-playing:] I just take a half and half.

LINDA: How about fifteen [dollars]?

ANN: You're leading into the money too fast, honey. Try: "What are you going to spend?" or "How much money are you going to spend?" or something like that.

LINDA: How much would you like to spend?

ANN: No! Not "like." 'Cause they don't *like* to spend anything.

LINDA: How much *would* you like to spend?

ANN: Make it a very definite, positive statement: "How much are you going to spend?"

Ann considers teaching hustling skills her most difficult and important task. In spite of her lengthy discussion on the tapes of the rules and techniques for dealing with her customer sexually, Ann states that it may take only a few minutes to "show a girl how to turn a trick." A substantially longer period is required, however, to teach her to hustle. To be adept at hustling, the woman must be mentally alert and sensitive to the client's response to what she is saying and doing and be able to act on those perceptions of his reactions. The hustler must maintain a steady patter of verbal coaxing, during which her tone of voice may be more important than her actual words.

In Ann's framework, then, hustling is a form of verbal sexual aggression. Referring to the problems in teaching novices to hustle, Ann notes that "taking the aggressive part is something women are not used to doing; particularly young women." No doubt, hustling is difficult to teach partly because the woman must learn to discuss sexual acts, whereas in her previous experience, sexual behavior and preferences had been negotiated nonverbally (see Gagnon and Simon, 1973:228). Ann feels that to be effective, each woman's "hustling rap" must be her own—one that comes naturally and will strike the clients as sincere. All of that takes practice. But Ann is aware that the difficulty in learning to hustle stems more from the fact that it involves inappropriate sex-role behavior. Bryan concludes that it is precisely this aspect of soliciting men on the telephone that causes the greatest distress to the novice call girl (1965:293). Thus, the call girl's income is affected by how much business she can bring in by her calls, that is, by how well she can learn to be socially aggressive on the telephone. The income of the house prostitute, in turn, depends heavily on her hustling skills in the bedroom. Ann's task, then, is to train the novice, who has recently come from a culture where young women are not expected to be sexually aggressive, to assume that role with a persuasive naturalness.

Following the first two major topics—client management through physical and verbal skills—the teaching of "racket" (prostitution world) values was the third-ranking topic of training and discussion on the teaching tapes. Bryan notes that the major value taught to call girls is "that of maximizing gains and minimizing effort, even if this requires transgressions of either a legal or moral nature" (1965:291). In her training, however, Ann avoids communicating the notion that the novices may exploit the customers in any way they can. For example, stealing or cheating clients is grounds for dismissal from the house. Ann cannot afford the reputation among her tricks that they risk being robbed when they visit her. Moreover, being honest with clients is extolled as a virtue. Thus, Ann urges the novices to tell the trick if she is nervous or unsure, to let him know she is new to the business. This is in direct contradiction to the advice pimps usually give their new women to hide their inexperience from the trick. Ann asserts that honesty in this case means that the client will be more tolerant of mistakes in sexual technique, be less likely to interpret hesitancy as coldness, and be generally more helpful and sympathetic. Putting her "basic principle" in the form of a simple directive, Ann declares: "Please the trick, but at the same time get as much money for pleasing him as you possibly can." Ann does not consider hustling to be client exploitation. It is simply the attempt to sell the customer the product with the highest profit margin. That is, she would defend hustling in terms familiar to the businessman or sales manager.

That Ann teaches hustling as a value is revealed in the following discussion between Ann and Sandy—a former hustler and longtime friend of Ann. Sandy, who married a former trick and still lives in town, has come over to the house to help instruct several novices in the hustling business.

Whores, Prostitutes and Hustlers

ANN: [To the turn-outs:] Don't get uptight that you're hesitating or you're fumbling, within the first week or even the first five years. Because it takes that long to become a good hustler. I mean you can be a whore in one night. There's nothing to that. The first time you take money you're a whore.

SANDY: This girl in Midtown [a small, midwestern city] informed me . . . I had been working there awhile . . . that I was a "whore" and she was a "prostitute." And I said: "Now what the hell does that mean?" Well the difference was that a prostitute could pick her customer and a whore had to take anybody. I said: "Well honey, I want to tell you something. I'm neither one." She said: "Well, you *work.*" I said: "I know, but I'm a *hustler.* I make *money* for what I do."

ANNE: And this is what I turn out—or try to turn out—hustlers. Not prostitutes. Not whores. But hustlers.

For Ann and Sandy the hustler deserves high status in the prostitution business because she has mastered a specific set of skills that, even with many repeat clients, earn her premiums above the going rate for sexual acts.

In the ideological training of call girls Bryan finds that "values such as fairness with other working girls, or fidelity to a pimp, may occasionally be taught" (1965: 291–2); the teaching tapes revealed Ann's affirmation of both these virtues. When a pimp brings a woman to Ann, she supports his control over that woman. For example, if during her stay at the house, the novices break any of the basic rules—by using drugs, holding back money (from either Ann or the pimp), lying or seeing another man—Ann will report the infractions to the woman's pimp. Ann notes: "If I don't do that and the pimp finds out, he knows I'm not training her right, and he won't bring his future ladies to me for training." Ann knows she is dependent on the pimps to help supply her with turn-outs. Bryan, likewise, finds a willingness among call girls' trainers to defer to the pimps' wishes during the apprenticeship period (1965:290).

Teaching fairness to other prostitutes is particularly relevant to the madam who daily faces the problem of maintaining peace among competing women at work under one roof. If two streetwalkers or two call girls find that they cannot get along, they need not work near one another. But if a woman leaves a house because of personal conflicts, the madam loses a source of income. To minimize potential negative feelings among novices, Ann stresses mutual support, prohibits "criticizing another girl," and denigrates the "prima donna"—the prostitute who flaunts her financial success before the other women.

In still another strategy to encourage fair treatment of one's colleagues in the establishment, Ann emphasizes a set of rules prohibiting "dirty hustling"—behavior engaged in by one prostitute that would undercut the business of other women in the house. Tabooed under the label of "dirty hustling" are the following: appearing in the line-up partially unclothed; performing certain disapproved sexual positions, such as anal intercourse; and allowing approved sexual extras without charging additional fees. The norms governing acceptable behavior vary from house to house and region to region, and Ann warns the turn-outs to ask about such rules when they begin work in a new establishment. The woman who breaks the work norms in a house, either knowingly or unknowingly, will draw the anger of the other women and can be fired by a madam eager to restore peace and order in the house.

Other topics considered on the tapes—in addition to physical skills, "hustling" and work values—were instruction on personal hygiene and grooming, role-playing of conversational skills with tricks on topics not related to sex or hustling ("living room talk"), house rules not related to hustling (such as punctuality, no perfume, no drugs), and guidelines for what to do during an arrest. There were specific suggestions on how to handle personal criticism, questions and insults from clients. In addition, the discussions on the tapes provided the novices with many general strategies for becoming "professionals" at their work, for example, the importance of personal style, enthusiasm ("the customer is always right"), and a sense of humor. In some ways these guidelines resemble a beginning course in salesmanship. But they also provide clues, particularly in combination with the topics on handling client insults and the emphasis on hustling, on how the house prostitute learns to manage a stable and limited clientele and cope psychologically with the repetition of the clients and the sheer tedium of the physical work (Hughes, 1971:342–5).

III. TRAINING HOUSE PROSTITUTES—
A PROCESS OF PROFESSIONAL SOCIALIZATION

Observing how Ann trains turn-outs is a study in techniques to facilitate identity change (see also Davis, 1971 and Heyl, 1975, chapter 2). Ann uses a variety of persuasive strategies to help give the turn-outs a new occupational identity as a "professional." One strategy is to rely heavily on the new values taught the novice to isolate her from her previous lifestyle and acquaintances. Bryan finds that "the value structure serves, in general, to create in-group solidarity and to alienate the girl, 'square' society" (1965:292). Whereas alienation from conventional society may be an indirect effect of values taught to call girls, in Ann's training of house prostitutes the expectation that the novice will immerse herself in the prostitution world ("racket life") is made dramatically explicit.

In the following transcription from one of the teaching tapes, the participants are Ann (age thirty-six at the time the tape was made), Bonnie (an experienced turn-out, age twenty-five) and Kristy (a new turn-out, age eighteen). Kristy has recently linked up with a pimp for the first time and volunteers to Ann and Bonnie her difficulty in adjusting to the racket rule of minimal contact with the square world—a rule her pimp is enforcing by not allowing Kristy to meet and talk with her old friends. Ann (A) and Bonnie (B) have listened to Kristy's (K) complaints and are making suggestions. (The notation "B-K" indicates that Bonnie is addressing Kristy.)

B-K: What you gotta do is sit down and talk to him and weed out your friends and find the ones he thinks are suitable companions for you—in your new type of life.
K-B: None of them.
A-K: What about *his* friends?
K-A: I haven't met many of his friends. I don't like any of 'em so far.
A-K: You are making the same mistake that makes me so goddamned irritated with square broads! You're taking a man and trying to train *him,* instead of letting the man train you.
K-A: What?! I'm not trying to train him, I'm just. . . .
A-K: All right, you're trying to force him to accept your friends.
K-A: I don't care whether he accepts them or not. I just can't go around not talking to anybody.
A-K: "Anybody" is your old man! He is your world. And the people he says you can talk to are the people that are your world. But what you're trying to do is force your square world on a racket guy. It's like oil and water. There's just no way a square and a racket person can get together. That's why when you turn out you've got to change your mind completely from square to racket. And you're still trying to hang with squares. You can't do it.

Strauss's (1969) concept of "coaching" illuminates a more subtle technique Ann employs as she helps the novice along, step by step, from "square" to "racket" values and lifestyle. She observes carefully how the novice progresses, elicits responses from her about what she is experiencing, and then interprets those responses for her. In the following excerpt from one of the teaching tapes, Ann prepares two novices for feelings of depression over their newly made decisions to become prostitutes.

Turn-out Blues

ANN: And while I'm on the subject—depression. You know they've got a word for it when you have a baby—it's called "postpartum blues." Now, I call it "turn-out blues." Every girl that ever turns out has 'em. And, depending on the girl, it comes about the third or fourth day. You'll go into a depression for no apparent reason. You'll wake up one morning and say: "Why in the hell am I doing this? Why am I here? I wanna go home!" And I can't do a thing to help you. The only thing I can do is to leave you alone and hope that you'll fight the battle yourself. But knowing that it will come and knowing that everybody else goes through it too does help. Just pray it's a busy night! So if you get blue and you get down, remember: "turn-out blues"—and everybody gets it. Here's when you'll decide whether you're going to stay or you're gonna quit.

Ann's description of "turn-out blues" is a good example of Strauss's account (1969:111–2) of how coaches will use prophecy to increase their persuasive power over their novices. In the case of "turn-out blues," the novice, if she becomes depressed about her decision to enter prostitution, will recall Ann's prediction that this would happen and that it happens to all turn-outs. This recollection may or may not end the woman's misgivings about her decision, but it will surely enhance the turn-out's impression of Ann's competence. Ann's use of her past experience to make such predictions is a form of positive leverage; it increases the probability that what she says will be respected and followed in the future.

In Bryan's study the call girls reported that their training was more a matter of observation than direct instruction from their trainer (1965:294). Ann, on the other hand, relies on a variety of teaching techniques, including lecturing and discussion involving other turn-outs who are further along in the training process and can reinforce Ann's views. Ann even brings in guest speakers, such as Sandy, the former hustler, who participates in the discussion with the novices in the role of the experienced resource person. "Learning the Hustling Rap," above, offers an example of role-playing—another teaching technique Ann frequently employs to help the turn-outs develop verbal skills. Ann may have to rely on more varied teaching approaches than the call-girl trainer because: (1) Ann herself is not working, thus her novices have fewer opportunities to watch their trainer interact with clients than do the call-girl novices; and (2) Ann's livelihood depends more directly on the success of her teaching efforts than does that of the call-girl trainer. Ann feels that if a woman under her direction does not "turn out well," not only will the woman earn less money while she is at her house (affecting Ann's own income), but Ann could also lose clients and future turn-outs from her teaching "failure."[5]

The dissolution of the training relationship marks the end of the course. Bryan claims that the sharp break between trainer and trainee shows that the training process itself is largely unrelated to the acquisition of a skill. But one would scarcely have expected the trainee to report "that the final disruption of the apprenticeship was the result of the completion of adequate training" (1965:296). Such establishments do not offer diplomas and terminal degrees. The present study, too, indicates that abrupt breaks in the training relationship are quite common. But what is significant is that the break is precipitated by personal conflicts exacerbated by both the narrowing of the skill-gap between trainer and trainee and the consequent increase in the novice's confidence that she can make it on her own. Thus, skill acquisition counts in such an equation, not in a formal sense ("completion of adequate training"), but rather in so far as it works to break down the earlier bonds of dependence between trainer and trainee.

IV. THE FUNCTION OF TRAINING AT THE HOUSE LEVEL OF PROSTITUTION

Bryan concludes that the training is necessitated by the novice's need for a list of clients in order to work at the call-girl level and not because the actual training is required to prepare her for such work. But turn-outs at the house level of prostitution do not acquire a clientele. The clients are customers of the house. In fact, the madam usually makes sure that only she has the names or phone numbers of her tricks in order to keep control over her business. If Ann's turn-outs (unlike call girls) do not acquire a clientele in the course of their training, why is the training period necessary?

Although Ann feels strongly that training is required to become a successful hustler at the house level, the function served by the training can be seen more as a spin-off of the structure of the occupation at that level: madams of establishments will often hire only trained prostitutes. Novices who pose as experienced hustlers are fairly easily detected by those proficient in the business working in the same house; to be found out all she need do is violate any of the expected norms of behavior: wear perfume, repeatedly fail to hustle any "over-money" or engage in dirty hustling. The exposure to racket values, which the training provides, may be more critical to the house prostitute than to the call girl. She must live and work in close contact with others in the business. Participants in house prostitution are more integrated into the prostitution world than are call girls, who can be and frequently are "independent"—working without close ties to pimps or other prostitutes. Becoming skilled in hustling is also less important for the call girl, and her minimum fee is usually high, making hustling for small increments less necessary. The house prostitute who does not know how to ask for more money, however, lowers the madam's income as well—another reason why madams prefer professional prostitutes.

The training of house prostitutes, then, reflects two problems in the social organization of house prostitution: (1) most madams will not hire untrained prostitutes; and (2) the close interaction of prostitutes operating within the confines of a house requires a common set of work standards and practices. These two factors differentiate house prostitution from call-girl and streetwalking operations and facilitate this madam's task of turning novices into professional prostitutes. The teaching madam employs a variety of coaching techniques to train turn-outs in sexual and hustling skills and to expose them to a set of occupational rules and values. Hers is an effort to prepare women with conventional backgrounds for work in the social environment of a house of prostitution where those skills and values are expected and necessary.

NOTES

1. This situation-specific induction into prostitution may be contrasted with the "smooth and almost imperceptible" transition to the status of poolroom "hustler" noted by Polsky (1969:80–1).

2. In the wider context of the national prostitution scene, Ann's situation reflects the "minor league" status of her geographical location. In fact, she trains women from other communities who move on to more lucrative opportunities in the big city. See the stimulating applications of the concept of "minor league" to the study of occupations in Faulkner (1974).

3. The term "hustling" has been used to describe a wide range of small-time criminal activities. Even within the world of prostitution, "hustling" can refer to different occupational styles; see Ross's description of the "hustler" who "is distinguished from ordinary prostitutes in frequently engaging in accessory crimes of exploitation," such as extortion or robbery (1959: 16). The use of the term here is thus highly specific, reflecting its meaning in Ann's world.

4. The dialogue sections (for example, "Ann as a Turn-out" and "Learning the Hustling Rap") are transcriptions from the teaching tapes. Redundant expressions have been omitted, and the author's comments on the speech tone or delivery are bracketed. Words italicized indicate emphasis by the speaker.

5. These data bear only on the skills and values to which Ann *exposes* the turn-outs; confirmation of the effects of such exposure awaits further analysis and is a study in its own right. See Bryan's (1966) study of the impact of the occupational perspective taught by call-girl trainers on the individual attitudes of call girls. See Davis (1971:315) for a description of what constitutes successful "in-service training" for streetwalkers.

REFERENCES

Barker, Roger G. (Ed.). 1963. *The Stream of Behavior: Explorations of Its Structure and Content*. New York: Appleton-Century-Crofts.

Bryan, James H. 1965. "Apprenticeships in prostitution." *Social Problems* 12 (Winter): 287–97.

———. 1966. "Occupational ideologies and individual attitudes of call girls." *Social Problems* 13 (Spring):441–50.

Davis, Nanette J. 1971. "The prostitute: Developing a deviant identity." Pp. 297–332 in James M. Henslin (ed). *Studies in the Sociology of Sex*. New York: Appleton-Century-Crofts.

Erwin, Carol. 1960. *The Orderly Disorderly House*. Garden City, N.Y.: Doubleday.

Faulkner, Robert R. 1974. "Coming of age in organizations: A comparative study of career contingencies and adult socialization." *Sociology of Work and Occupations* 1 (May): 131–73.

Gagnon, John H. and William Simon. 1973. *Sexual Conduct: The Social Sources of Human Sexuality*. Chicago: Aldine.

Gray, Diana. 1973. "Turning-out: A study of teenage prostitution." *Urban Life and Culture* 1 (January):401–25.

Hall, Susan. 1973. *Ladies of the Night*. New York: Trident Press.

Heyl, Barbara S. 1974. "The madam as entrepreneur." *Sociological Symposium* 11 (Spring): 61–82.

———. 1975. "The house prostitute: a case study." Unpublished Ph.D. dissertation, Department of Sociology, University of Illinois-Urbana.

Hughes, Everett C. 1971. "Work and self." Pp. 338–47 in *The Sociological Eye: Selected Papers*. Chicago: Aldine-Atherton.

Jackson, Bruce. 1969. *A Thief's Primer*. Toronto, Ontario: Macmillan.

Lewis, Gladys Adelina (ed.) 1942. *Call House Madam: The Story of the Career of Beverly Davis*. San Francisco: Martin Tudordale.

Polsky, Ned. 1969. *Hustlers, Beats and Others*. Garden City, N.Y.: Doubleday.

Ross, H. Laurence. 1959. "The 'Hustler' in Chicago." *Journal of Student Research* 1:13–19.

Sheehy, Gail. 1974. *Hustling: Prostitution in Our Wide-Open Society*. New York: Dell.

Stewart, George I. 1972. "On first being a john." *Urban Life and Culture* 1 (October): 255–74.

Strauss, Anselm L. 1969. *Mirrors and Masks: The Search for Identity*. San Francisco: Sociology Press.

Vogliotti, Gabriel R. 1975. *The Girls of Nevada*. Secaucus, N.J.: Citadel Press.

Winick, Charles and Paul M. Kinsie. 1971. *The Lively Commerce: Prostitution in the United States*. Chicago: Quadrangle Books.

QUESTIONS FOR DISCUSSION

1. Describe Ann's turn-out establishment. What is Ann's goal?

2. Aside from training women to become prostitutes, how can what Ann and others teach help women who are sexually active? How do the "turn-out blues" apply to your response?

3. Describe hustling for house prostitutes. Why does Ann consider teaching hustling skills difficult but the most important task?

On a Screen Near You: Cyberporn

Philip Elmer-Dewitt

Sex is everywhere these days—in books, magazines, films, television, music videos and bus-stop perfume ads. It is printed on dial-a-porn business cards and slipped under windshield wipers. It is acted out by balloon-breasted models and actors with unflagging erections, then rented for $4 a night at the corner video store. Most Americans have become so inured to the open display of eroticism—and the arguments for why it enjoys special status under the First Amendment—that they hardly notice it's there.

Something about the combination of sex and computers, however, seems to make otherwise worldly-wise adults a little crazy. How else to explain the uproar surrounding the discovery by a U.S. Senator—Nebraska Democrat James Exon—

that pornographic pictures can be downloaded from the Internet and displayed on a home computer? This, as any computer-savvy undergrad can testify, is old news. Yet suddenly the press is on alert, parents and teachers are up in arms, and lawmakers in Washington are rushing to ban the smut from cyberspace with new legislation—sometimes with little regard to either its effectiveness or its constitutionality.

If you think things are crazy now, though, wait until the politicians get hold of a report coming out this week. A research team at Carnegie Mellon University in Pittsburgh, Pennsylvania, has conducted an exhaustive study of online porn—what's available, who is downloading it, what turns them on—and the findings (to be published in the *Georgetown Law Journal*) are sure to pour fuel on an already explosive debate.

The study, titled *Marketing Pornography on the Information Superhighway*, is significant not only for what it tells us about what's happening on the computer networks but also for what it tells us about ourselves. Pornography's appeal is surprisingly elusive. It plays as much on fear, anxiety, curiosity and taboo as on genuine eroticism. The Carnegie Mellon study, drawing on elaborate computer records of online activity, was able to measure for the first time what people actually download, rather than what they say they want to see. "We now know what the consumers of computer pornography really look at in the privacy of their own homes," says Marty Rimm, the study's principal investigator. "And we're finding a fundamental shift in the kinds of images they demand."

What the Carnegie Mellon researchers discovered was:

There's an awful lot of porn online.   In an 18-month study, the team surveyed 917,410 sexually explicit pictures, descriptions, short stories and film clips. On those Usenet newsgroups where digitized images are stored, 83.5% of the pictures were pornographic.

It is immensely popular.   Trading in sexually explicit imagery, according to the report, is now "one of the largest (if not the largest) recreational applications of users of computer networks." At one U.S. university, 13 of the 40 most frequently visited newsgroups had names like *alt.sex.stories, rec.arts.erotica* and *alt.sex.bondage*.

It is a big moneymaker.   The great majority (71%) of the sexual images on the newsgroups surveyed originate from adult-oriented computer bulletin-board systems (BBS) whose operators are trying to lure customers to their private collections of X-rated material. There are thousands of these BBS services, which charge fees (typically $10 to $30 a month) and take credit cards; the five largest have annual revenues in excess of $1 million.

It is ubiquitous.   Using data obtained with permission from BBS operators, the Carnegie Mellon team identified (but did not publish the names of) individual consumers in more than 2,000 cities in all 50 states and 40 countries, territories and provinces around the world—including some countries like China, where possession of pornography can be a capital offense.

It is a guy thing.   According to the BBS operators, 98.9% of the consumers of online porn are men. And there is some evidence that many of the remaining 1.1% are women paid to hang out on the "chat" rooms and bulletin boards to make the patrons feel more comfortable.

It is not just naked women.   Perhaps because hard-core sex pictures are so widely available elsewhere, the adult BBS market seems to be driven largely by a demand for images that can't be found in the average magazine rack: pedophilia (nude photos of children), hebephilia (youths) and what the researchers call paraphilia—a grab bag of "deviant" material that includes images of bondage, sadomasochism, urination, defecation, and sex acts with a barnyard full of animals.

The appearance of material like this on a public network accessible to men, women and children around the world raises issues too important to ignore—or to oversimplify. Parents have legitimate concerns about what their kids are being exposed to and, conversely, what those children might miss if their access to the Internet were cut off. Lawmakers must balance public safety with their obligation to preserve essential civil liberties. Men and women have to come to terms with what draws them to such images. And computer programmers have to come up with more enlightened ways to give users control over a network that is, by design, largely out of control.

The Internet, of course, is more than a place to find pictures of people having sex with dogs. It's a vast marketplace of ideas and information of all sorts—on politics, religion, science and technology. If the fast-growing World Wide Web fulfills its early promise, the network could be a powerful engine of economic growth in the 21st century. And as the Carnegie Mellon study is careful to point out, pornographic image files, despite their evident popularity, represent only about 3% of all the messages on the Usenet newsgroups, while the Usenet itself represents only 11.5% of the traffic on the Internet.

As shocking and, indeed, legally obscene as some of the online porn may be, the researchers found nothing that can't be found in specialty magazines or adult bookstores. Most of the material offered by private BBS services, in fact, is simply scanned from existing print publications.

But pornography is different on the computer networks. You can obtain it in the privacy of your home—without having to walk into a seedy bookstore or movie house. You can download only those things that turn you on, rather than buy an entire magazine or video. You can explore different aspects of your sexuality without exposing yourself to communicable diseases or public ridicule. (Unless, of course, someone gets hold of the computer files tracking your online activities, as happened earlier this year to a couple dozen crimson-faced Harvard students.)

The great fear of parents and teachers, of course, is not that college students will find this stuff but that it will fall into the hands of those much younger—including some, perhaps, who are not emotionally prepared to make sense of what they see. Ten-year-old Anders Urmacher, a student at the Dalton School in New York City who likes to hang out with other kids in the Treehouse chat room on America Online, got E-mail from a stranger that contained a mysterious file with instructions for how to download it. He followed the instructions, and then he called his mom. When Linda Mann-Urmacher opened the file, the computer screen filled with 10 thumbnail-size pictures showing couples engaged in various acts of sodomy, heterosexual intercourse and lesbian sex. "I was not aware that this stuff was online," says a shocked Mann-Urmacher. "Children should not be subjected to these images."

This is the flip side of Vice President Al Gore's vision of an information super-highway linking every school and library in the land. When the kids are plugged in, will they be exposed to the seamiest sides of human sexuality? Will they fall prey to child molesters hanging out in electronic chat rooms?

It's precisely these fears that have stopped Bonnie Fell of Skokie, Illinois, from signing up for the Internet access her three boys say they desperately need. "They could get bombarded with X-rated porn, and I wouldn't have any idea," she says. Mary Veed, a mother of three from nearby Hinsdale, makes a point of trying to keep up with her computer-literate 12-year-old, but sometimes has to settle for monitoring his phone bill. "Once they get to be a certain age, boys don't always tell Mom what they do," she says.

"We face a unique, disturbing and urgent circumstance, because it is children who are the computer experts in our nation's families," said Republican Senator Dan Coats of Indiana during the debate over the controversial anti-cyberporn bill he co-sponsored with Senator Exon.

According to at least one of those experts—16-year-old David Slifka of Manhat-tan—the danger of being bombarded with unwanted pictures is greatly exaggerated. "If you don't want them you won't get them," says the veteran Internet surfer. Pri-vate adult BBSS require proof of age (usually a driver's license) and are off-limits to minors, and kids have to master some fairly daunting computer science before they can turn so-called binary files on the Usenet into high-resolution color pictures. "The chances of randomly coming across them are unbelievably slim," says Slifka.

While groups like the Family Research Council insist that online child molesters represent a clear and present danger, there is no evidence that it is any greater than the thousand other threats children face every day. Ernie Allen, executive director of the National Center for Missing and Exploited Children, acknowledges that there have been 10 or 12 "fairly high-profile cases" in the past year of children being se-duced or lured online into situations where they are victimized. Kids who are not on-line are also at risk, however; more than 800,000 children are reported missing every year in the U.S.

Yet it is in the name of the children and their parents that lawmakers are rac-ing to fight cyberporn. The first blow was struck by Senators Exon and Coats, who earlier this year introduced revisions to an existing law called the Communications Decency Act. The idea was to extend regulations written to govern the dial-a-porn industry into the computer networks. The bill proposed to outlaw obscene material and impose fines of up to $100,000 and prison terms of up to two years on anyone who knowingly makes "indecent" material available to children under 18.

The measure had problems from the start. In its original version it would have made online-service providers criminally liable for any obscene communications that passed through their systems—a provision that, given the way the networks operate, would have put the entire Internet at risk. Exon and Coats revised the bill but left in place the language about using "indecent" words online. "It's a frontal assault on the First Amendment," says Harvard law professor Laurence Tribe. Even veteran prose-cutors ridicule it. "It won't pass scrutiny even in misdemeanor court," says one.

The Exon bill had been written off for dead only a few weeks ago. Republican Senator Larry Pressler of South Dakota, chairman of the Commerce committee, which has jurisdiction over the larger telecommunications-reform act to which it is attached, told TIME that he intended to move to table it.

That was before Exon showed up in the Senate with his "blue book." Exon had asked a friend to download some of the rawer images available online. "I knew it was bad," he says. "But then when I got on there, it made *Playboy* and *Hustler* look like Sunday-school stuff." He had the images printed out, stuffed them in a blue folder and invited his colleagues to stop by his desk on the Senate floor to view them. At the end of the debate—which was carried live on C-SPAN—few Senators wanted to cast a nationally televised vote that might later be characterized as pro-pornography. The bill passed 84 to 16.

Civil libertarians were outraged. Mike Godwin, staff counsel for the Electronic Frontier Foundation, complained that the indecency portion of the bill would transform the vast library of the Internet into a children's reading room, where only subjects suitable for kids could be discussed. "It's government censorship," said Marc Rotenberg of the Electronic Privacy Information Center. "The First Amendment shouldn't end where the Internet begins."

The key issue, according to legal scholars, is whether the Internet is a print medium (like a newspaper), which enjoys strong protection against government interference, or a broadcast medium (like television), which may be subject to all sorts of government control. Perhaps the most significant import of the Exon bill, according to EFF's Godwin, is that it would place the computer networks under the jurisdiction of the Federal Communications Commission, which enforces, among other rules, the injunction against using the famous seven dirty words on the radio. In a TIME/CNN poll of 1,000 Americans conducted last week by Yankelovich Partners, respondents were sharply split on the issue: 42% were for FCC-like control over sexual content on the computer networks; 48% were against it.

By week's end the balance between protecting speech and curbing pornography seemed to be tipping back toward the libertarians. In a move that surprised conservative supporters, House Speaker Newt Gingrich denounced the Exon amendment. "It is clearly a violation of free speech, and it's a violation of the right of adults to communicate with each other," he told a caller on a cable-TV show. It was a key defection, because Gingrich will preside over the computer-decency debate when it moves to the House in July. Meanwhile, two U.S. Representatives, Republican Christopher Cox of California and Democrat Ron Wyden of Oregon, were putting together an anti-Exon amendment that would bar federal regulation of the Internet and help parents find ways to block material they found objectionable.

Coincidentally, in the closely watched case of a University of Michigan student who published a violent sex fantasy on the Internet and was charged with transmitting a threat to injure or kidnap across state lines, a federal judge in Detroit last week dismissed the charges. The judge ruled that while Jake Baker's story might be deeply offensive, it was not a crime.

How the Carnegie Mellon report will affect the delicate political balance on the cyberporn debate is anybody's guess. Conservatives thumbing through it for rhetorical ammunition will find plenty. Appendix B lists the most frequently downloaded files from a popular adult BBS, providing both the download count and the two-line descriptions posted by the board's operator. Suffice it to say that they all end in exclamation points, many include such phrases as "nailed to a table!" and none can be printed in TIME.

How accurately these images reflect America's sexual interests, however, is a matter of some dispute. University of Chicago sociologist Edward Laumann, whose 1994 *Sex in America* survey painted a far more humdrum picture of America's sex life, says the Carnegie Mellon study may have captured what he calls the "gaper phenomenon." "There is a curiosity for things that are extraordinary and way out," he says. "It's like driving by a horrible accident. No one wants to be in it, but we all slow down to watch."

Other sociologists point out that the difference between the Chicago and Carnegie Mellon reports may be more apparent than real. Those 1 million or 2 million people who download pictures from the Internet represent a self-selected group with an interest in erotica. The *Sex in America* respondents, by contrast, were a few thousand people selected to represent a cross section of all America.

Still, the new research is a gold mine for psychologists, social scientists, computer marketers and anybody with an interest in human sexual behavior. Every time computer users logged on to one of these bulletin boards, they left a digital trail of their transactions, allowing the pornographers to compile databases about their buying habits and sexual tastes. The more sophisticated operators were able to adjust their inventory and their descriptions to match consumer demand.

Nobody did this more effectively than Robert Thomas, owner of the Amateur Action BBS in Milpitas, California, and a kind of modern-day Marquis de Sade, according to the Carnegie Mellon report. He is currently serving time in an obscenity case that may be headed for the Supreme Court.

Thomas, whose BBS is the online-porn market leader, discovered that he could boost sales by trimming soft- and hard-core images from his database while front-loading his files with pictures of sex acts with animals (852) and nude prepubescent children (more than 5,000), his two most popular categories of porn. He also used copywriting tricks to better serve his customers' fantasies. For example, he described more than 1,200 of his pictures as depicting sex scenes between family members (father and daughter, mother and son), even though there was no evidence that any of the participants were actually related. These "incest" images were among his biggest sellers, accounting for 10% of downloads.

The words that worked were sometimes quite revealing. Straightforward oral sex, for example, generally got a lukewarm response. But when Thomas described the same images using words like choke or choking, consumer demand doubled.

Such findings may cheer antipornography activists; as feminist writer Andrea Dworkin puts it, "the whole purpose of pornography is to hurt women." Catharine MacKinnon, a professor of law at the University of Michigan, goes further. Women are doubly violated by pornography, she writes in *Vindication and Resistance,* one of three essays in the forthcoming *Georgetown Law Journal* that offer differing views on the Carnegie Mellon report. They are violated when it is made and exposed to further violence again and again every time it is consumed. "The question pornography poses in cyberspace," she writes, "is the same one it poses everywhere else: whether anything will be done about it."

But not everyone agrees with Dworkin and MacKinnon, by any means; even some feminists think there is a place in life—and the Internet—for erotica. In her new book, *Defending Pornography,* Nadine Strossen argues that censoring sexual expression would do women more harm than good, undermining their equality, their autonomy and their freedom.

The Justice Department, for its part, has not asked for new antiporn legislation. Distributing obscene material across state lines is already illegal under federal law, and child pornography in particular is vigorously prosecuted. Some 40 people in 14 states were arrested two years ago in Operation Longarm for exchanging kiddie porn online. And one of the leading characters in the Carnegie Mellon study—a former Rand McNally executive named Robert Copella, who left book publishing to make his fortune selling pedophilia on the networks—was extradited from Tijuana, and is now awaiting sentencing in a New Jersey jail.

For technical reasons, it is extremely difficult to stamp out anything on the Internet—particularly images stored on the Usenet newsgroups. As Internet pioneer John Gilmore famously put it, "The Net interprets censorship as damage and routes around it." there are border issues as well. Other countries on the Internet—France, for instance—are probably no more interested in having their messages screened by U.S. censors than Americans would be in having theirs screened by, say, the government of Saudi Arabia.

Historians say it should come as no surprise that the Internet—the most democratic of media—would lead to new calls for censorship. The history of pornography and efforts to suppress it are inextricably bound up with the rise of new media and the emergence of democracy. According to Walter Kendrick, author of *The Secret Museum: Pornography in Modern Culture,* the modern concept of pornography was invented in the 19th century by European gentlemen whose main concern was to keep obscene material away from women and the lower classes. Things got out of hand with the spread of literacy and education, which made pornography available to anybody who could read. Now, on the computer networks, anybody with a computer and a modem can not only consume pornography but distribute it as well. On the Internet, anybody can be Bob Guccione.

That might not be a bad idea, says Carlin Meyer, a professor at New York Law School whose *Georgetown* essay takes a far less apocalyptic view than MacKinnon's. She argues that if you don't like the images of sex the pornographers offer, the appropriate response is not to suppress them but to overwhelm them with healthier, more realistic ones. Sex on the Internet, she maintains, might actually be good for young people. "[Cyberspace] is a safe space in which to explore the forbidden and the taboo," she writes. "It offers the possibility for genuine, unembarrassed conversations about *accurate* as well as fantasy images of sex."

That sounds easier than it probably is. Pornography is powerful stuff, and as long as there is demand for it, there will always be a supply. Better software tools may help check the worst abuses, but there will never be a switch that will cut it off entirely—not without destroying the unbridled expression that is the source of the Internet's (and democracy's) greatest strength. The hard truth, says John Perry Barlow, cofounder of the EFF and father of three young daughters, is that the burden ultimately falls where it always has: on the parents. "If you don't want your children fixating on filth," he says, "better step up to the tough task of raising them to find it as distasteful as you do yourself."

QUESTIONS FOR DISCUSSION

1. Discuss the Carnegie Mellon research findings. Are there any results surprising or interesting? Explain your response.

2. Discuss senators Exon and Coats' bill, the Communications Decency Act. What was the disposition of the act? How helpful is the act? Explain your response.

3. How do feminists discuss the issue of cyberporn? Do you agree or disagree with Dworkin, MacKinnon, and Strossen's comments? Explain your response.

Check Mate: What I Teach My Daughter about Sex

Susie Bright

Yesterday, my 8-year-old daughter received her first lesson in chess from her godmother, Honey Lee. She has now played a total of three games with other people. (She's also been playing against herself—or some imaginary loser, I don't know who—but those games seem to end rather quickly.) I myself don't know a rook from a rooster, but I am captivated by watching her play. For one, she's very emotional and surprisingly rowdy. She whistles loudly while her opponent is contemplating the next move, and she had one big cry when her dad, Jon, "stole" her knight, as she put it. Instead of saying "Check," she sings out, "POP goes the weasel!" I picked up the beginners instruction book ("Chess Basics," by Nigel Short, highly recommended!) to see if this is really part of the chess vocabulary. "Oh yes," said Jon, "that's what Fischer said to Spassky in the Reykjavik match."

I didn't know whether to believe his story because I did find a story in the book's gamesmanship chapter about two adult champions, Petrosian and Korchnoi, who in a 1977 tournament kicked at each other under the table until a below-the-waist partition had to be built to separate them.

In any case, I am riveted by Aretha's competitive spirit and her desire to win and gloat. Early in her very first game with Honey Lee, she warned, "When I'm done with you, your life won't be worth a penny."

I was never encouraged to play like this when I was a child. I fantasized about knocking balls across the field to the amazement of my schoolmates, or beating a grown-up in a simple card game, but I was very afraid of being immodest or too competitive in public. The one contest of any significance I won was a spelling bee, although I was initially disqualified because the judge, our principal, made a mistake with the three-letter word "it's . . . As in, 'It's a beautiful day.'" I spelled it with the appropriate apostrophe, and he triumphantly rebuked me, "No, that's wrong, the correct answer is I-T-S."

I would never have contradicted him because that would be talking back, arrogant and a whole lot of trouble, but later a teacher pulled me into his office, and the principal grimly announced, "There has been an error. You have won the spelling bee." He handed me a gray piece of paper with the Riverside County Schools letterhead on top. I gathered there would be no publicity.

My fear of competition and my desire to "make nice" were the first aspects of my sexuality I confronted when I began to have sex with other people. If my lover touched me in a way that was displeasing, hurt or just didn't rock my boat, I would rather have died than said anything about it. I instantly grasped the technique of faking orgasm because it was as familiar as faking satisfaction about anything else when courtesy demanded I not speak up, but I also wouldn't show what I wanted with my hands. That, too, would have drawn too much attention to myself.

My salvation was that I read a lot, which introduced me to the notion that I would be happier if I asserted myself sexually. Germaine Greer said so! In addition, I had a couple of very blunt lovers who shattered my idea of what can and can't be said aloud between friends.

"The way you kiss, it's terrible," my first girlfriend told me. (I always fall for the direct type.) She was French, so I knew she must know what she was talking about. She gave me kissing lessons and explained the way she wanted me to caress her mouth. I was very grateful after the initial burn of embarrassment.

Another of my early boyfriends flat-out asked me, "Why don't you come with me? I don't like it as much if you don't. You masturbate, don't you? Show me!"

Well now, there he had me . . . if I was to please him, then I would have to come. It wasn't a matter of being self-centered.

Since I started writing so prolifically about sex, many readers have imagined me to be an erotic Catwoman on the prowl, taking what I want with grace, style and endless nerve. I think I'm more like a mouse who encountered some radical manifestos and was inspired to roar, overcoming my original character and training.

Is my daughter a tiger because of her feisty DNA or has my encouraging her to be strong and direct worked, miraculously? I've deliberately given her opportunities to do unladylike things. I've tolerated and defended her aggressiveness in speech and in movement. I never wrestled on the floor when I was a girl, but she loves it, and now asks for boxing gloves, too. I never talked back, but Aretha thinks nothing of arguing if she believes she's in the right. Perhaps it's the other girls around her. She's in some ways the prissiest of her friends, surfers and skateboarders who'd rather play soccer than Barbie.

What sexual advice have I given her? I've left it very open as to what sex is—that it's any two (or more, I'm so PC) people touching each other all over with their hands or mouths or genitals. She really doesn't know yet that when most people say "sex" they only mean penis-vagina intercourse. I can only hope that things change before she gets that message.

I've given her a few words about bodily fluids and STDs, but I don't want to harp on that. That's what she's going to hear from everyone else—about how adult sex is so dirty and disease-ridden and dangerous that it's amazing that anyone would want to go through puberty and attempt it. I tell her sex is supposed to feel good, really good, and that if it doesn't, stop what you're doing. I told her, "Don't worry about hurting the other person's feelings if you don't like the way it feels, because it's your body, after all."

She gives me a puzzled look, as if to say, "DUH, Mom! Why would I do anything that made me feel bad?" I hope she feels this much confidence as long as she lives because honestly, whose life is worth a penny without it?

QUESTIONS FOR DISCUSSION

1. In your opinion, what does the author teach her daughter about sex? Why is the article entitled "Check Mate?"

2. How is the author's sexual experience similar or dissimilar to someone you know? Explain your response.

3. You have been asked to speak to a group of fourth-graders about sex. From the reading, what information would you use?

C H A P T E R 14

SUICIDE

Gender Differences in the Suicide-Related Behaviors of Adolescents and Young Adults

Jennifer Langhinrichsen-Rohling, Peter Lewinsohn,
Paul Rohde, John Seeley, Candice M. Monson,
Kathryn A. Meyer, and Richard Langford

Gender differences in suicide-related behaviors were examined in an older adolescent and a young adult sample (primarily Caucasian). Suicide-related behaviors were assessed by the Life Attitudes Schedule (LAS) as well as by measures of depressive symptomatology and hopelessness. The LAS measures a broad continuum of potentially life-diminishing or life-enhancing behaviors. There are four LAS content-category subscales: overtly suicidal and death-related, self-related, risk and injury-related, and health-related behaviors. As hypothesized, in both samples, gender differences in the expression of suicide-related behaviors were obtained. Males from both samples endorsed substantially more risk-taking and injury-producing behaviors than females. Males in both samples also reported more negative health-related behaviors than females. In contrast, females reported more symptoms of depression than males. Hopelessness scores only differentiated male and female young adults; male and female adolescents did not differ significantly on the hopelessness measure. These findings

"Gender Differences in the Suicide-Related Behaviors of Adolescents and Young Adults" by Jennifer Langhinrichsen-Rohling, Peter Lewinsohn, Paul Rohde, John Seeley, Candice M. Monson, Kathryn A. Meyer, and Richard Langford. From *Sex Roles*, Dec. 1998, v39, i11–12, p. 839(1). Reprinted with permission of Plenum Publishing Corporation.

are primarily discussed in terms of gender-role socialization theory. Implications for the treatment of suicidality are drawn.

Controversy exists about how to define and measure suicidality in adolescents and young adults. Some researchers and clinicians have preferred to assess only overt suicidal behavior that is engaged in with intentionality. Other researchers have posited that there is a conceptual link among overt suicidal behaviors, less lethal suicidal behaviors such as self-mutilation and self-denigration, and life-threatening behaviors that may be unintentionally suicidal such as risk-taking, illness-enhancing or injury-producing behaviors (e.g., Lewinsohn et al., 1995; Shneidman, Farberow, & Litman, 1970). Consistent with this line of reasoning, a variety of diverse problem behaviors, many of which are self-destructive (e.g., cigarette smoking, alcohol use, risk taking), have been shown to be intercorrelated beginning in early adolescence in both males and females (Hawkins, Catalano, & Miller, 1992; Jessor & Jessor, 1975). Furthermore, as predicted theoretically, overtly suicidal behavior has been directly associated with a broad range of other potentially self-destructive and life-threatening behaviors. For example, individuals who have engaged in a suicide attempt have been found to be at greater risk of engaging in early sexual activity and substance use than individuals without a suicide behavior history (Ensminger, 1987). Data collected on 8th and 10th graders found that both alcohol use and risky behaviors were significant predictors of suicide ideation and attempts (Windle, Miller-Tutzauer, & Domenico, 1992). Furthermore, Shaffer and colleagues have determined that a variety of psychiatric disorders are related to suicide (Shaffer, Garland, Gould, Fisher & Trautman, 1988). For example, a study of adolescents in New York City who had committed suicide revealed that nearly ½ of the victims had been identified as needing treatment or had sought mental health services (Shaffer & Gould, 1987). Gender differences in the expression of and associations among various suicide-related behaviors have been a relatively neglected topic, however.

Consistent with the notion that diverse factors (e.g., depression, poor self-esteem, drug and alcohol abuse) are related to suicide, Lewinsohn and colleagues (1995) recently delineated a broad conceptualization of suicidality. Specifically, they posited a continuum of suicide-related behaviors. One end of their continuum consisted of life-enhancing behaviors; traditionally defined, overtly suicidal behaviors comprised the opposite pole. This group of researchers theorized that efforts to prevent fatal suicides will be enhanced if an individual's propensity to engage in life-enhancing behaviors is increased while her or his engagement in a variety of potentially life-threatening and suicide-related behaviors is decreased.

To test their theoretical construct, Lewinsohn et al. (1995) created a self-report measure, the Life Attitudes Schedule (LAS), that assesses engagement in a broad range of risky, life-threatening and potentially suicide-related behaviors. The instrument also assesses the degree to which the individual is engaging in life-promoting or life-enhancing behaviors. The LAS consists of four content category subscales which measure overtly suicidal and death-related (DR), illness and health-related (HR), risk and injury-related (IR), and self-related (SR) behaviors. Equal numbers of items were included on each subscale. Each subscale score is a combination of the number of negative items endorsed as well as the number of positive items that were not endorsed. To further delineate the construct of suicide proneness, items were selected that loaded highly on their respective subscales while showing only moderate correlations with measures of depressive symptomology, hopelessness, and social desirability. Initial data indicate that, as predicted, LAS total scores and all four LAS subscale scores are significantly associated with the occurrence of current suicide ideation and past suicide attempts in adolescents and young adults (Lewinsohn et al., 1995).

To date, no gender-specific results have been presented using the LAS as a measure of suicide-related behaviors. However, gender differences in overtly suicidal behaviors have been consistently demonstrated in the literature (e.g., Allgood-Merten, Lewinsohn, & Hops, 1990; McIntosh & Jewell, 1986; National Center for Health Statistics, 1994). For example, adolescent males engage in fatal suicidal behavior much more frequently than adolescent females. In 1992, among those aged 15 to 24, the rate of completed suicides was 21.9 per 100,000 for males and 3.7 per 100,000 for females (National Center for Health Statistics, 1994). Although young females do not complete suicide as often as young males, they frequently report more suicidal thoughts (Andrews & Lewinsohn, 1992; Simons & Murphy, 1985) and more suicide attempts (e.g., Andrews & Lewinsohn, 1992). Because gender differences in the expression of overt suicidal behavior have been documented, the current research was conducted to determine if there are also gender differences in the expression of the broad range of suicide-related behaviors that are assessed on the LAS.

It seems likely that there will be gender differences in LAS scores because gender differences in the prevalence of constructs that relate to suicidality have also been previously demonstrated. For example, adolescent females traditionally report more symptoms of depression and distress than do adolescent males (Allgood-Merten et al., 1990; Gore, Aseltine, & Colton, 1992). In contrast, male adolescents have been found to score higher than females on a factor labeled "foolhardiness" designed to reflect recklessness, dangerousness, and an interest in weapons (Clark, Sommerfeldt, Schwartz, Hedeker, & Watel, 1990). The activities that young males report engaging in for thrills and excitement have also been rated as riskier than the activities that females report doing (Lefkowitz, Kahlbaugh, & Sigman, 1994). While the links between depression and suicidality are well established, potential links between risk-taking, sensation-seeking, and suicide-related behavior warrant increased attention. Gender differences in the associations among these behaviors have also not been well established. Therefore, the second purpose of this study was to assess gender differences in expressions of suicidality as depicted by scores on the four LAS subscales. Gender differences in responses to measures of depressive symptomatology and hopelessness were concurrently examined. Specifically, the following hypotheses were proposed for the current study: (1) Females would score higher overall than males on the LAS. This expectation was based on past research findings that have reported higher scores for women than men on self-report measures of depression and suicide-related behavior. (2) The pattern of scores on the four LAS subscales was hypothesized to vary as a function of gender. Specifically, females were hypothesized to report more negative and fewer positive self-related behaviors than males. Females were also expected to score higher on the overtly suicidal and death-related LAS subscale than would males. In contrast, males were expected to have higher scores than females on the risk-taking and injury-related LAS subscale. Males were also expected to engage in more negative health-related behaviors than females. (3) Consistent with past research, females were expected to report more symptoms of depression and hopelessness than males.

This study reports on data that were collected from two large convenience samples. The first sample consisted of high school students, the second of college students. Data collection strategies and recruitment procedures differed across samples. In addition, the different-aged samples were collected in different parts of the United States. As a result, the two samples were not combined for data analysis. Instead, a priori, it was expected that any obtained gender differences would be replicated across the two samples.

METHOD

Participants

Sample One.   Data from two hundred and six adolescents comprised Sample One. All participants were enrolled in a large, urban, predominantly middle-class high school in Northern California. Ninety-nine of the participants (48.1%) were male and 106 (51.5%) were female. One participant did not report his or her gender, and was thus excluded from further data analysis. The majority of participants were Caucasian (73.3%). Of the remaining participants, 11.2% were Hispanic, 5.3% were Native American, 1.5% were Asian American, 1.5% were African American, and the remaining 7.3% classified themselves as "other." Their mean age was 16.3 years (SD = 1.08 years). All grade levels were represented in the sample.

Participants entered the project through a passive parental and active adolescent written consent procedure. Eighty percent of the potential subjects were eligible to participate after parents had declined participation and adolescents were administered the written informed consent procedure. For all participants, the pencil and paper assessment package was completed in a group setting under the supervision of a teacher and a research assistant in a 50-minute class period. To insure uniform data collection, research assistants administered the self-report packet following a prepared protocol. Payment consisted of donating one dollar to the school for every participant that completed the assessment packet. Several efforts were made to safeguard the welfare of participating adolescents. For instance, active adolescent consent to participate was required and referral numbers and follow-up clinical interviews were made available to all participants.

Sample Two.   There were 593 college students in Sample Two. All participants were enrolled in a large public university in the Midwest. Two hundred and eighty-nine of the participants (48.7%) were male and 304 (51.3%) were female. The majority of participants were freshmen and sophomores. The mean age of the sample was 19.6 years (SD = 1.79 years). The majority of participants were Caucasian (93.1%). Of the remaining participants, 2.5% were African American, 1.3% were Hispanic, .5% were Native American, .7% were Asian American, and the remaining 1.7% classified themselves as "other."

Participants in the college sample signed up for this study in order to receive research credit for one of several Introductory Psychology courses. The students completed the self-report packet in small groups that ranged from 5 to 25 participants. This was done in order to facilitate anonymity. As with the high school sample, ethical responsibilities were maintained in the data collection procedure. Active informed consent was required from all participants.

Measures

The following measures were included in both self-report packages and were analyzed for the current study. They were the Life Attitudes Schedule, a measure of depressive symptomatology (CES-D for adolescents and BDI for the college students), the Beck Hopelessness Scale, and the Crowne-Marlowe Social Desirability Scale.

Life Attitudes Schedule (LAS; Lewinsohn, Langhinrichsen, Langford, & Rohde, in press).   The LAS consists of 96-items designed to assess the broad construct of suicide-related behaviors. Half the items were designed to assess life-promoting behaviors; the other half of the items assess life-diminishing or life-threatening behaviors. LAS total scores range from 0 (no endorsement of any negative items and the endorsement of every positive item) to 96 (endorsement of every negative item and no endorsement of any positive item). Thus, higher LAS total scores indicate greater levels of suicide-related behaviors.

Four content category subscales scores are contained within the LAS. These subscales measure overtly suicidal and death-related, illness and health-related, risk and injury-related, and self-related behaviors. Each content category is assessed via 24 items. The overtly suicidal and death-related (DR) domain consists of death, traditional suicide, and longevity items (e.g., "I am hopeful that I will live to a ripe old age"; "I wrote a suicide note"). The illness and health-related (HR) domain is comprised of health, illness, lack of self-care, and wellness items (e.g., "Seeing a dentist for regular checkups is not important"; "I try to eat foods that are good for me"). The risk and injury-related (IR) category includes injury, risk-taking, and safety related behaviors (e.g., "I jumped on or off a moving vehicle"; "I am the type of person that thinks about how to protect myself"). The self-related (SR) domain assesses self-worth, self-efficacy, sense of accomplishment, and self-image. This subscale includes both self-denigrating and self-enhancing items (e.g., "Most of the time, I feel confident and assured"). Results of structural equation modeling confirmed the utility of the four LAS subscales. In general, as expected, the DR subscale loaded the highest on the overall construct while the IR subscale loaded the lowest (Lewinsohn et al., 1995).

The LAS has good reliability (i.e., coefficient alpha for total LAS score and four subscale scores = .92 and greater; test-retest reliability over a thirty day period ranged from .75 to .88). The LAS total score and the four content category subscale scores have all been shown to be significantly correlated with lifetime history of suicide attempts (r's ranging from .38 to .45, p [less than] .001), which provides some evidence for the validity of the measure. Lewinsohn and colleagues (1995) report that care was taken to select items that showed high LAS subscale correlations in conjunction with low to moderate correlations with measures of depression, hopelessness, and social desirability. Consistent with their selection criteria in both samples and for both genders, generally small to moderate correlations were obtained between LAS scores and the other psychosocial measures. Specifically, r's ranged from a low of .19 between IR and depression scores, to a high of .72 between SR and depression scores in adolescent males. Likewise, r's ranged from a low of .16 between HR and hopelessness scores for college males to a high of .78 between SR and hopelessness scores for adolescent females. In contrast, correlations between LAS scores and social desirability scores were lower. R's ranged from a low of .01 between LAS total and social desirability scores for college males to a high of $-.46$ between LAS total and social desirability scores for adolescent males. These results provide some evidence of the LAS's discriminant validity.

Gender differences in the obtained correlations were tested using Fisher's r to Z transformation. To protect against the increased probability of Type I error, only those correlations that differed at the p [less than] .01 level were considered significant. Using this criteria, no gender differences in the correlations between the LAS scores and the psychosocial variables were obtained.

The Center for Epidemiologic Studies-Depression Scale (CES-D; Radloff, 1977). This depression scale was administered to the adolescent sample. The CES-D is a 20-item self-report measure of depressive symptomatology. Participants rate the frequency with which they have experienced any symptoms during the past week. It is short, easy to read, and has been successfully used to assess depression in adolescent populations (Garrison, Shoenback, & Kaplan, 1983). The CES-D correlates highly (R = .70 to .80) with other self-report depression instruments (e.g., the Beck Depression Inventory) and has good psychometric properties (Radloff, 1977).

Beck Depression Inventory (BBDI; Beck 1961). This measure was used to assess symptoms of depression in Sample Two. This is a 21-item scale that measures symptoms of depression. Alphas range from .73 to .92 in nonpsychiatric samples (Beck, Rush, Shaw, & Emery, 1979). It is one of the most frequently used measures to assess depression in college students (Beck, Steer, & Garbin, 1988).

Beck Hopelessness Scale (BHS; Beck, Weissman, Lester, & Texler, 1974). The BHS is a 20-item, true-false inventory designed to measure lack of hope about the future. Beck et al. (1974) report internal consistency ratings of .93 for this measure. It has been used frequently with adolescent and young adult samples.

Crowne-Marlowe Social Desirability Scale (Crowne & Marlowe, 1960). Both samples were given an abbreviated version of this scale which measures the degree to which participants are responding in a socially desirable way. The six-item, true-false short form has been shown to have reasonable reliability and validity in an adolescent population (Andrews, Lewinsohn, Hops, & Roberts, 1993).

RESULTS
Gender Differences in Reports of Depressed, Hopeless, and Suicide-Related Behaviors

MANOVA was used to assess overall gender differences in reports of suicide-related behaviors, symptoms of depression, and hopelessness in the two samples. Participant gender was the independent variable while the depression measure (i.e., BDI or CES-D), BHS, and LAS total scores were the dependent variables in each analysis. As predicted, a main effect for participant gender was obtained with data collected from Sample One, Wilks Lambda (3, 181) = .90, p [less than] .001.

Gender Differences in Depressive Symptoms,   **TABLE 1**
Hopelessness, Social Desirability, and Suicide-Related Behaviors

Measure	Males	Females	Eta	F
Sample One—Adolescents (n: 99 Males, 106 Females)				
LAS	31.86	26.55	.04	7.02(b)
BHS	23.53	23.26	.00	[less than] 1
CES-D	7.20	8.89	.02	3.36 p = .07
SOC DES	5.13	5.33	.00	[less than] 1
Sample Two—Young Adults (n: 287 Males, 302 Females)				
LAS	29.50	23.48	.06	22.18(c)
BHS	26.09	24.89	.02	6.44(a)
BDI	6.26	9.08	.03	11.10(c)
SOC DES	4.29	4.55	.00	[less than] 1

(a) = p [less than] .05.

(b) = p [less than] .01.

(c) = p [less than] .001.

As shown in Table 1, follow-up ANOVAs revealed that the gender effect was significant for the LAS. Contrary to expectation, however, males reported significantly more suicide-related behavior than females, $F (1, 183) = 7.02$, p [less than] .01. Consistent with hypothesis, there was also a trend for female adolescents to report more symptoms of depression than male adolescents, $F (1, 183) = 3.36$, p = .068. No significant gender differences emerged on the hopelessness or social desirability scales.

When this analysis was repeated with data from Sample Two, the overall gender difference on the MANOVA was replicated, Wilks Lambda $(3, 329) = .77$, p [less than] .001. Follow-up ANOVAs indicated the college males also reported more suicide-related behaviors on the LAS than did college females, $F (1, 331) = 22.18$, p [less than] .001. As predicted, college women reported significantly more symptoms of depression on the BDI than college men, $F (1, 331) = 11.10$, p = .001. Contrary to expectation, however, a gender difference was also obtained on the BHS such that college men endorsed more hopelessness than did college women, $F (1, 331) = 6.44$, p = .012. No gender difference on the measure of social desirability was obtained.

To determine the robustness of the obtained gender differences on the LAS, two additional analyses were conducted. In both analyses, gender was the independent variable and LAS total scores from the two samples were the dependent variables. The measures of depressive symptoms, hopelessness, and social desirability were entered covariates in these analyses. As expected, for both samples, the significant gender differences on the LAS were maintained even when the measures of depressive symptoms, hopelessness, and social desirability were entered as covariates.

Gender Differences on the Four LAS Subscales

Two MANOVAs were conducted to determine if there were gender differences on the four LAS subscales. In each of these analyses, gender was the independent variable and the four LAS subscales were the dependent variables. As expected, a main effect

TABLE 2   Gender Differences on the Four LAS Subscales(a)

Subscale	Males	Females	Eta	F
	Sample One—Adolescents (n: 99 Males, 106 Females)			
DR	5.07	4.37	.01	1.64
HR	8.36	6.80	.03	6.7(b)
IR	14.10	9.78	.17	40.07(c)
SR	4.96	5.14	.00	[less than] 1
	Sample Two—Young adults (n: 287 Males, 302 Females)			
DR	4.06	3.88	.00	[less than] 1
HR	8.04	6.52	.04	14.14(c)
IR	13.54	9.41	.17	71.03(c)
SR	3.93	3.95	.00	[less than] 1

(a) DR = overtly suicidal and death-related, HR = health and illness related, IR = risk and injury related, SR = self-related.

(b) = p [less than] .01

(c) = p [less than] .001.

for gender was obtained for Sample One, Wilks Lambda (4, 193) = .80, p [less than] .001. As shown in Table 2, follow-up ANOVAs indicated that there were significant gender differences for the IR and HR subscales. Adolescent males reported substantially more risky and injury-related behaviors than females, F (1, 196) = 40.70, p [less than] .001. Males also reported more negative and fewer positive health-related behaviors than females, F (1, 196) = 6.70, p = .01. No gender differences on the DR and SR scales were obtained.

A main effect for gender was also obtained for Sample Two, Wilks Lambda (4, 346) = .81, p [less than] .001. As shown in Table 2, follow-up ANOVAs indicated that the findings obtained with Sample One were replicated with Sample Two. Specifically, there were significant gender differences for the IR and HR subscales. As predicted, college males reported substantially more risky and injury-related behaviors than females, F (1, 349) = 71.03, p [less than] .001. College males also reported more negative and fewer positive health-related behaviors than college females, F (1, 349) = 14.14, p [less than] .001. Again, no gender differences emerged on the DR and the SR subscales.

All the analyses reported above were conducted again with social desirability scores as a covariate. Although social desirability emerged as a significant covariate in each analysis, none of the reported findings were rendered insignificant. These analyses were repeated a third time using depressive symptomatology scores as a covariate. In these analyses, gender differences on the LAS became more pronounced, particularly in the college sample (LAS eta in the college sample = .20; LAS eta in the high school sample = .09). In fact, when BDI scores were used as a covariate in the college sample, all four LAS subscales were significantly differentiated by gender. Males scored higher than females on the SR and the DR subscales when variance in depression scores was removed statistically. Gender differences on the IR and HR subscales became even more pronounced. Similarly, in the high school sample, covarying out CES-D scores increased the gender differences on the IR and HR subscales and created a trend for a gender difference on the DR subscale. There were still no significant gender differences on the SR subscale in this analysis. Even when measures of depressive symptoms, hopelessness, and social desirability were simultaneously entered into the analyses as covariates, the significant gender differences on the LAS subscales were maintained in both samples.

DISCUSSION

As anticipated, gender differences in the expression of adolescents' and young adults' suicide-related behaviors were revealed in this study. Males from two independent samples were found to score higher overall on the LAS than females. In particular, in both groups, males reported substantially more risky and injury-related (IR) and negative health-related (HR) behaviors than females. These results support previous research findings that males have higher rates of sensation-seeking and risky behavior than females (e.g., Lefkowitz et al., 1994). The obtained LAS results are also consistent with adolescent males' higher rates of lethal suicide behavior, single car accidents, and externalizing disorders (e.g., Kandel & Davies, 1982; Leadbeater, Blatt, & Quinlan, 1995). In general, males appear to engage in more impulsive behavior than females and impulsivity has been linked to suicidal behavior (Garland & Zigler, 1993). Because it assesses a broad range of potentially risky, impulsive, suicide-prone, and life-threatening behaviors, the LAS may have particular utility as a measure to identify at-risk males. Certainly this is one of the few measures of suicidality in which males score higher than females to a degree that is consistent with the gender differential in fatal suicidal behavior. These gender-specific findings are particularly noteworthy as all items on the LAS were initially selected on the basis of their low or nonsignificant associations with gender (Lewinsohn et al., 1995).

In addition, as has been found in past research (Allgood-Merten et al., 1990; Gore et al., 1992), results from the current study revealed that young adult and adolescent females tended to report more symptoms of depression than young adult and adolescent males. These findings indicate that using only a measure of depression to assess suicidality may be more beneficial for identifying females than males. This finding has clinical implications because measures of depressive symptoms have often been used as markers of suicide risk (Garrison, Lewinsohn, Marsteller, Langhinrichsen, & Lann, 1991).

Theoretically, these gender-specific findings can be interpreted within gender-role socialization theory, which posits that males are taught to assert their independence and prove their physical prowess and masculinity during adolescence and young adulthood by engaging in risky, dangerous, and potentially self-destructive behavior (Clark et al., 1990). In contrast, females are socialized to express their dissatisfactions via internalizing behaviors and symptoms of depression. For example, females have been found to have a more ruminative and introspective style of responding to their distress than males (e.g., Nolen-Hoeksema, 1990; Nolen-Hoeksema, 1994). Females may also be taught to engage more body self-care and health-related behaviors than males because of the American culture's continued emphasis on females' physical appearance. This process might help to account for female's lower scores on the HR subscale. In contrast, males lack of engagement in positive health-related behaviors and high rates of injury-related behaviors may be somewhat normative for their gender. Thus, although men and women are equally capable of lethal suicidal behavior and may share some of the same precipitants for their suicidality, the particular type of suicidal behavior that is expressed is likely to be a function of culture-specific norms of gender and suicidal behavior (Canetto, 1992–93; Canetto & Lester, 1995).

In general, these results argue for the importance of assessing a broad array of suicide-related behaviors. The obtained findings are consistent with research demonstrating the interrelatedness of internalizing and disruptive behaviors (Loeber, Russo, Stouthamer-Loeber, & Lahey, 1994). Moreover, high LAS scores on all four content category subscales have previously been shown to be significantly correlated with reports of a past suicide attempt (Lewinsohn, et al., 1995), and both internalizing and externalizing problem behaviors have been used to significantly predict past suicide attempts in adolescent samples (Lewinsohn, Rohde, & Seeley, 1993). Therefore, use of the LAS total and subscale scores in both high school and college settings is likely to allow clinicians and researchers to increase their ability to successfully detect a larger variety of individuals who are at-risk for suicidality. However, use of the LAS as a general screener in school settings may not particularly advance suicide prevention efforts, as these programs may have limited effectiveness (Garland & Zigler, 1993). Instead, this instrument may be of benefit to schools coping with a student's recent suicide attempt or completion. High scorers on the LAS might be important targets for a mental health referral. The LAS may also be an important addition to a standard clinical assessment procedure with distressed and acting-out adolescents and young adults. Future research will be needed to assess the effectiveness of the LAS in applied settings and to determine whether it differentiates males and females who are undergoing treatment. It will also be important to determine if LAS scores can be used by clinicians to develop more effective and individualized treatment plans for adolescents with a history of suicidal behavior.

Another implication that can be drawn from this work is that suicide-related interventions may also need to be gender-specific. This is consistent with conclusions drawn by Canetto and Lester (1995). For example, at-risk males may benefit from interventions that are designed to interest males in activities that are less risky than current choices, and to learn paths to manhood and peer acceptance that do not necessitate engaging in risk-taking and health-compromising behaviors. In contrast, suicide interventions for women may be most effective when they address female's greater propensity for depressive thoughts and feelings and how depression, hopelessness and suicidality are related to each other and to gender-specific styles of coping. Future research can focus both on the processes that lead to higher levels of risk-taking behavior for males and higher levels of depressive behavior for females and the relation these behaviors have to cultural expressions of masculinity and femininity.

The relationships among the LAS, gender, and beliefs and attitudes about suicidality will also be important to examine in future research. Previous research has established that there are gender differences in how peers react to males' and females' suicidal behavior (White & Stillion, 1988). For example, females tend to be more sympathetic than males toward suicidal individuals; males evaluate suicidal males the most harshly. Furthermore, feminine males have been shown to express sympathy at rates that are equivalent to females, which suggests that gender role socialization is an important component of adolescent and young adult suicidal behavior (Stillion, McDowell, Smith & McCoy, 1986). It seems likely that gender differences in the tolerance and social support extended toward distressed peers in turn impacts how males and females express suicidal behavior. Future research will be needed to determine this empirically. Researchers have also established that there are gender differences in beliefs about the morality of suicide, and responses to a suicide awareness program (Overholser, Hemstreet, Spirito, & Vyse, 1989; Wellman & Wellman, 1986). Future research could focus on how these gender differences related to the etiology, expression, and maintenance of males' and females' suicidal behavior.

Limitations to the current study should be noted. All data were self-report and may thus reveal more about what girls and boys are willing to admit to doing, rather than how they are actually thinking, feeling, and behaving. This limitation is particularly important given that an association between LAS and social desirability scores has been established. Concerns about this limitation are allayed somewhat by the analyses that revealed that the gender-specific findings held even when social desirability scores were covaried out statistically. A second limitation to this study is that data from the two samples were collected using different data collection strategies, in different parts of the country, with different measures of depressive symptomatology. Hence, it was inappropriate to combine the two samples to consider whether age and gender interact in the development and expression of suicidal behavior. Research that is specifically designed to consider the influence of age and development on the expression of suicidality will be important. A final limitation to this research was the homogeneity of the participant's ethnicity across the two samples. Further research will be needed to consider the degree to which gender differences in the expression of suicide-related behaviors vary as a function of culture.

None the less, consistent gender differences in suicide-related behavior across two large convenience samples of adolescents and young adults were obtained. These findings lend strong support to the notion that suicidal behavior is gendered. Longitudinal research focused on understanding the development of the influence of gender on the expression of suicide-related behavior will be essential. A systematic consideration of the extent to which males and females have different developmental trajectories and, consequently, different forms and functions of psychopathology needs to be undertaken (Zahn-Waxler, 1993). The degree to which these gender differences are maintained throughout the life span will also be an important empirical topic. Finally, it will be clinically significant to determine if adding gender-specific components to suicide prevention and intervention programs will enhance their effectiveness.

REFERENCES

Allgood-Merten, B., Lewinsohn, P. M., & Hops, H. (1990). Sex differences and adolescent depression. *Journal of Abnormal Psychology, 99*, 55–63.

Andrews, J. A., & Lewinsohn, P. M. (1992). Suicidal attempts among older adolescents: Prevalence and co-occurrence with psychiatric disorders. *Journal of the American Academy of Child and Adolescent Psychiatry, 31,* 655–662.

Beck, A. T. (1969). Thinking and depression. *Archives of General Psychiatry, 9,* 324–333.

Beck, A. T., Rush, A. J., Shaw, B. F., & Emery, G. (1979). *Cognitive therapy for depression.* New York: Guilford Press.

Beck, A. T., Steer, R. A., & Garbin, M. G. (1988). Psychometric properties of the Beck Depression Inventory: Twenty-five years of evaluation. *Clinical Psychology Review, 8,* 77–100.

Beck, A. T., Weissman, A., Lester, D., & Texler, L. (1974). The measurement of pessimism: The Hopelessness Scale. *Journal of Consulting and Clinical Psychology, 42,* 861–865.

Canetto, S. S. (1992–93). She died for love and he for glory: Gender myths of suicidal behavior. *Omega, 26,* 1–17.

Canetto, S. S., & Lester, D. (1995). Gender and the primary prevention of suicide mortality. *Suicide and Life-Threatening Behavior, 25,* 58–69.

Clark, D. C., Sommerfeldt, L., Schwartz, M., Hedeker, D., & Watel, L. (1990). Physical recklessness in adolescence: Trait or by-product of depressive/suicidal states? *The Journal of Nervous and Mental Disease, 178,* 423–433.

Crowne, D. P., & Marlowe, D. A. (1960). A new scale of social desirability independent of psychopathology. *Journal of Consulting Psychology, 24,* 349–354.

Ensminger, M. (1987). Adolescent sexual behavior as it relates to other transition behaviors in youth. In S. Hofferth & C. Hayes (Eds.), *Risking the future: Adolescent sexuality, pregnancy and childbearing* (Vol. 2). Washington, DC: National Academy of Sciences.

Garland, A. F., & Zigler, E. (1993). Adolescent suicide prevention: current research and social policy implications. *American Psychologist, 48,* 169–184.

Garrison, C., Lewinsohn, P. M., Marsteller, F., Langhinrichsen, J., & Lann, I. (1991). The assessment of suicidal behaviour in adolescents. *Suicide and Life-Threatening Behavior, 21,* 217–230.

Garrison, C., Shoenback, V., & Kaplan, B. (1983). Depression symptoms in early adolescence. In A. Dead (Ed.), *Depression in multidisciplinary perspective.* New York: Brunner/Mazel.

Gore, S., Aseltine, R. H., Jr., & Colton, M. E. (1992). Social structure, life stress and depressive symptoms in a high school aged population. *Journal of Health and Social Behavior, 33,* 97–113.

Hawkins, J. D., Catalano, R. F., & Miller, J. Y. (1992). Risk and protective factors for alcohol and other drug problems in adolescence and early adulthood: Implications for substance abuse prevention. *Psychological Bulletin, 112,* 64–105.

Jessor, R., & Jessor, S. L. (1975). Adolescent development and the onset of drinking: A longitudinal study. *Journal of the Studies on Alcohol, 36,* 27–51.

Kandel, D. B., & Davies, M. (1982). Epidemiology of depressive modes in adolescence. *Archives of General Psychiatry, 39,* 1205–1212.

Langhinrichsen-Rohling, J., Sanders, A., Crane, M., & Monson, C. M. (in press). The influence of sex of participant and history of suicidality on college students' current suicide-related thoughts, feelings, and actions. *Suicide and Life-Threatening Behavior.*

Leadbeater, B. J., Blatt, S. J., & Quinlan, D. M. (1995). Gender-linked vulnerabilities to depressive symptoms, stress, and problem behaviors in adolescents. *Journal of Research on Adolescence, 5,* 1–29.

Lefkowitz, E. S., Kahlbaugh, P., & Sigman, M. D. (1994, February). Adolescent risk-taking and thrill-seeking: Relation to gender, AIDS beliefs, and family interactions. Poster presented at the Fifth Biennial Meeting of the Society for Research on Adolescence. San Diego, CA.

Lewinsohn, P. M., Langhinrichsen-Rohling, J., Langford, R., Rohde, P., Seeley, J. R., & Chapman, J. (1995). The Life Attitudes Schedule: A scale to assess adolescent life-enhancing and life-threatening behaviors. *Suicide and Life-Threatening Behavior, 25,* 458–474.

Lewinsohn, P. M., Langhinrichsen, J., Langford, R., & Rohde, P. (in press). *The Life Attitudes Schedule.* North Tonawand, NY: Multi-Health Systems, Inc.

Lewinsohn, P. M., Rohde, P., & Seeley, J. R. (1993). Psychosocial characteristics of adolescents with a history of suicide attempt. *Journal of the American Academy of Child and Adolescent Psychiatry, 32,* 60–68.

Loeber, R., Russo, M. F., Stouthamer-Loeber, M., & Lahey, B. B. (1994). Internalizing problems and their relation to the development of disruptive behaviors in adolescence. *Journal of Research on Adolescence, 4,* 615–637.

McIntosh, J. L., & Jewell, B. L. (1986). Sex difference trends in completed suicide. *Suicide and Life-Threatening Behavior, 16,* 16–27.

National Center for Health Statistics (1994). Monthly vital statistics report (Vol. 42). Public Health Service, Washington, DC: U.S. Government Printing Office.

Nolen-Hoeksema, S. (1990). *Sex differences in depression.* Stanford, CA: Stanford University Press.

Nolen-Hoeksema, S. (1994). An interactive model for the emergence of gender differences in depression in adolescence. *Journal of Research on Adolescence, 4,* 519–534.

Overholser, J. C., Hemstreet, A. H., Spirito, A., & Vyse, S. (1989). Suicide Awareness programs in the schools: Effects of gender and personal experience. *American Academy of Child and Adolescent Psychiatry, 28,* 925–930.

Radloff, L. (1977). A CES-D scale: A self-report depression scale for research in the general population. *Applied Psychological Measurement, 1,* 385–401.

Shaffer, D., & Gould, M. (1987). *Progress report: Study of completed and attempted suicide in adolescents (Contract No. R01-MH-38198).* Bethesda, MD: National Institute of Mental Health.

Shaffer, D., Garland, A., Gould, M., Fisher, P., & Trautman, P. (1988). Preventing teenage suicide: A critical review. *Journal of the American Academy of Child and Adolescent Psychiatry, 27,* 675–687.

Shneidman, E. S., Farberow, N. L., & Litman, R. E. (1970). *The psychology of suicide.* Scranton, PA: Science House.

Simons, R. L., & Murphy, P. I. (1985). Sex differences in the causes of adolescent suicide ideation. *Journal of Youth and Adolescence, 14,* 423–434.

Stillion, J. M., McDowell, E. E., Smith, R. T., & McCoy, P. A. (1986). Relationships between suicide attitudes and indicators of mental health among adolescents. *Death Studies, 10,* 289–296.

Wellman, M. M., & Wellman, R. J. (1986). Sex differences in peer responsiveness to suicide ideation. *Suicide and Life-Threatening Behavior, 16,* 360–378.

White, H., & Stillion, J. M. (1988). Sex differences in attitudes toward suicide. *Psychology of Women Quarterly, 12,* 357–366.

Windle, M., Miller-Tutzauer, C., & Domenico, D. (1992). Alcohol use, suicidal behavior, and risky activities among adolescents. *Journal of Research on Adolescence, 2,* 317–330.

Zahn-Waxler, C. (1993). Warriors and worriers: Gender and psychopathology. *Development and Psychopathology, 5,* 79–89.

QUESTIONS FOR DISCUSSION

1. Discuss three of the six hypotheses presented in the reading. Are the hypotheses supported? What does this mean for males and females in terms of suicide behavior?

2. Explain how gender-specific suicide is explained by gender-role socialization theory. How valid is the theory in explaining gender-specific suicide? Explain your response.

3. What is the best way to help suicidal individuals? How could you implement this at your college, university, or geographical location?

Anomic Suicide

Emile Durkheim

No living being can be happy or even exist unless his needs are sufficiently proportioned to his means. In other words, if his needs require more than can be granted, or even merely something of a different sort, they will be under continual friction and can only function painfully. Movements incapable of production without pain tend not to be reproduced. Unsatisfied tendencies atrophy, and as the impulse to live is merely the result of all the rest, it is bound to weaken as the others relax.

In the animal, at least in the normal condition, this equilibrium is established with automatic spontaneity because the animal depends on purely material conditions. All the organism needs is that the supplies of substance and energy constantly employed in the vital process should be periodically renewed by equivalent quantities; that replacement be equivalent to use. When the void created by existence in its own resources is filled, the animal, satisfied, asks nothing further. Its power of reflection is not sufficiently developed to imagine other ends than those implicit in its physical nature. On the other hand, as the work demanded of each organ itself depends on the general state of vital energy and the needs of organic equilibrium, use is regulated in turn by replacement and the balance is automatic. The limits of one are those of the other; both are fundamental to the constitution of the existence in question, which cannot exceed them.

This is not the case with man, because most of his needs are not dependent on his body or not to the same degree. Strictly speaking, we may consider that the quantity of material supplies necessary to the physical maintenance of a human life is subject to computation, though this be less exact than in the preceding case and a wider margin left for the free combinations of the will; for beyond the indispensable minimum which satisfies nature when instinctive, a more awakened reflection suggests better conditions, seemingly desirable ends craving fulfillment. Such appetites, however, admittedly sooner or later reach a limit which they cannot pass. But how determine the quantity of well-being, comfort or luxury legitimately to be craved by a human being? Nothing appears in man's organic nor in his psychological constitution which sets a limit to such tendencies. The functioning of individual life does not require them to cease at one point rather than at another; the proof being that they have constantly increased since the beginnings of history, receiving more and more complete satisfaction, yet with no weakening of average health. Above all, how establish their proper variation with different conditions of life, occupations, relative importance of services, etc.? In no society are they equally satisfied in the different stages of the social hierarchy. Yet human nature is substantially the same among all men, in its essential qualities. It is not human nature which can assign the variable limits necessary to our needs. They are thus unlimited so far as they depend on the individual alone. Irrespective of any external regulatory force, our capacity for feeling is in itself an insatiable and bottomless abyss.

But if nothing external can restrain this capacity, it can only be a source of torment to itself. Unlimited desires are insatiable by definition and insatiability is rightly considered a sign of morbidity. Being unlimited, they constantly and infinitely surpass the means at their command; they cannot be quenched. Inextinguishable thirst is constantly renewed torture. It has been claimed, indeed, that human activity naturally aspires beyond assignable limits and sets itself unattainable goals. But how can such an undetermined state be any more reconciled with the conditions of mental life than with the demands of physical life? All man's pleasure in acting, moving and exerting himself implies the sense that his efforts are not in vain and that by walking he has advanced. However, one does not advance when one walks toward no goal, or—which is the same thing—when his goal is infinity. Since the distance between us and it is always the same, whatever road we take, we might as well have made the motions without progress from the spot. Even our glances behind and our feeling of pride at the distance covered can cause only deceptive satisfaction, since the remaining distance is not proportionately reduced. To pursue a goal which is by definition unattainable is to condemn oneself to a state of perpetual unhappiness. Of course, man may hope contrary to all reason, and hope has its pleasures even when unreasonable. It may sustain him for a time; but it cannot survive the repeated disappointments of experience indefinitely. What more can the future offer him than the past, since he can never reach a tenable condition nor even approach the glimpsed ideal? Thus, the more one has, the more one wants, since satisfactions received only stimulate instead of filling needs. Shall action as such be considered agreeable? First, only on condition of blindness to its uselessness. Secondly, for this pleasure to be felt and to temper and half veil the accompanying painful unrest, such unending motion must at least aways be easy and unhampered. If it is interfered with only restlessness is left, with the lack of ease which it, itself, entails. But it would be a miracle if no insurmountable obstacle were never encountered. Our thread of life on these conditions is pretty thin, breakable at any instant.

To achieve any other result, the passions first must be limited. Only then can they be harmonized with the faculties and satisfied. But since the individual has no way of limiting them, this must be done by some force exterior to him. A regulative force must play the same role for moral needs which the organism plays for physical needs. This means that the force can only be moral. The awakening of conscience interrupted the state of equilibrium of the animal's dormant existence; only conscience, therefore, can furnish the means to re-establish it. Physical restraint would be ineffective; hearts cannot be touched by physicochemical forces. So far as the appetites are not automatically restrained by physiological mechanisms, they can be halted only by a limit that they recognize as just. Men would never consent to restrict their desires if they felt justified in passing the assigned limit. But, for reasons given above, they cannot assign themselves this law of justice. So they must receive it from an authority which they respect, to which they yield spontaneously. Either directly and as a whole, or through the agency of one of its organs, society alone can play this moderating role; for it is the only moral power superior to the individual, the authority of which he accepts. It alone has the power necessary to stipulate law and to set the point beyond which the passions must not go. Finally, it alone can estimate the reward to be prospectively offered to every class of human functionary, in the name of the common interest.

As a matter of fact, at every moment of history there is a dim perception, in the moral consciousness of societies, of the respective value of different social services, the relative reward due to each, and the consequent degree of comfort appropriate on the average to workers in each occupation. The different functions are graded in public opinion and a certain coefficient of well-being assigned to each, according to its place in the hierarchy. According to accepted ideas, for example, a certain way of living is considered the upper limit to which a workman may aspire in his efforts to improve his existence, and there is another limit below which he is not willingly permitted to fall unless he has seriously demeaned himself. Both differ for city and country workers, for the domestic servant and the day-laborer, for the business clerk and official, etc. Likewise the man of wealth is reproved if he lives the life of a poor man, but also if he seeks the refinements of luxury overmuch. Economists may protest in vain; public feeling will always be scandalized if an individual spends too much wealth for wholly superfluous use, and it even seems that this severity relaxes only in times of moral disturbance.[1] A genuine regimen exists, therefore, although not always legally formulated, which fixes with relative precision the maximum degree of ease of living to which each social class may legitimately aspire. However, there is nothing immutable about such a scale. It changes with the increase or decrease of collective revenue and the changes occurring in the moral ideas of society. Thus what appears luxury to one period no longer does so to another; and the well-being which for long periods was granted to a class only by exception and supererogation, finally appears strictly necessary and equitable.

Under this pressure, each in his sphere vaguely realizes the extreme limit set to his ambitions and aspires to nothing beyond. At least if he respects regulations and is docile to collective authority, that is, has a wholesome moral constitution, he feels that it is not well to ask more. Thus, an end and goal are set to the passions. Truly, there is nothing rigid nor absolute about such determination. The economic ideal assigned each class of citizens is itself confined to certain limits, within which the desires have free range. But it is not infinite. This relative limitation and the moderation it involves, make men contented with their lot while stimulating them moderately to improve it; and this average contentment causes the feeling of calm, active happiness, the pleasure in existing and living which characterizes health for societies as well as for individuals. Each person is then at least, generally speaking, in harmony with his condition, and desires only what he may legitimately hope for as the normal reward of his activity. Besides, this does not condemn man to a sort of immobility. He may seek to give beauty to his life; but his attempts in this direction may fail without causing him to despair. For, loving what he has and not fixing his desire solely on what he lacks, his wishes and hopes may fail of what he has happened to aspire to, without his being wholly destitute. He has the essentials. The equilibrium of his happiness is secure because it is defined, and a few mishaps cannot disconcert him.

But it would be of little use for everyone to recognize the justice of the hierarchy of functions established by public opinion, if he did not also consider the distribution of these functions just. The workman is not in harmony with his social position if he is not convinced that he has his desserts. If he feels justified in occupying another, what he has would not satisfy him. So it is not enough for the average level of needs for each social condition to be regulated by public opinion, but another, more precise rule, must fix the way in which these conditions are open to individuals. There is no society in which such regulation does not exist. It varies with times and places. Once it regarded birth as the almost exclusive principle of social classification; today it recognizes no other inherent inequality than hereditary fortune and merit. But in all these various forms its object is unchanged. It is also only possible, everywhere, as a restriction upon individuals imposed by superior authority, that is, by collective authority. For it can established only by requiring of one or another group of men, usually of all, sacrifices and concessions in the name of the public interest.

Some, to be sure, have thought that this moral pressure would become unnecessary if men's economic circumstances were only no longer determined by heredity. If inheritance were abolished, the argument runs, if everyone began life with equal resources and if the competitive struggle were fought out on a basis of perfect equality, no one could think its results unjust. Each would instinctively feel that things are as they should be.

Truly, the nearer this ideal equality were approached, the less social restraint will be necessary. But it is only a matter of degree. One sort of heredity will always exist, that of natural talent. Intelligence, taste, scientific, artistic, literary or industrial ability, courage and manual dexterity are gifts received by each of us at birth, as the heir to wealth receives his capital or as the nobleman formerly received his title and function. A moral discipline will therefore still be required to make those less favored by nature accept the lesser advantages which they owe to the chance of birth. Shall it be demanded that all have an equal share and that no advantage be given those more useful and deserving? But then there would have to be a discipline far stronger to make these accept a treatment merely equal to that of the mediocre and incapable.

But like the one first mentioned, this discipline can be useful only if considered just by the peoples subject to it. When it is maintained only by custom and force, peace and harmony are illusory; the spirit of unrest and discontent are latent; appetites superficially restrained are ready to revolt. This happened in Rome and Greece when the faiths underlying the old organization of the patricians and plebeians were shaken, and in our modern societies when aristocratic prejudices began to lose their own ascendancy. But this state of upheaval is exceptional; it occurs only when society is passing through some abnormal crisis. In normal conditions the collective order is regarded as just by the great majority of persons. Therefore, when we say that an authority is necessary to impose this order on individuals, we certainly do not mean that violence is the only means of establishing it. Since this regulation is meant to restrain individual passions, it must come from a power which dominates individuals; but this power must also be obeyed through respect, not fear.

It is not true, then, that human activity can be released from all restraint. Nothing in the world can enjoy such a privilege. All existence being a part of the universe is relative to the remainder; its nature and method of manifestation accordingly depend not only on itself but on other beings, who consequently restrain and regulate it. Here there are only differences of degree and form between the mineral realm and the thinking person. Man's characteristic privilege is that the bond he accepts is not physical but moral; that is, social. He is governed not by a material environment brutally imposed on him, but by a conscience superior to his own, the superiority of which he feels. Because the greater, better part of his existence transcends the body, he escapes the body's yoke, but is subject to that of society.

But when society is disturbed by some painful crisis or by beneficent but abrupt transitions, it is momentarily incapable of exercising this influence; thence come the sudden rises in the curve of suicides which we have pointed out above.

In the case of economic disasters, indeed, something like a declassification occurs which suddenly casts certain individuals into a lower state than their previous one. Then they must reduce their requirements, restrain their needs, learn greater self-control. All the advantages of social influence are lost so far as they are concerned; their moral education has to be recommenced. But society cannot adjust them instantaneously to this new life and teach them to practice the increased self-repression to which they are unaccustomed. So they are not adjusted to the condition forced on them, and its very prospect is intolerable; hence the suffering which detaches them from a reduced existence even before they have made trial of it.

It is the same if the source of the crisis is an abrupt growth of power and wealth. Then, truly, as the conditions of life are changed, the standard according to which needs were regulated can no longer remain the same; for it varies with social resources, since it largely determines the share of each class of producers. The scale is upset; but a new scale cannot be immediately improvised. Time is required for the public conscience to reclassify men and things. So long as the social forces thus freed have not regained equilibrium, their respective values are unknown and so all regulation is lacking for a time. The limits are unknown between the possible and the impossible, what is just and what is unjust, legitimate claims and hopes and those which are immoderate. Consequently, there is no restraint upon aspirations. If the disturbance is profound, it affects even the principles controlling the distribution of men among various occupations. Since the relations between various parts of society are necessarily modified, the ideas expressing these relations must change. Some particular class especially favored by the crisis is no longer resigned to its former lot, and, on the other hand, the example of its greater good fortune arouses all sorts of jealousy below and about it. Appetites, not being controlled by a public opinion, become disoriented, no longer recognize the limits proper to them. Besides, they are at the same time seized by a sort of natural erethism simply by the greater intensity of public life. With increased prosperity desires increase. At the very moment when traditional rules have lost their authority, the richer prize offered these appetites stimulates them and makes them more exigent and impatient of control. The state of de-regulation or anomy is thus further heightened by passions being less disciplined, precisely when they need more disciplining.

But then their very demands make fulfillment impossible. Overweening ambition always exceeds the results obtained, great as they may be, since there is no warning to pause here. Nothing gives satisfaction and all this agitation is uninterruptedly maintained without appeasement. Above all, since this race for an unattainable goal can give no other pleasure but that of the race itself, if it is one, once it is interrupted the participants are left empty-handed. At the same time the struggle grows more violent and painful, both from being less controlled and because competition is greater. All classes contend among themselves because no established classification any longer exists. Effort grows, just when it becomes less productive. How could the desire to live not be weakened under such conditions?

This explanation is confirmed by the remarkable immunity of poor countries. Poverty protects against suicide because it is a restraint in itself. No matter how one acts, desires have to depend upon resources to some extent; actual possessions are partly the criterion of those aspired to. So the less one has the less he is tempted to extend the range of his needs indefinitely. Lack of power, compelling moderation, accustoms men to it, while nothing excites envy if no one has superfluity. Wealth, on the other hand, by the power it bestows, deceives us into believing that we depend on ourselves only. Reducing the resistance we encounter from objects, it suggests the possibility of unlimited success against them. The less limited one feels, the more intolerable all limitation appears. Not without reason, therefore, have so many religions dwelt on the advantages and moral value of poverty. It is actually the best school for teaching self-restraint. Forcing us to constant self-discipline, it prepares us to accept collective discipline with equanimity, while wealth, exalting the individual, may always arouse the spirit of rebellion which is the very source of immorality. This, of course, is no reason why humanity should not improve its material condition. But though the moral danger involved in every growth of prosperity is not irremediable, it should not be forgotten.

If anomy never appeared except, as in the above instances, in intermittent spurts and acute crisis, it might cause the social suicide-rate to vary from time to time, but it would not be a regular, constant factor. In one sphere of social life, however—the sphere to trade and industry—it is actually in a chronic state.

For a whole century, economic progress has mainly consisted in freeing industrial relations from all regulation. Until very recently, it was the function of a whole system of moral forces to exert this discipline. First, the influence of religion was felt alike by workers and masters, the poor and the rich. It consoled the former and taught them contentment with their lot by informing them of the providential nature of the social order, that the share of each class was assigned by God himself, and by holding out the hope for just compensation in a world to come in return for the inequalities of this world. It governed the latter, recalling that worldly interests are not man's entire lot, that they must be subordinate to other and higher interests, and that they should therefore not be pursued without rule or measure. Temporal power, in turn, restrained the scope of economic functions by its supremacy over them and by the relatively subordinate role it assigned them. Finally, within the business world proper, the occupational groups by regulating salaries, the price of products and production itself, directly fixed the average level of income on which needs are partially based by the very force of circumstances. However, we do not mean to propose this organization as a model. Clearly it would be inadequate to existing societies without great changes. What we stress is its existence, the fact of its useful influence, and that nothing today has come to take its place.

Actually, religion has lost most of its power. And government, instead of regulating economic life, has become its tool and servant. The most opposite schools, orthodox economists and extreme socialists, unite to reduce government to the role of a more or less passive intermediary among the various social functions. The former wish to make it simply the guardian of individual contracts; the latter leave it the task of doing the collective bookkeeping, that is, of recording the demands of consumers, transmitting them to producers, inventorying the total revenue and distributing it according to a fixed formula. But both refuse it any power to subordinate other social organs to itself and to make them converge toward one dominant aim. On both sides nations are declared to have the single or chief purpose of achieving industrial prosperity; such is the implication of the dogma of economic materialism, the basis of both apparently opposed systems. And as these theories merely express the state of opinion, industry, instead of being still regarded as a means to an end transcending itself, has become the supreme end of individuals and societies alike. Thereupon the appetites thus excited have become freed of any limiting authority. By sanctifying them, so to speak, this apotheosis of well-being has placed them above all human law. Their restraint seems like a sort of sacrilege. For this reason, even the purely utilitarian regulation of them exercised by the industrial world itself through the medium of occupational groups has been unable to persist. Ultimately, this liberation of desires has been made worse by the very development of industry and the almost infinite extension of the market. So long as the producer could gain his profits only in his immediate neighborhood, the restricted amount of possible gain could not much overexcite ambition. Now that he may assume to have almost the entire world as his customer, how could passions accept their former confinement in the face of such limitless prospects?

Such is the source of the excitement predominating in this part of society, and which has thence extended to the other parts. There the state of crisis and anomy is constant and, so to speak, normal. From top to bottom of the ladder, greed is aroused without knowing where to find ultimate foothold. Nothing can calm it, since its goal is far beyond all it can attain. Reality seems valueless by comparison with the dreams of fevered imagination; reality is therefore abandoned, but so too is possibility abandoned when it in turn becomes reality. A thirst arises for novelties, unfamiliar pleasures, nameless sensations, all of which lose their savor once known. Henceforth one has no strength to endure the least reverse. The whole fever subsides and the sterility of all the tumult is apparent, and it is seen that all these new sensations in their infinite quantity cannot form a solid foundation of happiness to support one during days of trial. The wise man, knowing how to enjoy achieved results without having constantly to replace them with others, finds in them an attachment to life in the hour of difficulty. But the man who has always pinned all his hopes on the future and lived with his eyes fixed upon it, has nothing in the past as a comfort against the present's afflictions, for the past was nothing to him but a series of hastily experienced stages. What blinded him to himself was his expectation always to find further on the happiness he had so far missed. Now he is stopped in his tracks; from now on nothing remains behind or ahead of him to fix his gaze upon. Weariness alone, moreover, is enough to bring disillusionment, for he cannot in the end escape the futility of an endless pursuit.

We may even wonder if this moral state is not principally what makes economic catastrophes of our day so fertile in suicides. In societies where a man is subjected to a healthy discipline, he submits more readily to the blows of chance. The necessary effort for sustaining a little more discomfort costs him relatively little, since he is used to discomfort and constraint. But when every constraint is hateful in itself, how can closer constraint not seem intolerable? There is no tendency to resignation in the feverish impatience of men's lives. When there is no other aim but to outstrip constantly the point arrived at, how painful to be thrown back! Now this very lack of organization characterizing our economic conditions throws the door wide to every sort of adventure. Since imagination is hungry for novelty, and ungoverned, it gropes at random. Setbacks necessarily increase with risks and thus crises multiply, just when they are becoming more destructive.

Yet these dispositions are so inbred that society has grown to accept them and is accustomed to think them normal. It is everlastingly repeated that it is man's nature to be eternally dissatisfied, constantly to advance, without relief or rest, toward an indefinite goal. The longing for infinity is daily represented as a mark of moral distinction, whereas it can only appear within unregulated consciences which elevate to a rule the lack of rule from which they suffer. The doctrine of the most ruthless and swift progress has become an article of faith. But other theories appear parallel with those praising the advantages of instability, which, generalizing the situation that gives them birth, declare life evil, claim that it is richer in grief than in pleasure and that it attracts men only by false claims. Since this disorder is greatest in the economic world, it has most victims there.

Industrial and commercial functions are really among the occupations which furnish the greatest number of suicides (see Table 1). Almost on a level with the liberal professions, they sometimes surpass them; they are especially more afflicted than agriculture, where the old regulative forces still make their appearance felt most and where the fever of business has least penetrated. Here is best recalled what was once the general constitution of the economic order. And the divergence would be yet greater if, among the suicides of industry, employers were distinguished from workmen, for the former are probably most stricken by the state of anomy. The enormous rate of those with independent means (720 per million) sufficiently shows that the possessors of most comfort suffer most. Everything that enforces subordination attenuates the effects of this state. At least the horizon of the lower classes is limited by those above them, and for this same reason their desires are more modest. Those who have only empty space above them are almost inevitably lost in it, for no force restrains them.

Anomy, therefore, is a regular and specific factor in suicide in our modern societies; one of the springs from which the annual contingent feeds. So we have here a

TABLE 1   Suicides Per Million Persons of Different Occupations

	Trade	Trans-portation	Industry	Agri-culture	Liberal* Professions
France (1878–87)[†]	440	—	340	240	300
Switzerland (1876)	664	1,514	577	304	558
Italy (1866–76)	277	152.6	80.4	26.7	618[‡]
Prussia (1883–90)	754	—	456	315	832
Bavaria (1884–91)	465	—	369	153	454
Belgium (1886–90)	421	—	160	160	100
Wurttemburg (1873–78)	273	—	190	206	—
Saxony (1878)	—	341.59[§]	—	71.17	—

*When statistics distinguish several different sorts of liberal occupations, we show as a specimen the one in which the suicide-rate is highest.

[†]From 1826 to 1880 economic functions seem less affected (see *Compte-rendu* of 1880); but were occupational statistics very accurate?

[‡]This figure is reached only by men of letters.

[§]Figure represents Trade, Transportation and Industry combined for Saxony. (Ed.)

new type to distinguish from the others. It differs from them in its dependence, not on the way in which individuals are attached to society, but on how it regulates them. Egoistic suicide results from man's no longer finding a basis for existence in life; altruistic suicide, because this basis for existence appears to man situated beyond life itself. The third sort of suicide, the existence of which has just been shown, results from man's activity's lacking regulation and his consequent sufferings. By virtue of its origin we shall assign this last variety the name of *anomic suicide.*

Certainly, this and egoistic suicide have kindred ties. Both spring from society's insufficient presence in individuals. But the sphere of its absence is not the same in both cases. In egoistic suicide it is deficient in truly collective activity, thus depriving the latter of object and meaning. In anomic suicide, society's influence is lacking in the basically individual passions, thus leaving them without a check-rein. In spite of their relationship, therefore, the two types are independent of each other. We may offer society everything social to us, and still be unable to control our desires; one may live in an anomic state without being egoistic, and vice versa. These two sorts of suicide therefore do not draw their chief recruits from the same social environments; one has its principal field among intellectual careers, the world of thought—the other, the industrial or commercial world.

NOTE

1. Actually, this is a purely moral reprobation and can hardly be judicially implemented. We do not consider any reestablishment of sumptuary laws desirable or even possible.

QUESTIONS FOR DISCUSSION

1. What is Durkheim's basic explanation of suicide?

2. Compare Durkheim and Merton's use of the word anomie.

3. Assume that all people who attempt suicide are in some sense depressed. Also assume that only about 2 percent of all people who are depressed commit suicide. Does this correspond to what Durkheim would expect?

QUESTIONS FOR PART 4

1. How do the readings in Chapter 9 address rape and rape myths?

2. How do the readings in Chapter 10 address the involvement of white-collar crime?

3. How do the readings in Chapter 11 address the creation of crimes for certain drugs?

4. How do the readings in Chapter 12 address life for a drinker?

5. How do the readings in Chapter 13 address sexual deviance?

6. How do the readings in Chapter 14 address explanations for suicide?

PART 5

STUDIES IN STIGMA

Erving Goffman defines stigma as an attribute that disqualifies an individual from full social acceptance. He suggests that there are three principal types of stigma: (1) blemishes of character such as a mental disorder, (2) tribal or group membership such as homosexuality, and (3) physical limitations such as a physical disability. In Part 5 we address three types of stigma and how they are related to deviance.

What is a mental disorder? Why are people with mental disorders considered deviant? Almost any behavior that deviates from social norms can be classified as deviant behavior. Social attitudes toward individuals with mental disorders are sometimes sympathetic, but can also range from avoidance to ridicule to repulsion. "On Being Sane in Insane Places," David L. Rosenhan's classic experiment on deceiving mental health professions about the sanity of patients, questions the meaning of sanity. Rosenhan sent eight sane pseudopatients to psychiatric hospitals to find out whether the hospital could distinguish sanity from insanity. He discovers that even these patients experience countertherapeutic behaviors such as self-labeling, segregation, depersonalization, and powerlessness.

Recent media reports indicate an increasing public concern about the rise in the number of children diagnosed as having ADHD and being prescribed the drug ritalin. Peter Jensen, Lori Kettle, Margaret Roper, Michael Sloan, Mina Dulcan, Christina Hoven, Hector R. Bird, Jose J. Bauermeister, and Jennifer D. Payne discuss in their article how children are often misdiagnosed as having ADHD. Their study found that 12.5% of children meeting ADHD criteria had been treated with stimulants while 8 of 16 children who were prescribed a stimulant did not meet ADHD criteria. They concluded that medical treatments are often not used for children diagnosed as having ADHD and that there is not a substantial overtreatment with stimulants for ADHD children. Christy L. Picard's article compares NCAA Division I and III female college athletes and a group of non-athletes. Her results suggest that athletes at Division I schools showed more signs of pathological eating and were at a greater risk for an eating disorder due to higher levels of competition.

Homosexuality is a long-standing stigmatized deviant act. Individuals who are openly gay or lesbian or sexually different are often stigmatized and excluded from some social situations. Robert Meier and Gilbert Geis's article addresses the political visibilities and catalysts for the gay movement. They compare the gay movement's issues of social change with the women's movement. Robert Owens addresses the process of "coming out" in his article. Using data from personal interviews, Owens discusses the issues of feeling different, the awareness of same-sex attraction, coping strategies, first sexual contacts, dating, and self-identification as gay, lesbian, or bisexual. Do homosexual people have a substantially higher risk for emotional problems such as anxiety disorder, major depression, and suicide? Michael Bailey's article suggests that they do and discusses potential explanations for these disorders.

People with physical disabilities such as blindness and conjoined bodies can hardly be held accountable for their condition. Robert Scott's classic work addresses the socialization of blindness. Scott focuses on two principal mechanisms of personal encounters with the blind: preconceptions about blindness and reactions to the blind. He reveals four features of personal relationships that affect the socialization of blindness: stereotyped beliefs, blindness stigmas, encountering the blind, and social dependency. In an interview with the Schappell sisters, Lori and Reba, Natalie Angier discusses the life of conjoined twins. Angier addresses how happy the twins are in their situation, how they have different personalities, the twins' plans for college, how they do not see themselves as abnormal, and how they manage their nondeviant life.

CHAPTER 15

MENTAL DEVIANT BEHAVIOR

On Being Sane in Insane Places

David L. Rosenhan

If sanity and insanity exist, how shall we know them? The question is neither capricious nor itself insane. However much we may be personally convinced that we can tell the normal from the abnormal, the evidence is simply not compelling. It is commonplace, for example, to read about murder trials wherein eminent psychiatrists for the defense are contradicted by equally eminent psychiatrists for the prosecution on the matter of the defendant's sanity. More generally, there are a great deal of conflicting data on the reliability, utility, and meaning of such terms as "sanity," "insanity," "mental illness," and "schizophrenia."[1] Finally, as early as 1934, Benedict suggested that normality and abnormality are not universal.[2] What is viewed as normal in one culture may be seen as quite aberrant in another. Thus, notions of normality and abnormality may not be quite as accurate as people believe they are.

To raise questions regarding normality and abnormality is in no way to question the fact that some behaviors are deviant or odd. Murder is deviant. So, too, are hallucinations. Nor does raising such questions deny the existence of the personal anguish that is often associated with "mental illness." Anxiety and depression exist. Psychological suffering exists. But normality and abnormality, sanity and insanity, and the diagnoses that flow from them may be less substantive than many believe them to be.

At its heart, the question of whether the sane can be distinguished from the insane (and whether degrees of insanity can be distinguished from each other) is a simple matter: Do the salient characteristics that lead to diagnoses reside in the patients themselves or in the environments and contexts in which observers find them? From Bleuler, through Kretchmer, thought the formulators of the recently revised *Diagnostical and Statistical Manual* of the American Psychiatric Association, the belief has been strong that patients present symptoms, that those symptoms can be categorized, and, implicitly, that the sane are distinguishable from the insane. More recently, however, this belief has been questioned. Based in part on theoretical and anthropological considerations, but also on philosophical, legal, and therapeutic ones, the view has grown that psychological categorization of mental illness is useless at best and downright harmful, misleading, and pejorative at worst. Psychiatric diagnoses, in this view, are in the minds of the observers and are not valid summaries of characteristics displayed by the observed.[3,4,5]

Gains can be made in deciding which of these is more nearly accurate by getting normal people (that is, people who do not have, and have never suffered, symptoms of serious psychiatric disorders) admitted to psychiatric hospitals and then determining whether they were discovered to be sane and, if so, how. If the sanity of such pseudopatients were always detected, there would be *prima facie* evidence that a sane individual can be distinguished from the insane context in which he is found. Normality (and presumably abnormality) is distinct enough that it can be recognized wherever it occurs, for it is carried within the person. If, on the other hand, the sanity of the pseudopatients were never discovered, serious difficulties would arise for those who support traditional modes of psychiatric diagnosis. Given that the hospital staff was not incompetent, that the pseudopatient had been behaving as sanely as he had been outside of the hospital, and that it had never been previously suggested that he belonged in a psychiatric diagnosis betrays little about the patient but much about the environment in which an observer finds him.

This article describes such an experiment. Eight sane people gained secret admission to twelve different hospitals.[6] Their diagnostic experiences constitute the data of the first part of this article; the remainder is devoted to a description of their experiences in psychiatric institutions. Too few psychiatrists and psychologists, even those who have worked in such hospitals, know what the experience is like. They rarely talk about it with former patients, perhaps because they distrust information coming from the previously insane. Those who have worked in psychiatric hospitals are likely to have adapted so thoroughly to the settings that they are insensitive to the impact of the experience. And while there have been occasional reports of researchers who submitted themselves to psychiatric hospitalization,[7] these researchers have commonly remained in the hospitals for short periods of time, often with the knowledge of the hospital staff. It is difficult to know the extent to which they were treated like patients or like research colleagues. Nevertheless, their reports about the inside of the psychiatric hospital have been valuable. This article extends those efforts.

Pseudopatients and Their Settings

The eight pseudopatients were a varied group. One was a psychology graduate student in his twenties. The remaining seven were older and "established." Among them were three psychologists, a pediatrician, a psychiatrist, a painter, and a housewife. Three pseudopatients were women, five were men. All of them employed pseudonyms, lest their alleged diagnoses embarrass them later. Those who were in mental health professions alleged another occupation in order to avoid the special attentions that might be accorded by staff, as a matter of courtesy or caution, to ailing colleagues.[8] With the exception of myself (I was the first pseudopatient and my presence was known to the hospital administrator and chief psychologist, and, so far as I can tell, to them alone), the presence of pseudopatients and the nature of the research program were not known to the hospital staffs.[9]

The settings were similarly varied. In order to generalize the findings, admission into a variety of hospitals was sought. The twelve hospitals in the sample were located in five different states on the East and West coasts. Some were old and shabby, some were quite new. Some were research-oriented, others not. Some had good staff-patient ratios, others were quite understaffed. Only one was a strictly private hospital. All of the others were supported by state for federal funds or in one instance, by university funds.

After calling the hospital for an appointment, the pseudopatient arrived at the admissions office complaining that he had been hearing voices. Asked what the voices said, he replied that they were often unclear, but as far as he could tell they said "empty," "hollow," and "thud." The voices were unfamiliar and were of the same sex as the pseudopatient. The choice of these symptoms was occasioned by their apparent similarity to existential symptoms. Such symptoms are alleged to arise from painful concerns about the perceived meaninglessness of one's life. It is as if the hallucinating person were saying, "my life is empty and hollow." The choice of these symptoms was also determined by the *absence* of a single report of existential psychoses in the literature.

Beyond alleging the symptoms and falsifying name, vocation, and employment, no further alterations of person, history, or circumstances were made. The significant events of the pseudopatient's life history were presented as they had actually occurred. Relationships with parents and siblings, with spouse and children, with people at work and in school, consistent with the aforementioned exceptions, were described as they were or had been. Frustrations and upsets were described along with joys and satisfactions. These facts are important to remember. If anything, they strongly biased the subsequent results in favor of detecting sanity, since none of their histories or current behaviors were seriously pathological in any way.

Immediately upon admission to the psychiatric ward, the pseudopatient ceased simulating *any* symptoms of abnormality. In some cases, there was a brief period of mild nervousness and anxiety, since none of the pseudopatients really believed that they would be admitted so easily. Indeed, their shared fear was that they would be immediately exposed as frauds and greatly embarrassed. Moreover, many of them had never visited a psychiatric ward; even those who had, nevertheless had some genuine fears about what might happen to them. Their nervousness, then, was quite appropriate to the novelty of the hospital setting, and it abated rapidly.

Apart from that short-lived nervousness, the pseudopatient behaved on the ward as he "normally" behaved. The pseudopatient spoke to patients and staff as he might ordinarily. Because there is uncommonly little to do on a psychiatric ward, he attempted to engage others in conversation. When asked by staff how he was feeling, he indicated that he was fine, that he no longer experienced symptoms. He responded to instructions from attendants, to calls for medication (which was not swallowed), and to dining-hall instructions. Beyond such activities as were available to him on the admissions ward, he spent his time writing down his observations about the ward, its patients, and the staff. Initially these notes were written "secretly," but as it soon became clear that no one much cared, they were subsequently written on standard tablets of paper in such public places as the dayroom. No secret was made of these activities.

The pseudopatient, very much as a true psychiatric patient, entered a hospital with no foreknowledge of when he would be discharged. Each was told that he would have to get out by his own devices, essentially by convincing the staff that he was sane. The psychological stresses associated with hospitalization were considerable, and all but one of the pseudopatients desired to be discharged almost immediately after being admitted. They were, therefore, motivated not only to behave sanely, but to be paragons of cooperation. That their behavior was in no way disruptive is confirmed by nursing reports, which have been obtained on most of the patients. These reports uniformly indicate that the patients were "friendly," "cooperative," and "exhibited no abnormal indications."

THE NORMAL ARE NOT DETECTABLY SANE

Despite their public "show" of sanity, the pseudopatients were never detected. Admitted, except in one case, with a diagnosis of schizophrenia,[10] each was discharged with a diagnosis of schizophrenia "in remission." The label "in remission" should in no way be dismissed as a formality, for at no time during any hospitalization had any question been raised about any pseudopatient's simulation. Nor are there any indications in the hospital records that the pseudopatient's status was suspect. Rather, the evidence is strong that, once labeled schizophrenic, the pseudopatient was stuck with that label. If the pseudopatient was to be discharged, he must naturally be "in remission"; but he was not sane, nor in the institution's view, had he ever been sane.

The uniform failure to recognize sanity cannot be attributed to the quality of the hospitals, for, although there were considerable variations among them, several are considered excellent. Nor can it be alleged that there was simply not enough time to observe the pseudopatients. Length of hospitalization ranged from seven to fifty-two days, with an average of nineteen days. The pseudopatients were not, in fact, carefully observed, but this failure clearly speaks more to traditions within the psychiatric hospitals than to lack of opportunity.

Finally, it cannot be said that the failure to recognize the pseudopatients' sanity was due to the fact that they were not behaving sanely. While there was some tension present in all of them, their daily visitors could detect no serious behavioral consequences—nor, indeed, could other patients. It was quite common for the patients to "detect" the pseudopatients' sanity. During the first three hospitalizations, when accurate counts were kept, 35 of a total of 118 patients on the admissions ward voiced their suspicions, some vigorously. "You're not crazy. You're a journalist, or a professor [referring to the continual note-taking]. You're checking up on the hospital." While most of the patients were reassured by the pseudopatient's insistence that he had been sick before he came in but was fine now, some continued to believe that the pseudopatient was sane throughout his hospitalization.[11] The fact that the patients often recognized normality when staff did not raises important questions.

Failure to detect sanity during the course of hospitalization may be due to the fact that physicians operate with a strong bias toward what statisticians call the Type 2 error.[5] This is to say that physicians are more inclined to call a healthy person sick (a false positive, Type 2) than a sick person healthy (a false negative, Type 1). The reasons for this are not hard to find: it is clearly more dangerous to misdiagnose illness than health. Better to err on the side of caution, to suspect illness even among the healthy.

But what holds for medicine does not hold equally well for psychiatry. Medical illnesses, while unfortunate, are not commonly pejorative. Psychiatric diagnoses, on the contrary, carry with them personal, legal, and social stigmas.[12] It was therefore important to see whether the tendency toward diagnosing the sane insane could be reversed. The following experiment was arranged at a research and teaching hospital whose staff had heard these findings but doubted that such an error could occur in their hospital. The staff was informed that at some time during the following 3 months, one or more pseudopatients would attempt to be admitted into the psychiatric hospital. Each staff member was asked to rate each patient who presented himself at admissions or on the ward according to the likelihood that the patient was a pseudopatient. A 10-point scale was used, with a 1 and 2 reflecting high confidence that the patient was a pseudopatient.

Judgements were obtained on 193 patients who were admitted for psychiatric treatment. All staff who had sustained contact with or primary responsibility for the patient—attendants, nurses, psychiatrists, physicians, and psychologists—were asked to make judgements. Forty-one patients were alleged, with high confidence, to be pseudopatients by at least one member of the staff. Twenty-three were considered suspect by at least one psychiatrist. Nineteen were suspected by one psychiatrist and one other staff member. Actually, no genuine pseudopatient (at least from my group) presented himself during this period.

The experiment is instructive. It indicates that the tendency to designate sane people as insane can be reversed when the stakes (in this case, prestige and diagnostic acumen) are high. But what can be said of the 19 people who were suspected of being "sane" by one psychiatrist and another staff member? Were these people truly "sane," or was it rather the case that in the course of avoiding the Type 2 error the staff tended to make more errors of the first sort—calling the crazy "sane"? There is no way of knowing. But one thing is certain: Any diagnostic process that lends itself so readily to massive errors of this sort cannot be a very reliable one.

THE STICKINESS OF PSYCHODIAGNOSTIC LABELS

Beyond the tendency to call the healthy sick—a tendency that accounts better for diagnostic behavior on admission than it does for such behavior after a lengthy period of exposure—the data speak to the massive role of labeling in psychiatric assessment. Having once been labeled schizophrenic, there is nothing the pseudopatient can do to overcome the tag. The tag profoundly colors others' perceptions of him and his behavior.

From one viewpoint, these data are hardly surprising, for it has long been known that elements are given meaning by the context in which they occur. Gestalt psychology made this point vigorously, and Asch[13] demonstrated that there are "central" personality traits (such as "warm" versus "cold") which are so powerful that they markedly color the meaning of other information in forming an impression of a given personality. "Insane," "schizophrenic," "manic-depressive," and "crazy" are probably among the most powerful of such central traits.[14] Once a person is designated abnormal, all of his other behaviors were overlooked entirely or profoundly misinterpreted. Some examples may clarify this issue.

Earlier I indicated that there were no changes in the pseudopatient's personal history and current status beyond those of name, employment, and where necessary, vocation. Otherwise, a veridical description of personal history and circumstances was offered. Those circumstances were not psychotic. How were they made consonant with the diagnosis of psychosis? Or were those diagnoses modified in such a way as to bring them into accord with the circumstances of the pseudopatient's life, as described by him?

As far as I can determine, diagnoses were in no way affected by the relative health of the circumstances of a pseudopatient's life. Rather, the reverse occurred: The perception of his circumstances was shaped entirely by the diagnosis. A clear example of such translation is found in the case of a pseudopatient who had a close relationship with his mother but was rather remote from his father during his early childhood. During adolescence and beyond, however, his father became a close friend, while his relationship with his mother cooled. His present relationship with his wife was characteristically warm close and warm. Apart from occasional angry exchanges, friction was minimal. The children had rarely been spanked. Surely there is nothing especially pathological about such a history. Indeed, many readers may see a similar pattern in their own experiences, with no markedly deleterious consequences. Observe, however, how such a history was translated in the psychopathological context, this from the case summary prepared after the patient was discharged.

> This white 39-year-old male . . . manifests a long history of considerable ambivalence in close relationships, which begins in early childhood. A warm relationship with his mother cools during his adolescence. A distant relationship to his father is described as becoming very intense. Affective stability is absent. His attempts to control emotionality with his wife and children are punctuated by angry outbursts and, in the case of the children, spankings. And while he says that he has several good friends, one senses considerable ambivalence embedded in those relationships also . . .

The facts of the case were unintentionally distorted by the staff to achieve consistency with a popular theory of the dynamics of a schizophrenic reaction.[15] Nothing of an ambivalent nature had been described in relations with parents, spouse, or friends. To the extent that ambivalence could be inferred, it was probably not greater than is found in all human relationships. It is true the pseudopatient's relationships with his parents changed over time, but in the ordinary context that would hardly be remarkable—indeed, it might very well be expected. Clearly, the meaning ascribed to his verbalizations (that is, ambivalence, affective instability) was determined by the diagnosis: schizophrenia. An entirely different meaning would have been ascribed if it were known that the man was "normal."

All pseudopatients took extensive notes publicly. Under ordinary circumstances, such behavior would have raised questions in the minds of observers, as, in fact, it did among patients. Indeed, it seemed so certain that the notes would elicit suspicion that elaborate precautions were taken to remove them from the ward each day. But the precautions proved needless. The closest any staff member came to questioning these notes occurred when one pseudopatient asked his physician what kind of medication he was receiving and began to write down the response. "You needn't write it," he was told gently. "If you have trouble remembering, just ask me again."

If no questions were asked of the pseudopatients, how was their writing interpreted? Nursing records for three patients indicate that the writing was seen as an aspect of their pathological behavior. "Patient engages in writing behavior" was the daily nursing comment on one of the pseudopatients who was never questioned about his writing. Given that the patient is in the hospital, he must be psychologically disturbed. And given that he is disturbed, continuous writing must be a behavioral manifestation of that disturbance, perhaps a subset of the compulsive behaviors that are sometimes correlated with schizophrenia.

One tacit characteristic of psychiatric diagnosis is that it locates the sources of aberration within the individual and only rarely within the complex of stimuli that surrounds him. Consequently, behaviors that are stimulated by the environment are commonly misattributed to the patient's disorder. For example, one kindly nurse found a pseudopatient pacing the long hospital corridors. "Nervous, Mr. X?" she asked. "No, bored," he said.

The notes kept by pseudopatients are full of patient behaviors that were misinterpreted by well-intentioned staff. Often enough, a patient would go "berserk" because he had, wittingly or unwittingly, been mistreated by, say, an attendant. A nurse coming upon the scene would rarely inquire even cursorily into the environmental stimuli of the patient's behavior. Rather, she assumed that his upset derived from his pathology, not from his present interactions with other staff members. Occasionally, the staff might assume that the patient's family (especially when they had recently visited) or other patients had stimulated the outburst. But never were the staff found to assume that one of themselves or the structure of the hospital had anything to do with a patient's behavior. One psychiatrist pointed to a group of patients who were sitting outside the cafeteria entrance half an hour before lunchtime. To a group of young residents he indicated that such behavior was characteristic of the oral-acquisitive nature of the syndrome. It seemed not to occur to him that there were very few things to anticipate in a psychiatric hospital besides eating.

A psychiatric label has a life and an influence of its own. Once the impression has been formed that the patient is schizophrenic, the expectation is that he will continue to be schizophrenic. When a sufficient amount of time has passed, during which the patient has done nothing bizarre, he is considered to be in remission and available for discharge. But the label endures beyond discharge, with the unconfirmed expectation that he will behave as a schizophrenic again. Such labels, conferred by mental health professions, are as influential on the patient as they are on his relatives and friends, and it should not surprise anyone that the diagnosis acts on all of them as a self-fulfilling prophecy. Eventually, the patient himself accepts the diagnosis, with all of its surplus meanings and expectations, and behaves accordingly.[5]

The inferences to be made from these matters are quite simple. Much as Zigler and Phillips have demonstrated that there is enormous overlap in the symptoms presented by patients who have been variously diagnosed,[16] so there is enormous overlap in the behaviors of the sane and the insane. The sane are not "sane" all of the time. We lose our tempers "for no good reason." We are occasionally depressed or anxious, again for no good reason. And we may find it difficult to get along with one or another person—again for no reason that we can specify. Similarly, the insane are not always insane. Indeed, it was the impression of the pseudopatients while living with them that they were sane for long periods of time—that the bizarre behaviors upon which their diagnoses were allegedly predicated constituted only a small fraction of their total behavior. If it makes no sense to label ourselves permanently depressed on the basis of occasional depression, then it takes evidence that is presently available to label all patients insane or schizophrenic on the basis of bizarre behaviors or cognitions.

It is not known why powerful impressions of personality traits, such as "crazy" or "insane," arise. Conceivably, when the origins of and stimuli that give rise to a behavior are remote or unknown, or when the behavior strikes us as immutable, trait labels regarding the *behavior* arise. When, on the other hand, the origins and stimuli are known and available, discourse is limited to the behavior itself. Thus, I may hallucinate because I am sleeping, or I may hallucinate because I have ingested a peculiar drug. These are termed sleep-induced hallucinations, or dreams, and drug-induced hallucinations, respectively. But when the stimuli to my hallucinations are unknown, that is called craziness, or schizophrenia—as if that inference were somehow as illuminating as the others. . . .

The Consequences of Labeling and Depersonalization

Whenever the ratio of what is known to what needs to be known approaches zero, we tend to invent "knowledge" and assume that we understand more than we actually do. We seem unable to acknowledge that we simply don't know. The needs for diagnosis and remediation of behavioral and emotional problems are enormous. But rather than acknowledge that we are just embarking on understanding, we continue to label patients "schizophrenic," "manic depressive," and "insane," as if in fact in those words we had captured the essence of understanding. The facts of the matter are that we have known for a long time that diagnoses are often not useful or reliable, but we have nevertheless continued to use them. We now know that we cannot distinguish insanity from sanity. It is depressing to consider how that information will be used.

Not merely depressing, but frightening. How many people, one wonders, are sane but not recognized as such in our psychiatric institutions? How many have been needlessly stripped of their privileges of citizenship, from the right to vote and drive to that of handling their own accounts? How many have feigned insanity in order to avoid the criminal consequences of their behavior, and, conversely, how many would rather stand trial than live interminably in a psychiatric hospital—but are wrongly thought to be mentally ill? How many have been stigmatized by well-intentioned, but nevertheless erroneous, diagnoses? On the last point, recall again that a "Type 2 error" in psychiatric diagnosis does not have the same consequences it does in medical diagnosis. A diagnosis of cancer that has been found to be in error is cause for celebration. But psychiatric diagnoses are rarely found to be in error. The label sticks, a mark of inadequacy forever.

NOTES

1. P. Ash. *J. Abnorm. Soc. Psychol.* 44, 272 (1949); A. T. Beck, *Amer. J. Psychiat.* 119, 210 (1962); A. T. Boisen, *Psychiatry* 2, 233 (1938); N. Kreitman, *J. Ment. Sci.* 107, 876 (1961); N. Kreitman, P. Sainsbury, J. Morrisey, J. Towers, J. Scrivener, *ibid.,* p. 887; H. O. Schmitt and C. P. Fonda, *J. Abnorm. Soc. Psychol.* 52, 262 (1956); W. Seeman, *J. Nerv. Ment. Dis.* 118, 541 (1953). For an analysis of these artifacts and summaries of the disputes, see J. Zubin, *Annu. Rev. Psychol.* 18, 373 (1967); L. Phillips and J. G. Draguns, *ibid.* 22, 447 (1971).
2. R. Benedict, *J. Gen. Psychol.* 10, 59 (1934).
3. See in this regard H. Becker, *Outsiders: Studies in the Sociology of Deviance* (New York: Free Press, 1963); B. M. Braginsky, D. D. Braginsky, K. Ring, *Methods of Madness: The Mental Hospital as a Last Resort* (New York: Holt, Rinehart & Winston, 1969); G. M. Crocetti and P. V. Lemkau, *Amer. Sociol. Rev.* 30, 557 (1965); E. Goffman, *Behavior in Public Places* (New York: Free Press, 1964); R. D. Laing, *The Divided Self: A Study of Sanity and Madness* (Chicago: Quadrangle, 1960); D. L. Phillips, *Amer. Socio. Rev.* 28, 963 (1963); T. R. Sarbin, *Psychol. Today* 6, 18 (1972); E. Schur, *Amer. J. Sociol.* 75, 309 (1969); T. Szasz, *Law, Liberty and Psychiatry* (New York: Macmillan, 1963); *The Myth of Mental Illness: Foundations of a Theory of Mental Illness* (New York: Hoeber-Harper, 1963). For a critique of some of these views, see W. R. Gove, *Amer. Sociol. Rev.* 35, 873 (1970).
4. E. Goffman, *Asylums* (Garden City, NY: Doubleday, 1961).
5. T. J. Scheff, *Being Mentally Ill: A Sociological Theory* (Chicago: Aldine, 1966).
6. Data from a ninth pseudopatient are not incorporated in this report because, although his sanity went undetected, he falsified aspects of his personal history, including his marital status and parental relationships. His experimental behaviors therefore were not identical to those of the other patients.
7. A. Barry, *Bellevue Is a State of Mind* (New York: Harcourt Brace Jovanich, 1971); I. Belknap, *Human Problems of a State Mental Hospital* (New York: McGraw-Hill, 1956); W. Caudill, F. C. Redlich, H. R. Gilmore, E. B. Brody, *Amer. J. Orthopsychiat.* 22, 314 (1952); A. R. Goldman, R. H. Bohr, T. A. Steinberg, *Prof. Psychol.* 1, 427 (1970); unauthored, *Roche Report* 1 (No. 13), 8 (1971).
8. Beyond the personal difficulties that the pseudopatient is likely to experience in the hospital, there are legal and social ones that, combined, require considerable attention before entry. For example, once admitted to a psychiatric institution, it is difficult, if not impossible, to be discharged on short notice, state law to the contrary notwithstanding. I was not sensitive to these difficulties at the outset of the project, nor to the personal and situational emergencies that can arise, but later a writ of habeas corpus was prepared for each of the entering

pseudopatients and an attorney was kept "on call" during every hospitalization. I am grateful to John Kaplan and Robert Bartels for legal advice and assistance in these matters.

9. However distasteful such concealment is, it was a necessary first step to examining these questions. Without concealment, there would have been no way to know how valid these experiences were; nor was there any way of knowing whether whatever detections occurred were a tribute to the diagnostic acumen of the staff or to the hospital's rumor network. Obviously, since my concerns are general ones that cut across individual hospitals and staffs, I have respected their anonymity and have eliminated clues that might lead to their identification.

10. Interestingly, of the twelve admissions, eleven were diagnosed as schizophrenic and one, with the identical symptomatology, as manic-depressive psychosis. This diagnosis has a more favorable prognosis, and it was given by the only private hospital in our sample. On the relations between social class and psychiatric diagnosis, see A. B. Hollingshead and F. C. Redlich, *Social Class and Mental Illness: A Community Study* (New York: Wiley, 1958).

11. It is possible, of course, that patients have quite broad latitudes in diagnosis and therefore are inclined to call many people sane, even those whose behavior is patently aberrant. However, although we have no hard data on this matter, it was our distinct impression that this was not the case. In many instances, patients not only singled us out for attention, but came to imitate our behaviors and styles.

12. J. Cumming and E. Cumming, *Community Ment. Health 1*, 135 (1965); A. Farina and K. Ring, *J. Abnorm. Psychol. 70*, 47 (1965); H. E. Freeman and O. G. Simmons, *The Mental Patient Comes Home* (New York: Wiley, 1963); W. J. Johannsen, *Ment. Hygiene 53*, 218 (1969); A. S. Linsky, *Soc. Psychiat. 5*, 166 (1970).

13. S. E. Asch, *J. Abnorm. Soc. Psychol. 41*, 258 (1946); *Social Psychology* (New York: Prentice Hall, 1952).

14. See also I. N. Mensh and J. Wishner, *J. Personality 16*, 188 (1947); J. Wishner, *Psychol. Rev. 67*, 96 (1960); J. S. Bruner and R. Tagiuri, in *Handbook of Social Psychology*, G. Lindzey, ed. (Cambridge, MA: Addison-Wesley, 1954), vol. 2, pp. 634–54; J. S. Bruner, D. Shapiro, R. Tagiuri, in *Person Perception and Interpersonal Behavior*, R. Tagiuri and L. Petrullo, eds. (Stanford, CA: Stanford Univ. Press, 1958), pp. 277–288.

15. For an example of a similar self-fulfilling prophecy, in this instance dealing with the "central" trait of intelligence, see R. Rosenthal and L. Jacobson, *Pygmalion in the Classroom* (Holt, Rinehart & Winston, New York, 1968).

16. E. Zigler and L. Phillips, *J. Abnorm. Soc. Psychol. 63*, 69 (1961). See also R. K. Freudenberg and J. P. Robertson, *A. M. A. Arch. Neurol. Psychiatr. 76*, 14 (1956).

17. W. Mischel, *Personality and Assessment* (New York: Wiley, 1968).

QUESTIONS FOR DISCUSSION

1. Define your perception of normality and abnormality. Give seven examples of normality and its opposite, abnormality. Give seven examples of sanity and its opposite, insanity. Compare your normality responses to your insanity responses without being ethnocentric. How are they similar?

2. Describe the interaction in the hospital with the pseudopatients. How did some of the mental patients react to the pseudopatients? How does this apply to normality and abnormality in hospitals? Explain Type 2 error and its relationship to patients in hospitals.

3. Discuss how powerlessness, depersonalization, and labeling apply to the reading. Be sure to use examples from the reading.

Are Stimulants Overprescribed? Treatment of ADHD (Attention-Deficit Hyperactivity Disorder) in Four U.S. Communities

Peter S. Jensen, Lori Kettle, Margaret T. Roper, Michael T. Sloan, Mina K. Dulcan, Christina Hoven, Hector R. Bird, Jose J. Bauermeister, and Jennifer D. Payne

Recent media reports indicate that the public has become increasingly concerned about the apparent dramatic rise in the diagnosis of attention-deficit hyperactivity disorder (ADHD) and the prescription of psychostimulant medications, particularly methylphenidate (Ritalin®) (Hancock, 1996). There are, in fact, well-documented increases in the rate of medication treatment for hyperactivity among elementary and secondary school students over the past 18 years (Safer and Krager, 1994). Because of increased rates of prescribing, the Drug Enforcement Agency (DEA), responsible for regulating the level of methylphenidate production, regularly has had to increase the yearly allowable methylphenidate production quotas (Schmidt, 1987).

Despite the interest in the topic, little is actually known about why these increases are occurring. Skeptics have noted that the amount of methylphenidate prescribed is much higher in the United States than in any other country (Hancock, 1996), and they argue that the increases indicate inappropriate use of stimulants—that they are being used to treat all types of behavioral and academic problems (Schmidt, 1987). Others suggest that these increases are not cause for concern, but simply reflect the heightened professional and public awareness that has increased the level of identification and treatment of the disorder (Swanson et al., 1995). Regardless of the accuracy of either of these positions, under some circumstances physicians' evaluations and assessments of children with suspected ADHD may be inadequate, leading to inappropriate diagnosis and treatment of presumptive ADHD (Hancock, 1996; Jensen et al., 1989), while in other cases, assessment and treatment may be appropriate. In addition, some cases of ADHD might be undiagnosed and/or untreated. So what is actually known about the nature and frequency of various forms of ADHD treatments, as delivered in the community?

ADHD TREATMENT PRACTICES

Most treatments for ADHD fall into 2 categories: pharmacotherapy and various forms of counseling/psychotherapy. In addition, services provided within the school

"Are Stimulants Overprescribed? Treatment of ADHD (Attention-Deficit Hyperactivity Disorder) in Four U.S. Communities" by Peter S. Jensen, Lori Kettle, Margaret T. Roper, Michael T. Sloan, Mina K. Dulcan, Christina Hoven, Hector R. Bird, Jose J. Bauermeister, and Jennifer D. Payne. From *Journal of the American Academy of Child and Adolescent Psychiatry,* July 1999, v38 i7 p797(8). Reprinted by permission.

setting are frequently an essential part of the treatment plan. What is known about the frequency of provision of these services to children and adolescents with ADHD? We address each of these 3 components of ADHD treatment services below.

How Much Is Being Prescribed?

Data from Safer and Krager's (1985, 1988, 1994) series of studies over a 22-year period (1971–1993) suggest increased levels of prescribing, with the rate of medication treatment for elementary school students increasing from 1.07% in 1971 to 5.96% in 1987. Moreover, the prevalence of medication treatment for middle school students increased from 0.59% in 1975 to 2.98% in 1993; the rate for high school students increased from 0.22% in 1983 to 0.70% in 1993. Most recently, Safer and colleagues (1996) reported that the number of methylphenidate prescriptions for adolescents increased 2.5-fold from 1990 to 1995. They attributed the rise in methylphenidate use to an increase in the number of girls receiving a diagnosis of ADHD, the longer duration of medication treatment, and growing public acceptance of psychostimulant prescriptions.

While these data indicate increases in rates of prescribing, other evidence suggests that many children with apparent ADHD are not being identified and treated. For example, Szatmari and colleagues (1989) conducted an epidemiological survey of 2,701 children and their parents in the province of Ontario, Canada. Using parent-, teacher-, and child-completed behavior checklists, they estimated that 5.8% of children met criteria for ADHD, yet only a fraction of these same children (1 in 8) were taking any form of medication. More recently, Wolraich and colleagues (1996) conducted a countywide checklist-based survey of teachers to determine the number of children within the school system who were rated with high levels of hyperactive and inattentive symptoms, and whether they were receiving medication. Findings indicated that despite high levels of ADHD-like symptoms in 11.4% of children, only approximately one fourth of these children had been diagnosed or treated with stimulants for ADHD.

While findings from these 2 studies are informative, the extent to which they are more generally applicable to various communities across the United States is unclear. Moreover, the exclusive reliance on behavior checklists in both studies to obtain ADHD-relevant diagnostic information raises concerns about the validity of the ADHD diagnoses.

Frequency of Psychosocial Treatments

Bennett and Sherman (1983) found that in addition to using medication to treat hyperactivity, primary care physicians also reported using behavior modification, with significantly more pediatricians (94%) reporting its use than family physicians (71%) and general practitioners (61%). Similar findings have been reported by Copeland et al. (1987) and Moser and Kallail (1995). While these figures suggest relatively high levels of use of behavior therapies, they do not reflect what was done with individual children and are suspect on the grounds of likely overreporting due to social desirability factors and physicians' presenting their treatment practices in a favorable light. Data based on studies of the treatments that individual children actually receive suggest more problematic practices. Thus, Bosco and Robin (1980) reported that only 32% of identified hyperactive children received any form of counseling, and only 10% received behavior modification. The majority of children (74.5%) received methylphenidate, either alone or with the above 2 treatments. Similarly, Sandoval et al. (1980) found that 16.9% of children with ADHD were receiving some form of individual or parent counseling, and slightly more than one third (36.9%) were receiving some form of school-based intervention. Likewise, Jensen et al. (1989) found (based on medical records documentation) that only a small percentage of physicians implemented school interventions (16.2%) or psychotherapy (19.1%). And finally, in their survey of the province of Ontario, Szatmari and colleagues (1989) reported that less than one fifth of hyperactive children were receiving some form of mental health and/or social services intervention, while one third of these children were receiving school-based special educational services. In toto, these studies all suggest that only a minority of ADHD children receive some individual or family-based mental health services, and they often do not receive school-based supports. While these studies are of interest, studies of community-based samples of U.S. children are needed that demonstrate exactly how many children within given communities suffer from ADHD, and among these afflicted children, what types of services (medication, school-based services, or psychotherapeutic treatments) they do (or do not) receive. Do some children receive ADHD treatments (such as psychostimulants) who do not meet criteria for ADHD, and if so, how widespread is this phenomenon?

METHOD

During the first 6 months of 1992, we sampled youths aged 9 to 17 years and their primary caretakers in 4 communities (Atlanta, Georgia; New Haven, Connecticut; Westchester, New York; and San Juan, Puerto Rico). In what has become known as the Methods for the Epidemiology of Child and Adolescent Mental Disorders (MECA) Study (Lahey et al., 1996), we used epidemiological household sampling procedures to ascertain, enumerate, and recruit eligible children and families (one child per household) (see Lahey et al., 1996, for further description of samples and study methods). Children and their primary caretaker (usually the mother) were interviewed in their home by 2 lay interviewers (each blind to the other's findings), using a computer-assisted version (PC-DISC) of the National Institute of Mental Health Diagnostic Interview Schedule for Children (NIMH-DISC-2.3) (Shaffer et al., 1996). The analyses presented in this report include only those parent-child dyads for which DISC data allowed the diagnostic determination of ADHD in the child by either or both informants (1,285 total dyads). In most instances, children's primary caretakers were interviewed; 96% of caretakers were biological or adoptive parents. More than 98% of selected households were successfully enumerated, and 85% of families with an eligible child between ages 9 and 17 years participated in the survey.

Instruments and Measures

Version 2.3 of the NIMH DISC was used by all sites as the lay-administered structured diagnostic interview. The DISC generates diagnoses of major psychiatric diagnoses as defined by the DSM-III-R (American Psychiatric Association, 1987). The DISC has been shown to generate reliable and valid ADHD diagnoses (Jensen et al.,

1995; Schwab-Stone et al., 1996). The second component of the computer-assisted interview used by all sites was a multipart assessment battery covering demographic factors, intellectual ability of the youth, patterns of service utilization, barriers to service utilization, functional impairment, and potential risk and protective factors including school and family environments, family history of psychiatric disorder, parental supervision, life events, and physical maturity (Service Use and Risk Factors Interview). Details on the impairment measures are provided by Bird et al. (1996), risk factors are outlined by Goodman et al. (1998), and the service use measures are described by Leaf et al. (1996).

For the purposes of this article, we report services data concerning any medications prescribed (and by whom), school-based special educational services, and psychosocial treatments and/or counseling (and by whom). Type and intensity for each of these services within the past year were examined. In addition, parents completed a simple tally of yes/no questions asking whether they wanted or needed assistance for their child through various service options, e.g., school-based services, medicine for behavior problems, and counseling/psychotherapy. In this fashion, it could be determined whether those who wished for various treatment options actually obtained them.

To explore the possibility of over- versus underprescribing of psychostimulants, we determined whether children who were being treated with stimulants also met DISC criteria for ADHD, and conversely, we examined the extent to which children who met criteria for ADHD were receiving various forms of treatment, including psychostimulants.

Data Analyses

The total number of subjects who met DSM-III-R criteria for ADHD was determined for all 4 sites. We then examined the number of children who met criteria for ADHD and who were provided any of 3 types of treatment services, alone or in combination: medication (principally methylphenidate), any form of psychotherapeutic or behavioral treatment, or school-based services. Because availability of diagnostic and treatment services can vary substantially across different communities, we detail the ADHD prevalence rates and service use frequencies for each of the 4 communities. For comparative purposes, we examined the prevalence rates of services use in 3 groups: children with ADHD, children with other psychiatric conditions, and children with no psychiatric disorder. Because analyses were exploratory rather than hypothesis-driven (necessarily so, given the lack of information concerning the prevalence of ADHD treatments), power analyses were not conducted. However, confidence intervals of rates and proportions were computed to enable appropriate inferences about the strength of findings.

Thus this study takes advantage of community-based epidemiological samples drawn from 4 different U.S. communities to examine 3 vexing questions that have not been fully addressed in other studies to date: (1) To what extent are children with ADHD treated/under-treated across different communities? (2) What proportions of children with ADHD receive medication versus psychosocial treatments? (3) Within community samples, are "substantial" numbers of children with no evidence of ADHD being treated with stimulants?

RESULTS

Given the low medication prescribing rates overall, no significant differences were found across sites in proportions of children using stimulants or other medications. In general, children from Puerto Rico received fewer services/treatments of all types.

Only 12% of children with ADHD received stimulant treatment. One fourth of children with ADHD received some form of special services and assistance from the school, and almost one third of children with ADHD received some type of behavioral or psychotherapeutic help. Significant numbers of children with ADHD within these 4 communities were not receiving any services whatsoever.

Additional analyses (available from the authors upon request) indicated that primary care physicians do more than 85% of the prescribing. In terms of mental health and counseling treatments, 3% of children with ADHD were receiving mental health treatments from a psychiatrist; 12% were receiving help from a psychologist, while the remainder and the majority were receiving mental health treatments from other professionals, including general counselors, social workers, and others.

The cross-tabulation of diagnostic status by stimulant medication indicates that 8 of the 16 prescriptions for a psychostimulant were provided to children who did not meet full criteria for ADHD. To determine whether these prescriptions were being provided for children with ADHD but who no longer met full ADHD criteria (because they were being treated), we constructed a dimensional tally of all ADHD symptom criteria met by each subject. (The structure of the DISC interview allows the determination of subthreshold levels of symptoms, including criterion counts.) We then compared 4 groups of subjects: those who met ADHD criteria and were treated with stimulants; those who met ADHD criteria but were not treated with stimulants; those who did not meet ADHD criteria but were treated with stimulants; and those who neither met ADHD criteria nor were treated with stimulants (most of the sample).

ADHD subjects treated with medication had similar levels of ADHD symptoms compared with those not so treated. Also, children who were treated with stimulants and who did not meet ADHD criteria nonetheless had quite high levels of ADHD symptoms. Finally, fewer than one half of children who met criteria for ADHD received services deemed necessary by their parents. For example, 39 of 66 parents of children with ADHD noted their child's need for school services, but only 17 (43.6%) of these 39 children were actually receiving school services. Similar findings are noted for psychosocial treatments, in the general discrepancy between what parents think their children need versus what they actually receive. In relative contrast, regardless of diagnostic category, when parents believe that a child requires medication, it is readily available, regardless of whether the child meets full criteria for ADHD, other psychiatric conditions, or scores below the diagnostic threshold for any disorder.

DISCUSSION

Before discussion of our findings, several caveats are in order. First, it should be noted that our study sites' samples, while representative of 4 different communities, do not together constitute a nationally representative sample. Nonetheless, these 4 communities were quite diverse and interesting in their own right—a part-suburban, part-rural area in Georgia, a socioeconomically advantaged county in New York, an urban-suburban area from New Haven, and suburban San Juan of the island of Puerto Rico. To the extent that these differences are likely to shape access to services and patterns of care, the overall findings might reasonably be viewed as more broadly applicable to the range of communities and geographic areas across the United States than any single-site study. Of course, because the 4 communities were principally urban-suburban, studies of rural populations might yield quite different findings, especially given the scarcity of mental health resources and the greater reliance on non-specialist practitioners in such settings.

The second concern is related to the relatively small number of children on medication among those who meet criteria for ADHD. With a total of only 16 children being prescribed a stimulant (range: 1–5 across sites), the lack of significant differences among sites in 12-month prescribing rates is not surprising. While these small numbers could pose methodological problems for extensive analyses of the characteristics of the subgroup of children who are prescribed stimulants for ADHD, the relatively small number of children on medication is also quite informative, as it belies concerns about the presumed general overprescribing of stimulants for children with ADHD. Given other evidence about the large differences among physicians in the frequency of prescribing stimulant medications (Rappley et al., 1995; Sherman and Hertzig, 1991), a more cautious interpretation suggests that over- and under-prescribing may both occur, but are likely to be region-, community-, and provider-specific.

A third concern pertains to the dramatic differences in rates of ADHD across several of our communities, in particular Georgia (9.4%) and Puerto Rico (1.6%). While this difference could reflect problems with the translation of the DISC into Spanish, we are inclined to think that this is not the case, since the Spanish version of the DISC has been demonstrated to have very good psychometric properties (Jensen et al., 1995; Ribera et al., 1996; Rubio-Stipec et al., 1994; Schwab-Stone et al., 1996). A perhaps more plausible explanation has been advanced by other investigators (Ho et al., 1996; Mann et al., 1992), namely, that substantial differences in rates of reporting of ADHD symptoms occur across cultures, possibly because of different cultural thresholds to what constitutes acceptable versus deviant behaviors.

A fourth and final limitation concerns the tentative nature of our determination of what might constitute appropriate versus inappropriate prescription practices. The simple examination of whether a child who is receiving a psychostimulant also meets ADHD diagnostic criteria is obviously an imperfect criterion for determining "appropriate prescribing practices," but given the fact that these issues have remained relatively unexplored in the literature to date, such an approach is a reasonable first step.

In contrast to our study's weaknesses, unlike previously reported studies that in almost all instances described only the use of medications and medication-based practices, we were able to augment medication information with data about the range of school-based and mental health services that families might receive. Another strength in our study was our use of a standardized diagnostic interview to ascertain caseness of ADHD.

IMPLICATIONS

Concerns about dramatic levels of overprescribing are not supported by these data—fewer than 1 in 8 children with ADHD were actually taking medications. Of note, however, 8 of the 16 children who were prescribed a stimulant did not meet diagnostic criteria for ADHD. This could have been due to the fact that some of these children had treated ADHD and no longer met diagnostic criteria as a function of stimulant treatment. Symptoms (elevated well above those of nontreated children) of ADHD were found in all stimulant-treated children, regardless of whether they met full ADHD criteria. These data indicate that stimulant medications generally are being prescribed for children with ADHD or significant residual symptoms of such. Of course, because increases in stimulant prescriptions have continued subsequent to our study, were we to redo our study in 1998, altogether different results could well emerge.

A second implication from our findings concerns the types of treatments children with ADHD most frequently receive, compared to public perceptions and media reports. One third of children with ADHD received some form of counseling or mental health services, followed then by school-based services, compared with one eighth of the ADHD children receiving medication. Thus, in spite of the concerns that medication treatments are being substituted for other more appropriate treatments, such does not necessarily appear to be the case. More troubling, many children are not receiving needed services, regardless of whether they meet criteria for ADHD or some other condition. For example, among the 66 children with ADHD, only about one fourth to one third received school-based or psychotherapeutic services. These findings parallel previous reports that have suggested that of children needing mental health care, only about one third are actually receiving care (Institute of Medicine, 1989).

Reasons for these relatively low rates of prescribing for children with ADHD are unclear. If replicated, these findings may suggest that in the absence of specialists such as child psychiatrists, many pediatricians may be uncomfortable with prescribing, or when they do prescribe, use low, fixed doses that may not be maximally effective, leading families to explore other alternatives. Another possibility is that some of these children had received medication in earlier years. Regardless, these findings indicate that when a child is currently having significant home and school-based problems and is getting some form of behavioral or psychotherapeutic mental health care, parents', mental health providers', and possibly even physicians' concerns about or reluctance to use medication may require further education about the safety and efficacy of these current treatments and better information dissemination concerning their appropriateness.

It would have been optimal to have better process measures of mental health professionals' treatment practices, that is, the actual nature and quality of care rendered by providers and the extent to which services used empirically based treatments. Such in-depth measures are required to truly gauge what constitutes appropriate treatment, or to determine "over- versus underprescribing." While undertreatment and overtreatment, underdiagnosis and overdiagnosis certainly are real phenomena with any medical condition, alarmist or exaggerated reports can also do significant harm in that they discourage parents from seeking treatment for suffering children or otherwise increase the stigma and blame borne by families.

Critical questions to be answered in the near future concern the relative effectiveness of medication versus psychosocial treatments (Arnold et al., 1997; Richters et al., 1995), used alone or in combination, and how clinical outcomes differ among children who receive these various treatments. However, pending new data, the bulk of evidence to date suggests that stimulant medication is the most effective intervention for ADHD (e.g., Horn et al., 1991). Given the indication from our findings that children treated with stimulants have clearly elevated levels of ADHD symptoms, and that substantial numbers of children with ADHD are not receiving these efficacious interventions, our data do not support the notion that stimulants are overutilized for children with ADHD. Given the widespread concern about presumed inappropriate use of medication, better education appears warranted for parents, physicians, and the media about the appropriate assessments and treatments for ADHD.

REFERENCES

American Psychiatric Association (1987), *Diagnostic and Statistical Manual of Mental Disorders, 3rd edition-revised (DSM-III-R)*. Washington, DC: American Psychiatric Association.

Arnold LE, Abikoff HB, Cantwell DP et al. (1997), NIMH Collaborative Multimodal Treatment Study of Children with ADHD (MTA): design challenges and choices. *Arch Gen Psychiatry* 54:865–870.

Bennett FC, Sherman RA (1983), Management of childhood "hyperactivity" by primary care physicians. *J Dev Behav Pediatr* 4:88–93.

Bird HR, Andrews H, Schwab-Stone M et al. (1996), Global measures of impairment for epidemiologic and clinical use with children and adolescents. *Int J Methods Psychiatr Res* 6:295–307.

Bosco JJ, Robin SS (1980), Hyperkinesis: prevalence and treatment. In: *Hyperactive Children: The Social Ecology of Identification and Treatment*, Whalen CK, Henker B, eds. New York: Academic Press, pp 173–187.

Copeland L, Wolraich M, Lindgren S, Milich R, Woolson R (1987), Pediatricians' reported practices in the assessment and treatment of attention deficit disorders. *J Dev Behav Pediatr* 8:191–197.

Goodman SH, Hoven C, Narrow W et al. (1998), Measurement of risk for mental disorders and competence in a psychiatric epidemiologic community survey: the NIMH Methods for the Epidemiology of Child and Adolescent Mental Disorders (MECA). *Soc Psychiatry Psychiatr Epidemiol* 33:162–173.

Hancock L (1996), Mother's little helper. *Newsweek* March 18, pp 51–56.

Ho TP, Luk ES, Taylor E, Bacon-Shone J, Mak F (1996), Establishing the constructs of childhood behavioral disturbances in a Chinese population: a questionnaire study. *J Abnorm Child Psychol* 24:417–431.

Horn WF, Ialongo NS, Pascoe JM et al. (1991), Additive effects of psychostimulants, parent training, and self-control therapy with ADHD children. *J Am Acad Child Adolesc Psychiatry* 30:233–240.

Institute of Medicine (1989), Research on Children and Adolescents With Mental, Behavioral, and Developmental Disorders (Publication IOM-89-07). Washington, DC: National Academy Press.

Jensen PS, Xenakis SN, Shervette RS, Bain MW (1989), Diagnostic and treatment practices of attention deficit disorder in two general hospital clinics. *Hosp Community Psychiatry* 40:708–712.

Jensen PS, Roper M, Fisher P et al. (1995), Test-retest reliability of the Diagnostic Interview Schedule for Children (DISC 2.1): parent, child, and combined algorithms. *Arch Gen Psychiatry* 52:61–71.

Lahey B, Flagg E, Bird H et al. (1996), The NIMH Methods for the Epidemiology of Child and Adolescent Mental Disorders (MECA) Study: background and methodology. *J Am Acad Child Adolesc Psychiatry* 35:855–864.

Leaf PJ, Alegria M, Cohen P et al. (1996), Mental health service use in the community and schools: results from the four-community MECA study. *J Am Acad Child Adolesc Psychiatry* 35:889–897.

Mann EM, Ikeda Y, Mueller CW et al. (1992), Cross-cultural differences in rating hyperactive-disruptive behaviors in children. *Am J Psychiatry* 149:1539–1542.

Moser SE, Kallail KJ (1995), Attention-deficit hyperactivity disorder: management by family physicians. *Arch Fam Med* 4:241–244.

Rappley MD, Gardiner JC, Jetton JR, Houang RT (1995), The use of methylphenidate I Michigan. *Arch Pediatr Adolesc Med* 149:675–679.

Ribera JC, Canino G, Rubio-Stipec M et al. (1996), The Diagnostic Interview Schedule for Children (DISC-2.1) in Spanish: reliability in a Hispanic population. *J Child Psychol Psychiatry* 37:195–204.

Richters J, Arnold LEA, Jensen PS et al. (1995), NIMH Collaborative Multisite, Multimodal Treatment Study of Children With ADHD, I: background and rationale. *J Am Acad Child Adolesc Psychiatry* 34:987–1000.

Rubio-Stipec M, Canino GJ, Shrout P, Dulcan M, Freeman D, Bravo M (1994), Psychometric properties of parents and children as informants in child psychiatry epidemiology with the Spanish Diagnostic Interview Schedule for Children (DISC2). *J Abnorm Child Psychol* 22:703–720.

Safer DJ, Krager JM (1985), Prevalence of medication treatment for hyperactive adolescents. *Psychopharmacol Bull* 21:212–215.

Safer DJ, Krager JM (1988), A survey of medication treatment for hyperactive/inattentive students. *JAMA* 260:2256–2258.

Safer DJ, Krager JM (1994), The increased rate of stimulant treatment for hyperactive/inattentive students in secondary schools. *Pediatrics* 94:462–464.

Safer DJ, Zito JM, Fine EM (1996), Increased methylphenidate usage for attention deficit disorder in the 1990s. *Pediatrics* 98(6 part 1): 1084–1088.

Sandoval J, Lambert NM, Sassone D (1980), The identification and labeling of hyperactivity in children: an interactive model. In: *Hyperactive Children: The Social Ecology of Identification and Treatment,* Whalen CK, Henker B, eds. New York: Academic Press, pp 145–171.

Schmidt WE (1987), Sales of drug are soaring for treatment of hyperactivity. *New York Times* May 5, p C3.

Schwab-Stone M, Dulcan M, Jensen P et al. (1996), The NIMH Methods for the Epidemiology of Child and Adolescent Mental Disorders (MECA) Study: validity of the DISC-2. *J Am Acad Child Adolesc Psychiatry* 35:878–888.

Shaffer D, Fisher P, Dulcan M et al. (1996), The second version of the NIMH Diagnostic Interview Schedule for Children (DISC-2). *J Am Acad Child Adolesc Psychiatry.*

Sherman M, Hertzig ME (1991), Prescribing practices of Ritalin: the Suffolk County, New York study. In: *Ritalin: Theory and Patient Management*, Greenhill LL, Osman BB, eds. New York: MA Liebert, pp 187–194.

Swanson JM, Lerner M, Williams L (1995), More frequent diagnosis of attention deficit-hyperactivity disorder (letter). *N Engl J Med* 333:944.

Szatmari P, Offord DR, Boyle MH (1989), Correlates, associated impairments, and patterns of services utilization of children with attention deficit disorder: findings from the Ontario Child Health Study. *J Child Psychol Psychiatry* 30:205–217.

Wolraich ML, Hannah JN, Pinnock TY, Baumgaertel A, Brown J (1996), comparison of diagnostic criteria for attention-deficit hyperactivity disorder in a county-wide sample. *J Am Acad Child Adolesc Psychiatry* 35: 319–324.

QUESTIONS FOR DISCUSSION

1. Discuss your opinion of the results of the study. In your opinion, what actions should be taken to help children who are overdiagnosed?

2. Discuss the different types of treatments for ADHD children. Which method of treatment would you advocate for children in your geographical area? Explain your response.

3. Discuss the implications of the study. What does this mean for the children in your community area? Explain your answer.

The Level of Competition as a Factor for the Development of Eating Disorders in Female Collegiate Athletes

Christy L. Picard

INTRODUCTION

Instances of eating disorders have increased to epidemic proportions in the Western world. Anorexia and bulimia now rank among the major health problems in the U.S. (Taub and Blinde, 1992), and an even greater variety of disordered eating poses potential health problems (Yeager et al., 1993). While disordered eating is most common among adolescent women, it has also been observed in a variety of other populations. One study reported that girls as young as fourth grade were driven to diet out of dissatisfaction with their body shape (Thelen and Cormier, 1995). Surveys estimate that between 1% and 2% of college women have anorexia nervosa

(Pope et al., 1984), 6% to 8% suffer from bulimia (Katzman et al. 1984), and up to 50% display a variety of abnormal eating behaviors (Schotte and Stinkard, 1987).

Several factors may place an individual at higher risk than usual for the development of eating disorders. The need to maintain strong control over body shape has been identified as a risk factor for both anorexia and bulimia (Garner and Garfinkel, 1980; Taub and Blinde, 1992), and has been identified in many groups including, but not limited to, female adolescents, ballet dancers (Brooks-Gunn et al., 1988; Garner and Garfinkel, 1980; Garner et al., 1987), models (Garner and Garfinkel, 1980), and certain female athletes (Brooks-Gunn et al., 1988; Stoutjesdyk and Jevne, 1993; Taub and Blinde, 1992).

The research examining the level of risk for disordered eating in athletes has yielded mixed results. Some researchers have labeled college athletes as "high risk" (Greskoo and Karlsen, 1994; Rosen et al., 1986; Stoutjesdyk and Jevne, 1993), while others have not found support for such a label (Warren et al., 1990; Weight and Noakes, 1986; Wilkins et al., 1991). In one study of 182 female collegiate athletes, 32% reported practicing at least one pathogenic weight control behavior (Rosen et al., 1986). Conversely, Weight and Noakes (1986) found no significant difference in eating attitudes and the incidence of anorexia nervosa between competitive female runners and non-athlete controls.

One factor thought to account for a relationship between athletes and eating disorders is that athletes tend to exemplify several personality characteristics commonly seen in individuals with eating disorders. Some of these correlates include high self-expectations, competitiveness, perfectionism, compulsiveness, drive, self-motivation, and the intense pressure to be slim and perform (Garner and Garfinkel, 1980; Taub and Blinde, 1992; Thornton, 1990; Yates et al., 1983). Taub and Blinde (1992) added that factors such as pressure from coaches, parents, and peers, and emphasis placed on body form in certain sports may also contribute to the presence of eating disorders in athletes.

Explanations abound for the lack of relationship between athletes and disordered eating as well. Mallick et al. (1987) discovered that although athletes seemed to display anorexic-like behaviors, they differed from the non-athletes in that they had better self-images. Also, athletes may not have a morbid fear of being fat and are less likely than anorexics to perceive themselves as overweight (Wilkins et al., 1991). Finally, differences exist in motivation for weight control between athletes and anorexics (Stoutjesdyk and Jevne, 1993). Athletes are thought to lose weight in order to improve their performance (Rosen et al., 1986), whereas anorexics' weight loss efforts are driven by a desire to be thinner, suggesting different psychological mechanisms. It may, however, be necessary to consider the type of sport when making such comparisons. While improved performance may be a priority for many female athletes, sports that emphasize appearance (such as gymnastics and skating) also place aesthetic pressure on the athlete (Sesan, 1989). Certain subgroups of athletes may be more vulnerable to disordered eating than others (Stoutjesdyk and Jevne, 1993). Sports emphasizing leanness for the sake of better performance or appearance (e.g., gymnastics and figure skating), and sports possessing weight restrictions (e.g., wrestling and rowing), show a higher prevalence of disordered eating than sports without such characteristics (Borgen and Corbin, 1987; Stoutjesdyk and Jevne, 1993; Sundgot-Borgen, 1993; Sundgot-Borgen, 1994). Borgen and Corbin (1987) reported that 20% of athletes in sports favoring a lean physique were either exceptionally weight preoccupied or displayed tendencies toward eating disorders, as compared to only 6% and 10% of non-athletes and non-lean athletes respectively. Stoutjesdyk and Jevne (1993) found the highest percentage of females scoring in the anorexic range within lean type sports of gymnastics and diving. Along with the increased risk of disordered eating comes the risk for developing amenorrhea and osteoporosis (Yeager et al., 1993), the combination of which is referred to as the "female athlete triad" and is potentially fatal (Skolnick, 1993; Yeager et al., 1993). The triad is especially common among athletes competing in appearance or endurance sports (Yeager et al., 1993), and amenorrhea is a frequent symptom of eating disorders in female runners (Gadpaille et al., 1987).

The level of performance and competition in sports may be another factor contributing to the risk of disordered eating, but this relationship has not been systematically investigated. Stoutjesdyk and Jevne (1993) examined eating attitudes of athletes in different types of sports and as an aside mentioned that the only athletes who scored in the anorexic range were those engaged in national or international levels of competition. Examining the link between distance running and anorexia, Weight and Noakes (1986) also noticed that the majority of runners who displayed features of anorexia nervosa were highly competitive but again, this was not the focus of their study. Yeager et al. (1993) merely mentioned that a "pressure to excel" was common among athletes with one or more of the female athlete triad conditions.

In sum, the research on the risk for eating disorders among athletes is contradictory and inconclusive. The identification of athletes as a high-risk group for the development of eating disorders is debatable; however, there appears to be a consensus that certain subgroups of athletes have a higher tendency toward disordered eating than others (Borgen and Corbin, 1987; Rosen et al., 1986). The level of competition may also be linked to the level of risk of the athlete. This study examined the relationship between female collegiate athletes' levels of competition and their eating attitudes and behaviors (i.e., their disposition toward eating disorders). A second objective was to replicate previous findings that athletes of lean versus non-lean types of sport (distance running and light-weight rowing versus ice hockey and basketball, respectively) differed in risk of disordered eating.

Following the proposed link between disordered eating and competitive pressure (Stoutjesdyk and Jevne, 1993; Weight and Noakes, 1986), along with previous research on type of sport (Sundgot-Borgen, 1994; Stoutjesdyk and Jevne, 1993; Sundgot-Borgen, 1993), we predicted that the higher level competition and lean type sports would be associated with a higher prevalence of disordered eating.

METHOD

Participants

Women were recruited from four varsity teams at an NCAA Division I college and an NCAA Division III college, and a group of non-athletes served as a comparison group. Participation was voluntary and anonymous, as reinforced in a mandatory consent form. Of the 152 questionnaires distributed, 109 (72%) were returned. Table 1 presents the distribution of participants from the two schools representing either

Study Sample by Sport and Level of Competition (N = 109)   **TABLE 1**

Sports	N	Level of Competition	
		Division I	Division III
Non-lean athletes	45	23	22
Lean athletes	33	15	18
Non-athletes	31	16	15
N		54	55

non-athletes, athletes in lean-emphasizing or weight-restricting sports (cross-country running and lightweight crew), or athletes in non-lean sports (basketball and hockey). Aside from their athletic differences, the schools were nearly identical in terms of size, geographical setting, academic standing and academic interests, and both drew students from similar socioeconomic backgrounds and geographical regions.

Procedure

Letters were sent to the athletic directors and coaches at each school requesting their cooperation. Participants were told that the survey concerned eating, exercise, and health attitudes and behaviors in college students, and that the results were anonymous and would not be seen by coaches. Varsity athletes were approached at a team practice, and those willing to participate signed a consent form and received a packet of questionnaires to take home, complete, and return in a sealed envelope to a campus mail box. Non-athletes were drawn randomly from the student directory and were mailed an explanatory letter and a consent form. Non-athletes who reported participating in any sport at or above the varsity level were eliminated, and the rest were sent a packet of questionnaires to be returned via campus mail.

Measures

The Demographic and Health Questionnaire designed for this study provided data regarding age, height, weight, and ideal weight, as well as menstruation, diet, exercise, and training habits. Body Mass Index (weight divided by height squared in kg/m^2) and Ideal Weight Discrepancy (ideal weight subtracted from actual weight) were derived from responses. Five questions in Likert format examined the level of competition and stress the athletes experienced, including the pressure to maintain a lean body weight (0 = none to 7 = severe), subjective level of competitiveness (0 = very low to 7 = very high), ratings of fitness levels (0 = very unfit to 7 = very fit), amount of pressure from coaches and peers to perform (0 = none to 7 = severe), and nervousness before a competition (0 = not at all to 7 = extremely nervous).

The Eating Attitudes Test (EAT) screens for actual or initiatory cases of anorexia nervosa in both clinical and non-clinical populations. It has proven reliability (alpha = .90) and validity (r = .87, p [less than] .001) (Garner and Garfinkel, 1979). The abbreviated version (the EAT-26), which correlates highly with the original scale (r = 0.98), was used (Garner et al., 1982).

The Eating Disorder Inventory-2 (EDI-2) is used to assess attitudes, behaviors, and psychological characteristics that occur in individuals with eating disorders (Garner et al., 1983). It is both reliable (alpha = .83 to .93 for the EDI subscales), and valid (r = .44 to .68 for subscales, p [less than] .001), and together with the EAT is a useful screening tool for identifying individuals at risk of anorexia nervosa (Garner et al., 1983). While the EAT examines direct eating behaviors and attitudes surrounding food, the EDI examines other characteristics commonly seen in individuals with eating disorders, such as body image concerns. Therefore the combination of both tools was useful in examining the eating habits, behaviors, and concerns of the participants.

The entire packet of measures took about 10 minutes to complete.

RESULTS

Demographics

From the demographic and health questionnaire, a 2 × 3 analysis of variance, competition level (Division I vs. Division III) × sport (lean, non-lean, non-athlete), revealed main effects for the competition level. Higher subjective levels of competitiveness, $F(1,108) = 5.31$, p = .043, and pressure to perform, $F(1,108) = 4.39$, p = .039, were reported among the Division I versus the Division III athletes.

A main effect for sport was found in the Body Mass Index (BMI), $F(2,108) = 23.39$, p [less than] .001. The lean-sport athletes (runners and rowers) had lower BMI's, and were therefore, physically leaner than all other participants. There were no significant differences in ideal weight discrepancy. Every participant listed their ideal weight as less than, or in four cases equal to, their actual weight. The menstrual cycles of the lean athletes were also more irregular than both the non-lean athletes and the non-athletes, $F(2,108) = 3.87$, p = .024, indicating a greater chance of amenorrhea.

Eating Attitudes Test Scores

Results from a 2 × 3 ANOVA, competition level (Division I vs. Division III) × sport (lean, non-lean, non-athlete), indicated a main effect for competition. Division I participants had significantly higher EAT scores than the Division III participants $F(1,108) = 4.38$, p = .039, suggesting a higher prevalence of disordered eating patterns. Similarly, a main effect for the sport groups $F(2,108) = 17.45$, p [less than] .001, revealed that the lean athletes had significantly EAT scores than both the non-lean athletes and the non-athletes.

A significant competition × sport interaction, $F(2,108) = 3.23$, p = .044, further analyzed through simple effects and a Tukey post-hoc test (p [less than] .05), indicated higher EAT scores among the Division I athletes (both types) than the Division III, but no difference between the Division I non-athletes and the Division III non-athletes. Further, at the Division I level, lean athletes differed significantly from both the non-lean athletes and the non-athletes, but the non-lean athletes did not differ from the non-athletes. At the Division III level, lean athletes differed from the non-lean athletes, but not from the non-athletes, and again the non-lean athletes and non-athletes did not differ significantly.

Results From the MANOVAS of EDI Subscale Factors: Accounting for Differences Between Sport Groups and Level of Competition   TABLE 2

Subscale	Level of Competition		Sport Group	
	F	Sig. of F	F	Sig. of F
Drive for thinness	7.15	009(a)	28.01	000(b)
Bulimia	.969	.327	19.61	.000(b)
Body dissatisfaction	1.49	.225	13.37	.000(b)
Ineffectiveness	4.04	.047(a)	14.15	.000(b)
Perfectionism	3.35	.070	3.07	.271
Interpersonal distrust	3.99	.048(a)	3.19	.045(a)
Interoceptive awareness	2.04	157	20.67	000(b)
Maturity fears	4.40	.038(a)	1.14	.325
Asceticism	2.17	.144	8.45	.000(b)
Impulse regulation	1.62	.206	.993	.374
Social insecurity	4.25	.042(a)	1.32	.271

(a) p [less than] .05; (b) p [less than].001.

Eating Disorder Inventory Scores

The relationships between the EDI scores, competition level, and sport type resemble the results of the EAT. As with the EAT scores, in a 2 × 3 analysis of variance main effects were found for both the level of competition, $F(1,108) = 7.31$, p = .008, and the type of sport, $F(2,108) = 20.29$, p = .000. Once again, Division I athletes showed significantly higher scores than Division III athletes, and there was no difference between the non-athletes at each school. Simple effects and a Tukey post-hoc test (p [less than] .05) revealed that the lean athletes had higher scores than both the non-lean athletes and the non-athletes, but again the non-lean and non-athletes did not differ.

An overall interaction between Competition Level and Type of Sport for EDI scores was not significant, $F(2,108) = 2.26$, p = .110, because unlike the EAT scores, the EDI scores of the lean athletes were higher than both the non-lean and non-athletes at both levels of competition, and the non-lean and non-athletes did not differ from each other at either school.

EDI Subscales

For both the level of competition and type of sport, a MANOVA was used to examine which subscales accounted for the significant differences in EDI scores noted above. The results of the MANOVAS are listed in Table 2. There were five factors that contributed to the significant difference between Division I and Division III participants, the strongest being the drive for thinness subscale. The Division I subjects scored higher on the drive for thinness subscale than the Division III subjects.

For the Type of Sport analysis, there were six variables that accounted for the difference between groups on the EDI scale, including the drive for thinness, bulimia, body dissatisfaction, ineffectiveness, interoceptive awareness, and asceticism. The lean athletes scored higher than all other participants on all but the asceticism sub-scale, in which they scored higher than the non-lean athletes but not the non-athletes.

DISCUSSION

This study indicates that athletes of higher levels of competition may be more at risk for disordered eating than lower level athletes. The Division I athletes of both lean and non-lean sports scored significantly higher on both the EAT and EDI scales, in-dicating a higher prevalence of disordered eating, preoccupation with thinness, and fear of gaining weight (Garner et al., 1982). It could be argued that the differences between the Division I and Division III athletes may be attributed to other factors within each school; however, given the absence of differences among the non-athletes at each school on both measures the probability seems unlikely.

Division I athletes were more likely to display the characteristics that typically define eating disorder patients, the most important of which being the drive for thin-ness (DT) subscale of the EDI which indicates a preoccupation with weight and diet and a morbid fear of fat (Garner, 1991). These results contradict the findings of Wil-kins et al. (1991), who reported that athletes were not afraid of becoming fat and that this factor may separate athletes from anorexics and bulimics.

In keeping with previous research, data here provide support for the type of sport and its influence on disordered eating. Athletes who engaged in sports where lean-ness is favored (i.e., distance running) or in which weight restrictions are a factor (i.e., lightweight rowing) showed higher scores on tests of eating behaviors and the tendency toward eating disorders than either non-athletes or athletes of sports with-out such restrictions.

In general, the lean-sport athletes indicated all the signs and symptoms typical of eating disorders. More specifically, they had a distinct fear of fatness and a dissatis-faction with their overall shape, both of which are required for a diagnosis of an eat-ing disorder (Garner, 1991). They also had distinct feelings of self-discipline, denial, and control, all of which have been identified as risk factors for both anorexia and bulimia (Taub and Blinde, 1992). While scores on the EAT and the EDI may indicate maladaptive behaviors, they are not synonymous with a psychiatric diagnosis of an eating disorder and can only be used to infer the amount of risk for the development of one (Koenig and Wasserman, 1995).

Contradictions in previous research on athletes and disordered eating may stem from the effects of confounding factors such as the level of competition, and type of sport. Sundgot-Borgen (1994) used only elite athletes in his study and found signifi-cant differences between groups, but Weight and Noakes (1986) used female runners of all levels and found no significant differences. Studies controlling for other such factors may therefore produce more compatible results.

Moreover, it may well be the case that not all highly competitive athletes are at risk for disordered eating. Athletes in sports without weight restrictions or physical ap-pearance pressures were no more at risk for disordered eating than non-athletic peers.

Of course, other factors may also come into play with the groups of individuals at risk. The lean athletes in this study also had significantly more irregularity in their menstrual cycles. Amenorrhea is common among distance runners (Gadpaille et al.,

1987), other lean athletes (Yeager et al., 1993), and individuals with disordered eating patterns (Yeager et al., 1993). A link between high EAT scores and amenorrhea (Rippon et al., 1988) and between amenorrhea and eating disorders (Gadpaille et al., 1987) suggest a possibility that lean athletes with irregular menstrual cycles are at higher risk for eating disorders and for the female athlete triad (Skolnick, 1993).

A Note About Survey Responding

Truthful responding in the study was encouraged by guaranteeing anonymity and confidentiality of results; respondents knew that the information would not be given to coaches. Nevertheless, survey responses may have been underestimates of true prevalence because athletes with eating disorders may be reluctant to respond truthfully given the secretive nature of the disorders, and fear of the reactions of coaches and teammates (Wilmore, 1991). Wilmore (1991) found that none of the elite female athletes scored in the anorexic range of the EAT scale, yet 18% of them later sought treatment for eating disorders. If the responses to this survey are underestimates, they likely underestimate support of the hypothesis.

Implications and Further Research

The present study implicates the level of competition as a factor in the development of eating concerns and disordered eating patterns in athletes. Future research should examine in further detail the role of different levels of competition (e.g., national or international levels of athletic competition). In addition, a longitudinal study of different sport groups would help to decipher the direction of association and causal relation between athletes and eating disorders (i.e., whether anorexic-like individuals choose lean sport types, or lean sport types promote eating disorders).

    Certain collegiate athletes are at increased risk for the development of abnormal eating patterns and psychiatric eating disorders. Nutritional programs specific to the needs of the athletes should be designed for implementation in colleges, and coaches should be trained in early intervention. Some intervention efforts have been established (Greskoo and Karlsen, 1994; Powers and Johnson, 1996; Sesan, 1989; Thomsen and Sherman, 1993), yet such efforts need to be increased and geared toward a younger age group to help raise awareness and prevent such problems in the future.

REFERENCES

Borgen, J. S., and Corbin, C. B. P. (1987). Eating disorders among female athletes. *Psychol. Today* 15: 89–95.

Brooks-Gunn, J., Burrow, C., and Warren, M. P. (1988). Attitudes toward eating and body weight in different groups of female adolescent athletes, *Int. J. Eating Disord.* 7: 749–757.

Gadpaille, W. J., Sanborn, C. F., and Wagner, W. W. (1987). Athletic amenorrhea, major affective disorders, and eating disorders. *Am. J. Psychiatry* 144: 939–942.

Garner, D. M. (1991). Manual of Eating Disorder Inventory-2. *Psychological Assessments Resources,* Odessa, FL, pp. 1–18.

Garner, D. M., and Garfinkel, P. E. (1979). The Eating Attitudes Test: An index of the symptoms of anorexia nervosa. *Psychol. Med.* 9: 273–279.

Garner, D. M., and Garfinkel, P. E. (1980). Sociocultural factors in the development of anorexia nervosa. *Psychol. Med.* 10: 647–656.

Garner, D. M., Olmsted, Y. B., Bohr, Y., and Garfinkel, P. E. (1982). The Eating Attitudes Test: psychometric features and clinical correlates. *Psychol. Med.* 12: 871–878.

Garner, D. M., Olmsted, M. P., and Polivy, J. (1983). Development and validation of a multidimensional eating disorder inventory for anorexia nervosa and bulimia. *Int. J. Eating Disord.* 2: 1534.

Garner, D. M., Garfinkel, P. E., Rockert, W., and Olmsted, M. P. (1987). A prospective study of eating disturbances in the ballet. *Psychotherapy Psychosom.* 48: 170–175.

Greskoo, R. B., and Karlsen, A. (1994). The Norwegian program for primary, secondary, and tertiary prevention of eating disorders. *Eating Disord. J. Treatment Prey.* 2: 57–63.

Katzman, M. A., Wolchik, S. A., and Braver, S. L. (1984). The prevalence of frequent binge eating and bulimia in a nonclinical college sample, *Int. J. Eating Disord.* 3: 53–62.

Koenig, L. J., and Wasserman, E. L. (1995). Body image and dieting failure in college men and women: Examining links between depression and eating problems. *Sex Roles* 32: 225–249.

Mallick, M. J., Whipple, T. W., and Heurta, E. (1987). Behavioral and psychological traits of weight conscious teenagers: A comparison of eating disordered patients and high- and low-risk groups. *Adolescence* 22: 157–168.

Pope, H. G., Hudson, J. I., and Yurlegun-Todd, D. (1984). Anorexia nervosa and bulimia among 300 suburban shoppers. *Am. J. Psych.* 141: 292–294.

Powers, P. S., and Johnson, C. (1996). Small victories: Prevention of eating disorders among athletes. *Eating Disord. J. Treatment Prey.* 4: 364–377.

Rippon, C., Nash, J., Myburgh, K. H., and Noakes, T. D. (1988). Abnormal eating attitude test scores predict menstrual dysfunction in lean females. *Int. J. Eating Disord.* 7: 617–624.

Rosen, L. W., McKeag, D. B., Hough, D. O., and Curley, V. (1986). Pathogenic weight control behavior in female athletes. *Physical Sports Med.* 15: 79–86.

Schotte, D. E., and Stinkard, A. J. (1987). Bulimia vs. bulimic behaviors on a college campus. *J. Am. Med. Assoc.* 258: 1213–1215.

Sesan, R. (1989). Eating disorders and female athletes: A three level intervention program. *J. College Student Develop.* 30: 568–570.

Skolnick, A. A. (1993). 'Female athlete triad' risk for women. *J. Am. Med. Assoc.* 270: 921.

Stoutjesdyk, D., and Jevne, R. (1993). Eating disorders among high performance athletes. *J. Youth Adolesc.* 22: 271.

Sundgot-Borgen, J. (1993). Prevalence of eating disorders in female elite athletes. *Int. J. Sports Nutrition* 3: 29–40.

Sundgot-Borgen, J. (1994). Risk and trigger factors for the development of eating disorders in female elite athletes. *Med. Sci. Sports Exercise* 8: 414–419.

Taub, D. E., and Blinde, E. M. (1992). Eating disorders among adolescent female athletes: Influence of athletic participation and sport team membership. *Adolescence* 27: 833–848.

Thelen, M. H., and Cormier, J. F. (1995). Desire to be thinner and weight control among children and their parents. *Behav. Therapy* 26: 85–99.

Thompson, R. A., and Sherman, R. T. (1993). Reducing the risk of eating disorders in athletics. *Eating Disord. J. Treatment Prey.* 1: 65–78.

Thornton, J. S. (1990). Feast or famine: Eating disorders in athletes. *Physician Sports Med.* 18: 116–122.

Warren, B. J., Stanton, A. L., and Blessing, D. L. (1990). Disordered eating patterns in competitive female athletes. *Int. J. Eating Disord.* 9: 565–569.

Weight, L. M., and Noakes, T. D. (1986). Is running an analog of anorexia? A survey of the incidence of eating disorders in female distance runners. *Med. Sci. Sports Exercise* 19: 213–217.

Wilkins, J. A., Boland, F. J., and Albinson, J. (1991). A comparison of male and female university athletes and non-athletes on eating disorder indices: Are athletes protected? *J. Sport Behav.* 14: 129.

Wilmore, J. H. (1991). Eating and weight disorders in female athletes. *Int. J. Sports Nutrition* 1: 104–117.

Yates, A., Leehey, K., and Shisslak, C. M. (1983). Running—An analog of anorexia? *New England J. Med.* 308: 251–255.

Yeager, K. K., Agostini, R., Nattiv, A., and Drinkwater, B. (1993). The female athlete triad: Disordered eating, amenorrhea, osteoporosis. *Med. Sci. Sports Exercise* 25: 775–777.

QUESTIONS FOR DISCUSSION

1. Discuss factors that might place individuals at high risk of developing eating disorders. Have you noticed any of these factors at your college, university, or geographical location? Explain your response.

2. Discuss the results of the study. Should there be concerns to female athletes? Explain your response.

3. Division II sports were left out of the study. Where do you think Division II athletes would have been placed in the study? Explain your answer. How would big-name Division II programs compare to Division I programs?

CHAPTER 16

HOMOSEXUAL BEHAVIOR

The Gay Movement and Gay Communities

Robert F. Meier and Gilbert Geis

The gay movement became politically visible on June 28, 1969, when patrons of Stonewall, a gay bar in New York City's Greenwich Village, refused to cooperate with police who were carrying out a routine raid. The patrons, composed mostly of flamboyant drag queens and prostitutes, escalated their protests against the police into nearly five days of rioting that eventually involved hundreds of sympathetic supporters. The rioting appeared to accomplish little; no laws were changed, gays continued to be "bashed," and homosexuals continued to be regarded as socially and sexually marginal people. The significance of this resistance was in the imagination it sparked in gay people throughout the country and elsewhere. Many gays became eager to reject the social stigma and shame heaped upon them by conventional society (Bawer, 1996:4–15). Stonewall became synonymous with any resistance to that oppression.

Although resistance to social stigma and legal repression had existed prior to Stonewall, the rioting galvanized gay opinion like no other event. Gays had witnessed the success of the women's movement, which grew out of similarly felt oppression. But gays and lesbians had obstacles beyond traditionally held prejudices, reinforced by two of the most powerful institutions of social control in society: religion and the law. The women's movement had to confront antiquated tradition and stodgy

"The Gay Movement and Gay Communities" from *Victimless Crime? Prostitution, Drugs, Homosexuality, Abortion* by Robert F. Meier and Gilbert Geis, 1997, pp. 133–136. Reprinted with permission of Roxbury Publishing Company.

beliefs about gender roles but not moral condemnation or night sticks, as did the gay movement.

Only the cleverest and most energetic strategies would stand a chance against such puissant foes. Yet,

> gays developed a territorial base, with a matrix of bars, associations, publications, theaters, churches, writers, comedians, professional services, and eventually political representatives. Gayness became a sort of ethnicity with its own codes of recognition, rituals, parades, sacred days, even its own flag with a rainbow motif. (Gitlin, 1995:142–143)

The first generation of gays after Stonewall worked hard to produce such a community, but the effort was thought to require extremism and aggressiveness. It was to be a public community, which meant that homosexuals would have to be enticed to come out of the closet. Gay pride marches, celebrations of Stonewall, and organized events by such groups as Queer Nation were meant to shock, annoy, retaliate, and educate—all at the same time. There was a portion of the gay community that "developed a radical direct-action movement among men and women who are no longer interested in dwelling only within the safe ghettos of gaydom" (Browning, 1994:25). The closet was defined as only a temporary haven from the political realities of the movement and the drive for eventual freedom.

As in the women's movement, the gay movement produced a gulf among different generations, antagonisms among leaders and followers over points of ideology, and gender segregation. The movement was and continues to be far from monolithic. Some leaders oppose discrimination in any form; others preach the politics of sexual identity in which gayness must be affirmed as something special and distinctive. Some gays are willing to live peaceably in the absence of overt discrimination; others want no less than a social recanting of previous wrongs done to gays. Some wish only that gay bashing would be eliminated; others are more militant in demanding retributive—and in some cases—retaliatory justice.

The women's movement can appeal to both men and women and, hence, can lay theoretical claim to a large segment of potential supporters in the population, but the gay movement cannot assert itself on the basis of numbers only. Rather, it has to rely, ironically, on the moral strength and legal correctness of its position. Such a strategy involves a tricky balancing act, in which previously denigrated acts and conditions are reclaimed from the oppressors and reaffirmed. The first target has been language. The words "gay," "homosexual," and "lesbian" had served their purpose well for those before Stonewall, but they seemed old-fashioned in the 1990s. These were terms for the closet, not public discourse. More suitable to the militants were the previously hated expression of "queer," "faggot," and "dyke." By claiming such words as their own and providing them with positive meaning, the movement believed that it would liberate the terms from their oppressors. Frank Browning (1994:34) summarizes the strategy:

> Steal back all the hateful epithets thrown at gay people over the decades, turn them inside out, and celebrate them. If homophobes and fundamentalist preachers rant on about homosexuals recruiting the young because it's the only way to replenish their unholy ranks, then steal the language back. Yes, queer people

want to recruit the young, not by kidnapping young men as Chicago serial killer John Wayne Gacy did, but by being mentors and role models who would show gay and lesbian adolescents that they are not alone, that they are not freaks, that they need not continue committing suicides at three times the rate of straight teenagers.

So gay activists took one of two directions. The first direction was found in the first-generation gay activist after Stonewall, a militant who exaggerated gayness for effect. "Fag power, Dyke Power, Que-e-e-e-r Nation!" was the shout. Or "We're Queer! We're Here! Get Used to It!" signs, outrageous clothes, public displays of sexuality—anything that was acceptable to get across the message that the days of passivity—the closet—were over. The militant, dissatisfied with continued discrimination and social censure, would demand equality by highlighting the differences between gays and straights. One writer recounts that some gays were able to read a manifesto that advised:

> The next time some straight person comes down on you for being angry, tell them that until things change, you don't need any more evidence that the world turns at your expense. You don't need to see only hetero couples grocery shopping on your TV. . . . You don't want any more baby pictures shoved in your face until you can have or keep your own. No more weddings, showers, anniversaries, please, unless they were our own brothers and sisters celebrating. And tell them not to dismiss you by saying, "You have rights," "You have privileges," "You're overreacting," or "You have a victim's mentality." Tell them, "GO AWAY FROM ME, until YOU can change." (Browning, 1994:27)

While most gays would reject this statement as little more than an understandable but politically ineffective tantrum, rage is an important component of the gay agenda. Unmistakable gains have been attained politically and socially, and many gays have experienced what one writer called "virtual equality" (Vaid, 1995), a condition of being almost there, almost equal. By the mid-1990s, gays were said to be "virtually normal" (Sullivan, 1995). But "virtually" would be insufficient for many.

The second, more recent direction was that of the subdued activist who saw political extremism as part of a short-term agenda. More meaningful change would involve direct education, modeling, and quiet conformity. The gay educator didn't want to shake the boat, only to make sure there was room for everyone on board. He or she was interested in the more subtle, longer-term strategy of convincing straights that gays were very much like them in most of the ways that count. The moderate was convinced that the key to social change was political activism which, in turn, required a resolution of the basic conflict in the gay movement. The question needing resolution was whether gays were different from straights of whatever race and gender such that they required special status, or whether gays were just like everyone else except in sexual orientation.

The presentation of this conflict now determines the direction of the gay movement. What is required is not merely agreement on a political strategy but discussion and resolution of many issues, involving the extent to which homosexuality is deviant and according to whom, the degree to which gays can be accepted in straight society, and the role of law in liberating or repressing them.

REFERENCES

Akers, Ronald L. 1985. *Deviant Behavior: A Social Learning Approach,* 3rd edition. Belmont, CA: Wadsworth.

Appiah, K. Anthony. 1996. "The Marrying Kind." *New York Review of Books,* 43 (June 20): 48–54.

Baird, Robert M., and M. Katherine Baird, eds. 1995. "Introduction." *Homosexuality: Debating the Issues.* New York: Prometheus Books.

Barrett, Paul M. 1996. "How Hawaii Became Ground Zero in Battle over Gay Marriages." *Wall Street Journal,* June 17, 1996, pp. 1A, 5A.

Bawer, Bruce. 1996. *Beyond Queer: Challenging Gay Left Orthodoxy.* New York: Free Press.

Boswell, John. 1980. *Christianity, Social Tolerance, and Homosexuality.* Chicago: University of Chicago Press.

Browning, Frank. 1994. *The Culture of Desire.* New York: Simon and Schuster.

Davis, Kingsley. 1976. "Sexual behavior." Pp. 219–261 in Robert K. Merton and Robert Nisbet (eds.), *Contemporary Social Problems,* 4th ed. New York: Harcourt Brace Jovanovich.

Dworkin, Andrea. 1987. *Intercourse.* New York: Free Press.

Dworkin, Ronald. 1977. *Taking Rights Seriously.* New York: Oxford University Press.

Endleman, Robert. 1990. *Deviance and Psychopathology: The Sociology and Psychology of Outsiders.* Malabar, FL: Robert Krieger Publishing.

Eskridge, William N. 1996. *The Case for Same-Sex Marriage: From Sexual Liberty to Civilized Commitment.* New York: Free Press.

Ford, Clellan S., and Frank A. Beach. 1951. *Patterns of Sexual Behavior.* New York: Harper and Row.

Gallup Poll. 1986. "Sharp Decline Found in Support for Legalizing Gay Relations." *The Gallup Report,* Report Number 254, November: 24–26.

Geis, Gilbert. 1979. *Not the Law's Business.* New York: Schoken.

Gitlin, Todd. 1995. *The Twilight of Common Dreams.* New York: Henry Holt.

Goode, Erich, and Richard T. Troiden, eds. 1974. *Sexual Deviance and Sexual Deviants.* New York: Morrow.

Greenberg, David F. 1988. *The Construction of Homosexuality.* Chicago: University of Chicago Press.

Hamer, Dean, and Peter Copeland. 1994. *The Science of Desire: The Search for the Gay Gene and the Biology of Behavior.* New York: Simon and Schuster.

Harry, Joseph. 1982. *Gay Children Grown Up: Gender Culture and Gender Deviance.* New York: Praeger.

Hu, Stella, Angela M. L. Pattatucci, Chavis Patterson, Lin Li, David W. Fulker, Stacy S. Cherny, Leonid Kruglyak, and Dean H. Hamer. 1995. "Linkage Between Sexual Orientation And Chromosome Xq28 In Males But Not In Females". *Nature Genetics,* 11:248–256.

Katz, Jonathan, ed. 1976. *Gay American History: Lesbians and Gay Men in the U.S.A.* New York: Cromwell.

Katz, Jonathan. 1983. *Gay/Lesbian Almanac: A New Documentary.* New York: Harper and Row.

Kinsey, Alfred C., Wardell B. Pomeroy, and Charles E. Martin. 1948. *Sexual Behavior in the Human Male.* Philadelphia: Saunders.

Klassen, Albert D., Colin J. Williams, and Eugene E. Levitt. 1989. *Sex and Morality in the U.S.* Middletown, CT: Wesleyan University Press.

Langevin, Ron. 1985. "Introduction." Pp. 1–13 in *Erotic Preference, Gender Identity, and Aggression in Men: New Research Studies,* Ron Langevin (ed.). Hillsdale, NJ: Lawrence Erlbaum Associates.

Laumann, Edward O., John H. Gagnon, Robert T. Michael, and Stuart Michaels. 1994. *The Social Organization of Sexuality: Sexual Practices in the United States.* Chicago: University of Chicago Press.

LeVay, Simon, and Dean H. Hamer. 1994. "Evidence for a Biological Influence in Male Homosexuality." *Scientific American,* 270:44–45.

Luckenbill, David F. 1986. "Deviant Career Mobility: The Case of Male Prostitutes." *Social Problems,* 33:283–293.

Marsiglio, William. 1993. "Attitudes toward Homosexual Activity and Gays as Friends: A National Survey of Heterosexual 15- to 19-Year Old Males." *Journal of Sex Research,* 30:12–17.

McCall, William. 1996. "Judge: Same-Sex Partners of State Workers Must Get Benefits." *Corvallis Gazette-Times,* August 10, p. 1.

McWilliams, Peter. 1993. *Ain't Nobody's Business If You Do.* Los Angeles: Prelude Press.

New York Academy of Medicine Committee on Public Health. 1964. "Homosexuality," *Bulletin of the New York Academy of Medicine,* 40:576.

Plummer, Kenneth. 1975. *Sexual Stigma: An Interactionist Account.* London: Routledge and Kegan Paul.

Posner, Richard A. 1992. *Sex and Reason.* Cambridge, MA: Harvard University Press.

Quinn, D. Michael. 1996. *Same-Sex Dynamics among Nineteenth Century Americans: A Mormon Example.* Urbana: University of Illinois Press.

Roper, W. G. 1996. "The Etiology of Male Homosexuality." *Medical Hypotheses,* 46:85–88.

Shapiro, Joseph P. 1994. "Straight Talk about Gays." *U.S. News and World Report,* July 5:47.

Soards, Marion. 1995. *Scripture and Homosexuality: Biblical Authority and the Church Today.* Louisville, KY: Westminster John Knox Press.

Stephan, G. Edward, and Douglas R. McMullin. 1982. "Tolerance of Sexual Nonconformity: City Size as a Situational and Early Learning Determinant." *American Sociological Review,* 47:411–415.

Sullivan, Andrew. 1995. *Virtually Normal: An Argument about Homosexuality.* New York: Knopf.

Toobin, Jeffery. 1996. "Supreme Sacrifice." *The New Yorker,* July 8:43–47.

Vaid, Urvashi. 1995. *Virtual Equality: The Mainstreaming of Gay and Lesbian Liberation.* New York: Anchor Doubleday.

Vreeland, Carolyn N., Bernard J. Gallagher III, and Joseph A. McFalls, Jr. 1995. "The Beliefs of Members of the American Psychiatric Association on the Etiology of Male Homosexuality: A National Survey." *Journal of Psychology,* 129:507–517.

Williams, J. E. Hall. 1960. "Sex Offenses: The British Experience." *Law and Contemporary Problems,* 25:354–364.

QUESTIONS FOR DISCUSSION

1. Compare and contrast the gay movement and the women's movement.

2. How did the incident at Stonewall help jump-start the gay movement? Why would gays and lesbians like to be called previously hated expressions like "queer," "faggot," and "dyke"?

3. According to the authors, how is gayness like an ethnic status? Do you agree or disagree with gayness being an ethnicity? Explain your response.

Becoming Lesbian, Gay, and Bisexual

Robert E. Owens Jr.

Dear God, I am fourteen years old. . . .
[L]et me know what is happening to me.

Alice Walker, *The Color Purple*

Although many educators and professionals deny the existence of queer kids, the process of becoming lesbian, gay, or bisexual is very much an adolescent rite of passage. Sexual identification is not embraced immediately upon self-recognition and there is a gradual process of "coming out" to oneself. Most individuals pass from awareness to positive self-identity between ages thirteen and twenty, and a positive lesbian, gay, or bisexual identity is being established earlier today than in the past.[1] Relatively few middle school youths self-identify as lesbian, gay, or bisexual in contrast to as high as 6 to 7 percent of older high school males who describe themselves as primarily homosexual.[2]

This article describes a "generic" pattern of becoming. Many individual variations exist. The process seems to differ for men and women. In general, on issues of relational expectations, sexual awareness, and equality, young lesbians have more in common with young heterosexual women and young gays have more in common with young heterosexual men than young lesbians and gay men have with each other.[3] It is possible that being male or female is more important overall than being gay or lesbian. As has been observed, "Female homosexuality is to be understood as a unique female phenomenon, rather than a state which is either the same as or the reverse of male homosexuality."[4]

Even with these differences, the questions asked along the journey are surprisingly similar.

Am I really lesbian (or gay)?

Why me?

What will my parents think?

Am I the only one?

Should I tell my best friend?

In general, over a period of years and against a backdrop of stigmatization, lesbian, gay and bisexual youths gradually accept the label homosexual or bisexual for themselves as they interact with the sexual-minority community and increasingly disclose their sexual orientation.[5]

This chapter will describe a four-step process of "becoming." From their initial feelings of being different, gay, lesbian, and bisexual individuals gradually become aware of their same-sex attractions, engage in same-sex erotic behavior and dating, and finally self-identify as lesbian, gay, or bisexual.

FEELING DIFFERENT

Seventy percent or more of lesbian, gay, and bisexual adolescents and adults report feeling different at an early age, often as early as age four or five.[6] A fourteen-year-old gay male writes in his school paper,

> [T]here was always something I knew was a little bit different about me. I didn't know exactly what that might have been. It was just something that was there and I learned to accept it. . . .

Seventeen-year-old Kenneth recalls, "I've known I was different since I was five or six."[7] For some lesbian, gay, and bisexual adults the feeling of difference centered on a vague attraction to, or curiosity about, their own gender.

For most youths, these feelings are not sexual as we will see later in this section.[8] Early childhood experiences may later be interpreted in light of sexual orientation identification.

Many sexual-minority youths state that before they even knew what the difference was they were convinced of its importance. Linda recalls, "Quietly I knew."[9] From second grade on Tony knew instinctively that he was unlike other boys. They knew it too and targeted Tony for ridicule. Philip, a high schooler with deafness, recalls, "I didn't know what it was . . . , but I just knew I was different."[10] In contrast to reports from lesbians and gays, only about 10 percent of heterosexual adults report feeling different or odd as a child.[11]

Many lesbian and gay adults report that as children they felt like an "outsider" within their peer group and their family. They describe isolation, low popularity, scant dating, and lack of interest in the other sex, little participation in same-sex games, and gender nonconformity.[12] Derek, an African-American teen, had physical relations with other boys from age five on. He reports engaging in prepubescent games such as *doctor* and *I'll show you mine if you'll* . . . with boys, but never with girls. Furtive mutual fondling also occurred with other boys.[13]

As children, some but certainly not a majority of gay and bisexual male youths found men to be "enigmatic and unapproachable"[14] and felt more comfortable with women and girls. In general, these youth did not enjoy rough or athletic activities, especially team sports, and the coercion to participate, preferring instead books, art, and fantasy play. Doug, a very bright child, enjoyed problem-solving tasks such as puzzles and word games. Some gay male youths attribute their sexuality to a failure to develop "masculine" characteristics and to being "feminized" through ostracism by boys and association with girls with whom they shared more interests.[15] Jim, a young white gay male, was an isolate in his class except for the occasional female friend. He was labeled a *sissy* by the other children.

Many gay males report that they were more sensitive than other boys and had their feelings hurt more easily; cried more easily; had more aesthetic interests; were drawn to other "sensitive" boys, girls, and adults; and felt and acted less aggressively than their peers.[16] A young student recalls,

> I never felt like I fit in. I don't know why for sure. I feel different. I thought it was because I was more sensitive.[17]

"I had a keener interest in the arts," recalls a young gay man. "I never learned to fight; I just didn't feel I was like other boys. I was very fond of pretty things like ribbons and flowers and music; I was indifferent to boys' games, . . . I was more interested in watching insects and reflecting on certain things. . . ."[18] Male youths may experience the dichotomy of being attracted to the very bullies who are tormenting him.

Approximately 70 percent of lesbian and gay adults report gender non-conforming behaviors in contrast to 16 percent and 3 percent for heterosexual females and males, respectively.[19] These figures must be treated carefully because they are based on subject reports. In contrast, very few lesbian and gay adults exhibit gender or sex role inappropriate behaviors.[20]

The actual incidence of gender nonconformity is unknown and does not seem to be related to the amount of masculine or feminine behavior seen in an adult. "I was more masculine," recalls a young lesbian, "more independent, more aggressive, more outdoorish. . . ."[21] For the child who doesn't conform, whispers and innuendo about her or his sexual orientation may begin early. Parental admonitions to avoid another child who seems different can plant the seeds of homophobia early. Linda, a young lesbian, recalls her distaste for Barbie and learning in first grade to sneak on a pair of shorts under the dresses her mother made her wear to church.

Gender nonconformity seems to be related to socioeconomic status with the most exaggerated behavior found in lower income groups.[22] In general, those who are most different in gender behavior are the most pressured to change.[23] Malcolm, an African-American youth, was a devoutly religious child. Commenting on the boy's effeminate mannerisms, one congregant told Malcolm, "You're degrading God's name." To which the minister added, "You're a disgrace."[24] Often lesbian and gay children are teased for their gender nonconformity. The impact of teasing seems to be more severe for boys than for girls.[25] Miles, a very agile child, became awkward and less willing to try physical activity when at age five, someone stated that he ran "like a girl." "When the others called me names and stuff," recalls Malcolm, "I assumed they were right. I had very low self-esteem."[26] Boys may be viewed as weird and be rejected as undesirable playmates. For their part, some gay and bisexual boys reject play with other boys as unsafe and unenjoyable.[27]

At some level, both the child and the family recognize that a difference exists.[28] This recognition can lead to conflict within both the family and the individual. For me, it meant enrollment by my parents in Cub Scouts and Boy Scouts and an endless stream of failed attempts to play Little League. As for Malcolm, mentioned above, "I was embarrassing to them [his family], especially to my father." He continues,

> When we were in the projects and I would play with other kids, there were times when my mom would tell me to come in. She would say, "Those kids don't want you to play with them." She always made me feel like there was something wrong with me.[29]

For many individuals, feeling lesbian or gay is a natural part of themselves. Most lesbians, gays, and bisexuals state emphatically that they did not choose their sexual orientation and that they were not in control of their feelings. Kevin, now an adult, remembers that he had known he was gay, even before he heard the word or knew its meaning.[30] His older brothers' friends elicited very different feelings in him than in his brothers.

AWARENESS OF SAME-SEX ATTRACTIONS

Sexual awareness usually begins in early adolescence. Awareness is not a sudden event but a gradual sensitivity and consciousness, a growing realization that "I might be homosexual." Most individuals develop feelings and awareness before they ever have a label for them. For some youths, awareness is better described as confusion. Labeling of these feelings may become very frightening.[31] Edmund White in *A Boy's Own Story* (1982) recalled, "I see now that what I wanted was to be loved by men and to love them back but not to be a homosexual. . . ."[32] It is important to remember that while initial awareness may be met with some shock among self-identified youths, only 30 percent of lesbians and 20 percent of gay males report that they experienced negative feelings about themselves.[33] Dan realized he was gay when he was eleven. "It just gave me a sense of wholeness."[34]

The mean age for same-sex awareness and attractions is 10.9 to 13.2 years with a reported range of 10 to 18 years.[35] In a recent survey called *Sex on Campus*, 87 percent of lesbian college students and 63 percent of gay male college students report that they were aware of their sexual orientation by high school, although some knew their orientation in elementary school.

Recalling her confusion, Linda remembers that nobody felt as passionate about members of their own gender as she felt about two of her friends.[36] As a group, boys report being aware before girls. Approximately one-third of gay male youths report same-sex attractions prior to the onset of puberty.[37] In one study, a third of lesbian and gay teens claim that they knew they were homosexual prior to age ten.[38] Mickey, age eighteen, reports an awareness of such feelings at age four or five, concluding, "I've always wanted to touch and be touched by guys."[39] He began to realize that he was not heterosexual in seventh grade. Similarly, Andrew, a young white gay man, remembers noticing at age eight the beauty of his swimming instructor, an older boy of sixteen. Henry, a young college student, recalls, "What I wanted ever since I was five or six watching Marlo Thomas' boyfriend on television, was to have a man in my bed!"[40]

Often these early attractions are vague and impressionistic. One young gay teen recalled,

> My first memory of being attracted to men was a dream I had when I was six or seven. I was in a bathtub with a man in the middle of the forest. I remember this was a happy dream for me, and I dreamt it over and over again for years.[41]

More common is the fifth grade experience of Nathaniel, a middle-class African-American teen. "I was noticing guys," he recalls. "Not knowing that I was gay—just curious about guys."[42] Derek, also African American, recalls,

> In every grade, there was at least one boy that I had a certain fondness for. . . . And later on, . . . I recognized this as crushes. I wanted to spend as much time as I could with them.[43]

The vague same-sex attractions of childhood become eroticized in adolescence. The mean age for same-sex erotic fantasies among males is reported to be 13.9 years, for females somewhat later. "I had homosexual fantasies consistently," recalls Audrey, a young gay man. "If I had a heterosexual fantasy it was because I forced it upon myself."[44] At age fifteen, Mike began to collect photos of his best friend and to write his friend's name all over his notebooks. "It was like I was a junior high girl," he recalls. "I didn't know I had fallen in love with my best friend."[45] Norma Jean became aware of her sexual feelings while working as a store clerk.

> Some women runners kept coming by the store. . . . After they finished practicing, all of those gorgeous bodies would just pile into the store to get something to drink. I became very sexually excited about that. I made sure that I worked the nights they practiced.[46]

In his book *What the Dead Remember,* Harlan Green recalls from boyhood:

> I picked up the *Saturday Evening Post* . . . I turned the page and stopped. . . . I breathed out, transfixed at what I saw. . . . a picture of men and boys in black-and-white advertising Hanes or BVD's.[47]

Similarly, Darrell remembers scanning underwear ads closely in second grade, looking for an outline of what lay beneath. Erotic feelings can come from pictures, words, and voices that others might not consider erotic in the least.

Same-sex attractions are reported by many queer kids to have always been present, deep within the "natural self."[48] Richard, age nineteen, summarizes, "I have always been gay, although I did not know what that meant at the time."[49] In seventh grade, Amy, a young lesbian, recalls that she suddenly changed to a more gender-neutral manner of dress with the defense that "I just want to be myself."[50]

Gay male youths report that they first experienced an obsession to be near masculinity. Mickey, age eighteen, recalls,

> At eight I fell in love with Neal, this guy who rode my bus. . . . I guess he was fourteen. . . . I always wanted to sit with him or be next to him. . . . I spent my childhood fantasizing about men, not sexually . . . , but just being close to them and having them hold me or hug me.[51]

Former NFL running back David Kopay recalls a high school football captain two years older than himself. At the time, David did everything possible to be close to this teammate.[52] At age eleven, Tony developed nonsexual "crushes" on other boys in his scout troop. A seventeen-year-old Chinese-American youth recalls, "In high school . . . I had a crush on a guy I didn't even know."[53] Cory became obsessed with Jessie when both were in seventh grade. He recalls,

> I started going crazy. He looked better every day. I did not do anything during that time; I just looked a lot. . . . I just couldn't go up to him and say, "Hey, Jessie, I'm horny for you.[54]

Although occasionally a peer, the object of the obsession is more often an older male, such as a teacher, scout leader, coach, or older cousin. As a child, Denny's interest in big league sports masked his secretive crushes on the players he admired. A former neighbor, now an adult gay man, confided in me that he had wanted to have sex with me since we met when he was age thirteen and I was thirty-four. "These are often cases of unrequited love," explains Paul Gibson, author of the 1989 DHHS report on adolescent suicide, "with the youth never revealing their true feelings." [55]

Some lesbians report that they were attracted to women in authority positions. Maria had a crush on her third grade teacher. Tara, enamored with her softball coach recalls, "I would fantasize about being a man, so I could kiss her." [56] Other lesbian girls are attracted to peers. Lynn, a lesbian student, developed a crush on her best friend with whom she often danced to recorded music in the dark, took long walks, and slept over.

The reverse of the "typical" heterosexual pattern may occur in which a lesbian or gay teen is drawn to members of the other sex as friends but sexually attracted to members of the same sex. Gay male teens may have many female friends but not be sexually attracted to females. "I was real attracted to pretty women," explains Elisa, "but I identified more with men. But there was no attraction between us." [57]

An erotically aroused teen may or may not have engaged in same-sex sexual behavior. In fact, same-sex feelings and attractions almost always precede same-sex behavior.[58] Usually, erotic feelings appear in the early teens, although some adults report that they had same-sex fantasies and arousal as early as late preschool. Scott, an adolescent, recalls sexual interests at age six or seven. Approximately 70 percent of lesbians and 95 percent of gay men report same-sex arousal by age nineteen. In contrast, only 6 percent of heterosexual women and 20 percent of heterosexual men report same-sex arousal by this age.[59]

Most gay and bisexual male adolescents report that they initially believed that all boys felt as they did about other males.[60] Eventually they learned otherwise and the inner conflict of self versus society began, along with its accompanying confusion. Having defined himself along societal expectations, a youth may be concerned about the discrepancy that is now developing. An increasing doubt grows as both males and females become aware of their inability to fulfill heterosexual expectations. At age thirteen, Martin, an African-American student, was struck by the realization that his sexual feeling and being gay were one in the same. Chris, an eighteen-year-old male, adds, "I was under the impression that since I was gay, I wouldn't be able to do anything substantial with my life." [61]

Initial realization may be accompanied by intense anxiety and an identity crisis. Mike, a gay middle school student, spent hours in the counselor's office with vague school-related problems. He refused to face the sexual issue and recalls, "I was convinced they'd kick me out of school and send me to jail." [62] "[S]uddenly all the feelings . . . came together," recalls eighteen-year-old Joanne, "and pointed to the label, lesbian. As a result, I walked around like a shell shock victim for days." [63] "I was frightened," explains Linda. "Although I'd become somewhat comfortable with my label as 'the weird one,' lesbianism was *too* weird." [64] At age thirteen, Michelle kept a dream journal and she dreamt that she and a female friend kissed. "I immediately stopped keeping my journal," she recalls.[65] A sixteen-year-old Chinese-American Texan confides, "It was a total shock. . . . I simply could not accept myself. All the confidence I had in me disappeared." [66]

Positive self-regard and the plans for a bright future may appear to be lost as a teen recognizes that she or he is a member of a despised minority. Conflict can occur between a teen's positive self-esteem and her or his own internalized homophobia with its negative connotations.

Although this conflict easily resolves for some, others incorporate familial and societal values of the homosexual as sick, wrong, and undesirable, a member of a despised minority. Paul became aware of his attraction to other boys and to male TV stars at age eleven. "When the feelings did not go away I became distraught; the problem seemed beyond my control," he recalls. "I spent whole days crying alone in my room, and my family and friends didn't know what to make of me." [67] A fourteen-year-old bisexual girl reports, "I'd been thinking about it but I didn't want to believe what I was thinking." [68] She had first begun to be attracted to both boys and girls at age eleven. Having "bought into" society's negative values and social conditioning, queer kids may begin to hate themselves intensely.

Many gay and bisexual male youths report that sexual thoughts and feelings intrude on everything else. [69] They report being frightened by their awareness, threatened by the possibilities, and energized by the intensity of their feelings and the sense of the forbidden. "I knew I was clearly checking out the guys in the shower after soccer practice," a young gay man recalls. "This scared the shit out of me." [70] With little except negative feedback from home or school, these adolescents have no context within which to make sense of their feelings.

Myths and stereotypes of lesbians, gays, and bisexuals are a source of much of their confusion. A nineteen-year-old lesbian recalls, "I heard so many times, 'You look straight.' I thought that was stupid. . . . What's looking straight, what's looking gay?" [71] A young gay male or lesbian may try to fulfill the stereotype or be repelled by it. Lack of appropriate role models only exacerbates the situation.

The extreme loneliness of this period, as described by lesbian, gay, and bisexual adults, may be even more acute for younger adolescents who do not have the maturity to explore these complex feelings. This isolation may be made more harsh by a youth's active avoidance of other students suspected of being lesbian, gay, or bisexual. A queer kid can become the loneliest person in the high school. [72] Vulnerable and afraid of being revealed, a sexual-minority teen may be incapable of withstanding peer pressure to conform to a heterosexual standard that does not fit. Anti-homosexual jokes or ridicule are especially painful.

Youths may learn to hide their desires as wrong, believing that they will change or decrease. "Over the years, these tiny denials have a cumulative effect." [73] The youth may hate herself or himself for feeling a certain way but the sexual feelings continue to come. The result may be acting out, rebellion, dangerous sexual behavior, depression, and/or suicide.

In similar fashion, those youth who are very open about their sexual orientation may also experience isolation. Former "friends" may ridicule or stop associating. When Jill began to dress in a more "butch" manner, her phone fell silent as more and more girls she had considered friends avoided her. At school, she became an isolate.

The sense of isolation increases with the fear of discovery and rejection, especially for teens who belong to racial minorities. Many honestly believe that they are the only lesbian or gay student in their high school. [74]

Early recognition of sexual orientation by today's teens contrasts sharply with the lengthy process of self-acceptance and identification reported by many older lesbian, gay, and bisexual adults. Few resources or guides exist to facilitate this process for the young adolescent.

Coping Strategies

The three most common coping strategies for defending one's self against internalized and externalized homophobia, from least to most satisfactory for psychological adjustment, are repression of desires, suppression of homosexual impulses, and acceptance and disclosure to others.[75] Although each is discussed below as a distinct coping strategy, it is rare that distinctions are so clear in real life. Instead, the behavior of each lesbian, gay, and bisexual adolescent evolves to best serve the individual and the immediate environment.

Few youths enter treatment to change their sexual orientation unless the family makes this demand. Most attempt home remedies, especially ones that deny same-sex attractions and emphasize heterosexual roles.[76]

Repression of Desires

By repressing unacceptable or disconcerting desires, the lesbian or gay teen attempts to prevent these desires from entering her or his consciousness. Unfortunately, this strategy offers the youth no opportunity to integrate sexual desires and sexual identity. Behavior and identity also become disconnected. Repressed or hidden feelings may work for a while but eventually emerge, often unexpectedly, resulting in panic, coping disruption, and disorganization. Some repressed teens may display acting out behavior.[77] The most common forms of repression are rationalization, relegation to insignificance, and compartmentalization.

Rationalization.   The rationalization strategy uses the claim that the behavior was only for gratification and was a special case or situation.[78] Events are characterized as isolated incidents not to be repeated. Common rationalizations include:

> We were both drunk.
>
> We just got high and. . . .
>
> It's just a phase; I'll grow out of it.
>
> I just needed some money.
>
> I was just lonely.
>
> All guys do it once.

Relegation to insignificance.   The insignificance strategy can be summarized as "No big deal."[79] Common types include:

> It was just experimenting; so what.
>
> It was no big deal; we hardly touched.
>
> I was curious, that's all.
>
> It was just a favor for a friend, nothing more.
>
> He's the one who gave me head.

Compartmentalization.   In the compartmentalization strategy, sex and relation-ships become disconnected.[80] Sexual behavior is set aside as if it is unrelated to the person who engages in it. Common compartmentalization phrases include:

It just happened once.

It's just something that happened; I'm not like that.

I mess around; it don't mean nothing.

We're really good friends; that's it.

I love her, not all women.

Suppression of Homosexual Impulses

Unlike repression which tries to prevent same-sex desires, suppression tries to over-ride them. The result is a moratorium on development that merely delays positive same-sex sexual identification until age thirty or forty but does not "cure" an individ-ual.[81] Sexual orientation does not change. In general, the more heterosexual expe-rience a person has had, the older the age at which she or he self-identifies as lesbian or gay.[82]

Youths employing this coping strategy are heavily invested in "the big lie" or "the big secret." Real fear of exposure and/or rejection exists. On the incorrect assump-tion that homosexuals possess the behavioral characteristics of the other sex, youth may attempt to remedy the situation by accentuating gender-typical behavior. This compensation can be noted in an accentuated male swagger or in male bodybuilding or female interest in clothing and cosmetics. Spontaneity may be suppressed as the youth attempts to control all behaviors and agonizes about all uncertainties.

A teen may become sexually active with the other sex, to the point of pregnancy in some extreme cases among women.[83] Meanwhile, fear may prevent same-sex experimentation, sexual maturation, and exploration of intimate relationships by a youth suppressing same-sex feeling. Peter, age seventeen, did not begin to question his sexuality until age sixteen and attempted to suppress his feelings, thus delaying identification of his true sexual feelings.[84] Withholding and suppressing personal in-formation and interests in order to gain peer acceptance results in a false persona that is kept in place with vigilance and elaborate defenses.

As a result of delayed sexual identity, some individuals display adolescent-like behaviors when they finally come out even if well beyond adolescence. These might include intense but brief romantic involvements, frequent sexual experimentation, and overconcern for one's own physical appearance.[85]

Withdrawal to celibacy or asexuality.   Although relatively infrequent, a small per-centage of gay male youths report that puberty was sexless and that they were deeply involved in masculine activities such as sports.[86] These teens may look and act like other males and participate in male activities. They may crave male friendships deeply, but terminate or avoid close relations because of the temptations encoun-tered. Likewise, masturbation may be avoided because of same-sex erotic feelings and fantasies. Often these young adults deny that they had sexual feelings and attrac-tions as children.

Tran, who lettered in three sports, believed he was too busy for sex. Carlton competed in high school track, gymnastics, and swimming but had to be cautious with the feelings he could not totally suppress.

> I felt like I had to be ultra-careful in the dressing room. I couldn't let my eyes wander; I couldn't let anyone suspect the slightest thing. I found myself putting up more of a front in sports than anywhere else. I felt like it was such a proving ground—proving my manhood to my father, to the other guys, to myself.[87]

Tom had his paper route, football, Boy Scouts, and the band.

Females may also choose to remain asexual under the guise of saving themselves for marriage or for just the right man. These individuals may even avoid information pertaining to sexuality. Those who choose to remain asexual, whether male or female, may delay awareness and gradual self-acceptance.[88]

In some cases, individuals, particularly those from strict religious backgrounds, may use crossdressing or transexualism in an attempt to confront their true feelings of homoeroticism. Sometimes this behavior disappears when individuals become exposed to a gay peer group.[89] In some ethnic groups, especially among Hispanics, cross-dressing behavior is more common among both lesbian and gay youths.

Overcoming obstructionism requires overwhelming alternative information or a traumatic event that forces the individual to be honest with her or his feelings. It is difficult to help a youth with sexual issues when for the youth there is no issue.

Denial and heterosexual dating and sex.   "Passing" or "learning to hide" as a heterosexual is the most common adjustment.[90] Responding to peer and societal pressure, a teen may use heterosexual dating as an attempt to fit in or to change her or his sexual orientation. Approximately two-thirds of gay men and three-fourths of lesbians have engaged in heterosexual dating.[91] The increasing pressure on teens to date may have raised these values to 85 percent of today's youth.[92] These figures may be even higher in conservative areas, such as the South. Norma Jean, a poor, small town, Southern teen, felt she had no choice but to adopt a heterosexual persona. As for being an open lesbian, "It wasn't an option."[93] Vince was muscular, played football, made fag jokes, had a regular girlfriend, attended a fundamentalist church, and sang in an evangelical choir. The outward signs were a perfect cover. He tried to be "The Best Heterosexual of the Year." Eric, a white twenty-one-year-old, recalls, "Hiding became an art."[94]

A youth may try to control any mannerisms or dress that might be perceived as gender inappropriate by peers. Lesbian and gay adolescents monitor themselves:

Is my voice too high?

Did I appear too happy when she entered the room?

Am I standing too close?

Malcolm describes the process:

> I had to become more masculine. . . . I would make myself walk[,]. . . . I would make my voice sound[,]. . . . I would make myself sit a certain way. It was total insanity. I was not being me.[95]

Greg, a young African-American man, recalls, "Once when we were in Sunday school, the guys were ganging up on me about my eyelashes, saying they were entirely too long and beautiful for me to really be normal. I was so appalled . . . that I went home and cut my eyelashes off."[96] In addition to perpetuation of self-denial, this strategy can lead to problems of self-esteem because it fails to change sexual interests and desires.

Some teens attempt to cultivate a heterosexual role and to engage in antigay jokes and teasing. By teasing others suspected of similar feelings, a youth hopes to deflect suspicion from herself or himself.[97] In other words, the *best defense is a good offense*. Audrey told fag jokes and played pranks: "My social objective was to fit in," he explains, "and homosexuality definitely was not fitting in. I wanted to deny it in front of other people because I denied it to myself."[98] John, age sixteen, continues, "And the more nasty comments I made about gay people, the less gay I felt, like further and further away from this horrible thing."[99] "[O]nce I beat up a guy for being a faggot," confesses another young gay man. "No one suspected me [of being gay]," he adds, "because I did sports and had a girlfriend."[100]

Heterosexual girlfriends and boyfriends become a screen for hiding the true self and a wall between a youth and self-awareness. This wall enabled at least one gay youth to confide, "I guess I'm homosexual until I find the right woman."[101] Nathaniel, a middle-class African-American high schooler, explains, "When I got Delta [his girlfriend], it was like a cover. I was having sex with a man but for security I got a girlfriend. . . ."[102] As part of the cover, members of the other gender may be pursued as "sport" but with little serious intentions. Trying desperately to fit in, a young man explains, "It [pursuing women] gave me something to do to tell the other guys who were always bragging."[103] Monica acted "boy-crazy," although she had no sexual interest in boys. For Lee, a lesbian high schooler, the most difficult part was pretending when "I really didn't care."[104]

A lesbian or gay youth actually may believe that she or he is heterosexual but uncomfortable. Sex may fell unnatural, lacking an emotional component.[105] "I was just going through the motions," explains Kimberly. "It was expected of me, so I did it."[106] Drew reports that heterosexual dating anxiety results in nausea and diarrhea before each date.

Heterosexual relationships are often of short duration for gay and bisexual male youths but over half involve some sexual contact with young women.[107] Sex with one or two girls is often sufficient to satisfy curiosity. Gay and bisexual males express a preference for friendship over romance with females. As one adolescent explains, "I was disappointed because it [dating and sex] was such hard work—not physically, I mean, but emotionally."[108]

In similar fashion, lesbians may have two or three sexual contacts in the context of heterosexual dating. "I never really wanted to be intimate with any guy," explains Georgina, in a comment that echoes the sentiment of many lesbian teens. "I always wanted to be their best friend."[109]

Young women may engage in promiscuous heterosexual behavior in an attempt to make themselves heterosexual or to prove to themselves that they really are not lesbian.[110] Lisa, a young lesbian, states that she never enjoyed sex with boys but did it "to prove I wasn't gay."[111] This tactic can be summed up as *I can't be a lesbian; I have a boyfriend.*

The benefits of this coping strategy tend to be short-lived. A fourteen-year-old white gay male writes in his school paper,

> If there is anything I hate, it is having to be a fake. Unfortunately, I'm forced
> to do this every day of my life when I go to school.

"I fervently tried to take a more active interest in girls," notes Paul, "but I could tell that it was contrived." [112] Although she strenuously tried to be heterosexual in high school, Bonnie, a twenty-one-year-old bisexual, remembers, "I developed painful crushes on female teachers and straight girlfriends that left me feeling so pathetic and impotent." [113] Passing as a heterosexual negates a lesbian or gay youth's feelings and ultimately herself or himself. John, age nineteen recalls,

> I used to stand around with the guys and try to look interested in all their gas
> about this girl and that. . . . All the time I'd be thinking about one or the other
> of them. It seemed like I didn't belong. [114]

Fear of exposure becomes very real. Jack, age twenty-two and described as "straight-acting," recalls the panic at age thirteen or fourteen when "I saw a Bloomingdale ad for Calvin Klein underwear that I could not take my eyes off of." [115]

The overall result of this charade is psychological tension which may lead to depression, shame, fear of disclosure, and anxiety, although her or his surface demeanor may seem calm. [116] For example, covert gay college men experience more psychological tension, social problems, and isolation than openly gay men. [117] In short, those in hiding "have the most concern over self-esteem, self-acceptance, and status and feel the most social isolation, powerlessness, normlessness, and personal incompetence." [118] Paul, a gay college student, tried desperately as a teen to be heterosexual. "From seventh grade to tenth grade," he recalls, "while everyone's hormones were running rampant, I was attracted to no one; emotionally I was numb. I hadn't a clue about what it meant to be sexually attracted to someone." [119]

Healthy personalities develop when they are shared openly and honestly with others. Those in hiding have little opportunity to date or to develop same-sex relationships in a socially sanctioned context similar to that of heterosexual youth.

Redirection of energies into other areas (Compensation).   Some queer kids become too busy to bother with sex. [120] The student who is class president, yearbook editor, school play lead, and tennis team captain is too busy for any dating, heterosexual or otherwise. Although effeminate-acting and disinterested in girls, Jacob, an African-American high school overachiever, was never questioned by his family about his sexual orientation. "He don't have time for girls . . ." his family rationalized. "He's doing his books." [121] A positive correlation of both delayed same-sex erotic behavior and self-identification with better grades exists among gay men. [122] In other words, those with later behavior and self-identification have higher grades.

In similar fashion, in ninth grade, Malcolm, a young African-American male, became a religious true-believer, an indefatigable "pioneer" in the Jehovah's Witnesses. He would hide in his room for hours reading the Bible and proselytize door to door for over 100 hours a month. "If I prayed all the time and stayed active in the church," he reasoned, "maybe somehow I could appease God . . . and He wouldn't be so angry at me. . . ." [123] When this strategy failed, Malcolm devoted himself to his studies and became an honor student.

Self-Acceptance and Disclosure to Others ("Coming Out")

Accepting one's homosexuality is the optimal strategy.[124] Unfortunately, society encourages lesbian and gay teens to adopt a repressive or suppressive strategy. As a consequence, lesbian and gay youths become subject to psychological stresses that can affect their well-being.

Awareness of sexual orientation may encourage some adolescents to gain limited exposure to the larger lesbian, gay, and bisexual community. First steps may be very tentative, such as calling a hot line or attending a youth group. A teen may attempt to get to know another lesbian or gay youth or adult. For this youth, the goal is usually explorational, not sexual or relational.[125]

Accepting oneself need not always be a traumatic process. Michael, age sixteen, remarks:

> I can bring love and happiness to someone's life . . . Different is not bad. . . .[126]

Another teen, a cocky fourteen-year-old, adds that although the thought of being gay was frightening, the possible terror of living life as a "pseudo-heterosexual" was indescribable.

FIRST SEXUAL CONTACTS AND DATING

Same-sex erotic contact, such as body rubbing, manual genital contact, or oral genital contact, may begin around thirteen years for gay boys and fifteen for lesbian girls.[127] The mean age of first consensual orgasm is approximately sixteen years for today's young gay males, although some studies have placed the age much earlier.[128] Data vary with the definition of "first sexual contact" and with the age of both the subjects and the study itself. It is important to remember that many children, who, as adults, do not identify as lesbian or gay, engage in adolescent same-sex erotic behavior, and that such behavior is not a cause of homosexuality.

Frequently, the first same-sex kiss, whether sexual in nature or not, is significant. Nancy describes the "incredible feeling" when she and Connie brushed their lips together lightly for the first time. Neither Dan nor Steven, swim team buddies, said anything as they turned from their homework toward each other. "I remember thinking, oh God, don't let me be wrong; don't let me be wrong about this one thing in my entire life," Dan recalls.[129] The unplanned kiss was like a rush of air through his body. Others describe a feeling of "being home at last."

Both boys and girls engage in same-sex erotic behavior prior to self-identifying but the time between these two events differs greatly from about two-and-a-half years for males to four months on average for females.[130] Approximately two-thirds of gay male youths who reported that their sexuality felt natural very early welcomed puberty as a link between their attractions and sexual behavior. "Until then," reports a male student, "I had felt I could never fall in love, that I had no sexual feelings. . . ."[131] Often, there is an "ah-ha" or "eureka" when sexual arousal, imagery, fantasy, romantic notions, and sexual behavior come together. According to Linda, a young adult, she had "an undeniable feeling that this was . . . best for me."[132] A young gay man explains, "I didn't feel like I was cheating on Beth [his girlfriend] because the sex felt so different, so right." [133] John, a high school student continues, "I felt everything that I hadn't felt with a woman." [134] Although some lesbian, gay, and bisexual youths will still persist in their notion that they are similar sexually to others of their gender, they have great difficulty denying their attractions, and most explore and experiment with same-sex behaviors.

For some youths, same-sex erotic behavior is accompanied by shame and guilt. For example, although he had been engaging in sexual behavior for at least four years, Mike continued to promise not to transgress again and to ask God's help in making him stop. Those males with a history of rejection may begin to fear that such rejection will continue. [135] Same-sex sexual encounters are one more thing for which they may be ostracized by their peers. When Rob, age thirteen, agreed to perform oral sex on a very handsome athlete in his junior high, he had no idea that the ensuing "trap" would reveal his homosexuality to many of the boys in his scout troop and result in extreme embarrassment and isolation.

Lesbian and gay youths are more likely to engage in early sexual behavior of all types than are heterosexuals. As a group, lesbian and gay adults report earlier and more frequent same-sex contacts than do heterosexuals and more adolescent sexual behavior involving both sexes. [136] These trends are most evident among gay male youth. [137] Almost two-thirds of gay men masturbated with another boy before adolescence, more than double the figure reported by nongay men. [138]

In general, gay males behave sexually much like heterosexual males. Both are more likely than females to give in to sexual urges and to have serial partners. Unlike their heterosexual counterparts, however, gay males don't have socially sanctioned opportunities to learn a dating "script." Freed from the constraints that women place on male sexual assertiveness, however, gay males may experience a series of sexual encounters with little emotional attachment. In contrast, women, regardless of their sexual orientation, tend to value intimacy and attachment in their intimate relationships with others.

Sex role behaviors may become evident within same-sex relationships although they are not unique to lesbian and gay couples. [139] Male relationships may lack intimacy if both partners exhibit male sex role characteristics of competitiveness and independence. In contrast, female relationships may be very intimate while lacking individuality and autonomy. These tendencies can be counterbalanced over time with practice and patience.

As many as 70 percent of lesbian and gay teens may have had some *heterosexual* experience by adulthood. [140] In contrast, over 30 percent of self-identified lesbian and gay youths have no *same*-sex experience. [141] Heterosexual erotic behavior may continue despite a personal recognition of a lesbian, gay, or bisexual orientation by an individual. For some sexual-minority teens, heterosexual activity is part of their "cover."

In general, lesbians are more likely to continue to engage in heterosexual behaviors and relationships even after they come out to themselves. This difference reflects the positive relationship between self-worth and *success-with-boyfriends* found among junior and senior high school girls.

Emotionally vulnerable youth, especially males, may make sexual contact with other gay teens or adults under less than ideal conditions. Unfortunately, these furtive, solely sexual contacts may set a pattern for interactions with other guys. Dangerous settings, such as parks or rest areas after sunset, and the exploitive nature of the sexual encounters reinforce the worst gay stereotypes and may be internalized by a teen. For those queer kids still denying their feelings, such sexual encounters allow them to participate in sexual behavior devoid of an affectional component.

On the other hand, a naive youth may also misinterpret the intense feelings aroused by sex, the supportive environment, and physical affection as romantic love. Joseph, a gay student, concluded, "There's no simple answers; feelings are more important than sex." [142] Subsequent nonfulfillment of these feelings and the indifference of the other participant may convince the youth that love and intimacy are not possible within same-sex relationships. [143]

Other youths may welcome a mutual emotional commitment, nurture it, and develop mature feelings. Jacob, an African-American high school junior, met Warren, a senior, in a community choir. Jacob recalls,

> [W]e started to have sex. It started to be an emotional thing. He got to the point of telling me he loved me. That was the first time anybody ever said anything like that. It was kind of hard to believe that *even after sex* there are really feelings. We became good friends. [144]

For Christopher, age nineteen, expressing his sexual orientation for the first time confirmed and validated his feelings.

> Having repressed my sexuality for so long, it was an amazing experience to physically express it . . . It was . . . a giant step toward accepting who I am. [145]

For still others, sex and love may come together and be terrifying. "And that [emotion] scared the shit out of me . . ." confides a young gay man. "As I moved into eighth grade, it became more and more clear to me what it was all about, and it was sex, but with a twist, romance." [146] Still, it is important to remember that only 15 percent of self-identified lesbian teens and 25 percent of gay teens report negative feelings regarding their first same-sex erotic activity. [147]

Same-sex experiences are often the product of crushes, hero-worship and/or intimate friendships, especially for heterosexual youth. [148] Jeff's first experience at age twelve occurred at the urging of a high school neighbor. Another young man explains,

> Derek was my best friend. After soccer practice the fall of our junior year we celebrated both making the "A" team by getting really drunk. We were just fooling around and suddenly our pants were off. . . . I was so scared I stayed out of school for three days. . . . [149]

Beyond enjoyment, these erotic behaviors serve an information-gathering and comparison, reassurance, or experimentation function for queer kids. [150]

Among all males, the reported differences in the percentage of individuals engaging in same sex activity varies, especially with increasing age (Table 1). As would be expected, gay youths increasingly engage in same-sex sexual behavior while nongay youth demonstrate an opposite trend. [151]

Possibly as high as 60 percent of *all* pre*p*ubescent males and 33 percent of *all* pre*p*ubescent girls have some type of same-sex experience, while 33 percent and 17 percent, respectively, experience postpubescent orgasm with a same-sex partner. [152]

TABLE 1   Percentage of Males Engaging in Same-Sex Experiences

	Sexual Orientation	
Age Level	Nongay	Gay
During childhood (ages 5–9 yrs.)	5	40
During preadolescence (ages 10–12 yrs.)	25	60
During adolescence (ages 13–17 yrs.)	15	70
During early adulthood (ages 18–24 yrs.)	5	95

Adapted from Savin-Williams and Lenhart, 1990.

Although other studies have reported lower percentages, the values are still higher than the percentage who will later identify as lesbian or gay.[153] Matthew, a student, observed, "Everyone is a bit gay, especially when you're young."[154] Only 2 percent of heterosexual men and almost no heterosexual women report predominantly homosexual activity as adolescents. In contrast 56 percent of gay men and 41 percent of adult lesbians had engaged in predominantly homosexual behavior.[155] The majority of heterosexual youth who engage in same-sex erotic behavior do not continue to do so in adulthood.

The quality of same-sex erotic behavior appears to differ among adolescents based on sexual orientation. Same-sex sexual activity by lesbians and gays may be more emotional, more planned, and less playful than same-sex activity among heterosexuals.[156] Lesbian and gay teens report a strong affectional component and feelings of love and desire. In contrast, the heterosexual activity by lesbian and gay youths may have an experimental quality.

Among gay and bisexual male youths, there is a positive relationship between the age of onset of puberty and the age of first same-sex erotic activity.[157] In contrast, the age of first heterosexual activity among the same youths is more closely related to chronological age. Heterosexual encounters seem to begin around age fifteen regardless of the age of onset of puberty.

Most lesbian and gay youths state that they want committed same-sex relationships.[158] The more a youth has engaged in same-sex erotic behavior, the more she or he desires a romantic relationship. In our culture, the process of dating and forming relationships is an important developmental experience through which we define ourselves and gain self-confidence. Falling in love and forming long-term romantic relationships is related to higher self-esteem and self-acceptance.

Although societal supports are nonexistent, 90 percent of lesbians and nearly 70 percent of gay youth between the ages of fourteen and twenty-three have been involved in at least one same-sex relationship. Unfortunately, these are often secretive, short-lived, and covert.[159] Social pressures make it difficult to form and maintain same-sex romantic relationships. "[L]ove of a woman," explains Diane, "was never a possibility that I even realized could be."[160] "[L]ove was something I watched other [heterosexual] people experience and enjoy," continues Lawrence. "I was expected to be part of a world with which I had nothing in common."[161] John, age sixteen, adds, "I've gone through my whole life not getting to know about relationships, not learning about any of this stuff."[162] Vic, a college-age lesbian, concludes,

> I think that whole straight ceremony of dating is not accessible to queer youth.
> Dating implies a certain amount of choice, the freedom to pick and choose.
> Well, pick and choose from who? A bunch of people who have to flash each
> other secret signals to be recognized.[163]

SELF-IDENTIFICATION

Self-identification usually occurs in late adolescence between the ages of sixteen
and twenty-one.[164] The mean age for openly gay young adult males to self-identify
is 16.2 years, although some self-identifying occurs as early as age 14.[165] As some
segments of society become more open, self-identification is occurring even earlier,[166]
especially in large urban areas. There is much individual variation. Self-identification
does not imply self-acceptance which is dependent on many things, including romantic relationships and acceptance by others.

The relative lateness of sexual identification for lesbians and gays as compared
to heterosexuals can be attributed to homophobia, discrimination, and societal prohibitions. No doubt, confusion and fear account for the lapse of time between first
becoming aware of erotic feelings and labeling them. Sixteen-year-old Andy was
caught between self-hatred for cowardly remaining in the closet and the threat of harassment should he come out.

Although Victor, a seventeen-year-old bisexual, reminds us that, "Bisexuals are
not waiting to make up their minds about their sexual identity," [167] for many *youths,*
self-identification as bisexual is a rest stop on the way to claiming a lesbian or gay
identity. Bisexuality seems more accepted and maintains a link to the heterosexual majority. More African Americans than European Americans choose this route
because African-American communities seem to be more tolerant of bisexuality.[168]
African-American teens may adopt a less heterosexual lifestyle than their white peers
while being less willing to identify as lesbian or gay.

Initial feelings may gradually evolve into a sense of relief, of well-being, and
"rightness." Naomi, age twenty, explains,

> [T]his feeling was so natural, I guess I trusted my own feelings enough not to
> believe anybody's negative ones. Because I felt that if what I'm doing is what
> they're saying is sick and bad, well they [the critics] must be sick and bad.[169]

This conscious recognition of sexuality is the beginning of self-acceptance.[170] A lesbian or gay teen may express self-anger at her or his earlier pattern of conformity.

Self-identification can be viewed as a two-step process in which a youth first accepts her or his sexual identity and then integrates that identity with her or his personality and self-concept. A twenty-year-old female student describes how it feels to
be an integrated whole:

> I feel that I am the terrific person I am today because I'm a lesbian. I decided
> that I was gay when I was very young. After making that decision, which was
> the hardest thing I could ever face, I feel like I can do anything.[171]

Through acceptance, an adolescent begins to view the notion of "lesbianness," "gayness," or bisexuality in a positive way. "It's a real love and trust of women, and respect," states Brenda, age seventeen. "It's something inside me that I can't explain." [172]

During this phase, a youth may make an initial disclosure or "come out" to a very trusted friend or family member. Disclosure seems very important for positive self-identification as lesbian or gay.[173]

Acceptance is followed by integration in which a young person identifies as lesbian or gay and proud. "My soul feels more comfortable," explains Shannon, age seventeen. "It feels right."[174] Integration and pride often are accompanied by a public self-disclosure known as "Coming out of the closet" or simply as "Coming out." Coming out is an adjusting between the real and the social self and as such, is a necessary process for healthy personality integration. Being known as lesbian or gay is an important step in identity formation.[175] Early self-disclosure seems to be related positively to high self-esteem. Michael, a student, explains the process:

> Contrary to . . . opinion, I didn't wake up one morning and say, "Gee, I think I'll be gay for the rest of my life. That'll be fun." Why would anyone choose a life filled with discrimination? The only choice I made was to come out of the closet.[176]

Coming out and the often accompanying anger and pride are important for identity stabilization.[177] Integration of sexual identity is a lifelong process.

Integration is enhanced through interaction with other lesbian and gay adults and involvement in the sexual-minority community.[178] Typically, youths come to prefer social interactions with other lesbians and/or gays. Self-identification with others becomes positive.

When compared to those still in sexual-identity turmoil, the well-integrated lesbian or gay male may have higher self-esteem and greater well-being, a greater capacity for and more confidence in love both in sexual relationships and in friendships, and increased productivity.[179] More energy is spent in living an open life and less on trying to hide it. Honest heterosexuals occasionally will express awe for the lesbian or gay adult who has come through this process and now knows who she or he is and faces the world confident and unafraid. The battle with homophobia and heterosexism and with their own internalized "demons" can leave the winners proud to proclaim "I am who I am!"

Unfortunately, not everyone can accept her or his own lesbian, gay, or bisexual identity. The road to a well-integrated, healthy, positive, cohesive identity is strewed with those who can not really accept themselves and continue to use all manner of subterfuge to hide their true identity.

CONCLUSION

Sexual self-identification is part of the larger process of adolescent development, a confusing, sometimes contradictory, process in itself. For many teens their sexuality is ambiguous and not clearly delineated. They may begin to understand their feelings but not to clarify them. All the while these processes are occurring within an atmosphere of adolescent conformity and burgeoning independence. The adolescent question "Who am I?" becomes entangled with a second question "What does it mean to be lesbian or gay?"[180]

Most homosexuals accept their sexual orientation and "lead successful, productive, nonneurotic lives as self-acknowledged gay men and lesbians."[181] More than half would not change their sexual orientation if such were possible, despite the negative attitudes of the larger society.[182] The process of becoming has its own rewards. Questioning and exploring can make an individual more sensitive to difference and more accepting of it. Every issue from coming out to having children is there to be explored, discussed, and decided. Erna, a young Navajo woman, concludes that, as a result, "[W]e are special, because we're able to deal with . . . life."[183]

Through the process of "becoming" lesbian, gay, or bisexual, the individual's range of possibilities increases, and the coping skills acquired to transcend adversity give an individual the ability to find fulfillment. "I have learned a lot more about myself," states Rachel, a sixteen-year-old Midwesterner. "If I had a choice, I wouldn't change my sexual orientation."[184]

NOTES

1. G. J. McDonald, "Individual differences in the coming-out process for gay men: Implications for theoretical models," *Journal of Homosexuality,* (Volume 8, 1982). P. K. Rector, "The acceptance of homosexual identity in adolescence: A phenomenological study," Dissertation Abstracts International, (Volume 43, 1982).

2. G. J. Remafedi, M. Resnick, R. Blum, and L. Harris, L., "Demography of sexual orientation in adolescents," *Pediatrics,* (Volume 89, 1992).

3. Dailey, 1979; Peplau, Cochran, Rook, and Pedesky, 1978. M. T. Saghir and E. Robins, "Clinical aspects of female homosexuality." In J. Marmor, Ed., *Homosexual behavior: A modern reappraisal.* (New York: Basic Books, 1980).

4. M. Kirkpatrick and C. Morgan, "Psychodynamic psychotherapy of female homosexuality." In J. Marmor, Ed., *Homosexual behavior: A modern reappraisal.* (New York: Basic Books, 1981), p. 372–373.

5. R. R. Troiden, "Homosexual identity formation," *Journal of Adolescent Health Care,* (Volume 9, 1988).

6. A. P. Bell, M. S. Weinberg, and S. K. Hammersmith, *Sexual preference: Its development in men and women.* (Bloomington, IN: Indiana University Press, 1981a). R. C. Savin-Williams, "Memories of childhood and early adolescent sexual feelings among gay and bisexual boys: A narrative approach." In R. C. Savin-Williams and K. M. Cohen, Eds., *The lives of lesbians, gays, and bisexuals.* (Fort Worth, TX: Harcourt & Brace College Publishers, 1996c). R. R. Troiden, "Becoming homosexual: A model of gay identity acquisition," *Psychiatry,* (Volume 42, 1979).

7. A. Heron, Ed., *One teenager in ten.* (Boston, MA: Alyson, 1983), p. 115.

8. Durby, 1994.

9. L. Heal, "It happened on Main Street." In B. L. Singer, Ed., *Growing up gay/growing up lesbian.* (New York: The New Press, 1994), p. 9.

10. P. J. Gorton, "Different from the others." In R. Luczak, Ed., *Eyes of desire: A deaf gay and lesbian reader.* (Boston, Alyson, 1993), p. 20.

11. Bell, Weinberg, and Hammersmith, 1981a.

12. R. Green, "Gender identity in childhood and later sexual orientation," *American Journal of Psychiatry,* (Volume 142, 1985). G. B. MacDonald, "Exploring sexual identity:

Gay people and their families," *Sex Education Coalition News,* (Volume 5, 1983). R. Robertson, "Young gays." In J. Hart and J. Richardson, Eds., *The theory and practice of homosexuality.* (New York: Routledge & Kegan Paul, 1981). B. Zuger, "Early effeminate behavior in boys," *Journal of Nervous and Mental Disorders,* (Volume 172, 1984). R. C. Savin-Williams, *Gay and lesbian youth: Expressions of identity.* (Washington, DC: Hemisphere, 1990). R. C. Savin-Williams, "Ethnic- and sexual-minority youth." In R. C. Savin-Williams and K. M. Cohen, Eds., *The lives of lesbians, gays, and bisexuals.* (Fort Worth, TX: Harcourt & Brace College Publishers, 1996b).

13. K. Chandler, *Passages of pride: Gay and lesbian youth come of age.* (New York: Times Books, 1995).

14. Savin-Williams, 1996c, p. 100.

15. Ibid.

16. Bell, Weinberg, and Hammersmith, 1981a. R. C. Friedman and L. O. Stern, "Juvenile aggressivity and sissiness in homosexual and heterosexual males," *Journal of the Academy of Psychoanalysis,* (Volume 8, 1980). R. Green, "Childhood cross-gender behavior and subsequent sexual preference," *American Journal of Psychiatry,* (Volume 36, 1979).

17. R. R. Troiden, "Becoming homosexual: A model of gay identity acquisition," *Psychiatry,* (Volume 42, 1979), p. 363.

18. Bell, Weinberg, and Hammersmith, 1981a, p. 74.

19. M. T. Saghir and E. Robins, *Male and female homosexuality.* (Baltimore, MD: Williams & Wilkins, 1973).

20. I. Bieber, *Homosexuality: A psychoanalytic study.* (New York: Basic Books, 1962). M. Eisner, *An investigation of the coming-out process, lifestyle, and sex-role orientation of lesbians.* Unpublished doctoral dissertation, York University, Toronto (1982). R. B. Evans, "Childhood parental relationships of homosexual men," *Journal of Consulting and Clinical Psychology,* (Volume 33, 1969). E. A. McCauley and A. A. Ehrhardt, "Role expectations and definitions: A comparison of female transsexuals and lesbians," *Journal of Homosexuality,* (Volume 3, 1977). C. Van Cleave, "Self-identification, self-identification discrepancy, an environmental perspective of women with same-sex preference." *Dissertation Abstracts International,* (Volume 38, 1978).

21. Bell, Weinberg, and Hammersmith, 1981a, p. 148.

22. S. O. Murray, *Social theory, homosexual realities.* (New York: Gai Sabre Books, 1984). R. Parker, "Youth, identity, and homosexuality: The changing shape of sexual life in contemporary Brazil." In G. Herdt, Ed., *Gay and lesbian youth.* (Binghamton, NY: Harrington Park Press, 1989).

23. R. Green, *The "sissy boy syndrome" and the development of homosexuality.* (New Haven, CT: Yale University Press, 1987). Whitan, 1983.

24. J. T. Sears, *Growing up gay in the South: Race, gender, and journeys of the spirit.* (Binghamton, NY: Harrington Park Press, 1991a), p. 50.

25. R. C. Savin-Williams and R. E. Lenhart, "AIDS prevention among gay and lesbian youth: Psychosocial stress and health care intervention guidelines." In D. G. Ostrow, Ed., *Behavioral aspects of AIDS.* (New York: Plenum Medical Book Co., 1990).

26. Sears, 1991a, p. 49.

27. Savin-Williams, 1996c.

28. E. Coleman, "Developmental stages of the coming-out process," *Journal of Homosexuality,* (Volume 7, 1982b).

29. Sears, 1991a, p. 49.

30. K. Jennings, "American dreams." In B. L. Singer, Ed., *Growing up gay/growing up lesbian.* (New York: The New Press, 1994).

31. Sanders, 1980. C. J. Straver, "Research on homosexuality in the Netherlands," *The Netherlands' Journal of Sociology,* (Volume 12, 1976).

32. E. White, *A boy's own story.* (New York: E. P. Dutton, 1982), p. 169.

33. A. M. Boxer, *Betwixt and between: Developmental discontinuities of gay and lesbian youth.* Paper presented at the Society for Research on Adolescence, Alexandria, VA (1988, March).

34. E. Bass and K. Kaufman, *Free your mind: The book for gay, lesbian, and bisexual youth — and their allies.* (New York: Harper Perennial, 1996), p. 24.

35. Coleman, 1982b. K. Jay and A. Young, *The gay report: Lesbians and gay men speak out about sexual experiences and lifestyles.* (New York: Simon & Schuster, 1979). A. C. Kinsey, W. B. Pomeroy, and C. E. Martin, *Sexual behavior in the human male.* (Philadelphia, PA: W. B. Saunders, 1948). H. Kooden, S. Morin, D. Riddle, M. Rogers, B. Sang, and F. Strassburger, *Removing the stigma. Final Report of the Task Force on the Status of Lesbian and Gay Male Psychologists.* (Washington, DC: American Psychological Association, 1979). MacDonald, 1983; Rector, 1982. R. A. Rodriguez, *Significant events in gay identity development: Gay men in Utah.* Paper presented at the annual meeting of the American Psychological Association, Atlanta, GA (1988, August). G. Sanders, "Homosexuals in the Netherlands," *Alternative Lifestyles,* (Volume 3, 1980). J. Spada, *The Spada Report: The newest survey of gay male sexuality.* (New York: New American Library, 1979). Troiden, 1979.

36. Heal, 1994.

37. G. J. Remafedi, "Male homosexuality: The adolescent's perspective," *Pediatrics,* (Volume 79, 1987b).

38. S. K. Telljohann and J. H. Price, "A qualitative examination of adolescent homosexuals' life experiences: Ramifications for secondary school personnel," *Journal of Homosexuality,* (Volume 26, 1993).

39. Savin-Williams, 1996c, p. 97.

40. Sears, 1991a, p. 112.

41. Remafedi, 1987b, p. 328.

42. J. T. Sears, "Black-gay or gay-black: Choosing identities and identifying choices." In G. Unks, Ed., *The gay teen.* (New York: Routledge, 1995), p. 142.

43. Chandler, 1995, p. 6.

44. Sears, 1991a, p. 352.

45. P. Singer, "Breaking through," Rochester, NY, *Democrat and Chronicle,* (1993, July 4), p. D-1.

46. Sears, 1991a, p. 84.

47. J. Green, "This school is out," *The New York Times Magazine,* (1991, October 13), p. 18.

48. Savin-Williams, 1996c.

49. Ibid., p. 100.

50. Chandler, 1995, p. 16.

51. Savin-Williams, 1996c, p. 98.

52. D. Kopay and P. D. Young, *The David Kopay Story: An extraordinary self-revelation.* (New York: Donald Fine, 1988).

53. "Starting over: A Chinese teenager comes to a new home and comes out," *Crossroads,* (1996, Winter/Spring), p. 6.

54. Sears, 1991a, p. 205.

55. P. Gibson, "Report of the Secretary's Task Force on Youth Suicide." In M. Feinleib, Ed., *Prevention and intervention in youth suicide.* Washington, DC: U.S. Department of Health and Human Services, Public Health Services; Alcohol, Drug Abuse and Mental Health Administration, (1989), pp. 3–131.

56. Chandler, 1995, p. 82.

57. Sears, 1991a. p. 290.

58. Boxer, 1988. A. H. Buss, *Self-consciousness and social anxiety.* (San Francisco, CA: W. H. Freeman, 1980). McDonald, 1982; Remafedi, 1987b. T. Roesler and R. Deisher, "Youthful male homosexuality," *Journal of the American Medical Association,* (Volume 219, 1972). Savin-Williams, 1990.

59. Bell, Weinberg, and Hammersmith, 1981a, 1981b.

60. Savin-Williams, 1996c.

61. S. Maguen, "Gay rural youth lack support from the community," *The Advocate,* (1992, November 17), p. 54.

62. Singer, 1993, p. D-1.

63. Heron, 1983, pp. 9–10.

64. Heal, 1994, p. 10.

65. Bass and Kaufman, 1996, p. 17.

66. "Starting over," 1996, p. 6.

67. P. D. Toth, "Realizing it's OK to be gay." Rochester, NY, *Times Union,* (1993, October 12), p. 15.

68. S. Parsavand, "Discussion groups ease acceptance for gay high schoolers," *Schenectady Gazette,* (1993, December 5), p. A-1.

69. D. A. Anderson, "Family and peer relations of gay adolescents." In S. C. Geinstein, Ed., *Adolescent psychiatry: Developmental and clinical studies,* Volume 14. (Chicago: The University of Chicago Press, 1987). Savin-Williams, 1996c.

70. R. C. Savin-Williams, "Dating and romantic relations among gay, lesbian, and bisexual youths." In R. C. Savin-Williams and K. M. Cohen, Eds., *The lives of lesbians, gays, and bisexuals.* (Forth Worth, TX: Harcourt & Brace College Publishers, 1996a), p. 170.

71. M. Schneider, "Sappho was a right-on adolescent: Growing up lesbian," *Journal of Homosexuality,* (Volume 17, 1989), p. 118.

72. J. L. Norton, "The homosexual and counseling," *Personnel and Guidance Journal,* (Volume 54, 1976).

73. R. Fisher, *The gay mystique: The myth and reality of male homosexuality.* (New York: Stein & Day, 1972), p. 249.

74. B. L. Singer, Ed., *Growing up gay/growing up lesbian.* (New York: The New Press, 1994).

75. Bell, Weinberg, and Hammersmith, 1981a. V. C. Cass, "Homosexual identity formation: Testing a theoretical model," *Journal of Homosexuality,* (Volume 4, 1979). J. A. Lee, "Going public: A study of the sociology of homosexual liberation," *Journal of Homosexuality,* (Volume 3, 1977). A. K. Malyon, "The homosexual adolescent: Developing issues and social bias," *Child Welfare,* (Volume 60, 1981). A. D. Martin, "Learning to hide: The socialization of the gay adolescent," *Adolescent Psychiatry,* (Volume 10, 1982). K. Plummer, *The making of the modern homosexual.* (London: Hutchinson, 1981). Reiche and Dannecker, 1977; Sanders, 1980; Troiden, 1979.

76. Anderson, 1987.

77. G. P. Mallon, "Gay and no place to go: Assessing the needs of gay and lesbian adolescents in out-of-home settings," *Child Welfare,* (Volume 71, 1992a).

78. Cass, 1979; Plummer, 1981. R. C. Savin-Williams and R. G. Rodriguez, "A developmental, clinical perspective on lesbian, gay male, and bisexual youths." In T. P. Gulotta, G. R. Adams, and R. Montemayor, Eds., *Adolescent sexuality.* (Newbury Park, CA: Sage, 1993). C. A. Tripp, *The homosexual matrix.* (New York: McGraw-Hill, 1975). Troiden, 1979.

79. Cass, 1979; Reiche and Dannecker, 1977; Savin-Williams and Rodriguez, 1993; Troiden, 1979.

80. C. DeMontflores and S. J. Schultz, "Coming out: Similarities and differences for lesbians and gay men," *Journal of Social Issues,* (Volume 34, 1978). Malyon, 1981; Martin, 1982; Tripp, 1975.

81. Malyon, 1981.

82. Troident and Goode, 1980.

83. D. A. Anderson, "Lesbian and gay adolescents: Social and developmental considerations." In G. Unks, Ed., *The gay teen.* (New York: Routledge, 1995).

84. Savin-Williams, 1996c.

85. Ross-Reynolds, 1988.

86. Cass, 19779; Lee, 1977. G. Sanders, 1980.

87. Sears, 1991a, p. 194.

88. Savin-Williams, 1996c.

89. E. S. Hetrick and A. D. Martin, "Developmental issues and their resolution for gay and lesbian adolescents," *Journal of Homosexuality,* (Volume 14, 1987).

90. Bell, Weinberg, and Hammersmith, 1981; Cass, 1979; Martin, 1982; Lee, 1977. R. Reich and M. Dannecker, "Male homosexuality in West Germany—A sociological investigation," *Journal of Sex Research,* (Volume 13, 1977). G. Sanders, 1980. M. S. Weinberg and C. J. Williams, *Male homosexualities: Their Problems and adaptations.* (New York: Penguin, 1974).

91. M. S. Weinberg and C. J. Williams, *Male homosexualities: A study of diversity among men and women.* (New York: Simon & Shuster, 1978). S. Schafer, "Sexual and social problems of lesbians," *Journal of Sex Research,* (Volume 12, 1976). Spada, 1979. R. R. Troiden and E. Goode, "Variables related to acquisition of gay identity," *Journal of Homosexuality,* (Volume 5, 1980). Weinberg and Williams, 1974.

92. B. S. Newman and P. G. Muzzonigro, "The effects of traditional family values on the coming out process of gay male adolescents," *Adolescence,* (Volume 28, 1993).

93. Sears, 1991a, p. 82.

94. L. D. Brimmer, *Being different: Lambda youth speak out.* (New York Franklin Watts, 1995), p. 37.

95. Sears, 1991a, p. 52.

96. L. Due, *Joining the Tribe,* (New York: Anchor Books, 1995), p. 242.

97. C. A. Rigg, "Homosexuality and adolescence," *Pediatric Annual,* (Volume 11, 1982).

98. Sears, 1991a, p. 353.

99. Due, 1995, p. 77.

100. K. M. Cohen and R. C. Savin-Williams, "Developmental perspectives on coming out to self and others. In R. C. Savin-Williams and K. M. Cohen, Eds., *The lives of lesbians, gays, and bisexuals.* (Fort Worth, TX: Harcourt & Brace College Publishers, 1996), p. 126.

101. D. Boyer, "Male prostitution and homosexual identity." In G. Herdt, Ed., *Gay and Lesbian Youth.* (Binghamton, NY: Harrington Park Press, 1989), p. 169.

102. Sears, 1991a, p. 132.

103. Savin-Williams, 1996a, p. 173.

104. Due, 1995, p. 108.

105. G. Herdt and A. M. Boxer, *Children of horizons: How gay and lesbian teens are leading a new way out of the closet.* (Boston: Beacon Press, 1993).

106. Sears, 1991a, p. 327.

107. Herdt and Boxer, 1993. G. J. Remafedi, "Adolescent homosexuality: Psycho-social and medical implications," *Pediatrics,* (Volume 79, 1987a). Remafedi, 1987b; Roesler and Deisher, 1972; Savin-Williams, 1990; Sears, 1991a.

108. Savin-Williams, 1996b, p. 172.

109. Sears, 1991a, p. 327.

110. M. S. Schneider and B. Tremble, "Training service providers to work with gay and lesbian adolescents: A workshop," *Journal of Counseling and Development,* (Volume 65, 1986).

111. Heron, 1983, p. 76.

112. Toth, 1993, October 12, p. 15.

113. Bass and Kaufman, 1996, p. 33.

114. J. Gover, "Gay youth in the family," *Journal of Emotional and Behavioral Problems,* (Volume 2, Number 4, 1993), p. 36.

115. Savin-Williams, 1996c, p. 105.

116. Weisberg and Williams, 1974.

117. L. J. Braaten and C. D. Darling, "Overt and covert homosexual problems among male college students," *Genetic Psychology Monographs,* (Volume 71, 1965).

118. F. L. Myrick, "Homosexual types: An empirical investigation," *Journal of Sex Research,* (Volume 10, 1974a), p. 234.

119. Toth, 1993, October 12, p. 15.

120. G. Sanders, 1980.

121. Sears, 1995, p. 140.

122. J. Harry, "Adolescent sexuality: Masculinity-femininity, and educational attainment," ERIC Document No. 237395, (1983a).

123. Sears, 1991a, p. 52.

124. Martin, 1982.

125. Savin-Williams, 1990.

126. Michael, "Different is not bad." In B. L. Singer, Ed., *Growing up gay/growing up lesbian.* (New York: The New Press, 1994), p. 60.

127. J. A. Cook, A. M. Boxer, and G. Herdt, *First homosexual and heterosexual experiences reported by gay and lesbian youth in an urban community.* Paper presented at the annual meeting of the American Sociological Association. San Francisco, CA, (1989).

128. S. M. Brady, "The relationship between differences in stages of homosexual identity formation and background characteristics, psychological well-being and homosexual adjustment," *Dissertation Abstracts International,* (Volume 45, 1985). Rodriguez, 1988.

129. Bass and Kaufman, 1996, p. 92.

130. Sears, 1991a.

131. R. A. Isay, "The development of sexual identity in homosexual men." In S. I. Greenspan and G. H. Pollack, Eds., *The course of life: Volume IV, Adolescence.* (Medison, CT: International Universities Press, 1991), p. 477.

132. Heal, 1994, p. 12.

133. Savin-Williams, 1996a, p. 175.

134. Due, 1995, p. 74.

135. Savin-Williams, 1996c.

136. M. Manosevitz, "Early sexual behavior in adult homosexual and heterosexual males," *Journal of Abnormal Psychology,* (Volume 76, 1970).

137. DeMonteflores and Schultz, 1978; Shafer, 1977; Remafedi, 1987a.

138. Saghir and Robins, 1973.

139. J. C. Gonsiorek, "Mental health issues of gay and lesbian adolescents," *Journal of Adolescent Health Care,* (Volume 9, 1988).

140. Cook, Boxer, and Herdt, 1989.

141. G. Ross-Reynolds, "Issues in counseling the 'homosexual' adolescent." In J. Grimes, Ed., *Psychological approaches to problems of children and adolescents.* (Des Moines, IA: Iowa Department of Education, 1982).

142. M. Mac an Ghaill, "Schooling, sexuality and male power: Towards an emancipatory curriculum," *Gender and Education,* (Volume 3, 1991), p. 298.

143. Gover, 1993.

144. Sears, 1991a, p. 127.

145. Brimmer, 1995, p. 64.

146. Savin-Williams, 1996c, p. 104.

147. Boxer, 1988.

148. C. L. Chng, "Adolescent homosexual behavior and the health educator," *Journal of School Health,* (Volume 61, 1980).

149. Savin-Williams, 1996a, p. 171.

150. M. Glasser, "Homosexuality in adolescence," *British Journal of Medical Psychology,* (Volume 50, 1977).

151. Savin-Williams and Lenhart, 1990.

152. J. Diepold and R. D. Young, "Empirical studies of adolescent sexual behavior: A critical review," *Adolescence,* (Volume 14, 1979). Kinsey, Pomeroy, and Martin, 1948. A. C. Kinsey, W. B. Pomeroy, C. E. Martin, and P. H. Gebhard, *Sexual behavior in the human female.* (Philadelphia, PA: W. B. Saunders, 1953).

153. R. E. Fay, C. F. Turner, A. D. Klassen, and J. H. Gagnon, "Prevalence and patterns of same-gender sexual contact among men," *Science,* (Number 243, 1989). E. Goode and L. Haber, "Sexual correlates of homosexual experience: An exploratory study of college women," *Journal of Sex Research,* (Volume 13, 1977). R. Sorenson, *Adolescent sexuality in contemporary society.* (New York: World Book, 1973).

154. Mac an Ghaill, 1991, p. 298.

155. Bell, Weinberg, and Hammersmith, 1981a. A. P. Bell, M. S. Weinberg, and S. K. Hammersmith, *Sexual preference: Its development in men and women, (Statistic appendix).* Bloomington, IN: Indiana University Press, 1981b).

156. Isay, 1991.

157. R. C. Savin-Williams, "An exploratory study of pubertal maturation timing and self-esteem among gay and bisexual male youths," *Developmental Psychology,* (Volume 31, 1995).

158. D'Augelli, 1991. J. Harry and W. B. DeVall, *The social organization of gay males.* (New York: Praeger, 1978). Remafedi, 1987a; G. Sanders, 1980; Savin-Williams, 1990.

159. Sears, 1991a.

160. Stanley and Wolf, 1980, p. 47.

161. W. Curtis, Ed., *Revelations: A collection of gay male coming-out stories.* (Boston, MA: Alyson, 1988), p. 109–110.

162. Due, 1995, p. 74.

163. Ibid., pp. 123–124.

164. Rodriguez, 1988; Savin-Williams and Lenhart, 1990.

165. Remafedi, 1987b; Rodriguez, 1988.

166. D. Offer and A. M. Boxer, "Normal adolescent development: Empirical research findings." In M. Lewis, Ed., *Child and adolescent psychiatry: A comprehensive textbook.* (Baltimore, MD: Williams & Wilkins, 1991). Rector, 1982.

167. Chandler, 1995, p. 146.

168. Sears, 1991a.

169. Schneider, 1989, pp. 121–122.

170. Isay, 1991.

171. Schneider, 1989, p. 123.

172. Ibid., p. 128.

173. S. M. Jourard, *The transparent self.* (New York: Van Nostrand, 1971).

174. Schneider, 1989, p. 128.

175. Malyon, 1981.

176. M. Sluchan, "Whose world is it anyway?" *The Weekly Pennsylvanian,* (1993, March 30), p. 4.

177. Savin-Williams, 1990.

178. Troiden, 1979.

179. Isay, 1991.

180. Schneider, 1989.

181. Savin-Williams, 1990, p. 182.

182. Remafedi, 1987b.

183. E. Pahe, "Speaking up." In B. L. Singer, Ed., *Growing up gay/growing up lesbian.* (New York: The New Press, 1994), p. 234.

184. A. Heron, Ed., *Two teenagers in twenty.* (Boston, MA: Alyson, 1994), p. 14.

QUESTIONS FOR DISCUSSION

1. According to the author, discuss the four-step process of becoming lesbian, gay, or bisexual. Give illustrations from the reading to support your response.

2. How does the process of becoming lesbian, gay, or bisexual differ for men and women? Give illustrations from the reading to support your response.

3. Compare and contrast becoming gay and lesbian (homosexuality) to heterosexuality.

Homosexuality and Mental Illness

Michael J. Bailey

No topic has caused the field of psychiatry more controversy than homosexuality, and 2 articles in this issue of the archives are likely to reopen past controversies and begin new ones. [1,2] These studies contain arguably the best published data on the association between homosexuality and psychopathology, and both converge on the same unhappy conclusion: homosexual people are at a substantially higher risk for some forms of emotional problems, including suicidality, major depression, and anxiety disorder. Preliminary results from a large, equally well-conducted Dutch study [3] generally corroborate these findings.

METHODOLOGICAL ADVANCES AND LIMITATIONS

The strength of the new studies is their degree of control. All too often, prior studies marshaled to examine the mental illness or health of homosexual people used samples seemingly selected to prove the point the researchers hoped to make. [4] Gay men

"Homosexuality and Mental Illness" by Michael J. Bailey. From *Archives of General Psychiatry,* Oct. 1999, v56 i10 p883. Reprinted by permission.

undergoing therapy seemed dysfunctional, while volunteers from homophile organizations seemed well. The current studies are not susceptible to this criticism.

The study by Fergusson et al. [1] focused on 1007 children from New Zealand who were observed until the age of 21 years. This sample represents 80% of a birth cohort; hence, results are exceedingly unlikely to be owing to unrepresentative sampling or differential dropout. Subjects whom they classified as gay, lesbian, or bisexual were at an increased lifetime risk for suicidal ideation and behavior, major depression, generalized anxiety disorder, conduct disorder, and nicotine dependence (odds ratios, 2.8–6.2 [compared with the heterosexual sub-sample]).

The study by Herrell et al. [2] used a powerful technique: the co-twin control method. Specifically, these investigators studied male twins in which one was homosexual and the other heterosexual (by the authors' definitions of these respective categories). It is difficult to imagine how findings of mental health differences between homosexual and heterosexual co-twins might be spurious. Herrell et al. found that gay twins had higher lifetime rates on 4 measures of suicidality compared with their heterosexual co-twins (odds ratios, 2.4–6.5). (The heterosexual co-twins of homosexual twins scored higher on the suicidal indicators compared with twins from pairs concordant for heterosexuality, although the difference was significant for only one suicidal symptom.) Results of logistic regression suggested that much, but not all, of the increased risk for suicide among homosexual subjects was owing to increased depression.

Although the new studies represent notable methodological advances compared with most prior research, they also have their limitations. The most important limitation, shared by both studies (as well as their Dutch counterpart [3]), concerns the definition of homosexuality. Both studies included in the definition of a homosexual person any subject who had had a same-sex sexual experience as an adult. In contrast, homosexual orientation is usually assessed by patterns of sexual attraction and fantasy. It is conceivable that some of the subjects who had engaged in homosexual behavior were not even attracted to people of their own sex. For example, 8 of 28 subjects classified as homosexual by Fergusson et al. [1] labeled themselves heterosexual. The problem is that heterogeneity among the homosexual subjects complicates interpretations of the results. For example, perhaps experimentation with homosexuality among heterosexually oriented people is associated with impulsivity, and this trait, rather than homosexual orientation, is associated with psychopathology. The decision to label these subjects homosexual by the authors was probably guided by both constraints on available data (which were not collected primarily to study this question) and concerns about statistical power. Regarding the latter, homosexuality was rare even by the lenient behavioral definition (2% in the study by Herrell et al. [2]; 3% in the other studies [1,3]). The low prevalence of homosexuality undoubtedly also was the reason why Fergusson et al. elected to combine gay men and lesbians into one group in their analyses. Most sexual orientation researchers believe that the causes of male and female sexual orientation differ, and if so, the correlates may differ as well. Thus, it would be optimal to perform separate analyses on gay men and lesbians. For all of these reasons, future studies should be even larger than the new ones and should include direct measures of sexual orientation.

POTENTIAL EXPLANATIONS

Several reactions to the new studies are predictable. First, some mental health professionals who opposed the successful 1973 referendum to remove homosexuality from DSM-III [5] will feel vindicated. Second, some social conservatives will attribute the findings to the inevitable consequences of the choice of a homosexual lifestyle. Third, and in stark contrast to the other 2 positions, many people will conclude that widespread prejudice against homosexual people causes them to be unhappy or worse, mentally ill. Commitment to any of these positions would be premature, however, and should be discouraged. In fact, a number of potential interpretations of the findings need to be considered, and progress toward scientific understanding will be achieved only by eliminating competing explanations.

Consider first the idea that increased depression and suicidality among homosexual people are caused by societal oppression. This is an eminently reasonable hypothesis. Surely it must be difficult for young people to come to grips with their homosexuality in a world where homosexual people are often scorned, mocked, mourned, and feared, and there is considerable anecdotal evidence that the "coming out" process is emotionally difficult. [6] The hypothesis would be strengthened by findings that issues related to self-acceptance, or acceptance by others, often trigger homosexual people's depressive and suicidal episodes. Furthermore, homosexual people should not, by this model, be more suicidal than heterosexual people in reaction to stressors of equal magnitude. It would indeed be surprising if antihomosexual attitudes were not part of the explanation of increased suicidality among homosexual people, but this remains to be demonstrated.

A second possibility is that homosexuality represents a deviation from normal development and is associated with other such deviations that may lead to mental illness. One need not believe that homosexuality is a psychopathologic trait (i.e., a behavioral or emotional trait that necessarily creates problems for the individual or for society) to believe that evolution has worked to ensure heterosexuality in most cases and that homosexuality may represent a developmental error. This hypothesis would be supported by findings that homosexual people (and people disposed to suicidality and depression) have higher rates of indicators of developmental instability, such as fluctuating asymmetry, left-handedness, and minor physical anomalies. [7] Although research has linked left-handedness to both male [8] and female [9] sexual orientation, considerably more research would be necessary to validate the general hypothesis.

Another developmental hypothesis concerns gender. On average, homosexual people are sex-atypical with respect to some traits, both during childhood [10] and adulthood. [11] The most influential etiologic hypothesis of homosexuality implicates sex atypical levels of prenatal androgens. [12,13] Perhaps these influences also make gay men more susceptible to types of psychopathology more commonly found in women and affect lesbians analogously. This hypothesis is consistent with the association between male homosexuality and both depression and suicidality found by Herrell et al., [2] as well as prior reports [14] that gay men, like women, score higher on psychological tests of neuroticism than heterosexual men. It would also imply that gay men should have lower rates than heterosexual men of diagnoses such as antisocial personality disorder, which more commonly affects men than women. Lesbians should have opposite vulnerabilities to gay men's. Unfortunately, Fergusson et al., [1] who had both male and female subjects, did not report results separately.

Another possible explanation is that increased psychopathology among homosexual people is a consequence of lifestyle differences associated with sexual orientation. For example, gay men are probably not innately more vulnerable to the human immunodeficiency virus, but some have been more likely to become infected because of 2 behavioral risk factors associated with male homosexuality: receptive anal sex and promiscuity. It is unclear how an analogous model would account for homosexual people's increased rates of suicidality and depression, although at least one other disorder may be explicable in this way. Gay men appear to be vastly over-represented among male patients with eating disorders. [15] One explanation is that the gay male culture emphasizes physical attractiveness and thinness, just as the heterosexual culture emphasizes female physical attractiveness and thinness. [16]

It is unlikely that any one of these models will explain all of the differences in the psychopathology between homosexual and heterosexual people. Perhaps social ostracism causes gay men and lesbians to become depressed, but why would it cause gay men to have eating disorders? Two things are certain, however. First, more research is needed to understand the fascinating and important findings of Fergusson et al. [1] and Herrell et al. [2] Second, it would be a shame—most of all for gay men and lesbians whose mental health is at stake—if sociopolitical concerns prevented researchers from conscientious consideration of any reasonable hypothesis.

REFERENCES

1. Fergusson DM, Horwood LJ, Beautrais AL. Is sexual orientation related to mental health problems and suicidality in young people? *Arch Gen Psychiatry.* 1999; 56: 876–880.

2. Herrell R, Goldberg J, True WR, Ramakrishnan V, Lyons M, Eisen S, Tsuang MT. Sexual orientation and suicidality: a co-twin control study in adult men. *Arch Gen Psychiatry.* 1999; 56:867–874.

3. Sandtort TGM, Graaf RD, Biji RV, Schnabel P. *Sexual Orientation and Mental Health: Data from the Netherlands Mental Health Survey and Incidence Study (NEMESIS).* Stony Brook, NY: International Academy of Sex Research; 1999.

4. Bayer RB. *Homosexuality and American Psychiatry: The Politics of Diagnosis.* New York, NY: Basic Books Inc Publishers; 1981.

5. American Psychiatric Association. *Diagnostic and Statistical Manual of Mental Disorders, Third Edition.* Washington, DC: American Psychiatric Association; 1980.

6. Isay R. *Becoming Gay: The Journey to Self-Acceptance.* New York, NY: Pantheon Books; 1996.

7. Thornhill R, Moller AP. Developmental stability, disease and medicine. *Biol Re Camb Philos Soc.* 1997;72:497–548.

8. Lindesay J. Laterality shift in homosexual men. *Neuropsychologia.* 1987;25:965–969.

9. McCormick CM, Witelson SF, Kingstone E. Left-handedness in homosexual men and women: neuroendocrine implications. *Psychoneuroendocrinology.* 1990; 15:69–76.

10. Bailey JM, Zucker KJ. Childhood sex-typed behavior and sexual orientation: a conceptual analysis and quantitative review. *Dev Psychol.* 1995; 31:43–55.

11. Lippa RA. Gender-related traits in gay men, lesbians, and heterosexual men and women: the virtual identity of homosexual-heterosexual diagnosticity and gender diagnosticity. *J Pers*. In press.

12. Byne W. Parsons B. Human sexual orientation: the biologic theories reappraised. *Arch Gen Psychiatry*. 1993; 50:228–239.

13. Meyer-Bahlburg HFL. Psychoendocrine research on sexual orientation: current status and future options. In: Vries GJD, Bruin JPCD, Uylings HBM, Corner MA, eds. *Sex Differences in the Brain*. New York, NY: Elsevier Science Inc; 1984:375–398.

14. Van den Aardweg GJ. Male homosexuality and the neuroticism factor: an analysis of research outcome. *Dynamic Psychother*. 1985; 3:79–87.

15. Carlat DJ, Camargo CA Jr. Review of bulimia nervosa in males. *Am J Psychiatry*. 1991; 148:831–843.

16. Siever MD. Sexual orientation and gender as factors in socioculturally acquired vulnerability to body dissatisfaction and eating disorders. *J Consult Clin Psychol*. 1994; 62: 252–260.

QUESTIONS FOR DISCUSSION

1. Compare and contrast homosexual and heterosexual suicidal behaviors.

2. According to the author, why do homosexuals commit suicide?

3. How can society help homosexuals decrease their suicidal behavior?

PHYSICAL DISABILITIES

The Making of the Blind in Personal Interaction

Robert A. Scott

One of the ways in which a person who has difficulty seeing learns how to be a blind man is by interacting with those who see. When normals come face-to-face with someone who cannot see, their preconceptions about, and reactions to, blindness are expressed as expectations of how the blind man ought to behave. The blind man is poignantly reminded of the social identity imputed to him by others. All blind men respond to this identity in some way, even if only to dispute it; for those who internalize it, this putative social identity becomes a personal identity in fact.

There are two principal mechanisms of personal encounters through which the blind are socialized. The first relates to preconceptions about blindness that people who can see bring to encounters with blind men, the second to the reactions of the sighted during the encounter. Normal people will react to a blind man as a blind man only if they perceive him as such. The interpersonal context of the socialization of the blind applies, therefore, only to those blind people whose blindness is readily apparent to others. Most laymen know little or nothing about the technical definitions of blindness that professionals use, and even those who do cannot make the fine discriminations that are required in order to determine where someone whose vision is severely impaired falls relative to the line set by the definition. The socialization processes described in this article apply, therefore, to people who are totally or virtually blind, and whose impairment is readily identifiable as such.

PRECONCEPTIONS ABOUT BLINDNESS

The preconceptions that the sighted bring to situations of interaction with the blind are of two sorts. On the one hand, there are stereotypic beliefs about blindness and the blind that they have acquired through the ordinary processes of socialization in our culture; and, on the other, there is the fact that blindness is a stigmatizing condition. Each factor makes its special contribution to the social identity that is reserved for blind men. As I suggested in the introduction, the notions of a blindness stereotype and of blindness as stigma have already received a good deal of attention in the literature.[1] Little would be gained, therefore, by a simple reiteration of these ideas. My purpose here is to place the stereotype and stigma notions into the perspective of this book by explaining the contributions each makes to the self-concept that a blind man acquires.

Stereotypic Beliefs about Blindness

In my introduction I described a number of the general beliefs that laymen have about the blind. These beliefs involve notions of helplessness, docility, dependency, melancholia, aestheticism, and serious-mindedness. While few laymen accept all these beliefs, most of them do adhere to at least a few of them. These misconceptions are brought by them to situations of interaction with the blind and are expressed as expectations of the behavior and attitudes of the blind person. Because of them, deep and stubborn "grooves and channels" are created into which all the blind man's actions and feelings are pressed. Their existence makes it extremely difficult for meaningful communication and unstrained relationships to occur between the seeing and the blind.

   The effects such beliefs have upon blind people when they interact with sighted people have been succinctly described by Gowman:

> An individual taking up the role of blind man is conceptually relocated along the margins of the dominant social structure and a peripheral social role is assigned to him. His rights and obligations are redefined in a manner which is believed to mesh with the character of the disability. The newly blinded person is reevaluated in all his aspects, and the evaluative scale shifts from the measurement of specific individual qualities or capabilities to the assessment of the global condition of blindness. What distinguishes the blind role from other types of roles is its all pervasive character. Blindness is not an attribute to be put on or cast off as the situation demands, but a constant characteristic which affects the quality of each of the individual's relationships in occupational, recreational, and other contexts. When evaluation is thus expanded to cover an individual's entire personality structure, a stereotype is operative. The blind may be assigned a social role which so transforms them that they emerge as a labeled segment of society. Social interaction becomes stunted and artificial under the impress of the stereotype.[2]

   It is impossible for blind men to ignore these beliefs; they have no choice but to respond to them. These responses vary, but in a highly patterned way. Some blind people come to concur in the verdict that has been reached by those who see. They adopt as a part of their self-concept the qualities of character, the feelings, and the behavior patterns that others insist they must have. Docility, helplessness, melancholia, dependency, pathos, gratitude, a concern for the spiritual and the aesthetic, all become a genuine part of the blind man's personal identity. Such blind men might be termed "true believers"; they have become what others with whom they interact assume they must become because they are blind.

Not all blind people are true believers; there are many who explicitly, indeed insistently, reject the imputations made of them by others. They thereby manage to insulate a part of the self-concept from the assaults made on it by normals. The personal identity of such a person is not that of a blind man, but of a basically normal person who cannot see. For the blind man who responds in this way, there remains the problem that most people who see do not share the view he has reached about himself. Some blind men respond simply by complying with the expectations of the sighted, in a conscious and deliberate way. They adopt an external facade that is consistent with the normals' assumptions about them, but they are aware that it is a facade, and they are ready to drop it whenever occasions permit them to do so. Ordinarily, the reason for acquiescence is expedience; in fact, every blind man, whether he accepts or rejects the social identity imputed to him, will be found to acquiesce at least some of the time. For example, several blind people have told me that when they use public transportation, fellow passengers will occasionally put money into their hands. When this occurs, a blind man cannot very well give a public lecture on the truth about blindness; in fact, to do anything but acquiesce and accept the gift will leave him open to charges of ingratitude and bitterness. There are other blind people who use the acquiescent facade not for expedience but as a weapon. Those who beg, for example, deliberately cultivate it in order to encourage people to give them money. The beggar strikes up an unstated pact with the world: he agrees to behave exactly as others insist that he must, and in exchange he makes the normal person pay with money. How dear this price is, of course, depends upon the beggar's ability to exploit the emotions that lead his victims to expect him to acquiesce in the first place.

Another way that blind men cope with discrepancies between putative and personal identity is to resist and negate the imputations of others at every turn. By so doing, personal integrity is preserved, but the cost is very high. It requires an enormous commitment and expenditure of energy to resist these forces, and the blind man who does so inevitably alienates himself from other people. Even those who follow this road successfully are left with a certain bitterness and frustration that is the inevitable residue of any attempt to break the stubborn molds into which the blind man's every action is pressed.

Clearly, then, these stereotypic beliefs about the blind have profound consequences for the self-concept of every blind man. He may internalize them as a part of his self-image, or he may reject them as completely false and misleading, but he cannot ignore them. The fact that he cannot is one of the several reasons why homogeneity develops in this otherwise heterogenous group of individuals.

Blindness as Stigma

Blindness is a stigma, carrying with it a series of moral imputations about character and personality. The stereotypical beliefs I have discussed lead normal people to feel that the blind are different; the fact that blindness is a stigma leads them to regard blind men as their physical, psychological, moral, and emotional inferiors. Blindness is therefore a trait that discredits a man by spoiling both his identity and his respectability.

When a person with a stigma encounters a normal person, barriers are created between them.[3] These barriers, though symbolic, are often impenetrable. They produce a kind of "moving away," much like the action of two magnetized particles of metal whose similar poles have been matched. These avoidance reactions are often induced by a fear that direct contact with a blind person may be contaminating, or that the stigmatized person will somehow inflict physical or psychic damage. Such reactions and fears are completely emotional and irrational in character.

The effects of these reactions on a blind man are profound. Even though he thinks of himself as a normal person, he recognizes that most others do not really accept him, nor are they willing or ready to deal with him on an equal footing. Moreover, as Goffman has observed, "the standards he has incorporated from the wider society equip him to be intimately alive to what others see as his failure, inevitably causing him, if only for moments, to agree that he does indeed fall short of what he really ought to be."[4] As a result, he may feel shame because he knows he possesses a defiling attribute. It is when the blind person finds himself in the company of the sighted that these self-derogating feelings are aroused. "The central feature of the stigmatized individual's situation in life," writes Goffman, "is a question of what is often, if vaguely, called 'acceptance.' Those who have dealings with him fail to accord him the respect and regard which the uncontaminated aspects of his social identity have led them to anticipate extending and have led him to anticipate receiving; he echoes this denial by finding that some of this own attributes warrant it."[5]

The stigma of blindness makes problematic the integrity of the blind man as an acceptable human being. Because those who see impute inferiority, the blind man cannot ignore this and is forced to defend himself. If, as sometimes occurs, the blind man shares the values of the sighted, the process becomes even more insidious; for when this is the case, a man's personal identity is open to attack from within as well as from without.

THE SITUATION OF INTERACTION

Preconceptions about blindness are not the only elements of personal encounters that determine a blind man's socialization experiences; certain features of the actual encounter play an important role as well. First of all, the norms governing ordinary personal interaction cannot, as a rule, be applied when one of the actors is unable to see. Furthermore, blind people, because they cannot see, must rely for assistance upon the seeing with whom they interact. As a result, many of the interactions that involve the sighted and blind men become relationships of social dependency. Each of these factors is intimately related to the kind of self-concept the blind man develops, and, because of them, this group of people is made even more homogeneous.

Blindness and the Conduct of Personal Relationships

Vision plays an extremely important role in the face-to-face encounters of everyday life. The initial impressions we have of people are acquired largely through vision, and the success of our subsequent relationships with them depends to a considerable extent on our ability to see. It is when one of the actors is blind that we recognize how central a part vision plays in our relationships with other people.

Establishing an Identity

One of the things we do upon meeting someone for the first time is to impute to him a familiar social identity. We label him elderly, handsome, debonair, cultured, timid, or whatever. From the identity we have imputed to him, we anticipate what his tastes and interests will be, the kinds of attitudes he will have, and how he is likely to behave. We search for clues to help us to classify the person as a type of individual, and we then apply norms of conduct associated with "his type" in order to guide us in our subsequent interaction with him. Unless we can do this quickly and accurately, we are at a loss as to how to proceed, and experience the situation as embarrassing and stressful.

What people wear, how they look, the way they stand, and the gestures they use are important clues in helping us to reach some decision about the type of person we have encountered. Indeed, most of the initial impressions we form of persons are based on clues that must be seen to be detected. For this reason, initial encounters between the seeing and the blind are set awry; since one actor is blind, each is deprived of significant information about the other.

The blind person will be at a loss to know precisely with whom he is dealing, and what kind of behavior to expect from him. He will be slow to piece together the information he needs correctly to infer the person's social identity. This produces an ambiguity and uneasiness that can become so intense that the relationship never develops.

The inability to gather accurate information on which to form impressions is not a problem only for the blind person; it leads to uneasiness on the part of the sighted individual as well. He is uncertain that the image he tries to project will be received, or, if received, accurately interpreted. He will realize immediately that his general appearance is no longer useful for conveying information about himself. His uneasiness intensifies if he does not know which nonvisual clues the blind man is using to "size him up." Is it the tone of voice or the content of words? Does his tone of voice convey something about himself that he is unaware of or wishes to conceal? He does not know how to convey to the blind man an impression equivalent to the one conveyed through sight. When visual clues to a person's social identity are missing, a difficult and awkward situation is created for both the blind man, who does not develop a complete impression, and the sighted man, who is unsure of the impression he has made.

Conversely, there is the process of social identification of the blind person by the sighted individual. In addition to the stigmatizing quality of blindness, blind people may be insensitive to the role of visual factors in projecting accurate impressions, or they may not know how to create the kind of impression they want to. In this regard, the blind person is dependent upon others. They select his clothes and advise him on his posture. To some extent, then, his appearance is dependent on the tastes of his helpers. Clearly, ambiguities and uncertainties surround the identity of the blind man as well as that of those who see.

Norms Governing Personal Interaction

These ambiguities and uncertainties will sometimes cause the encounter to terminate prematurely. If not, initial uncertainties are carried into the next stage of the relationship, for it is on the basis of initial impressions that we know what to expect of others, and how we should behave toward them. The rules that are applied to an individual on the basis of our identification of him, and the rules that he in turn applies to us, lend structure and substance to the relationship. Since a person's blindness tends to overwhelm those with whom he interacts, it is the blindness more than anything else that identifies him. Blindness is a comparatively rare event in any population, so that only a few sighted individuals ever interact with a blind man in any sustained way. What most of us know about blindness comes from the mass media, religious and other writings, common sense, hearsay, or from occasional contacts with a salesman of products made by the blind, or the solicitor from the local blindness agency. When we encounter a blind man, the rules we apply to him are extremely vague; they tell us what to expect in only the most tentative way. Our lack of direct experience makes the situation more uncertain. This normative ambiguity, indeed normlessness, applies as much to the blind person's understanding of the sighted individual's behavior as it applies to the sighted individual's understanding of how to behave with a blind person. In this sense, the difficulty lies with the relationship and not with the individual partners to it.

One facet of this problem deserves special emphasis. Easy social interaction is contingent upon the possession of certain skills and information that can be used as a conversation goes along. I have in mind here the most elementary kinds of things. For example, when a person meets me in my office, I may invite him to take a seat. He may wish to have a cup of coffee and a cigarette. Perhaps we will share lunch together either in my office or in a restaurant. He may wish to use the telephone while we visit. I may want to show him written materials, or he may have to excuse himself to use the bathroom. Easy, uninterrupted communication hinges on the ability of both of us to carry out the activities necessary to the encounter with ease and independence. Let us go back over the encounter as I have described it and assume that this person is blind. When he enters my office, I find I cannot simply invite him to take a seat but must conduct him to it. As I approach him, I realize that I do not know how to direct him easily, and he does not know if I can do so either. Consequently, we share an awkward moment as I try to lead him to a chair and back him into it. He in turn tries to accept my assistance gracefully while trying not to be impaled by the arms of the chair. After he is seated, I offer him coffee. When it arrives, I realize that he may want cream and sugar. It is unclear if he can manage this himself. If he cannot, and he asks me to do it for him, I may place the coffee in front of him only to find that he knocks it over when reaching for it. When lighting a cigarette, he may put a match to it yet fail to ignite it. If he continues to puff away, do I tell him or let him discover it for himself? Suppose that he flicks ashes on himself; do I point this out to him and therefore bring his disability into the conversation or do I let it pass unnoticed? When he asks to use the bathroom, what do I do? How do I direct him to a toilet or urinal or get him to the washstand? We enter a restaurant for lunch, and I realize that he cannot read the menu. How do I help him to get seated at the table? He orders meat and the question arises as to whether or not he is able to cut it. And what about things on the table of which he is unaware, such as butter and rolls? How do I help him to get these things? From its inception to its conclusion, the interaction is filled with uncertainty, awkwardness, and ambiguity, making such meetings frustrating, embarrassing, and tense.[6]

Sighted people may try to avoid situations in which such encounters are likely to occur, but there will always be occasions when this is impossible. No one can very well walk away from a blind man who is obviously lost in a crowded street; he cannot ignore the blind man who is about to walk into the side of a building or into an open ditch; he cannot shun the blind person during intimate social gatherings. When normals and blind men are coerced into relationships with one another, most normals fall back on common sense. The trouble with commonsense ideas about the blind is that they are often ridiculous. One man, for example, asked a student in a school for the blind how he knew when he was awake. Another once reprimanded a blind man who had walked into the side of a building for failing to "look out" where he was going. Still another expressed amazement that all a blind man had to do in order to get around a city was to tell his seeing-eye dog the address he wanted and hang on. A blind person can expect anything to happen when he meets a sighted person for the first time. It is almost impossible for him to know how the sighted person will react to him. This fact makes for frustrating uncertainties on the part of both parties to the encounter. It has an added consequence for the blind person. He interprets sighted peoples' responses to him as additional evidence that, in their world, the blind are different and lesser people. Why else would sighted people be so insensitive to the blind man's problems, or so anxious to terminate encounters with him? This interpretation by the blind only serves to reinforce the sense of their differentness that is already implied in the preconceptions about blindness.

The effects of uncertainty and ambiguity do not end when the encounter does. The memory of them lingers, giving rise to the impulse to avoid another such experience. Normative ambiguity is, therefore, one of the several reasons many sighted people impulsively act to avoid contact with blind people. It is to be noted in passing that such impulses are accompanied by considerable guilt.

Communication Problems

Vision plays an enormously important role in personal communication. When we speak to someone, it is customary for us to maintain eye contact. This is learned from the earliest age, so that by about the time a child begins school, this very important lesson has been learned. To turn away or focus on a distant object when addressing another person can be attributed to rudeness, shyness, or guilt. Frequently, the lack of visual contact is one of the factors responsible for the statement, "We simply could not communicate." Eye contact signifies honesty, directness, attentiveness, respect, and a variety of other virtues that are the important ingredients of successful human communication.

The important role of vision in personal interaction has many implications for the blind. The blind person is often able to follow a conversation more closely by turning his ear to the speaker. He may therefore develop the disconcerting habit of turning his head slightly away from the speaker's face. The speaker reacts as he would to a sighted individual turning away—he is not sure the listener is attentive.

But the problem is more complicated than this. In addition to the fact that the blind person may incline his head this way or that to maximize sound, the appearance of his eyes may be disconcerting. When blindness is due to accident, the face and eyes may be disfigured. Sometimes the eyes bulge or are set at peculiar angles; in other cases, they maybe opaque or gray. When a person is losing vision, he may also lose the ability to control the eye muscles, so that the eyes constantly flutter and roll about in their sockets. These deformities present major difficulties in communication, because eye contact may be not only disturbing but repulsive to the observer.

Blindness prevents the blind person from getting visual feedback for his own body gestures. This may result in the inadvertent development of gestures and bodily movements that are disruptive of communication. A person who cannot observe his own expressions or others' reactions to them becomes insensitive to the importance of facial and bodily gestures in communication. The blind man often appears to be smiling the smile of a simpleton, or gesturing in a way that makes him appear retarded or mentally deranged. Once again, his unusual facial and body gestures are interpreted as they would be if he could see. This is especially a problem among the congenitally blind, who, out of a need for stimulation, develop peculiar body movements called "blindisms." They probe at the face, tilt the body and roll it, move up and down and back and forth, and often have ticks and twitches of the face. If the sighted person interprets these gestures as being responsive to what he has said, further misunderstandings in communication arise. Conversely, the gesture as a mechanism of communication by the sighted person is also eliminated, which limits the range of expressions he can use. These problems with communication serve to heighten the already difficult misunderstandings that have been created by normative ambiguity.

These three problems of establishing the desired personal identity in the mind of the other actor, of uncertainty as to how to interact with a blind man, and of miscommunication all work together to produce that peculiar blend of annoyance, frustration, ambiguity, anger, tension, and irritation that describes human interaction that is spoiled. The important point is that the source of this unhappy outcome is not to be found so much in the erroneous beliefs that the seeing hold about blindness, although such beliefs clearly are not entirely innocent; rather, it lies in what might be called "the mechanics" of interpersonal conduct. In this sense, the problem lies more with the relationship itself than with the erroneous conceptions held by those who are parties to it.

All of this has two important effects on the blind man's self-concept. First, he is once again reminded that he is different from most people and that the satisfying personal relationships that are commonplace to them are, for the most part, denied to him. Second, because so many of his relationships with other people are spoiled, he is denied the kind of honest and direct feedback that is so essential for maintaining clear and realistic conceptions about the kind of person he is.[7] Often the blind man gets no feedback at all, and when he does it is usually badly distorted. As a result, the blind man can easily acquire either an unduly negative or an unreasonably positive conception of his own abilities.

Even though many of the problems that characterize encounters between sighted and blind men arise from the mechanics of interpersonal conduct, it does not always follow that blind men explain these problems to themselves in this way. On the contrary, many of them apparently assume that a normal person's behavior is caused by his beliefs. Thus, when a sighted person behaves assertively toward a blind man so as to eliminate uncertainty, the blind man infers that the other's actions are caused by a belief that blindness makes him helpless. It is for this reason, I think, that blind people have placed so much stock in the notion of an elaborate, rigid stereotype of the blind.

Blindness and Social Dependency

Interactions between the blind and the sighted become relationships of social dependency. To explain why this occurs, it is necessary to digress for a moment in order to clarify some of the factors that enter into ordinary personal relationships.[8] A useful starting point is to ask the question, "Why do people enter into associations with one another?" One answer is, because most of the things that men find pleasurable in life can be obtained only through associations with other people. Some of these associations, such as those which occur between close friends, are rewarding in and of themselves. Other associations are extrinsically rewarding. By this I mean that the partners to the interaction derive particular benefits from social relations because those with whom they interact provide them with a service or fulfill a personal need. Extrinsically rewarding relationships are among the most troublesome of all relationships in which blind people engage; for this reason, I will be concerned primarily with them.

A characteristic of most persons is their liking to help others and do things for them. When we do a favor for someone, he usually feels grateful, and the gratitude he expresses becomes the reward we obtain for having helped him. This is particularly the case if he is vocal in his gratitude, since public expressions of gratitude go a long way toward establishing a good reputation for the person who has rendered the favor. In addition to the fact that we receive gratitude in exchange for favors, we know from experience that rendering a favor will often result in receiving one. When we do someone a favor, he is not only grateful to us; he feels obligated to us. In order to discharge his obligation, he reciprocates by doing something for us. There develops through this circular reciprocity a social bond that is the heart of lasting associations between people. Another reward that is sought in association with others is social approval. The social approval of others, especially of those whose opinions we most value, is of great significance to us. It acts as a curb on our tendency to become too selfish in associations with others. If, in our relations with others, we are too egoistic, we lose social approval. To gain social support requires us to give to others a bit more than we get, or at least not to try to gain more than we give.

Given the fact, then, that men associate with one another out of a desire to obtain the social rewards of gratitude, reciprocity, and social approval, we may now wonder why people associate differentially with one another. What determines who will associate with whom?

Social attraction is the force that impels men to establish social relations with one another. One person will be attracted to another if he anticipates that associating with him will result in some rewards for him. His desire for social rewards and his perception of the probability of their realization motivate one person to associate with another.

Attraction of one person to another will result in social exchange. This is so because the person who is attracted to another also wants to prove himself attractive to the other, since his ability to associate with him and reap the benefits expected from the association is contingent on the other finding him an attractive associate.

Mutual attraction depends upon the anticipation that the association between persons will be mutually rewarding, and social attraction ultimately leads to social exchange. The people involved, however, are not always social equals. There are situations in which A needs what B is able to supply, but A has nothing that B needs. To be sure, the supplier in this exchange may feel rewarded by expressions of gratitude that will most certainly be forthcoming from A; but, if the association is to be durable and long-lasting, it will be necessary for A to provide a compensatory reward in exchange for what B is supplying. What are the alternatives that A might follow? If he is capable of it, he may be able to force B to help him by employing either physical force or social coercion. But this implies that A has a high degree of control over a situation, something he seldom, in fact, has. A second alternative is to seek the rewards he wishes elsewhere, perhaps through someone with whom he can develop an equitable exchange. The final alternative is to forego the satisfaction of the reward. Which alternatives are viable or even possible will depend upon the individuals involved, the intensity of the need, and the structure of the situation. There will be many situations in which none of these alternatives is feasible, and in these cases A has no choice but to exchange the favors given by B for compliance with B's requests. In this case, the compliance of one person is the reward a supplier receives for the services he renders. Compliance with the demands of others is the substance of power and, for this reason, willingness to comply with the wishes of others is often a very generous reward. It can be seen, incidentally, that exchange processes give rise to differentiation of power among persons, since there will always be some people who are capable of providing services that others need. Power is attained when the supplier makes the satisfaction of those in need contingent on their compliance with his wishes.

In forming extrinsic social relationships, we continually evaluate one another in terms of potential attractiveness. We assess visible and inferred qualities to get some impression about whether or not other persons are potential providers of the services we require. If we decide that the other will be able to meet our needs, sustained social relations become possible. If we decide otherwise, our relationships will either terminate or become casual. Since people seek services from others in exchange for things they want, it follows that people with roughly equivalent, but somewhat different, potentials will be attracted to one another. People will be reluctant to enter into encounters they believe will force them either into too great a degree of compliance (when they are receivers) or (when they are suppliers) into a position in which the compliance of the receiver is not worth the services rendered. For these reasons, "social likes" tend to attract "social likes."

A number of basic problems in personal associations between the blind and the sighted are clarified by viewing them from the perspective of social exchange theory. A person's evaluation of the potential social attractiveness of another will be radically affected if the person evaluated cannot see. Most sighted people will assume that they will have to offer more services to the blind man than the blind man will be able to offer them. Moreover, since blind persons require assistance in getting about in their environment, they will be automatically put into the sighted person's debt. This implies that persons cannot realistically expect to receive payment in kind for favors rendered. That this is recognized in our culture is evident from the fact that most encounters involving the blind and the sighted are defined as charitable. In charitable relationships, the donor person is expected to give generously to the one who is stigmatized, and not to expect to receive anything in return. Actually, this commonly held belief is only partly true, since the charitable person is repaid in part by the fact that his giving is usually public and therefore results in social approval and in part by receiving gratitude. But social approval and gratitude are not in themselves sufficient to sustain relationships. Reciprocity, to be genuine, must involve both socially valued compliance and the capacity to perform socially valued favors.

The blind person is, therefore, by virtue of his dependency, the subordinate in a power relationship. As a rule, none of the alternatives available to subordinates in power relationships are open to him. He cannot forego the service required, since performing important activities of daily life depends on the cooperation of sighted persons. It is unlikely that he will turn elsewhere, partly because he cannot always do that on his own and partly because his situation will be unlikely to change greatly if he does. Finally, he cannot very well rely on force to have favors done for him. He is, therefore, backed into a position of compliance.

Persons who are in positions of power must weigh the cost of granting a service to someone against the potential value of his compliance. Because of his marginal status, a blind man's compliance is only of limited worth to those who seek to gain powerful positions in the mainstream of society. Furthermore, the value of a blind person's compliance may be offset by the investment of time and effort required to render the service. For these reasons, many sighted people avoid encounters with the blind because they anticipate that the compliance of a blind person will be of little value to them. Once again, such avoidance is not achieved without feeling guilt. These guilt feelings become apparent whenever the blind and the sighted are thrust into one another's company.

These factors have several consequences for the blind person's socialization experiences in personal relationships. For one thing, since blindness is a social debit, it follows that blind persons will find it difficult to develop enduring associations with sighted persons who are otherwise their intellectual, psychological, and social equals. The fact that many blind people are not treated as "normal" in this respect is the motive behind a common reaction pattern in the blind. Some blind people disavow their blindness entirely by learning how to perform activities that are normally reserved for people who can see. These include such activities as skiing, golfing, driving automobiles, or doing elaborate repairs on the house. While such activities can be performed by certain blind people, competence in them is attained at the cost of a tremendous personal effort.

Even more fundamental than this, however, are the demoralizing and humiliating effects upon the self of continuously being treated charitably. The blind person comes to feel that he is not completely accepted as a mature, responsible person. As a second-class citizen, he must deal with the eroding sense of inadequacy that inevitably accompanies that status. Incidentally, it is important to note that this problem does not stem from the preconceptions others have about blindness; it is an effect of introducing the factor of blindness into the equation which describes the mechanics of interpersonal conduct.

There is one condition under which a blind person is able to escape this dilemma—the possession of a valued quality, trait, or attribute that he can use to compensate for his blindness. It is no accident that the blind persons who become most completely integrated into the larger society possess wealth, fame, or exceptional talent. These people can exchange prestige or money for the favors they must accept in order to function in daily life. They are unusual. A majority of blind people are elderly and poor, two traits that also have very low potential attractiveness to others.

In summary, four features of personal relationships affect the socialization of the blind. These are (1) the stereotyped beliefs that those who see bring to the interaction; (2) the fact that blindness is a stigma; (3) the fact that the conduct of such interactions is profoundly disturbed when one of the actors to the encounter cannot see; and (4) the fact that, by their nature, these are relationships of social dependency. The first two factors affect socialization outcomes in two ways; they force the blind man to recognize that he is a different, and lesser, person and they create a social identity that he either internalizes as a part of his self-concept or reacts to by rejecting. The stereotyped beliefs and the stigma, being contingencies no blind man can ignore, impose certain uniform behavioral patterns on those whom society labels blind.

The last two factors, which relate to the mechanics of interpersonal conduct, affect socialization outcomes in three ways; they force upon blind people further evidence of their difference; they deny them the kind of honest, uncluttered feedback about self that is commonplace to the sighted; and they place them in a subordinate position, making it difficult for them to form intimate relationships with those they regard as their intellectual and psychological equals. Together, these processes feed on one another and, from the initial heterogeneity of the blindness population, homogeneous patterns begin to emerge.

NOTES

1. See Harry Best, *Blindness and the Blind in the United States,* Macmillan Company, New York, 1934, p. 279; Hector Chevigny and Sydell Braverman, *The Adjustment of the Blind,* Yale University Press, New Haven, 1950, p. 26; and Alan G. Gowman, *The War Blind in American Social Structure,* American Foundation for the Blind, New York, 1957, p. 104.

2. Gowman, *op. cit.,* p. 46.

3. Fred Davis, "Deviance Disavowal: The Management of Strained Interaction by the Visibly Handicapped," in Howard S. Becker, ed., *The Other Side,* The Free Press of Glencoe, New York, 1964, pp. 119–138, and Robert Kleck, Ono Hiroshi, and Albert H. Hastorf, "The Effects of Physical Deviance upon Face-to-Face Interaction," *Human Relations,* Vol. 19, No. 4, 1966, pp. 425–436.

4. Erving Goffman, *Stigma: Notes on the Management of Spoiled Identity,* Prentice-Hall, Inc., Englewood Cliffs, N.J., 1963, p. 7.

5. *Ibid.,* pp. 8–9.

6. Alan G. Gowman, "Blindness and the Role of the Companion," *Social Problems,* Vol. 4, No. 1, 1956, pp. 68–75.

7. Stephen A. Richardson, "The Effects of Physical Disability on the Socialization of a Child," in David A. Goslin, ed., *Handbook of Socialization Theory and Research,* Rand McNally & Company, Chicago, 1969.

8. See Peter M. Blau, *Exchange and Power in Social Life,* John Wiley & Sons, Inc., New York, 1964; and George C. Homans, *Social Behavior,* Harcourt, Brace & World, New York, 1961.

QUESTIONS FOR DISCUSSION

1. Discuss what the author means by the socialization of the blind in personal interaction. Give illustrations from the reading to support your response.

2. How have you interacted with a blind person? Discuss preconceptions about blindness. Be sure to address stereotypes and stigma about blind people.

3. Discuss, from a blind person's perspective, what problems sighted people have when interacting with blind people. Give illustrations from the reading to support your response.

Thinking Twice on Splitting Twins

Natalie Angier

Like many twins descended from a single egg whom the world has deemed "identical," Lori and Reba Schappell prefer to emphasize their differences over their similarities. Lori is warm and boisterous and maternal. She wants to get married and have babies, she says, and at the age of 36, she wants to do it soon. Reba is quiet and self-contained, and she squirms whenever her sister hugs her in public or tells her that she loves her. Reba is focused on her fledgling career as a country singer. Last month, she flew to California to accept an L.A. Music Award for best new country artist of the year. Lori keeps her brown hair short, speaks with the broad-voweled accent of Pennsylvania's Reading area, where the sisters have always lived, and loves strawberry daiquiris. Reba colors her wavy hair copper, has adopted a Nashville twang and is a teetotaler.

But there are certain things the sisters undeniably share. They are, literally, of one flesh. They are conjoined at the head, portions of their skull, scalp and blood vessels fused at the side in a mirror-image configuration, so that they face in opposite directions. And though they have two distinct brains, they are of one mind in their opinion about whether they would ever consider undertaking the risks of surgical separation.

"Our point of view is no, straight-out no," said Reba, for the moment not the quiet one. "Why would you want to do that? For all the money in China, why? You'd be ruining two lives in the process."

"And we'd miss the other one horribly if she were to die," Lori added. Reba and Lori live a life that no singleton can imagine, and one that looks unbearably difficult. Where one goes, so must the other. Reba is short and cannot walk for herself, and so her sister wheels her around on a bar stool. They venture out in the world fearlessly, and the world never stops staring at the sight of them. But as they see it, their lives are no more difficult than all lives.

"There are good days and bad days—so what?" says Reba. "This is what we know. We don't hate it. We live it every day. I don't sit around questioning it, or asking myself what I could do differently if I were separated."

"Thinking Twice on Splitting Twins" by Natalie Angier from *The San Diego Union-Tribune,* January 7, 1998, E-1. Reprinted with permission of The New York Times.

Nor does Lori appreciate being held up as an exemplar of fortitude. "People come up to me and say, 'You're such an inspiration. Now I realize how minor my own problems are compared to yours.' But they have no idea what problems I have or don't have, or what my life is like." The attitudes of the Schappell sisters are strongly felt, but not unique. Alice D. Dreger, a historian of anatomy at Michigan State University in East Lansing, argues that the common assumption that life as a conjoined twin is not worth living is an outsider's premise—one in need of some serious scrutiny. Dreger has written a detailed analysis of the medical treatment of conjoined twins in Studies in History and Philosophy of Science.

She reviews cases in which surgeons have aggressively sought to separate conjoined twins even when the distribution of shared organs resulted in the extreme disability or death of one. She points out that throughout history, conjoined twins who reached adulthood have expressed satisfaction with their linked lives.

THE BIDDENDEN MAIDS

Consider the case of Mary and Eliza Chulhurst, one of the first documented examples of conjoined twins. Born fused at the lower back and buttocks in the year 1100, the Chulhurst sisters—also known as the Biddenden Maids—lived for 34 years in Kent, England. After the death of one sister, doctors urged the survivor to allow them to attempt surgical separation to save her. She refused, declaring "As we came together, we will go together." She died several hours later. Despite the testimony of twins themselves, Dreger argues, doctors persist in regarding conjoined twins—also known by the now-disfavored term "Siamese twins"—as monstrously abnormal beings who must be individuated surgically, even if such standard mores of medicine as "first do no harm" are set aside to do so.

She explores at length the disturbing ethics that surrounded the widely publicized case of Angela and Amy Lakeberg, born in 1993. The girls were attached breast to belly, sharing a liver and a single, six-chambered heart, rather than two four-chambered ones. Because the infants had no chance of survival in their conjoined state, the parents opted for a long-shot effort to save one of the twins at the expense of the other. In an extraordinary operation, a team of surgeons at Children's Hospital of Philadelphia cut off circulation to Amy, deliberately sacrificing her to salvage the heart for Angela. The effort ultimately failed, and Angela died 10 months later, but the doctors involved, as well as many others, have maintained that it was better to try to rescue one twin rather than allow both to die. Yet, as Dreger argues in her new report, if the twins had been born separate, and the health of both were failing, nobody would have proposed that one sickly twin should be killed and her organs harvested for the sake of the other's survival.

> "That we would go so far as to intentionally asphyxiate a conscious head, and kill one of the twins in order to make a singleton out of them, just astonished me," Dreger said. "There was much ethical discussion at the time about the cost of the operation, but very little about whether or not this was morally right."

As Dreger sees it, the willingness in this case to forfeit one young life for another bespeaks society's extreme discomfort with its outliers, those who in the past would have been called freaks and exhibited at circus sideshows, but who today are often the objects of elaborate and heroicized surgical interventions. The implications of how conjoined twins are treated, say Dreger, go beyond the rare condition, which occurs in about one in 50,000 to one in 100,000 births.

TREATMENT TRAIN

Dreger and others who share her views see parallels between medical attitudes toward conjoined twins, and toward those children born with other anomalies, including ambiguous genitals, dwarfism, congenital deafness and the like. Such conditions have invited aggressive attempts at fixing, often through a long series of operations, medications and rehabilitations; and most have required that the therapies be performed on children too young to have a say in whether they want to be treated or not. "There's an allurement called the treatment train," said Dr. David C. Thomasma, director of the medical humanities program at Loyola University Chicago Medical Center. "You hop on, but it's very hard to get off. The power of the medical model today to shape our decisions is enormous." Thomasma is an author of a paper that appeared last year in the Hastings Center report about the Lakeberg twin case. Like Dreger, he was disturbed by the ease with which Amy Lakeberg's life was taken. "I remember during all the debates how quickly people who wanted to intervene could move into the idea that one twin was like a parasite or an appendage—language that makes one morally able to divide them."

FISSION FAILURE

Nobody knows what causes conjoining, but most experts assume it arises when a fertilized egg begins to split in two, as happens with identical twins, but then fails to complete the fission. The conjoining can occur at many points of the body, and result in a variety of configurations, from a major sharing of brain, heart, liver or intestines, as well as external limbs, to the development of two nearly complete individuals connected only by a minor bridge of flesh on the torso. For unknown reasons, about three-quarters of conjoined twins are female. All but 25 percent of conjoined twins are either stillborn or die soon after birth. Those who survive have always been objects of fascination. The most renowned of all conjoined twins were Chang and Eng Bunker, Chinese brothers who were born in Siam in the early 19th century and thus were called Siamese twins by P. T. Barnum, who was their manager for a time. Attached by a round band of flesh extending from breastbone to navel, the Bunkers profited handsomely by exhibiting themselves to the public, and were able to become gentlemen farmers and slave owners in North Carolina. They married two daughters of a clergyman, set up two separate households, and between them fathered 22 children. A musical now on Broadway called "Side Show" celebrates the lives of another famous set of conjoined twins, Daisy and Violet Hilton, who appeared in the classic 1932 movie "Freaks" and later ended up working in a supermarket.

What people find difficult to imagine about conjoined twins is how they can stand doing everything together. The Bunker twins managed by taking turns every three days on who was in charge. Reba Schappell capitulated to Lori for a few years while Lori attended college and then worked in a hospital. Now Lori says it is Reba's turn to dominate, and she accompanies her sister to recording sessions and performances, at which she practices a kind of Zen detachment. "I'm just there, I'm not doing anything," said Lori. "I say hi to people, but then I let Reba get down to business."

Yet the sisters also engage in a constant and perhaps largely subconscious dance of intimacy: fixing a sister's stray lock of hair, or picking a bit of lint off the other's shirt. When they move through a shopping mall they look like young girls, best friends, heads bowed together, murmuring to each other, swept up in a realm of their own. Doctors who have performed separation surgeries say that the apparent contentment of the Schappell sisters is just as well, for in their case separation is probably impossible, and would lead to mutual death by hemorrhaging. But surgeons say that each case of conjoined twins is unique, and that in many instances separation is possible, if difficult. They insist that even risky attempts are usually worth the effort, and that the hard truth is, yes, it may be better to end up with one reasonably normal singleton than to accept conjoined twins, whose shared organs may not be up to supporting them and whose lives will be severely restricted. For parents, the desire to give their children as normal a life as possible is so profound that it can override any fears they have about the dangers of separation surgery. Michelle Roderick said that after she gave birth in May 1996 to Shawna and Janelle, attached at the abdomen, she and her husband never considered keeping the girls conjoined. "We felt that since the chances were so good that they'd come out of it as healthy individuals, we didn't think it would be fair not to try," she said. "If they'd stayed conjoined, they wouldn't have been able to walk normally or develop normally and it would have been pretty sad." Dreger said she was not immutably opposed to separation surgery, nor does she romanticize the state of being conjoined. What she does question is the presumption behind the most radical surgical pyrotechnics that a life conjoined is hopeless, absurd, no life at all. She argues that decisions about who is in pain and who should be "fixed"—whatever their purported abnormality must go beyond mechanical or economic or even philosophical considerations, to include the voices of those who know best. "We need to have people who live with a condition be allowed to describe for themselves what it's like," she said. "Otherwise we'll assume it's a living hell, and that we're doing the right thing by eliminating them."

QUESTIONS FOR DISCUSSION

1. How are Lori and Reba Schappell similar to other humans and how are they different?

2. From the reading, explain how conjoined (Siamese) twins are different from two siblings or a set of identical twins?

3. Discuss why the Schappell sisters might be considered deviant. Discuss why onlookers of conjoined twins might be deviant.

QUESTIONS FOR PART 5

1. How do the readings in Chapter 15 address how individuals are labeled and how individuals internalize mental deviant behavior?

2. How do the readings in Chapter 16 address the development and life of gays and lesbians?

3. How do the readings in Chapter 17 address the development and life of physically disabled individuals?